Selected Titles in This Series

23 **Alberto Candel and Lawrence Conlon,** Foliations I, 2000

22 **Günter R. Krause and Thomas H. Lenagan,** Growth of algebras and Gelfand-Kirillov dimension, 2000

21 **John B. Conway,** A course in operator theory, 2000

20 **Robert E. Gompf and András I. Stipsicz,** 4-manifolds and Kirby calculus, 1999

19 **Lawrence C. Evans,** Partial differential equations, 1998

18 **Winfried Just and Martin Weese,** Discovering modern set theory. II: Set-theoretic tools for every mathematician, 1997

17 **Henryk Iwaniec,** Topics in classical automorphic forms, 1997

16 **Richard V. Kadison and John R. Ringrose,** Fundamentals of the theory of operator algebras. Volume II: Advanced theory, 1997

15 **Richard V. Kadison and John R. Ringrose,** Fundamentals of the theory of operator algebras. Volume I: Elementary theory, 1997

14 **Elliott H. Lieb and Michael Loss,** Analysis, 1997

13 **Paul C. Shields,** The ergodic theory of discrete sample paths, 1996

12 **N. V. Krylov,** Lectures on elliptic and parabolic equations in Hölder spaces, 1996

11 **Jacques Dixmier,** Enveloping algebras, 1996 Printing

10 **Barry Simon,** Representations of finite and compact groups, 1996

9 **Dino Lorenzini,** An invitation to arithmetic geometry, 1996

8 **Winfried Just and Martin Weese,** Discovering modern set theory. I: The basics, 1996

7 **Gerald J. Janusz,** Algebraic number fields, second edition, 1996

6 **Jens Carsten Jantzen,** Lectures on quantum groups, 1996

5 **Rick Miranda,** Algebraic curves and Riemann surfaces, 1995

4 **Russell A. Gordon,** The integrals of Lebesgue, Denjoy, Perron, and Henstock, 1994

3 **William W. Adams and Philippe Loustaunau,** An introduction to Gröbner bases, 1994

2 **Jack Graver, Brigitte Servatius, and Herman Servatius,** Combinatorial rigidity, 1993

1 **Ethan Akin,** The general topology of dynamical systems, 1993

Foliations I

Foliations I

Alberto Candel
Lawrence Conlon

Graduate Studies in Mathematics
Volume 23

American Mathematical Society
Providence, Rhode Island

2000 *Mathematics Subject Classification.* Primary 57R30.

The first author was partially supported by NSF Grant #DMS-9973086.

ABSTRACT. This book is the first of two volumes on foliations. It treats the foundations of the theory, foliated manifolds of codimension one, and selected topics about higher codimension and abstract foliated spaces.

Library of Congress Cataloging-in-Publication Data

Candel, Alberto, 1961–
Foliations I / Alberto Candel, Lawrence Conlon.
p. cm.—(Graduate studies in mathematics, ISSN 1065-7339 ; v. 23)
Includes bibliographical references and index.
ISBN 0-8218-0809-5 (alk. paper)
1. Foliations (Mathematics) I. Title. II. Series. III. Conlon, Lawrence, 1933–
QA613.62.C37 1999
514′.72 21–dc21 99-045694

∞ The paper used in this book is acid-free and falls within the guidelines
established to ensure permanence and durability.
Visit the AMS home page at URL: http://www.ams.org/

10 9 8 7 6 5 4 3 2 1 05 04 03 02 01 00

To our wives, Juana and Jackie,
and in loving memory of
Edna Conlon

Contents

Preface

This is the first of two volumes on the qualitative theory of foliations. It is divided into three parts, the first being a primer on foliated manifolds. The material here could be the basis of a short graduate course or learning seminar, preparing students to begin exploring the literature. The remainder of this volume is a "sampler" of more advanced topics. In Part 2, we focus on foliated manifolds of codimension one. Here, hands–on geometric methods yield a structure theory (the theory of levels) which is reminiscent of the classical Poincaré–Bendixson theory of flows on surfaces. In arbitrary codimension (Part 3) these methods break down and we turn to the techniques of ergodic theory. Again the theory of flows, now in higher–dimensional manifolds, is the model. Specifically, we consider foliations to be generalized dynamical systems and extend to these systems the theory of invariant measures (following J. F. Plante [**107**], D. Sullivan [**127**], *et al.*) and the theory of topological entropy (following E. Ghys, R. Langevin and P. Walczak [**50**]). For the most part, these methods apply equally well to laminations (partial foliations of manifolds) and even to abstract laminations (foliated metric spaces). In Chapter 11, we introduce this abstract setting and most of the subsequent results are proven for foliated metric spaces.

The measures studied in Part 3 are invariant under holonomy. There are interesting conditions guaranteeing the existence of such measures (*cf.* Theorem 12.3.1 and Theorem 13.4.2) and the corresponding ergodic theory has beautiful applications. The foliation cycles of D. Sullivan (Chapter 10) exemplify this. The limitation of this approach is that holonomy–invariant measures do not always exist. On the other hand, the harmonic measures of L. Garnett [**46**] always exist and the ergodic theory for foliations based on them also has profound applications. We plan to treat this in the second volume.

Foliation theory has developed to the point that an encyclopedic treatment is out of the question. Our choice of topics is quite subjective, depending largely on the authors' tastes and expertise. As to topics omitted, we call special attention to the quantitative theory of foliations, culminating in the stunning existence and classification theorems of W. Thurston in the 1970s [**134, 135**]. Likewise, the applications of noncommutative geometry to foliations, pioneered by A. Connes (see [**31**]), and the beautiful study of transverse geometry (see P. Molino [**91**] and P. Tondeur [**137, 138**]) are not treated. Our hope is that the material we do present will whet the reader's appetite for more. In the second volume, in addition to harmonic measures, we will offer introductory treatments of a few other specialized topics, including the exotic characteristic classes of foliations and foliated 3–manifold theory.

Part 1

The Foundations

Foreword to Part 1

This part is intended to be a primer in foliation theory. A good understanding of its contents should make much of the literature on foliations accessible and will be presupposed in the rest of the book.

We have tried to present a wealth of examples in the conviction that they are essential for understanding abstract theory. Wherever feasible, we have even preceded formal definitions with informal examples, occasionally allowing rigor to be compromised for the sake of clarity. This is especially true in the earlier sections, where foliated manifolds are sometimes constructed by "cut and paste" methods. Careful justifications for these and subsequent constructions are given in Chapter 3.

It will be assumed that the reader has a fairly good background in manifold theory. For example, the material in references [**11**], [**29**], and [**124**] should be adequate for most purposes.

A few remarks on notation are in order. The notation $C^r(M)$ will stand for the space of all real–valued C^r functions on the smooth manifold M, $0 \leq r \leq \infty$ or $r = \omega$ (analytic). The term "smooth", without further qualification, generally means "smooth of class C^∞". If $\pi : E \to M$ is a bundle of a given smoothness class C^r, then $\Gamma(E)$ denotes the set of C^r sections of the bundle. The total space of the tangent bundle of M will be denoted by $T(M)$ and the space of smooth vector fields is $\mathfrak{X}(M) = \Gamma(T(M))$. If $\pi : E \to M$ is a vector bundle, E^* denotes the total space of the dual bundle. The notation $\Lambda^k(E)$ denotes the total space of the kth exterior power of $\pi : E \to M$ and $A^k(M) = \Gamma(\Lambda^k(T(M)^*))$ is the space of k–forms on M. The graded exterior algebra of differential forms on M will be denoted by $A^*(M)$.

We will follow the convention that all manifolds are 2nd countable and Hausdorff. Unless otherwise stipulated, all manifolds are smooth, possibly with boundary and corners.

Chapter 1

Foliated Manifolds

1.1. What Is a Foliation?

Roughly speaking, a foliation $\mathcal{F}$ is an equivalence relation on an n–manifold M, the equivalence classes being connected, immersed submanifolds, all of the same dimension k. (By the term "immersed submanifold" we mean the image of a one–to–one immersion $L \hookrightarrow M$.) One further requires that, locally, the decomposition into equivalence classes be modeled on the decomposition of $\mathbb{R}^n$ into the cosets $x + \mathbb{R}^k$ of the standardly imbedded subspace $\mathbb{R}^k$. In particular, this decomposition of $\mathbb{R}^n$ is itself a foliation. The equivalence classes of $\mathcal{F}$ are called the *leaves* of the foliation.

Before making these notions precise, we consider examples.

Example 1.1.1. Let M be an n–manifold, B a manifold of dimension $n-k$, $\partial M = \emptyset = \partial B$, and let

$$\pi : M \to B$$

be a smooth *fiber bundle*. That is, there is a k–manifold F and, for each $x \in B$, an open neighborhood U of x in B and a commutative diagram

$$\begin{array}{ccc} \pi^{-1}(U) & \xrightarrow{\varphi} & U \times F \\ {\scriptstyle\pi}\downarrow & & \downarrow{\scriptstyle p} \\ U & \xrightarrow[\text{id}]{} & U \end{array}$$

with φ a diffeomorphism and p the canonical projection onto the first factor. Each subspace $\pi^{-1}(x)$, $x \in B$, is clearly an imbedded k–manifold diffeomorphic to F. The manifold B is the *base* of the fiber bundle and F is the *fiber*. The *total space* of the fiber bundle is M. As x ranges over B, the connected components of the fibers $\pi^{-1}(x)$ range over the leaves of a foliation $\mathfrak{F}$. We say that $\mathfrak{F}$ has *codimension* $q = n - k = \dim B$ and *dimension* $k = \dim F$. Remark that the leaves are also fibers of a bundle whose base $\widetilde{B}$ is a covering space of B. Indeed, $\widetilde{B} = M/\mathfrak{F}$ is the manifold obtained by collapsing each leaf of $\mathfrak{F}$ to a point.

Example 1.1.2. A smooth submersion

$$f : M \to B$$

from an n–manifold to a q–manifold (both without boundary) provides a mild generalization of the previous class of examples. The submersion theorem asserts that, for each point $y \in M$, there is a coordinate neighborhood $(U, y^1, \dots, y^n)$ of y in M and a coordinate neighborhood $(V, x^1, \dots, x^q)$ of $f(y)$ in B, relative to which the formula for $f|U$ becomes

$$f(y^1, \dots, y^n) = (y^1, \dots, y^q).$$

It follows that the level sets $f^{-1}(x)$ are properly imbedded submanifolds of M of dimension $k = n - q$ and that, locally, these submanifolds fit together exactly like parallel copies of $\mathbb{R}^k$ in $\mathbb{R}^n$. Thus, the connected components of the nonempty level sets of f are the leaves of a foliation $\mathfrak{F}$ of codimension $q = \dim B$. If B is connected and each of these level sets is compact (*e.g.*, if M itself is compact), then

$$f : M \to B$$

is actually a fiber bundle (Exercise 1.1.3).

While a fiber bundle is always a submersion, the converse is false. As an example, consider

$$f : \mathbb{R}^2 \to \mathbb{R},$$
$$f(x, y) = (x^2 - 1)e^y.$$

Here,

$$f_x(x, y) = 2xe^y,$$
$$f_y(x, y) = (x^2 - 1)e^y.$$

The derivative f_x only vanishes along the y–axis, while f_y only vanishes along the lines $x = \pm 1$; so f is a submersion. The reader should check that the level sets of f give the foliation $\mathfrak{F}$ in Figure 1.1.1. The vertical lines $x = \pm 1$ are the components of the level set $f = 0$, hence are leaves. Each curve between these lines is a level set $f = c$ with $c < 0$ and is asymptotic to $x = \pm 1$. Each level set $f = c$, $c > 0$, falls into two components, one lying in the region $x > 1$ and asymptotic to $x = 1$, one in the region $x < -1$ and asymptotic to $x = -1$. The leaf space $\mathbb{R}^2/\mathfrak{F}$ (again formed by collapsing each leaf to a point and imposing the quotient topology) is locally Euclidean

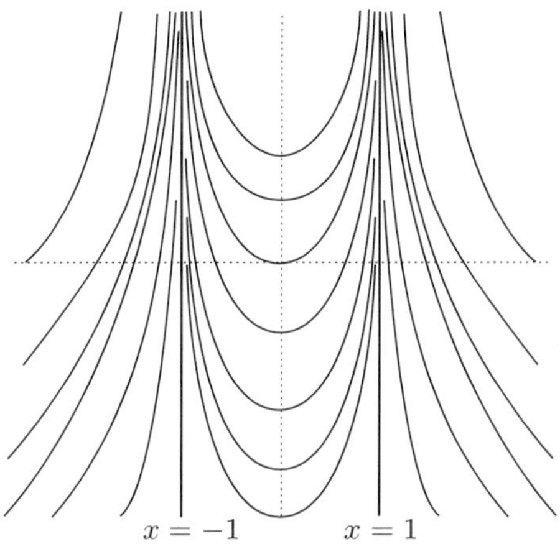

Figure 1.1.1. The foliation by the submersion $(x^2 - 1)e^y$

of dimension 1, but is not Hausdorff (Exercise 1.1.4); so this cannot be the base manifold of a bundle.

We extend the above submersion to a submersion

$$f : \mathbb{R}^{n+1} \to \mathbb{R}$$

by writing $\mathbb{R}^{n+1} = \mathbb{R}^n \times \mathbb{R}$ and using cylindrical coordinates (r, z, t). Here, z ranges over the unit sphere $S^{n-1} \subset \mathbb{R}^n$ and the coordinates of any point $v \neq 0$ in $\mathbb{R}^n$ are $(r, z) = (\|v\|, v/\|v\|)$. These coordinates are bad at the origin, where r is not smooth (not even of class C^1) and every pair $(0, z)$ represents the origin. However, r^2 is smooth on all of $\mathbb{R}^n$ and the formula

$$f(r, z, t) = (r^2 - 1)e^t$$

defines the desired submersion. For $n \geq 2$, each level set $f = c$ is connected, hence is a leaf of a foliation $\mathcal{F}$ of $\mathbb{R}^{n+1}$. The level $f = 0$ is exactly the cylinder $S^{n-1} \times \mathbb{R}$, each level $f = c < 0$ is a hypersurface of revolution diffeomorphic to $\mathbb{R}^n$, and each level $f = c > 0$ is a hypersurface of revolution diffeomorphic to $S^{n-1} \times \mathbb{R}$. The foliation of $\mathbb{R}^3$ is depicted in Figure 1.1.2. Since the leaves fall into two distinct diffeomorphism classes, it remains evident that this foliation cannot be a fiber bundle.

Exercise 1.1.3. If $\partial M = \emptyset = \partial B$ and B is connected, prove that a submersion $f : M \to B$ with compact level sets is a fiber bundle.

Exercise 1.1.4. Prove that the leaf space $\mathbb{R}^2/\mathcal{F}$ for the foliation in Figure 1.1.1 is locally Euclidean of dimension 1, but not Hausdorff.

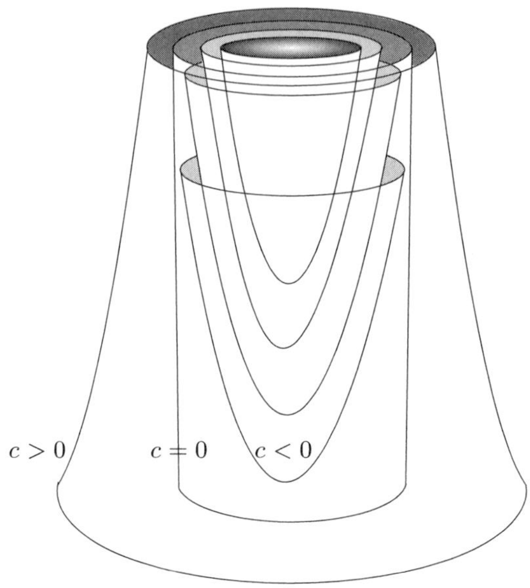

Figure 1.1.2. Foliating $\mathbb{R}^3$ by the level sets $(r^2 - 1)e^t = c$

Example 1.1.5. If $X \in \mathfrak{X}(M)$ is a nonsingular (*i.e.*, nowhere zero) vector field, then the local flow defined by X patches together to define a foliation of dimension 1. Indeed, given an arbitrary point $x \in M$, the fact that X is nonsingular allows us to find a coordinate neighborhood $(U, x^1, \dots, x^n)$ about x such that

$$-\epsilon < x^i < \epsilon, \quad 1 \le i \le n,$$

and

$$\frac{\partial}{\partial x^1} = X|U.$$

Geometrically, the flow lines of $X|U$ are just the level sets

$$x^i = c^i, \quad 2 \le i \le n,$$

where all $|c^i| < \epsilon$. Since we follow the convention that manifolds are 2nd countable, the careful reader may worry that a leaf of this foliation might be the "long line", a notorious example of a connected manifold that is not 2nd countable [**124**, Appendix A]. We will see in the next section that such anomalies are precluded by the 2nd countability of M itself. For the moment, the difficulty can be sidestepped by requiring that X be a complete field (*e.g.*, that M be compact), hence that each leaf be a flowline.

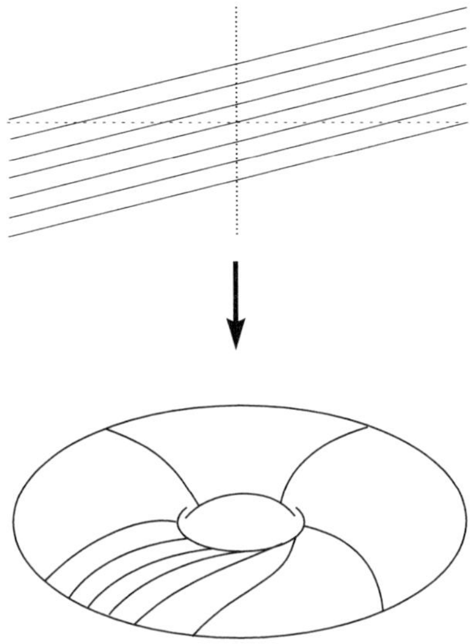

Figure 1.1.3. A linear foliation of T^2

An interesting class of 1–dimensional foliations on the torus T^2 is given by the *linear* foliations. A constant vector field

$$\widetilde{X} \equiv \begin{bmatrix} a \\ b \end{bmatrix}$$

on $\mathbb{R}^2$ is invariant by all translations in $\mathbb{R}^2$, hence passes to a well–defined vector field X on $T^2 = \mathbb{R}^2/\mathbb{Z}^2$. We assume that $a \neq 0$. The foliation $\widetilde{\mathcal{F}}$ on $\mathbb{R}^2$ produced by $\widetilde{X}$ has as leaves the parallel lines of slope b/a. This foliation is also invariant under translations and passes to the foliation $\mathcal{F}$ on T^2 produced by X (Figure 1.1.3). We consider separately the cases that the slope $\rho = b/a$ is rational or irrational.

Each leaf of $\widetilde{\mathcal{F}}$ is of the form

$$\widetilde{L} = \{(x_0 + ta, y_0 + tb)\}_{t\in\mathbb{R}}.$$

When the slope is rational, we can take $a, b \in \mathbb{Z}$. For fixed $t_0 \in \mathbb{R}$, the points of $\widetilde{L}$ corresponding to values of $t \in t_0 + \mathbb{Z}$ all project to the same point of T^2; so the corresponding leaf L of $\mathcal{F}$ is an imbedded circle in T^2. Since L is arbitrary, $\mathcal{F}$ is a foliation of T^2 by circles. It follows rather easily that this foliation is actually a fiber bundle $\pi : T^2 \to S^1$ (Exercise 1.1.6).

When the slope $\rho = b/a$ is irrational, the foliation on T^2 looks radically different. Indeed, each leaf L of $\mathcal{F}$ is a one–to–one immersion of $\mathbb{R}$ and is everywhere dense in T^2 (Kronecker's theorem). To see this, note first that, if a leaf $\widetilde{L}$ of $\widetilde{\mathcal{F}}$ does not project one–to–one into T^2, there must be a real number $t \neq 0$ such that ta and tb are both integers. But this would imply that $b/a \in \mathbb{Q}$. In order to show that each leaf L of $\mathcal{F}$ is dense in T^2, it is enough to show that, for every $v \in \mathbb{R}^2$, each leaf $\widetilde{L}$ of $\widetilde{\mathcal{F}}$ achieves arbitrarily small positive distances from suitable points of the coset $v + \mathbb{Z}^2$. A suitable translation in $\mathbb{R}^2$ allows us to assume that $v = 0$; so we are reduced to showing that $\widetilde{L}$ passes arbitrarily close to suitable points $(n, m) \in \mathbb{Z}^2$. The line $\widetilde{L}$ has slope–intercept equation

$$y = \rho x + c.$$

So it will suffice to find, for arbitrary $\eta > 0$, integers n and m such that

$$|\rho n + c - m| < \eta.$$

Equivalently, $c \in \mathbb{R}$ being arbitrary, we are reduced to showing that the set $\{\rho n - m\}_{m,n \in \mathbb{Z}}$ is dense in $\mathbb{R}$. This is essentially the criterion of Eudoxus that ρ and 1 be incommensurable (*i.e.*, that ρ be irrational). A more modern proof is suggested in Exercise 1.1.7.

Exercise 1.1.6. Let $\mathcal{F}$ be a linear foliation of T^2, as in Example 1.1.5, with slope b/a rational. Prove that each leaf has a neighborhood U diffeomorphic to $(-\epsilon, \epsilon) \times S^1$ in such a way that, for $-\epsilon < t < \epsilon$, the imbedded circle $\{t\} \times S^1$ is identified with a leaf. Use this to prove that the leaves of the foliation are the fibers of a fiber bundle $\pi : T^2 \to S^1$.

Exercise 1.1.7. If $G \subset \mathbb{R}$ is a nontrivial additive subgroup and if there is a neighborhood $U \subset \mathbb{R}$ of 0 such that $U \cap G = \{0\}$, prove that G is infinite cyclic. Next, show that, if $\rho \in \mathbb{R}$ is irrational, the additive subgroup $G_\rho \subset \mathbb{R}$, generated by $\{1, \rho\}$, is isomorphic to $\mathbb{Z} \oplus \mathbb{Z}$. Use these results to prove that G_ρ is dense in $\mathbb{R}$. Conclude that each leaf of a linear foliation of T^2, with slope b/a irrational, is dense in T^2.

Example 1.1.8. If G is a Lie group and $H \hookrightarrow G$ a connected Lie subgroup, then the collection $\{gH\}_{g \in G}$ of left cosets of H is the set of leaves of a foliation $\mathcal{H}$. If H is a *closed* subgroup, then G/H is a manifold and $\mathcal{H}$ is a foliation with leaves the fibers of a fiber bundle $\pi : G \to G/H$. If H is not closed, the closed subgroup theorem guarantees that $\overline{H}$ will also be a Lie subgroup and, indeed, the closure of each leaf gH is the coset $g\overline{H}$; so the closures of the leaves of $\mathcal{F}$ are the fibers of a bundle $\pi : G \to G/\overline{H}$.

The linear foliations of T^2 are special cases of this. Similarly, the connected Lie subgroups of $T^n = \mathbb{R}^n/\mathbb{Z}^n$ correspond one–to–one to the vector subspaces $V \subset \mathbb{R}^n$. According to the "position" of the k–dimensional subspace $V \subset \mathbb{R}^n$ with respect to $\mathbb{Z}^n$, the leaves in T^n will all be diffeomorphic to $\mathbb{R}^{k-r} \times T^r$, for some integer r, $0 \leq r \leq k$, and the closure of each leaf will be an imbedded torus in T^n of dimension q, $k \leq q \leq n$. The integers

r and q are determined by V and all the possibilities $0 \le r < k < q \le n$ can occur. It is clear, however, that $r = k$ if and only if $q = k$, the case in which $V \cap \mathbb{Z}^n$ is a k–dimensional lattice subgroup of V.

Example 1.1.9. Let G be a Lie group, $H \hookrightarrow G$ a connected Lie subgroup, and let $\mathcal{H}$ be the foliation of G by the left cosets of H. Suppose that $\Gamma \subset G$ is a discrete subgroup. Then $M = \Gamma\backslash G$, the space of *right* cosets of Γ, is a manifold of the same dimension as G. If Γ is *cocompact* in G, the manifold M is compact. At any rate, since $\mathcal{H}$ is invariant under left translations and Γ is divided out from the left, the foliation $\mathcal{H}$ passes to a well–defined foliation $\mathcal{H}_\Gamma$ on M. While $\mathcal{H}$ is a fairly tame foliation, $\mathcal{H}_\Gamma$ can be rather wild.

Of some interest is the case in which $G = \mathrm{Sl}(2,\mathbb{R})$, the *special linear group* consisting of all real 2×2 matrices A with $\det(A) = 1$. This is a 3–dimensional Lie group. Let $H \subset \mathrm{Sl}(2,\mathbb{R})$ be the subgroup given by

$$H = \left\{ \begin{bmatrix} a & b \\ 0 & a^{-1} \end{bmatrix} \right\}_{a>0}.$$

This is a connected, 2–dimensional subgroup. Actually, we are going to work in the group

$$G = \mathrm{PSl}(2,\mathbb{R}) = \mathrm{Sl}(2,\mathbb{R})/\{\pm I\}$$

where H survives intact as a subgroup.

Let $\mathbf{H}$ denote the *hyperbolic plane,* the simply connected Riemannian 2–manifold of constant curvature -1. A standard model of $\mathbf{H}$ is the open upper half plane

$$\mathbf{H} = \{x + iy \in \mathbb{C} \mid y > 0\}.$$

In this model, the metric is

$$\frac{dx^2 + dy^2}{y^2}.$$

This metric is conformal to the Euclidean metric $dx^2 + dy^2$. Thus angles are the same in both geometries but, as one approaches $y = 0$, hyperbolic distances become enormous compared to Euclidean distances. Indeed, a vector tangent to $\mathbf{H}$ at $x + iy$ has hyperbolic length 1 if and only if it has Euclidean length y.

If

$$\begin{bmatrix} a & b \\ c & d \end{bmatrix} \in \mathrm{Sl}(2,\mathbb{R}),$$

one easily checks that the linear fractional transformation

$$z \mapsto \frac{az + b}{cz + d}$$

carries the upper half plane onto itself. In Exercise 1.1.10, you will show that this transformation is an orientation–preserving isometry of the hyperbolic metric and that every orientation–preserving isometry is of this form. Since composition of linear fractional transformations corresponds to multiplication of the corresponding coefficient matrices, we obtain a surjective homomorphism of $\mathrm{Sl}(2,\mathbb{R})$ onto the group $\mathrm{Iso}_+(\mathbf{H})$ of orientation–preserving isometries. The kernel of this projection is clearly the center $\{\pm I\}$ of $\mathrm{Sl}(2,\mathbb{R})$, and so $\mathrm{Iso}_+(\mathbf{H}) = \mathrm{PSl}(2,\mathbb{R})$. In Exercise 1.1.10, you will also show that this isometry group acts (via its differentials) as a smooth transformation group of the manifold

$$T^1(\mathbf{H}) \cong \mathbf{H} \times S^1$$

of unit tangent vectors to $\mathbf{H}$ and that this action is simply transitive. Thus, fixing a basepoint $v_0 \in T^1(\mathbf{H})$, we can identify $T^1(\mathbf{H})$ with

$$\mathrm{PSl}(2,\mathbb{R}) \cdot v_0 \cong \mathrm{PSl}(2,\mathbb{R}).$$

It will be convenient to take the basepoint to be

$$v_0 = \begin{bmatrix} 0 \\ 1 \end{bmatrix}_i \in T_i^1(\mathbf{H}),$$

the unit vector tangent at $i \in \mathbf{H}$ that points in the positive y direction. We are going to describe H and its left cosets in $\mathrm{PSl}(2,\mathbb{R})$ as sections of the S^1–bundle $T^1(\mathbf{H})$ which can be easily visualized. Indeed, a matrix

$$A = \begin{bmatrix} a & b \\ 0 & a^{-1} \end{bmatrix} \in H$$

carries $i \mapsto a^2 i + ba$ and one readily checks that the correspondence

$$A \mapsto a^2 i + ba$$

is a diffeomorphism between H and $\mathbf{H}$. The differential of this transformation sends

$$\begin{bmatrix} 0 \\ 1 \end{bmatrix}_i \mapsto \begin{bmatrix} 0 \\ a^2 \end{bmatrix}_{a^2 i + ab},$$

imbedding H as the smooth section of $T^1(\mathbf{H})$ which, to each point $z \in \mathbf{H}$, assigns the unit tangent vector at z that points in the positive y direction. This is the unit velocity field for the family of parallel geodesics in $\mathbf{H}$ given by the vertical lines. Under our identification of $T^1(\mathbf{H})$ with $\mathrm{PSl}(2,\mathbb{R})$, the subgroup H "is" this section.

In order to see the cosets gH as sections of $T^1(\mathbf{H})$, we need to recall that the (oriented but otherwise unparametrized) geodesics in hyperbolic geometry are just the intersections of the open upper half plane with oriented circles that meet S_∞, the "circle at infinity", orthogonally. The circle S_∞ is the real axis, compactified by the point ∞. A geodesic, "extended to infinity", meets this circle orthogonally in exactly two points, its "source" and its "target". The geodesics with target $\infty \in S_\infty$ are just the vertical straight line segments, oriented in the positive y direction. This will be called

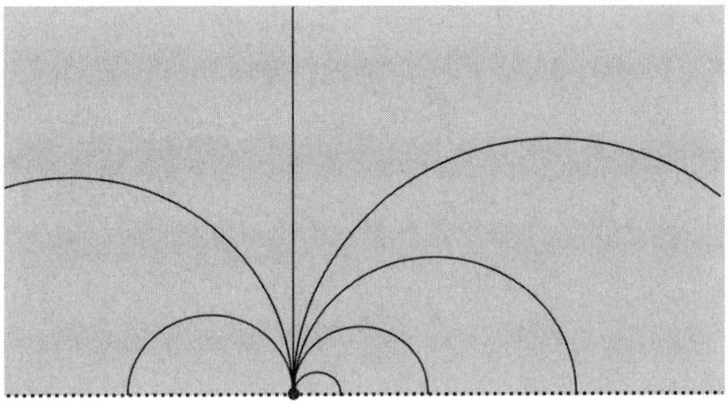

Figure 1.1.4. The pencil $\mathcal{P}_x$

the "geodesic pencil through ∞" and denoted by $\mathcal{P}_\infty$. In Figure 1.1.4, we draw the pencil $\mathcal{P}_x$ of geodesics with finite target $x \in \mathbb{R} \subset S_\infty$. The linear fractional transformation $z \mapsto g(z)$, defined by $g \in \mathrm{Psl}(2, \mathbb{R})$, extends by the same formula to the circle at infinity and, being an isometry of $\mathbf{H}$, it sends $\mathcal{P}_x$ to $\mathcal{P}_{g(x)}$. The subgroup H is exactly the stabilizer of $\infty \in S_\infty$, hence is exactly the subgroup that preserves $\mathcal{P}_\infty$. It follows that the elements of the coset gH are exactly the elements of $\mathrm{Iso}_+(\mathbf{H})$ that send $\mathcal{P}_\infty$ to $\mathcal{P}_x$, where $x = g(\infty)$. Viewing H as the section of $T^1(\mathbf{H})$ consisting of the unit velocity vectors along the pencil $\mathcal{P}_\infty$, we see that gH is identified as the section consisting of the unit velocity vectors along the pencil $\mathcal{P}_x$ (Figure 1.1.5). Every $x \in S_\infty$ occurs, and so we have completely described the leaves of $\mathcal{H}$ in $\mathrm{Psl}(2, \mathbb{R})$.

This foliation is carried to a foliation $\mathcal{H}_\Gamma$ of a compact manifold

$$M = \Gamma \backslash \mathrm{Psl}(2, \mathbb{R})$$

as follows. The compact, orientable surface Σ_g of genus $g > 1$ has a Riemannian metric of constant curvature -1, hence has universal cover $\mathbf{H}$, the group Γ of covering transformations being a group of orientation–preserving isometries of $\mathbf{H}$. This is the subgroup $\Gamma \subset \mathrm{Psl}(2, \mathbb{R})$ and M is the total space of the unit tangent bundle $T^1(\Sigma_g)$, a circle bundle over the surface. The foliation $\mathcal{H}$ is invariant under left translations by Γ, hence descends to a foliation $\mathcal{H}_\Gamma$, the leaves of which are locally sections of $T^1(\Sigma_g)$. That is, this foliation is transverse to the circle fibers (*cf.* Definition 1.1.11). It can be shown that each leaf of $\mathcal{H}_\Gamma$ is dense in M, that exactly a countable infinity of them are diffeomorphic to the cylinder and that all the other leaves are diffeomorphic to the plane.

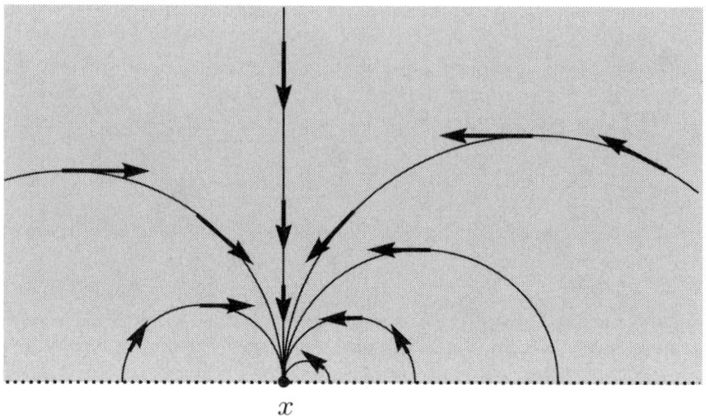

Figure 1.1.5. The section of unit velocity vectors along $\mathcal{P}_x$

Exercise 1.1.10. You are to prove that the group $\mathrm{Psl}(2,\mathbb{R})$, acting as linear fractional transformations of the upper half plane, is exactly the group $\mathrm{Iso}_+(\mathbf{H})$ and that its action on $T^1(\mathbf{H})$, via the differentials of these linear fractional transformations, is simply transitive. Here is a suggested way to proceed:

1. Show that the 2–dimensional subgroup $H \subset \mathrm{Psl}(2,\mathbb{R})$ (as defined in Example 1.1.9) and the 1–dimensional subgroup
$$L = \left\{ \pm \begin{bmatrix} 1 & 0 \\ c & 1 \end{bmatrix} \right\}_{c\in\mathbb{R}}$$
both determine linear fractional transformations of $\mathbf{H}$ that preserve the hyperbolic metric and the orientation of $\mathbf{H}$.
2. Show that L and H generate $\mathrm{Psl}(2,\mathbb{R})$, concluding that this is a subgroup of $\mathrm{Iso}_+(\mathbf{H})$.
3. Show that the action of H on $\mathbf{H}$ is *transitive,* hence that $\mathbf{H}$ is a Riemannian homogeneous space. A standard theorem in Riemannian geometry asserts that such homogeneous spaces are *complete* in their metric. Use these facts to show that an arbitrary isometry $\varphi : \mathbf{H} \to \mathbf{H}$ is uniquely determined by $\varphi(i)$ and φ_{*i}.
4. Show that the subgroup of $\mathrm{Psl}(2,\mathbb{R})$ that fixes i is exactly
$$\mathrm{PSO}(2) = \mathrm{SO}(2)/\{\pm I\}.$$
Use this and the above to prove both that each $\varphi \in \mathrm{Iso}_+(\mathbf{H})$ is an element of $\mathrm{Psl}(2,\mathbb{R})$ and that this group acts simply transitively on $T^1(\mathbf{H})$.

One also allows foliations of manifolds with boundary. If N is a component of ∂M, we demand either that $\mathcal{F}$ be *transverse* to N or *tangent* to N.

Definition 1.1.11. Let $N \subset M$ be a smooth submanifold. We say that $\mathcal{F}$ is transverse to N (and write $\mathcal{F} \pitchfork N$) if, for each leaf L of $\mathcal{F}$ and each point $x \in L \cap N$, $T_x(L)$ and $T_x(N)$ together span $T_x(M)$. At the other extreme, we say that $\mathcal{F}$ is tangent to N if, for each leaf L of $\mathcal{F}$, either $L \cap N = \emptyset$ or $L \subseteq N$.

Remark. We will see that, whenever $\mathcal{F} \pitchfork N$, there is naturally induced on N a foliation of the same codimension as $\mathcal{F}$, each leaf of which is a path component of $N \cap L$, L some leaf of $\mathcal{F}$.

Example 1.1.12. Let

$$D^n = \{v \in \mathbb{R}^n \mid \|v\| \leq 1\},$$

the unit disk in $\mathbb{R}^n$. The manifold $M = D^n \times S^1$ (called a *solid torus*) has boundary $\partial M = S^{n-1} \times S^1$. The foliation $\mathcal{F}_0$ of M by leaves $D^n \times \{y\}$, $y \in S^1$, is transverse to ∂M. In this case, the foliation is modeled locally on the foliation of Euclidean half space

$$\mathbb{H}^{n+1} = \{(x^1, \ldots, x^{n+1}) \mid x^1 \leq 0\}$$

by the subspaces parallel to

$$\mathbb{H}^n = \{(x^1, \ldots, x^n, 0) \mid x^1 \leq 0\}.$$

This is a fairly trivial example, but it leads to a more interesting foliation, one tangent to the boundary, by a process of "spinning" the leaves asymptotically along ∂M. This is the famous *Reeb foliation* of the solid torus.

The easiest way to visualize the Reeb foliation is to lift it to the solid cylinder $\widetilde{M} = D^n \times \mathbb{R}$ via the universal covering map. The submersion

$$f : \mathbb{R}^n \times \mathbb{R} \to \mathbb{R},$$
$$f(r, z, t) = (r^2 - 1)e^t$$

of Example 1.1.2 restricts to a submersion

$$f : D^n \times \mathbb{R} \to \mathbb{H}^1.$$

Here, $0 \leq r \leq 1$, and so $f \leq 0$. The boundary $S^{n-1} \times \mathbb{R}$ is the level set $f = 0$ and each level $f = c < 0$ is a "cup–shaped" manifold diffeomorphic to $\mathbb{R}^n$. The resulting foliation $\widetilde{\mathcal{F}}$ is tangent to $\partial\widetilde{M}$. The cases $n = 1, 2$ are illustrated in Figure 1.1.6.

Translation in the t–coordinate

$$(r, z, t) \mapsto (r, z, t + a)$$

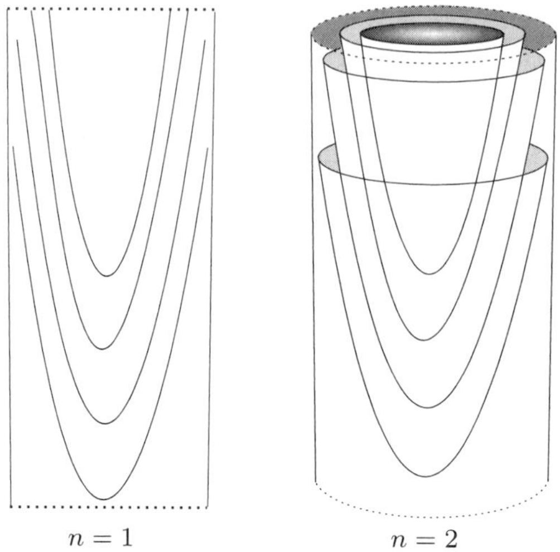

Figure 1.1.6. The foliation $\widetilde{\mathcal{F}}$ of $D^n \times \mathbb{R}$

carries the level set $f = c$ diffeomorphically onto the level set $f = ce^a$. This carries the boundary $f = 0$ diffeomorphically onto itself and, if $a \neq 0$, moves every interior leaf to a different interior leaf. In particular, the $\mathbb{Z}$–action

$$\mathbb{Z} \times (D^n \times \mathbb{R}) \to D^n \times \mathbb{R},$$
$$(k, (r, z, t)) \mapsto (r, z, t + k)$$

preserves $\widetilde{\mathcal{F}}$, inducing a foliation $\mathcal{F}$ on the quotient space

$$D^n \times (\mathbb{R}/\mathbb{Z}) = D^n \times S^1 = M.$$

Each boundary leaf of $\widetilde{\mathcal{F}}$ quotients to a component of ∂M and each interior leaf of $\widetilde{\mathcal{F}}$ is carried diffeomorphically onto an interior leaf of $\mathcal{F}$. The interior leaves still look like cup–shaped Euclidean spaces, but they now wind around inside the solid torus like infinite snakes repeatedly swallowing their own tails (and everyone else's tails as well). These leaves all wind out asymptotically toward the boundary $S^{n-1} \times S^1$. The Reeb foliation of the solid torus is not defined by a submersion, but the foliation of the *interior* $M_0 = \operatorname{int} D^n \times S^1$ is defined by a submersion $g : M_0 \to S^1$ (Exercise 1.1.13). The case $n = 2$ is illustrated in Figure 1.1.7.

Exercise 1.1.13. Let $\mathcal{G}$ be the foliation of $M_0 = \operatorname{int} D^n \times S^1$ obtained by restricting the Reeb foliation (Example 1.1.12) to M_0. Prove that $\mathcal{G}$ is defined by a submersion $g : M_0 \to S^1$. On the other hand, show that, if a submersion f of a manifold to a 1–manifold has a compact level set $f^{-1}(a)$, then every level set that comes sufficiently close to $f^{-1}(a)$ is also compact. Conclude that the Reeb foliation itself cannot be defined by a submersion.

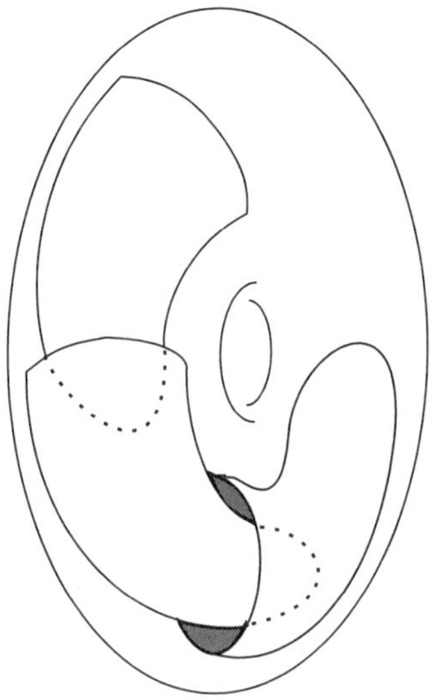

Figure 1.1.7. The Reeb foliation of $D^2 \times S^1$

The submersion $f : D^n \times \mathbb{R} \to \mathbb{H}^1$ admits a deformation f_s, $0 \le s \le 1$, through submersions

$$f_s : D^n \times \mathbb{R} \to \mathbb{H}^1,$$
$$f_s(r, z, t) = f(sr, z, t).$$

Let $\widetilde{\mathcal{F}}_s$ denote the foliation of $D^n \times \mathbb{R}$ defined by f_s. We view this as a deformation of $\widetilde{\mathcal{F}}_0$ to $\widetilde{\mathcal{F}}_1 = \widetilde{\mathcal{F}}$. This deformation projects in a well–defined way to a deformation $\mathcal{F}_s$ of $\mathcal{F}_0$ to $\mathcal{F}_1 = \mathcal{F}$. It is clear that $\mathcal{F}_0$ is the foliation of the solid torus M by the disks $D^n \times \{x\}$, the foliation transverse to ∂M with which we began this example. The intermediate foliations $\mathcal{F}_s$, $0 < s < 1$, are diffeomorphic to $\mathcal{F}_0$, but "look" more and more like $\mathcal{F}_1$ as $s \to 1$. This deformation is one interpretation of "spinning" the leaves of $\mathcal{F}_0$ along ∂M.

Example 1.1.14. Let $\Sigma = \Sigma_2$ denote the 2–holed torus, $C_1, C_2 \subset \Sigma$ the two imbedded circles depicted in Figure 1.1.8. Let $\mathcal{F}$ be the product foliation of the 3–manifold $M = \Sigma \times S^1$ with leaves $\Sigma \times \{y\}$, $y \in S^1$. Note that $N_i = C_i \times S^1$ is an imbedded torus and that $\mathcal{F} \pitchfork N_i$, $i = 1, 2$. Let $\mathrm{Diff}_+(S^1)$ denote the group of orientation–preserving diffeomorphisms of S^1 and choose $f_1, f_2 \in \mathrm{Diff}_+(S^1)$. Cut M apart along N_1 and N_2, letting N_i^+ and N_i^- denote the resulting copies of N_i, $i = 1, 2$. At this point we have a manifold $M' = \Sigma' \times S^1$ with four boundary components $\{N_i^\pm\}_{i=1,2}$. The foliation $\mathcal{F}$ has passed to a foliation $\mathcal{F}' \pitchfork \partial M'$, each leaf of which is of the form $\Sigma' \times \{y\}$, $y \in S^1$ (Figure 1.1.9).

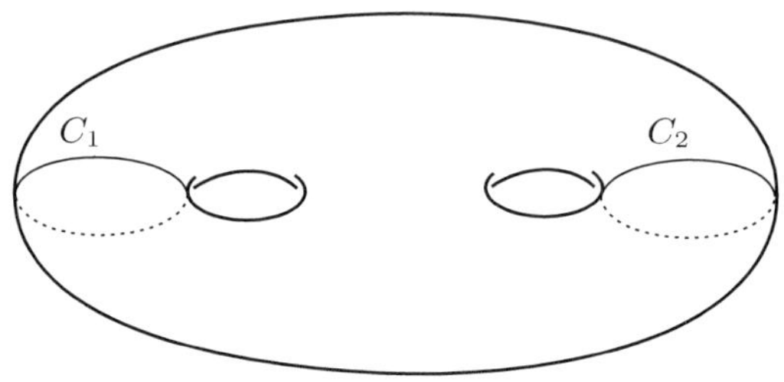

Figure 1.1.8. The 2–holed torus Σ

This leaf meets $\partial M'$ in four circles $C_i^\pm \times \{y\} \subset N_i^\pm$. If $z \in C_i$, we denote the corresponding points in $C_i^\pm$ by $z^\pm$ and "reglue" N_i^- to N_i^+ by the identification

$$(z^-, y) \equiv (z^+, f_i(y)), \quad i = 1, 2.$$

Since f_1 and f_2 are orientation–preserving diffeomorphisms of S^1, they are isotopic to the identity and the manifold obtained by this regluing operation is homeomorphic to M (Exercise 1.1.15). The leaves of $\mathcal{F}'$, however, reassemble to produce a new foliation $\mathcal{F}(f_1, f_2)$ of M. If a leaf L of $\mathcal{F}(f_1, f_2)$ contains a piece $\Sigma' \times \{y_0\}$, then

$$L = \bigcup_{g \in G} \Sigma' \times \{g(y_0)\},$$

where $G \subset \mathrm{Diff}_+(S^1)$ is the subgroup generated by $\{f_1, f_2\}$. These copies of Σ' are attached to one another by identifications

$$(z^-, g(y_0)) \equiv (z^+, f_1(g(y_0))), \qquad \forall\, z \in C_1,$$
$$(z^-, g(y_0)) \equiv (z^+, f_2(g(y_0))), \qquad \forall\, z \in C_2,$$

where g ranges over G. The leaf is completely determined by the G–orbit of $y_0 \in S^1$ and can be simple or immensely complicated. For instance, a leaf will be compact precisely if the corresponding G–orbit is finite. As an extreme example, if G is trivial ($f_1 = f_2 = \mathrm{id}_{S^1}$), then $\mathcal{F}(f_1, f_2) = \mathcal{F}$. If an orbit is dense in S^1, the corresponding leaf is dense in M. As an example, if f_1 and f_2 are rotations through rationally independent multiples of 2π, every leaf will be dense. In other examples, which we will see later, some leaf L has closure $\overline{L}$ that meets each factor $\{w\} \times S^1$ in a Cantor set. Similar constructions can be made on $\Sigma \times I$, where I is a compact, nondegenerate interval. Here we take $f_1, f_2 \in \mathrm{Diff}_+(I)$ and, since ∂I is fixed pointwise by all orientation–preserving diffeomorphisms, we get a foliation having the

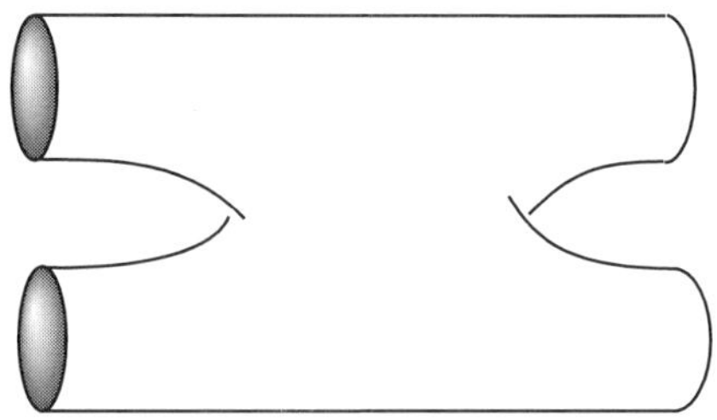

Figure 1.1.9. The typical leaf $\Sigma' \times \{y\}$ of $\mathcal{F}'$

two components of ∂M as leaves. Note that, when we form M' in this case, we get a foliated manifold with corners. In either case, this construction is called the *suspension* of a pair of diffeomorphisms and is a fertile source of interesting examples of codimension–one foliations.

Exercise 1.1.15. Let $f \in \mathrm{Diff}_+(S^1)$ and show that there is a smooth homotopy f_t, such that $f_1 = f$, $f_0 = \mathrm{id}$, and $f_t \in \mathrm{Diff}_+(S^1)$, $0 \le t \le 1$. (We say that f is *isotopic* to the identity.) Use this to justify the assertion in Example 1.1.14 that the regluing of M' described there gives a manifold homeomorphic to M. (In fact, the reglued manifold is *diffeomorphic* to M when its differentiable structure is appropriately defined.)

These examples were somewhat informal. We turn to the precise definitions. In what follows, the symbol $\mathbb{F}^p$ denotes either the full Euclidean space $\mathbb{R}^p$ or Euclidean half space

$$\mathbb{H}^p = \{(x^1, x^2, \dots, x^p) \mid x^1 \le 0\}.$$

Definition 1.1.16. A *rectangular neighborhood* in $\mathbb{F}^n$ is an open subset of the form $B = J_1 \times \cdots \times J_n$, where each J_i is a (possibly unbounded) relatively open interval in the ith coordinate axis. If J_1 is of the form $(a, 0]$, we say that B has boundary

$$\partial B = \{(0, x^2, \dots, x^n) \in B\}.$$

In the following, we will consider coordinate charts that have values in $\mathbb{F}^{n-q} \times \mathbb{F}^q$, allowing the possibility of manifolds with boundary and (convex) corners.

Definition 1.1.17. Let M be an n–manifold. A *foliated chart* on M of codimension q is a pair (U, φ), where $U \subseteq M$ is open and $\varphi : U \to B_\tau \times B_\pitchfork$ is a diffeomorphism, $B_\pitchfork$ being a rectangular neighborhood in $\mathbb{F}^q$ and B_τ

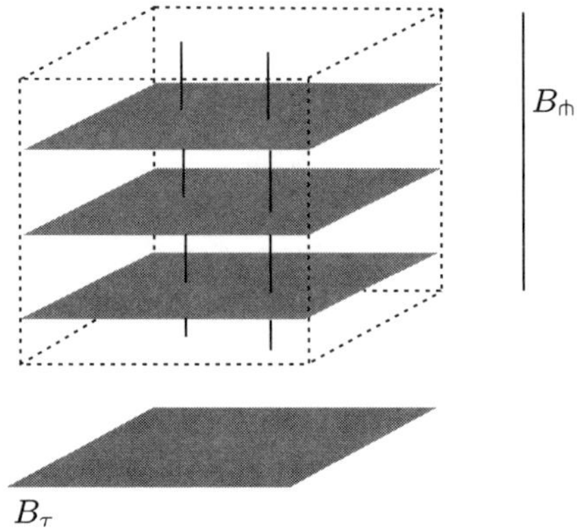

Figure 1.1.10. A 3–dimensional foliated chart

a rectangular neighborhood in $\mathbb{F}^{n-q}$. The set $P_y = \varphi^{-1}(B_\tau \times \{y\})$, where $y \in B_\pitchfork$, is called a *plaque* of this foliated chart. For each $x \in B_\tau$, the set $S_x = \varphi^{-1}(\{x\} \times B_\pitchfork)$ is called a *transversal* of the foliated chart. The set $\partial_\tau U = \varphi^{-1}(B_\tau \times (\partial B_\pitchfork))$ is called the *tangential boundary* of U and $\partial_\pitchfork U = \varphi^{-1}((\partial B_\tau) \times B_\pitchfork)$ is called the *transverse boundary* of U.

In Figure 1.1.10, we show a foliated chart with $n = 3$ and $q = 1$. The plaques are 2–dimensional and the transversals are 1–dimensional.

Remark. A foliated chart will be the basic model for all foliations, the plaques being the leaves. One should read B_τ as "B–tangential" and $B_\pitchfork$ as "B–transverse". One should also pay attention to various possibilities. If both $B_\pitchfork$ and B_τ have empty boundary, the foliated chart models codimension–q foliations of n–manifolds without boundary (Figure 1.1.10). If one, but not both of these rectangular neighborhoods has boundary, the foliated chart models the various possibilities for foliations of n–manifolds with boundary and without corners. Specifically, if $\partial B_\pitchfork \neq \emptyset = \partial B_\tau$, then $\partial U = \partial_\tau U$ is a union of plaques and the foliation by plaques is tangent to the boundary. If $\partial B_\tau \neq \emptyset = \partial B_\pitchfork$, then $\partial U = \partial_\pitchfork U$ is a union of transversals and the foliation is transverse to the boundary. For the case $n = 3$ and $q = 1$, these possibilities are illustrated in Figures 1.1.11 and 1.1.12. Finally, if $\partial B_\pitchfork \neq \emptyset \neq \partial B_\tau$, we obtain a model of a foliated manifold with a corner separating the tangential boundary from the transverse boundary (Figure 1.1.13).

Definition 1.1.18. Let M be an n–manifold, possibly with boundary and corners, and let $\mathcal{F} = \{L_\lambda\}_{\lambda \in \mathfrak{L}}$ be a decomposition of M into connected, topologically immersed submanifolds of dimension $k = n - q$. Suppose that M admits an atlas $\{U_\alpha, \varphi_\alpha\}_{\alpha \in \mathfrak{A}}$ of foliated charts of codimension q such

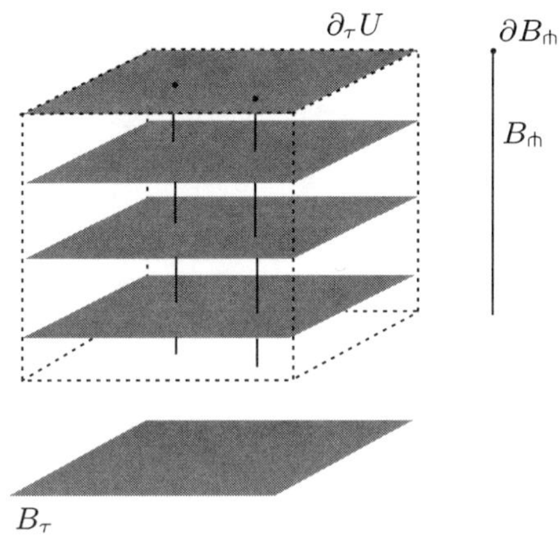

Figure 1.1.11. Foliation tangent to the boundary ($\partial B_\pitchfork \neq \emptyset = \partial B_\tau$)

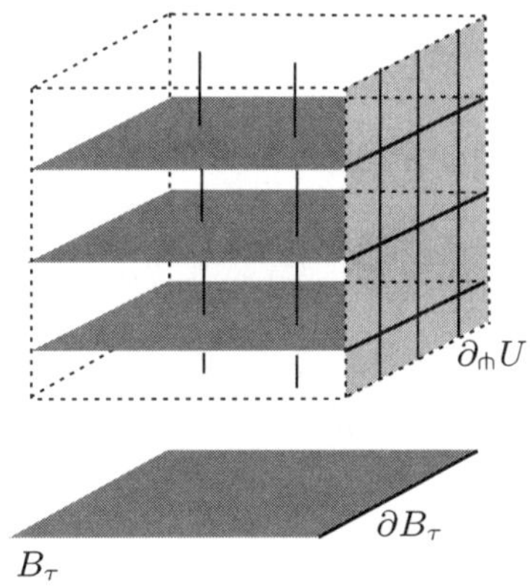

Figure 1.1.12. Foliation transverse to the boundary ($\partial B_\tau \neq \emptyset = \partial B_\pitchfork$)

that, for each $\alpha \in \mathfrak{A}$ and each $\lambda \in \mathfrak{L}$, $L_\lambda \cap U_\alpha$ is a union of plaques. Then $\mathfrak{F}$ is said to be a foliation of M of codimension q (and dimension k) and $\{U_\alpha, \varphi_\alpha\}_{\alpha\in\mathfrak{A}}$ is called a foliated atlas associated to $\mathfrak{F}$. Each L_λ is called a leaf of the foliation and the pair $(M, \mathfrak{F})$ is called a foliated manifold. If the foliated atlas is of class C^r ($0 \leq r \leq \infty$ or $r = \omega$), then the foliation $\mathfrak{F}$ and the foliated manifold $(M, \mathfrak{F})$ are said to be of class C^r.

Remarks. (1) Implicit in this definition is the fact that plaques in distinct foliated charts intersect (if at all) in open subsets of one another. There is no such compatibility requirement for the transversals in distinct foliated charts.

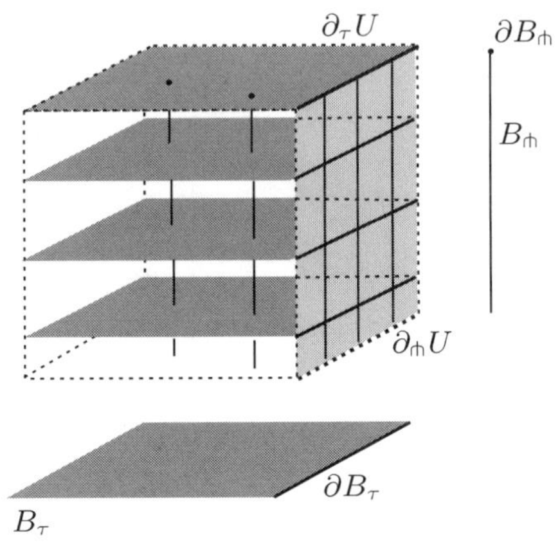

Figure 1.1.13. Foliation with corner ($\partial B_\tau \neq \emptyset \neq \partial B_\pitchfork$)

(2) By a *decomposition* of a set X into a family of subsets $\{X_\lambda\}_{\lambda \in \mathfrak{L}}$, we mean that X is the disjoint union of these subsets. Equivalently, a decomposition is an equivalence relation on X with equivalence classes X_λ. In the case of a foliation $\mathfrak{F}$, we can describe this equivalence relation as follows. Let $\{U_\alpha, \varphi_\alpha\}_{\alpha \in \mathfrak{A}}$ be an associated foliated atlas. If $x, y \in M$, we write $x \sim y$ if and only if there is a finite sequence $\{P_{\alpha_i} \subset U_{\alpha_i}\}_{i=0}^{p}$ of plaques such that $x \in P_{\alpha_0}$, $y \in P_{\alpha_p}$, and $P_{\alpha_{i-1}} \cap P_{\alpha_i} \neq \emptyset$, $1 \leq i \leq p$. It is trivial that this is an equivalence relation on M. The fact that the equivalence classes are exactly the leaves of $\mathfrak{F}$ is left to the reader. In particular, the equivalence relation depends only on $\mathfrak{F}$, not on the choice of the associated foliated atlas. Finally, if $\mathfrak{F}$ is of class C^r, each leaf is a C^r–immersed submanifold (Exercise 1.1.19).

Exercise 1.1.19. Let L be a leaf in a foliated manifold $(M, \mathfrak{F})$ of class C^r. Show that L, in its manifold topology, has a natural C^r structure relative to which the inclusion map $i : L \hookrightarrow M$ is a C^r immersion.

Definition 1.1.20. Let $(M, \mathfrak{F})$ be a foliated manifold and let $\{U_\alpha, \varphi_\alpha\}_{\alpha \in \mathfrak{A}}$ be an associated foliated atlas. Then the *tangential boundary* of M is

$$\partial_\tau M = \bigcup_{\alpha \in \mathfrak{A}} \partial_\tau U_\alpha$$

and the *transverse boundary* is

$$\partial_\pitchfork M = \bigcup_{\alpha \in \mathfrak{A}} \partial_\pitchfork U_\alpha.$$

Evidently, $\partial_\tau M$ is a union of leaves of $\mathfrak{F}$ and $\mathfrak{F} \pitchfork \partial_\pitchfork M$. If $\partial_\tau M \neq \emptyset$ and $\partial_\pitchfork M \neq \emptyset$, then the corners of M are the components (if any) of $\partial_\tau M \cap \partial_\pitchfork M$.

While foliated manifolds with corners are not the primary focus of this book, they occur as intermediate steps in certain constructions of foliated manifolds without corners.

1.2. Foliated Atlases

An equivalent way to define a foliation is to give only the foliated atlas, without reference to a decomposition of M into leaves. Since we have only defined the term "foliated atlas" with reference to such a decomposition, we will need a general definition of this term. In the following, r is allowed to take all values $0 \le r \le \infty$ or $r = \omega$.

Definition 1.2.1. A foliated atlas of codimension q and class C^r on the n–manifold M is a C^r–atlas $\mathfrak{U} = \{U_\alpha, \varphi_\alpha\}_{\alpha\in\mathfrak{A}}$ of foliated charts of codimension q which are *coherently foliated* in the sense that, whenever P and Q are plaques in distinct charts of $\mathfrak{U}$, then $P \cap Q$ is open both in P and Q.

Example 1.2.2. Let $f : \mathbb{R}^{n-1} \to \mathbb{R}$ be smooth and, for each $c \in \mathbb{R}$, set $f_c = f + c$. Let

$$L_c = \{(x^1, \dots, x^{n-1}, f_c(x^1, \dots, x^{n-1}))\}_{(x^1,\dots,x^{n-1})\in\mathbb{R}^{n-1}},$$

the graph of f_c in $\mathbb{R}^n$. To simplify notation, set $x = (x^1, \dots, x^{n-1})$ and $y = x^n$, the last coordinate in $\mathbb{R}^n$. Then,

$$\varphi : \mathbb{R}^n \to \mathbb{R}^n,$$
$$\varphi(x, y) = (x, y - f(x))$$

is a diffeomorphism with inverse

$$\varphi^{-1}(x, y) = (x, y + f(x)).$$

Since

$$\varphi(L_c) = \{(x, c)\}_{x\in\mathbb{R}^{n-1}}, \quad \forall c \in \mathbb{R},$$

the single coordinate chart $(\mathbb{R}^n, \varphi)$ is a foliated atlas having the graphs L_c as plaques. Since there is only one foliated chart, the condition that any two of the charts are coherently foliated is satisfied vacuously. In this example, the codimension is 1.

Example 1.2.3. For a more interesting case, we return to Example 1.1.5. For simplicity, assume that $\partial M = \emptyset$ and let $X \in \mathfrak{X}(M)$ be a nonsingular vector field. This field is not necessarily complete, hence may not generate a global flow. For each $z \in M$, we can find a coordinate chart $(U, x^1, y^1, \dots, y^{n-1})$ about z in which $X|U = \partial/\partial x^1$. We can choose U so that this $\mathbb{R}^n$–chart is rectangular, obtaining a foliated chart of codimension $n-1$, where $x = x^1$ ranges over B_τ and $y = (y^1, \dots, y^{n-1})$ ranges over $B_\pitchfork$. The 1–dimensional plaques, defined by $y \equiv$ const., are local flow lines of X. We cover M with such charts, obtaining an atlas $\mathfrak{U}$. By the uniqueness of solutions of ordinary differential equations, it follows that the intersection of any two local flow lines is an open subset of each, and so $\mathfrak{U}$ is a foliated atlas. Intuitively, it should follow that the full system of plaques links together into a family of connected, immersed submanifolds of dimension 1 which will be the leaves of a foliation. This is not immediately obvious, the main problem being to show that these "leaves" are 2nd countable. In Example 1.1.5, we bypassed the technical problems by assuming that the vector field was complete. The discussion in this section will make clear how the 2nd countability of leaves is forced by the 2nd countability of M.

Example 1.2.4. View $\mathbb{R}^2$ as an additive group, fix $\rho \in \mathbb{R}$, and consider the subgroup $\Lambda_\rho \subset \mathbb{R}^2$ generated by $(0,1)$ and $(1,-\rho)$. These vectors are linearly independent. So $\Lambda_\rho \cong \mathbb{Z}^2$ is a full lattice subgroup and $M = \mathbb{R}^2/\Lambda_\rho$ is diffeomorphic to the torus T^2. The corresponding projection $\mathbb{R}^2 \to M$ is a covering map and, if $B = J_1 \times J_2 \subset \mathbb{R}^2$ is a small enough open rectangular neighborhood, it projects diffeomorphically onto an open set $U \subset M$. The inverse $\varphi : U \to B$ of this diffeomorphism defines a coordinate chart (U, φ) which we view as a foliated chart (U, x, y) of codimension 1. Cover M by an atlas $\{U_\alpha, x_\alpha, y_\alpha\}_{\alpha\in\mathfrak{A}}$ consisting entirely of charts constructed this way. In Exercise 1.2.5, you are asked to verify directly that this is a foliated atlas of class C^ω and is associated to a foliation $\mathcal{F}_\rho$ of M. In Exercise 1.2.6, you will show that there is an analytic diffeomorphism $\theta : M \to T^2$ carrying $\mathcal{F}_\rho$ onto the linear foliation of slope ρ (*cf.* Example 1.1.5).

Exercise 1.2.5. Consider the atlas $\{U_\alpha, \varphi_\alpha\}_{\alpha\in\mathfrak{A}}$ on $M = \mathbb{R}^2/\Lambda_\rho$, constructed as in Example 1.2.4 relative to a choice of $\rho \in \mathbb{R}$. Show that this is a foliated atlas of codimension 1 and of class C^ω. Verify directly (without appeal to the results of this section) that this foliated atlas is associated to a foliation $\mathcal{F}_\rho$ of M by 1–dimensional leaves.

Exercise 1.2.6. If $M = \mathbb{R}^2/\Lambda_\rho$ is constructed as in Example 1.2.4, find an analytic diffeomorphism $\theta : M \to T^2$ that carries $\mathcal{F}_\rho$ to the linear foliation of T^2 of slope ρ as constructed in Example 1.1.5.

Exercise 1.2.7. We say that a nonsingular (*i.e.*, nowhere vanishing) 1–form $\omega \in A^1(M)$ *defines* a foliation $\mathcal{F}$ of codimension 1 if it vanishes exactly on vectors tangent to the leaves of $\mathcal{F}$. By the previous exercise, view the foliation $\mathcal{F}_\rho$ as a foliation of T^2 and show that it is defined by a *closed* nonsingular 1–form ω_ρ on T^2. The *periods* of this closed form are the numbers that can be obtained by line integrals $\int_\sigma \omega_\rho$ over piecewise smooth closed loops σ on T^2. Show that the set of periods is the additive subgroup $G_\rho \subset \mathbb{R}$ generated by 1 and ρ.

Remarks. Let $\mathfrak{U} = \{U_\alpha, \varphi_\alpha\}_{\alpha \in \mathfrak{A}}$ be a C^r foliated atlas of codimension q on the n–manifold M. For $w \in U_\alpha \cap U_\beta$, write

$$\varphi_\alpha(w) = (x_\alpha(w), y_\alpha(w)) \in B^\alpha_\tau \times B^\alpha_\pitchfork,$$
$$\varphi_\beta(w) = (x_\beta(w), y_\beta(w)) \in B^\beta_\tau \times B^\beta_\pitchfork.$$

We often write $(U_\alpha, x_\alpha, y_\alpha)$ for $(U_\alpha, \varphi_\alpha)$, with

$$x_\alpha = (x^1_\alpha, \dots, x^{n-q}_\alpha),$$
$$y_\alpha = (y^1_\alpha, \dots, y^q_\alpha).$$

On $\varphi_\beta(U_\alpha \cap U_\beta)$, consider the change of coordinates formula

$$g_{\alpha\beta}(x_\beta, y_\beta) = \varphi_\alpha \circ \varphi_\beta^{-1}(x_\beta, y_\beta) = (x_\alpha(x_\beta, y_\beta), y_\alpha(x_\beta, y_\beta)).$$

The condition that $(U_\alpha, x_\alpha, y_\alpha)$ and $(U_\beta, x_\beta, y_\beta)$ be coherently foliated means that, if $P \subset U_\alpha$ is a plaque, the connected components of $P \cap U_\beta$ lie in (possibly distinct) plaques of U_β. Equivalently, since the plaques of U_α and U_β are level sets of the transverse coordinates y_α and y_β, respectively, each point $z \in U_\alpha \cap U_\beta$ has a neighborhood in which the formula

$$y_\alpha = y_\alpha(x_\beta, y_\beta) = y_\alpha(y_\beta)$$

is independent of x_β. This is a useful way to reformulate the notion of "coherently foliated" charts.

1.2.A. Coherence and regular foliated atlases. As already mentioned, one wants to link overlapping plaques of a foliated atlas to form the leaves of a foliation. For this and many other purposes, the general definition of "foliated atlas" is a bit clumsy. One problem is that a plaque of $(U_\alpha, \varphi_\alpha)$ can meet multiple plaques of (U_β, φ_β) (Figure 1.2.1). It can even happen that a plaque of one chart meets infinitely many plaques of another chart. We are going to show that no generality is lost in assuming the situation to be much more regular.

Definition 1.2.8. Two foliated atlases $\mathfrak{U}$ and $\mathcal{V}$ on M of the same codimension and smoothness class C^r are *coherent* ($\mathfrak{U} \approx \mathcal{V}$) if $\mathfrak{U} \cup \mathcal{V}$ is a foliated C^r–atlas.

Lemma 1.2.9. *Coherence of foliated atlases is an equivalence relation.*

Proof. Reflexivity and symmetry are immediate. Let $\mathfrak{U} \approx \mathcal{V}$ and $\mathcal{V} \approx \mathcal{W}$. Let $(U_\alpha, x_\alpha, y_\alpha) \in \mathfrak{U}$ and $(W_\lambda, x_\lambda, y_\lambda) \in \mathcal{W}$ and suppose that there is a point $w \in U_\alpha \cap W_\lambda$. Choose $(V_\delta, x_\delta, y_\delta) \in \mathcal{V}$ such that $w \in V_\delta$. By the above remarks, there is a neighborhood N of w in $U_\alpha \cap V_\delta \cap W_\lambda$ such that

$$y_\delta = y_\delta(y_\lambda) \quad \text{on } \varphi_\lambda(N),$$
$$y_\alpha = y_\alpha(y_\delta) \quad \text{on } \varphi_\delta(N),$$

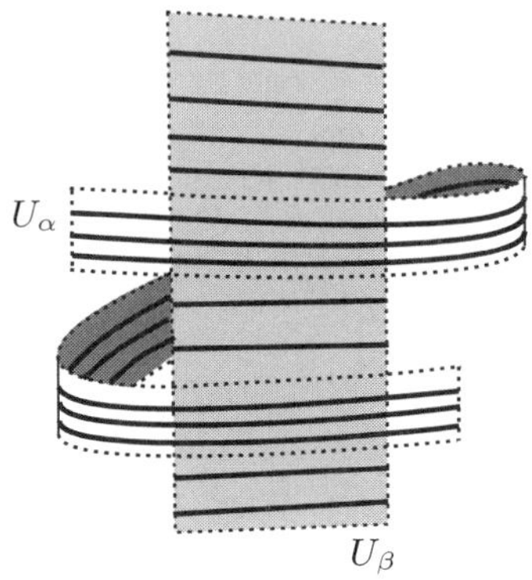

Figure 1.2.1. Plaques of U_α each meet two plaques of U_β.

and hence

$$y_\alpha = y_\alpha(y_\delta(y_\lambda)) \quad \text{on } \varphi_\lambda(N).$$

Since $w \in U_\alpha \cap W_\lambda$ is arbitrary, we conclude that $y_\alpha(x_\lambda, y_\lambda)$ is locally independent of x_λ. We have proven that $\mathcal{U} \approx \mathcal{W}$, hence that coherence is transitive. □

Lemma 1.2.10. *Let $\mathcal{U}$ and $\mathcal{V}$ be foliated atlases on M and suppose that $\mathcal{U}$ is associated to a foliation $\mathcal{F}$. Then $\mathcal{U}$ and $\mathcal{V}$ are coherent if and only if $\mathcal{V}$ is also associated to $\mathcal{F}$.*

Proof. If $\mathcal{V}$ is also associated to $\mathcal{F}$, every leaf L is a union of $\mathcal{V}$–plaques and of $\mathcal{U}$–plaques. These plaques are open subsets in the manifold topology of L, hence intersect in open subsets of each other. Since plaques are connected, a $\mathcal{U}$–plaque cannot intersect a $\mathcal{V}$–plaque unless they lie in a common leaf; so the foliated atlases are coherent.

Conversely, if we only know that $\mathcal{U}$ is associated to $\mathcal{F}$ and that $\mathcal{V} \approx \mathcal{U}$, let Q be a $\mathcal{V}$–plaque. If L is a leaf of $\mathcal{F}$ and $w \in L \cap Q$, let $P \subset L$ be a $\mathcal{U}$–plaque with $w \in P$. Then $P \cap Q$ is an open neighborhood of w in Q and $P \cap Q \subset L \cap Q$. Since $w \in L \cap Q$ is arbitrary, it follows that $L \cap Q$ is open in Q. Since L is an arbitrary leaf, it follows that Q decomposes into disjoint open subsets, each of which is the intersection of Q with some leaf of $\mathcal{F}$. Since Q is connected, $L \cap Q = Q$. Finally, Q is an arbitrary $\mathcal{V}$–plaque, and so $\mathcal{V}$ is associated to $\mathcal{F}$. □

Remark. Let (U, φ) and (W, ψ) be foliated charts such that $\overline{U} \subset W$ and $\varphi = \psi | U$. Then, if $\varphi(U) = B_\tau \times B_\pitchfork$, we see that $\psi|\overline{U}$, written $\overline{\varphi}$, carries $\overline{U}$

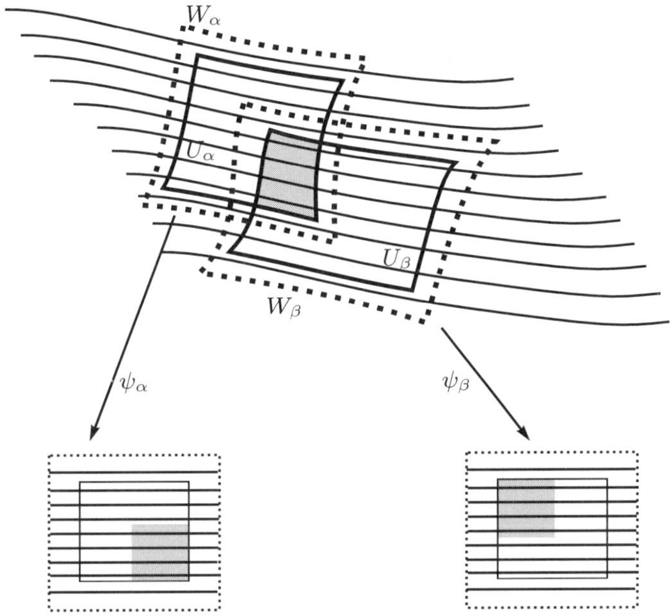

Figure 1.2.2. Sample charts in a regular foliated atlas

diffeomorphically onto $\overline{B}_\tau \times \overline{B}_\pitchfork$. Thus, we can speak of the (closed) plaques and transversals of $\overline{U}$.

Definition 1.2.11. A foliated atlas $\mathfrak{U} = \{U_\alpha, \varphi_\alpha\}_{\alpha \in \mathfrak{A}}$ of class C^r is said to be *regular* if

1. for each $\alpha \in \mathfrak{A}$, $\overline{U}_\alpha$ is a compact subset of a foliated chart (W_α, ψ_α) and $\varphi_\alpha = \psi_\alpha | U_\alpha$;
2. the cover $\{U_\alpha\}_{\alpha \in \mathfrak{A}}$ is locally finite;
3. if $(U_\alpha, \varphi_\alpha)$ and (U_β, φ_β) are elements of $\mathfrak{U}$, then the interior of each closed plaque $P \subset \overline{U}_\alpha$ meets at most one plaque in $\overline{U}_\beta$.

Remarks. Properties (1) and (3) in this definition are illustrated in Figure 1.2.2. By (1), the coordinates x_α and y_α extend to coordinates $\overline{x}_\alpha$ and $\overline{y}_\alpha$ on $\overline{U}_\alpha$ and we write $\overline{\varphi}_\alpha = (\overline{x}_\alpha, \overline{y}_\alpha)$. Property (3) is equivalent to requiring that, if $U_\alpha \cap U_\beta \neq \emptyset$, the transverse coordinate changes

$$\overline{y}_\alpha = \overline{y}_\alpha(\overline{x}_\beta, \overline{y}_\beta)$$

be independent of $\overline{x}_\beta$. That is,

$$\overline{g}_{\alpha\beta} = \overline{\varphi}_\alpha \circ \overline{\varphi}_\beta^{-1} : \overline{\varphi}_\beta(\overline{U}_\alpha \cap \overline{U}_\beta) \to \overline{\varphi}_\alpha(\overline{U}_\alpha \cap \overline{U}_\beta)$$

has the formula

$$\overline{g}_{\alpha\beta}(\overline{x}_\beta, \overline{y}_\beta) = (\overline{x}_\alpha(\overline{x}_\beta, \overline{y}_\beta), \overline{y}_\alpha(\overline{y}_\beta)).$$

A fortiori, similar assertions hold without the overlines. Note that the transverse coordinate map y_α can be viewed as a submersion

$$y_\alpha : U_\alpha \to \mathbb{F}^q = \mathbb{R}^q \text{ or } \mathbb{H}^q$$

and that the formulas $y_\alpha = y_\alpha(y_\beta)$ can be viewed as diffeomorphisms

$$\gamma_{\alpha\beta} : y_\beta(U_\alpha \cap U_\beta) \to y_\alpha(U_\alpha \cap U_\beta).$$

These satisfy the *cocycle conditions*. That is, on $y_\delta(U_\alpha \cap U_\beta \cap U_\delta)$,

(1.2.1) $$\gamma_{\alpha\delta} = \gamma_{\alpha\beta} \circ \gamma_{\beta\delta}$$

and, in particular,

(1.2.2) $$\gamma_{\alpha\alpha} = \mathrm{id}_{y_\alpha(U_\alpha)},$$

(1.2.3) $$\gamma_{\alpha\beta} = \gamma_{\beta\alpha}^{-1}.$$

Definition 1.2.12. The set $\gamma = \{\gamma_{\alpha\beta}\}_{\alpha,\beta\in\mathfrak{A}}$ is the *holonomy cocycle* of the regular foliated atlas.

Remark. The holonomy cocycle plays a central role in foliation theory. Note that, by property (1) in the definition of regular foliated atlas, the set $y_\beta(U_\alpha \cap U_\beta)$ has compact closure in $\mathbb{F}^q$ and, by property (3), that $\gamma_{\alpha\beta}$ has a well–defined, smooth (of class C^r) extension $\overline{\gamma}_{\alpha\beta}$ to this compact set. In particular, the component functions of $\gamma_{\alpha\beta}$ and their derivatives of all orders $\leq r$ are bounded functions.

Exercise 1.2.13. Choosing the atlas in Exercise 1.2.5 with a little care, show that it can be assumed to be regular. Show that the associated holonomy cocycle consists of transformations of the form

$$\gamma_{\alpha\beta}(y) = y + c_{\alpha\beta}$$

and that this regular atlas can be chosen so that the numbers $c_{\alpha\beta}$ generate G_ρ.

Definition 1.2.14. Let $(M, \mathcal{F})$ be smooth of class C^1 and let $\{\gamma_{\alpha\beta}\}_{\alpha\beta\in\mathfrak{A}}$ be the associated holonomy cocycle of a regular foliated atlas for $\mathcal{F}$. We say that $\mathcal{F}$ is *transversely orientable* if $\det J\gamma_{\alpha\beta} > 0$, $\forall\, \alpha, \beta \in \mathfrak{A}$, where $J\gamma$ denotes the Jacobian matrix of the C^1 diffeomorphism γ.

Exercise 1.2.15. Let $(M, \mathcal{F})$ be C^1, let $T(\mathcal{F}) \subset T(M)$ be the subbundle tangent to the leaves, and define the *normal bundle* of $\mathcal{F}$ to be the quotient bundle $Q = T(M)/T(\mathcal{F})$ over M. Prove that $\mathcal{F}$ is transversely orientable if and only if Q is an orientable bundle. In case $\mathcal{F}$ is smooth of codimension one, show that this is equivalent to the existence of a nonsingular 1–form that defines $\mathcal{F}$ (*cf.* Exercise 1.2.7). In particular, transverse orientability is independent of the choice of regular foliated atlas for $\mathcal{F}$.

Lemma 1.2.16. *Every regular foliated atlas of codimension q is associated to a unique foliation $\mathfrak{F}$ of codimension q.*

Proof. Let $\mathfrak{U} = \{U_\alpha, \varphi_\alpha\}_{\alpha\in\mathfrak{A}}$ be a regular foliated atlas of codimension q. Define an equivalence relation on M by setting $x \sim y$ if and only if either there is a $\mathfrak{U}$–plaque P_0 such that $x, y \in P_0$ or there is a sequence $\{P_0, P_1, \dots, P_p\}$ of $\mathfrak{U}$–plaques such that

$$\begin{aligned} &x \in P_0, \\ &y \in P_p, \\ &P_i \cap P_{i-1} \neq \emptyset, \qquad 1 \le i \le p. \end{aligned}$$

This will be called a *plaque chain of length p* connecting x and y. In the case that $x, y \in P_0$, we say that $\{P_0\}$ is a plaque chain of length 0 connecting x and y. The fact that $\sim$ is an equivalence relation is clear. It is also clear that each equivalence class L is a union of plaques. Since $\mathfrak{U}$–plaques can only overlap in open subsets of each other, L is locally a topologically immersed submanifold of dimension $n - q$. The open subsets of the plaques $P \subset L$ form the base of a locally Euclidean topology on L of dimension $n - q$ and L is clearly connected in this topology. It is also trivial to check that L is Hausdorff. The main problem is to show that L is 2nd countable. Since each plaque is 2nd countable, the same will hold for L if we show that the set of $\mathfrak{U}$–plaques in L is at most countably infinite. Fix one such plaque P_0. By the definition of a regular, foliated atlas, P_0 meets only finitely many other plaques. That is, there are only finitely many plaque chains $\{P_0, P_1\}$ of length 1. By induction on the length p of plaque chains that begin at P_0, we prove similarly that there are only finitely many of length $\le p$. Since every $\mathfrak{U}$–plaque in L is, by the definition of $\sim$, reached by a finite plaque chain starting at P_0, the assertion follows. □

Lemma 1.2.17. *Every foliated atlas has a coherent refinement that is regular.*

Proof. Fix a metric on M and a foliated atlas $\mathcal{W}$. Consider first the case that M is compact. Passing to a subcover, if necessary, we can assume that $\mathcal{W} = \{W_j, \psi_j\}_{j=1}^{\ell}$ is finite. Let $\epsilon > 0$ be a Lebesgue number for $\mathcal{W}$. That is, any subset $X \subseteq M$ of diameter $< \epsilon$ lies entirely in some W_j. For each $x \in M$, choose j such that $x \in W_j$ and choose a foliated chart (U_x, φ_x) such that

$$\begin{gathered} x \in U_x \subseteq \overline{U}_x \subset W_j, \\ \varphi_x = \psi_j | U_x, \\ \operatorname{diam}(U_x) < \epsilon/2. \end{gathered}$$

Suppose that $U_x \subset W_k$, $k \neq j$, and write $\psi_k = (x_k, y_k)$ as usual, where

$$y_k : W_k \to \mathbb{F}^q$$

is the transverse coordinate map. This is a submersion having the plaques in W_k as level sets. Thus, y_k restricts to a submersion

$$y_k : U_x \to \mathbb{F}^q.$$

This is locally constant in x_j; so choosing U_x smaller, if necessary, we can assume that $y_k|\overline{U}_x$ has the plaques of $\overline{U}_x$ as its level sets. That is, each plaque of W_k meets (hence contains) at most one (compact) plaque of $\overline{U}_x$. Since $1 \leq k \leq \ell < \infty$, we can choose U_x so that, whenever $U_x \subset W_k$, distinct plaques of $\overline{U}_x$ lie in distinct plaques of W_k. Pass to a finite subatlas $\mathcal{U} = \{U_i, \varphi_i\}_{i=1}^N$ of $\{U_x, \varphi_x\}_{x\in M}$. If $U_i \cap U_j \neq \emptyset$, then $\operatorname{diam}(U_i \cup U_j) < \epsilon$, and so there is an index k such that $\overline{U}_i \cup \overline{U}_j \subseteq W_k$. Distinct plaques of $\overline{U}_i$ (respectively, of $\overline{U}_j$) lie in distinct plaques of W_k. Hence each plaque of $\overline{U}_i$ has interior meeting at most one plaque of $\overline{U}_j$ and *vice versa.* By construction, $\mathcal{U}$ is a coherent refinement of $\mathcal{W}$ and is a regular foliated atlas.

If M is not compact, local compactness and 2nd countability allows us to choose a sequence $\{K_i\}_{i=0}^\infty$ of compact subsets such that

$$K_i \subset \operatorname{int} K_{i+1}, \qquad \forall i \geq 0,$$

$$M = \bigcup_{i=0}^{\infty} K_i.$$

Passing to a subatlas, we assume that

$$\mathcal{W} = \{W_j, \psi_j\}_{j=0}^\infty$$

is countable and we find a strictly increasing sequence $\{n_\ell\}_{\ell=0}^\infty$ of positive integers such that

$$\mathcal{W}_\ell = \{W_j, \psi_j\}_{j=0}^{n_\ell}$$

covers K_ℓ. Let δ_ℓ denote the distance from K_ℓ to $\partial K_{\ell+1}$ and choose $\epsilon_\ell > 0$ so small that

$$\epsilon_\ell < \min\left\{\frac{\delta_\ell}{2}, \epsilon_{\ell-1}\right\}, \qquad \ell \geq 1,$$

$$\epsilon_0 < \frac{\delta_0}{2},$$

and ϵ_ℓ is a Lebesgue number for $\mathcal{W}_\ell$ (as an open cover of K_ℓ) and for $\mathcal{W}_{\ell+1}$ (as an open cover of $K_{\ell+1}$). More precisely, if $X \subset M$ meets K_ℓ (respectively, $K_{\ell+1}$) and $\operatorname{diam} X < \epsilon_\ell$, then X lies in some element of $\mathcal{W}_\ell$ (respectively, $\mathcal{W}_{\ell+1}$). For each $x \in K_\ell \smallsetminus \operatorname{int} K_{\ell-1}$, construct (U_x, φ_x) as for the compact case, requiring that $\overline{U}_x$ be a compact subset of W_j and that $\varphi_x = \psi_j|U_x$, some $j \leq n_\ell$. Also, require that $\operatorname{diam} \overline{U}_x < \epsilon_\ell/2$. As before, pass to a finite subcover $\{U_i, \varphi_i\}_{i=n_{\ell-1}+1}^{n_\ell}$ of $K_\ell \smallsetminus \operatorname{int} K_{\ell-1}$. (Here, we take $n_{-1} = 0$.)

This creates a regular foliated atlas $\mathcal{U} = \{U_i, \varphi_i\}_{i=1}^{\infty}$ that refines $\mathcal{W}$ and is coherent with $\mathcal{W}$. □

The following is now obvious.

Theorem 1.2.18. *The correspondence between foliations on M and their associated foliated atlases induces a one–to–one correspondence between the set of foliations on M and the set of coherence classes of foliated atlases.*

We now have an alternative definition of the term "foliation".

Definition 1.2.19. A foliation $\mathcal{F}$ of codimension q and class C^r on M is a coherence class of foliated atlases of codimension q and class C^r on M.

By Zorn's lemma, it is obvious that every coherence class of foliated atlases contains a unique maximal foliated atlas. Thus,

Definition 1.2.20. A foliation of codimension q and class C^r on M is a maximal foliated C^r–atlas of codimension q on M.

In practice, we generally use a relatively small foliated atlas to represent a foliation. Usually, we also require this atlas to be regular.

Example 1.2.21. A C^r foliation $\mathcal{F}$ induces a C^r foliation $\mathcal{F}|\partial_{\pitchfork} M$ on $\partial_{\pitchfork} M$ having the same codimension q as $\mathcal{F}$. The leaves of this foliation are the connected components of the intersections $L \cap \partial_{\pitchfork} M$, as L ranges over the leaves of $\mathcal{F}$. Indeed, if $\{U_\alpha, x_\alpha, y_\alpha\}_{\alpha \in \mathfrak{A}}$ is a regular foliated atlas for $\mathcal{F}$, let

$$\mathfrak{A}' = \{\alpha \in \mathfrak{A} \mid W_\alpha = U_\alpha \cap \partial_{\pitchfork} M \neq \emptyset\}.$$

By notational abuse, let

$$\begin{aligned} x_\alpha &= x_\alpha | W_\alpha, \\ y_\alpha &= y_\alpha | W_\alpha, \end{aligned}$$

$\forall\, \alpha \in \mathfrak{A}'$. It is then elementary to check that $\{W_\alpha, x_\alpha, y_\alpha\}_{\alpha \in \mathfrak{A}'}$ is a regular foliated atlas of codimension q and smoothness class C^r. Evidently, the leaves of the associated C^r foliation of $\partial_{\pitchfork} M$ are as asserted. Similarly, $\mathcal{F}$ induces a C^r–foliation $\mathcal{F}|\partial_\tau M$ on $\partial_\tau M$ of codimension $q - 1$.

In the case of foliations $\mathcal{F}$ of codimension 1, the components of $\partial_\tau M$ are leaves of $\mathcal{F}$ and the corresponding foliation $\mathcal{F}|\partial_\tau M$ is of little interest, being of codimension 0.

1.2.B. Foliations of class C^0. The case $r = 0$ is rather special. Those C^0 foliations that arise in practice are usually "smooth–leaved". More precisely, they are of class $C^{r,0}$, in the following sense.

Definition 1.2.22. A foliation $\mathfrak{F}$ is of class $C^{r,k}$, $r > k \geq 0$, if the corresponding coherence class of foliated atlases contains a regular foliated atlas $\{U_\alpha, x_\alpha, y_\alpha\}_{\alpha\in\mathfrak{A}}$ such that the change of coordinate formula

$$g_{\alpha\beta}(x_\beta, y_\beta) = (x_\alpha(x_\beta, y_\beta), y_\alpha(y_\beta))$$

is of class C^k, but x_α is of class C^r in the coordinates x_β and its mixed x_β partials of orders $\leq r$ are C^k in the coordinates (x_β, y_β).

Remark. Definition 1.2.22 suggests the more general concept of a *foliated space* or *abstract lamination* $(M, \mathfrak{F})$. One relaxes the condition that the transversals be open, relatively compact subsets of $\mathbb{R}^q$, allowing the transverse coordinates y_α to take their values in some more general topological space Z. The plaques are still open, relatively compact subsets of $\mathbb{R}^p$, the change of transverse coordinate formula $y_\alpha(y_\beta)$ is continuous and $x_\alpha(x_\beta, y_\beta)$ is of class C^r in the coordinates x_β and its mixed x_β partials of orders $\leq r$ are continuous in the coordinates (x_β, y_β). One usually requires M and Z to be locally compact, second countable and metrizable. This may seem like a rather wild generalization, but there are contexts in which it is useful. In this book, foliated spaces will become important in Part 3.

Example 1.2.23. For foliations of class $C^{r,0}$ and, more generally, for foliated spaces $(M, \mathfrak{F})$, the holonomy cocycle is C^0, but the individual leaves come with a C^r structure. The tangent spaces to the leaves are well defined and form a continuous vector bundle $T(\mathfrak{F})$ on M, called the tangent bundle of $\mathfrak{F}$. Indeed, this bundle is determined by the continuous structure cocycle

$$g_{\alpha\beta} : U_\alpha \cap U_\beta \to \mathrm{Gl}(p),$$

$$g_{\alpha\beta}(x_\beta, y_\beta) = \left[\frac{\partial x^i_\alpha}{\partial x^j_\beta}(x_\beta, y_\beta)\right].$$

In the case of a $C^{r,0}$–foliated manifold, the atlas in Definition 1.2.22 only establishes M as a C^0 manifold and one does not have a tangent bundle $T(M)$. Even if there is a differentiable structure on M, there seems to be no natural way to identify $T(\mathfrak{F})$ as a p–plane subbundle of $T(M)$. The following slightly stronger requirement on $\mathfrak{F}$ remedies this.

Definition 1.2.24. Let $\mathfrak{F}$ be of class $C^{r,0}$ on a manifold M. If M has a differentiable structure relative to which each leaf is C^1–immersed and the resulting inclusion $T(\mathfrak{F}) \hookrightarrow T(M)$ imbeds $T(\mathfrak{F})$ as a C^0 p–plane subbundle of $T(M)$, then $(M, \mathfrak{F})$ is said to be of class $C^{r,0+}$. A foliation is said to be *integral to a C^0 plane field* if and only if it is of class at least $C^{1,0+}$.

Example 1.2.25. Let $(M, \mathfrak{F})$ be a foliated n–manifold of codimension 1 and suppose that M is smooth and $\mathfrak{F}$ is integral to a C^0 hyperplane field $E \subset T(M)$. It follows that $\mathfrak{F}$ is of class $C^{\infty,0+}$. We outline the main ideas here, leaving details until Section 5.1. A Riemannian metric on M defines a C^0 line field $E^\perp$ which, by a small homotopy, becomes a smooth subbundle of $T(M)$ transverse to E. With care, one can guarantee that this line field is tangent to $\partial_\pitchfork M$. By the Frobenius theorem (Theorem 1.3.8), smooth line fields on M are integrable, giving rise to a smooth, 1–dimensional foliation $\mathcal{L}$ of M that is transverse to $\mathcal{F}$ and tangent to $\partial_\pitchfork M$. In this situation, it is possible to choose an atlas $\{U_\alpha, x_\alpha, y_\alpha\}_{\alpha\in\mathfrak{A}}$ that is simultaneously a regular foliated atlas for $\mathcal{F}$ and $\mathcal{L}$. That is, x_α is $\mathbb{R}^{n-1}$–valued, y_α is $\mathbb{R}$–valued, the level sets of x_α are plaques of $\mathcal{L}$ and the level sets of y_α are plaques of $\mathcal{F}$. The change of coordinates on $U_\alpha \cap U_\beta$ takes the form

$$\begin{aligned} x_\alpha &= x_\alpha(x_\beta), \\ y_\alpha &= y_\alpha(y_\beta), \end{aligned}$$

where $x_\alpha(x_\beta)$ is of class C^∞ but $y_\alpha(y_\beta)$ may only be continuous. This proves that the foliated manifold is of class $C^{\infty,0}$. The hypothesis that $\mathcal{F}$ is integral to E means that each leaf of $\mathcal{F}$ is a C^1–immersed submanifold of the smooth manifold M, the vector bundle $E \subset T(M)$ being identified via these immersions with $T(\mathcal{F})$. Furthermore, if the holonomy cocycle $\gamma_{\alpha\beta}(y_\beta) = y_\alpha(y_\beta)$ is of class C^r, $1 \le r \le \infty$, $\forall\, \alpha, \beta \in \mathfrak{A}$, then $(M, \mathcal{F})$ is a C^r–foliated manifold.

1.2.C. Transverse structures. Let $\gamma = \{\gamma_{\alpha\beta}\}_{\alpha,\beta\in\mathfrak{A}}$ be the holonomy cocycle associated to a regular foliated atlas on M and let $\mathcal{F}$ denote the corresponding foliation. The cocycle consists of C^r diffeomorphisms between open subsets of $\mathbb{R}^q$. Suppose that $\mathcal{S}$ is a geometric structure on an open subset U of $\mathbb{R}^q$ containing the domains and ranges of all $\gamma_{\alpha\beta} \in \gamma$ and let $G \subseteq \operatorname{Diff}^r(U)$ be the group of automorphisms of $\mathcal{S}$.

Definition 1.2.26. If each $\gamma_{\alpha\beta} \in \gamma$ is the restriction to $\operatorname{dom}\gamma_{\alpha\beta}$ of an element of G, then the holonomy cocycle γ is said to be a transverse G–structure for the foliated manifold $(M, \mathcal{F})$.

We have already seen an example of this in Definition 1.2.14. A transversely orientable foliation is one admitting a transverse G–structure with $G = \operatorname{Diff}^r_+(U)$.

Example 1.2.27. Let $\mathcal{S}$ be a Riemannian metric on an open subset $U \subseteq \mathbb{R}^q$, $G = \operatorname{Iso}(U)$ the group of isometries of this metric. A foliated manifold $(M, \mathcal{F})$ admitting a transverse $\operatorname{Iso}(U)$ structure is said to be (transversely) Riemannian. While every q–manifold admits a Riemannian metric, transverse Riemannian structures for foliations are rare, imposing rigid symmetries on the foliated manifold. The theory of Riemannian foliations is highly developed [**91**] and very interesting, but will not be pursued in this book.

Example 1.2.28. Another important example is the class of (transversely) Lie foliations. Let G be a q–dimensional Lie group. Using local coordinate charts on G, one can suppose that the domain $y_\beta(U_\alpha \cap U_\beta)$ of $\gamma_{\alpha\beta}$ is identified with an open subset $V_{\alpha\beta} \subset G$, $\forall\, \alpha, \beta \in \mathfrak{A}$. If this can be done in such a way that each $\gamma_{\alpha\beta}$ is identified with (the restriction to $V_{\alpha\beta}$ of) left translation in G by an element of G, the foliation is (transversely) Lie, modeled on the Lie group G. Again, this places an enormous restriction on the foliated manifold $(M, \mathcal{F})$. Remark that there is a left–invariant Riemannian metric on G and that this induces a transversely Riemannian structure on $(M, \mathcal{F})$.

Example 1.2.29. Similarly, one can define (transversely) homogeneous foliations modeled on a q–dimensional homogeneous space G/K. Here, the domains $V_{\alpha\beta}$ of the cocycle elements $\gamma_{\alpha\beta}$ are identified with open subsets of G/K in such a way that $\gamma_{\alpha\beta}$ becomes (the restriction to $V_{\alpha\beta}$ of) a left translation in G/K by an element of G.

By this point, readers can extend the list of examples for themselves. Every example of a geometric structure on q–manifolds yields an analogous transverse structure for foliated manifolds of codimension q. This important topic is one of many that will not be pursued systematically in this book. The interested reader is referred to [**52**, Chapitre III].

1.3. The Frobenius Theorem

Foliations are sometimes given by infinitesimal data, either by a C^∞ k–plane distribution $E \subseteq T(M)$ or by the annihilator ideal $I^*(E) \subseteq A^*(M)$ of such a distribution. This infinitesimal data amounts to the global analogue of a system of first–order partial differential equations of constant rank. Generally, such a system is *overdetermined* and corresponds to a foliation if and only if it is *completely integrable.* The integrability condition is the classical result known as the Frobenius theorem. Here, we only review this theorem. Proofs will be found, for example, in [**11**, pp. 159–161], [**29**, pp. 118–123 and pp. 178–180], [**15**, Appendix], [**126**, pp. 130–136], and [**147**, pp. 42–45].

Let E be a k–plane distribution on a C^∞ n–manifold M. That is, E is the total space of a C^∞, k–dimensional subbundle of the tangent bundle $\pi : T(M) \to M$. A C^∞ foliation $\mathcal{F}$ of dimension k gives rise to such a distribution, the vector subspace $E_x \subseteq T_x(M)$ being the tangent space to the leaf of $\mathcal{F}$ through x, $\forall\, x \in M$. We have called E the *tangent bundle to the foliation* and have denoted it by $T(\mathcal{F})$ (*cf.* Example 1.2.23). We are interested in the inverse problem: given the k–plane distribution E, can we find a foliation $\mathcal{F}$ with k–dimensional leaves such that $E = T(\mathcal{F})$?

Example 1.3.1. Let $E \subset T(\mathbb{R}^3)$ be a 2–plane distribution spanned by C^∞ vector fields of the form

$$X = \frac{\partial}{\partial x} + g(x, y)\frac{\partial}{\partial z},$$
$$Y = \frac{\partial}{\partial y} + h(x, y)\frac{\partial}{\partial z}.$$

Let $\rho : \mathbb{R}^3 \to \mathbb{R}^2$ be the projection $\rho(x, y, z) = (x, y)$ and assume that $E = T(\mathcal{F})$ for some 2–dimensional foliation $\mathcal{F}$. Since

$$\rho_{*(x,y,z)}(X_{(x,y,z)}) = \left.\frac{\partial}{\partial x}\right|_{(x,y)},$$
$$\rho_{*(x,y,z)}(Y_{(x,y,z)}) = \left.\frac{\partial}{\partial y}\right|_{(x,y)},$$

the inverse function theorem implies that a leaf of $\mathcal{F}$ through any point (x_0, y_0, z_0) has a neighborhood of (x_0, y_0, z_0) that is carried by ρ diffeomorphically onto an open subset of $\mathbb{R}^2$. That is, a leaf of $\mathcal{F}$ is locally a graph $z = f(x, y)$ of a C^∞ function f and

$$X_{(x,y,f(x,y))} = \begin{bmatrix} 1 \\ 0 \\ g(x, y) \end{bmatrix} = \begin{bmatrix} 1 \\ 0 \\ \frac{\partial f}{\partial x}(x, y) \end{bmatrix},$$
$$Y_{(x,y,f(x,y))} = \begin{bmatrix} 0 \\ 1 \\ h(x, y) \end{bmatrix} = \begin{bmatrix} 0 \\ 1 \\ \frac{\partial f}{\partial y}(x, y) \end{bmatrix}.$$

The locally–defined function $f(x, y)$ solves the (overdetermined) system of partial differential equations

$$\frac{\partial f}{\partial x} = g(x, y),$$
$$\frac{\partial f}{\partial y} = h(x, y).$$

This implies that

$$\frac{\partial g}{\partial y} = \frac{\partial^2 f}{\partial y\,\partial x} = \frac{\partial^2 f}{\partial x\,\partial y} = \frac{\partial h}{\partial x}.$$

That is, a necessary condition for the existence of the foliation such that $E = T(\mathcal{F})$ is that

$$(*) \qquad \frac{\partial g}{\partial y} = \frac{\partial h}{\partial x}.$$

Conversely, this condition implies that $\omega = g\,dx + h\,dy$ is closed, hence locally exact. Local solutions of $df = \omega$ provide locally–defined functions f, unique up to an additive constant, whose graphs are local leaves tangent to E. In this way, one easily produces a foliated atlas on $\mathbb{R}^3$, hence a foliation $\mathcal{F}$ such that $E = T(\mathcal{F})$. Condition $(*)$, which we call an *integrability condition*, can also be written in terms of brackets. Indeed,

$$[X, Y] = \left(\frac{\partial h}{\partial x} - \frac{\partial g}{\partial y}\right)\frac{\partial}{\partial z},$$

and so equation $(*)$ becomes

$$[X, Y] = 0$$

(the fields *commute*). Arbitrary fields $Z_1, Z_2 \in \Gamma(E)$ (the space of C^∞ sections $s : \mathbb{R}^3 \to E$) are $C^\infty(\mathbb{R}^3)$–linear combinations of X and Y, and hence commutativity of X and Y implies that $[Z_1, Z_2] \in \Gamma(E)$. Our integrability condition becomes: *the space* $\Gamma(E)$ *of* C^∞ *sections of* E *is a Lie subalgebra of* $\mathfrak{X}(\mathbb{R}^3)$. This exemplifies the *vector field* version of the Frobenius theorem.

Remark. In the above example, the words "local" and "locally" can be dropped throughout. One proves this using the simple connectivity of $\mathbb{R}^3$ and $\mathbb{R}^2$.

Throughout this section, all manifolds, distributions, vector fields, forms, *etc.*, will be of class C^∞. Thus, $\mathfrak{X}(M) = \Gamma(T(M))$ is closed under the bracket, and hence is a Lie algebra. Also, the exterior algebra $A^*(M)$ is closed under exterior differentiation, and hence is a differential graded algebra. For simplicity of exposition, we will also assume that $\partial M = \emptyset$, although the reader can easily adapt our treatment to the more general case.

Definition 1.3.2. A k–plane distribution E on the C^∞ manifold M is the total space of a C^∞, k–dimensional subbundle of the tangent bundle $T(M)$.

Definition 1.3.3. A k–plane distribution E on M is *involutive* if the space $\Gamma(E)$ of C^∞ sections of E is a Lie subalgebra of $\mathfrak{X}(M)$.

Definition 1.3.4. Let E be a k–plane distribution on M, $w \in M$. An *integral manifold* to E through w is a smoothly immersed submanifold N of M of dimension k such that $w \in N$ and $E_x = T_x(N)$, $\forall\, x \in N$.

Definition 1.3.5. A k–plane distribution E on M is *completely integrable* if, through each point $w \in M$, there passes an integral manifold to E.

In the following, $\Lambda^p(\Gamma(E))$ denotes the pth exterior power of $\Gamma(E)$ as a $C^\infty(M)$–module.

Definition 1.3.6. Let E be a k–plane distribution on M. The *annihilator* $I^*(E) \subseteq A^*(M)$ is the graded ideal $I^*(E) = \{I^p(E)\}_{p=0}^\infty$ defined by the condition: $\omega \in I^p(E) \Leftrightarrow \omega(v) \equiv 0$, $\forall\, v \in \Lambda^p(\Gamma(E))$.

Remark that $\Lambda^0(\Gamma(E)) = C^\infty(M)$, and hence $I^0(E) = \{0\}$.

Definition 1.3.7. A *differential graded ideal* $I^* \subseteq A^*(M)$ is a graded ideal such that $d(I^*) \subseteq I^*$.

Theorem 1.3.8 (Frobenius Theorem). *For a k–plane distribution E on M, the following are equivalent:*

1. *E is completely integrable;*
2. *$I^*(E)$ is a differential graded ideal;*
3. *$d(I^1(E)) \subseteq I^2(E)$;*
4. *E is involutive;*
5. *there is a C^∞ foliated atlas on M of codimension $q = n - k$, every plaque of which is an integral manifold to E.*

Remarks. The implications (1) $\Rightarrow$ (2), (2) $\Rightarrow$ (3), and (5) $\Rightarrow$ (1) are trivial. The implication (3) $\Rightarrow$ (4) is an elementary consequence of the well–known formula

$$d\omega(X \wedge Y) = X\omega(Y) - Y\omega(X) - \omega([X,Y]), \tag{1.3.1}$$

where $\omega \in A^1(M)$ and $X, Y \in \mathfrak{X}(M)$. The difficult step is to prove (4) $\Rightarrow$ (5). As remarked earlier, this is a standard result in manifold theory and will not be proven in this book.

As it stands, the Frobenius theorem is a local result. But Theorem 1.2.18 and property (5) of Theorem 1.3.8 imply the following global consequence.

Corollary 1.3.9. *The k–plane distribution on M is completely integrable if and only if $E = T(\mathcal{F})$ for a C^∞ foliation $\mathcal{F}$ of codimension $q = n - k$.*

The following examples show some of the ways that the Frobenius theorem is used in foliation theory.

Example 1.3.10. We return to Example 1.1.8. If G is a Lie group and $H \hookrightarrow G$ is a connected Lie subgroup, let $\mathfrak{g}$ and $\mathfrak{h} \subset \mathfrak{g}$ be the corresponding Lie algebras of left–invariant vector fields on G. Then $\mathfrak{h}$ spans a distribution $E \subset T(G)$ and the fact that $\mathfrak{h}$ is closed under the Lie bracket implies that E is involutive. Hence $E = T(\mathcal{H})$ for a C^∞ foliation $\mathcal{H}$. Evidently, the distribution E is invariant under left translations in the group; so $\mathcal{H}$ is also invariant under left translations. It is standard from elementary Lie theory that the leaf through the identity $e \in G$ is a connected Lie group, which is easily seen to be an open subgroup of H. Since H is connected, it is the only open subgroup of itself; so the leaves of $\mathcal{H}$ are the left cosets of H.

Example 1.3.11. Let $\mathcal{F}$ be a C^∞, transversely–orientable foliation of codimension 1, $E = T(\mathcal{F})$, and let $\omega \in A^1(M)$ be a nonsingular form that defines $\mathcal{F}$ (*cf.*, Exercise 1.2.7). Thus, $\omega \in I^1(E)$ and, by Theorem 1.3.8, $d\omega \in I^2(E)$.

In Exercise 1.3.12, you will show that $I^*(E)$ is the principal ideal generated by ω; so there exists a form $\eta \in A^1(M)$ such that

$$d\omega = \omega \wedge \eta.$$

But

$$0 = d^2\omega = d(\omega \wedge \eta) = \omega \wedge \eta \wedge \eta - \omega \wedge d\eta = -\omega \wedge d\eta.$$

Since ω is nonsingular, it follows that there is a form $\gamma \in A^1(M)$ such that

$$d\eta = \omega \wedge \gamma.$$

Then

$$d(\eta \wedge d\eta) = d\eta \wedge d\eta = \omega \wedge \gamma \wedge \omega \wedge \gamma = 0.$$

The closed 3–form $\eta \wedge d\eta$ on M determines a de Rham cohomology class

$$\operatorname{gv}(\mathcal{F}) = [\eta \wedge d\eta] \in H^3(M),$$

called the *Godbillon–Vey class* of $\mathcal{F}$. Although ω is only determined by $\mathcal{F}$ up to multiplication by nowhere 0 functions $f \in C^\infty(M)$, and η is not uniquely determined by ω, you will show (Exercise 1.3.13) that $\operatorname{gv}(\mathcal{F})$ depends only on $\mathcal{F}$. Much has been learned about this "exotic characteristic class" since its introduction in the early 1970s [**53**], but the experts are still exploring its mysteries.

Exercise 1.3.12. Let $\omega \in A^1(M)$ be a nonsingular form and let E be the $(n-1)$–plane distribution annihilated by ω. Prove that $I^*(E)$ is the principal ideal in $A^*(M)$ generated by ω.

Exercise 1.3.13. Let $\omega \in A^1(M)$ be a nonsingular form defining the foliation $\mathcal{F}$.

1. If

$$d\omega = \omega \wedge \eta = \omega \wedge \widetilde{\eta},$$

prove that $\eta \wedge d\eta$ and $\widetilde{\eta} \wedge d\widetilde{\eta}$ are cohomologous 3–forms.

2. If $f \in C^\infty(M)$ is nowhere zero and

$$\widetilde{\omega} = f\omega,$$
$$d\omega = \omega \wedge \eta,$$

prove that

$$d\widetilde{\omega} = \widetilde{\omega} \wedge \widetilde{\eta},$$

where η and $\widetilde{\eta}$ differ by an exact form.

3. Using these computations, prove that $\operatorname{gv}(\mathcal{F})$ depends only on $\mathcal{F}$, not on the allowable choices of η and ω.

Example 1.3.14. Topologists were only convinced that $\operatorname{gv}(\mathcal{F})$ was an interesting characteristic class when foliations were found for which $\operatorname{gv}(\mathcal{F}) \neq 0$. The first example was due to R. Roussarie [**53**]. Later, in [**132**], W. Thurston showed that $\operatorname{gv}(\mathcal{F}) \in H^3(S^3) = \mathbb{R}$ takes all real values as $\mathcal{F}$ varies over all foliations of S^3 of codimension 1. We sketch the Roussarie example.

Let $M = \Gamma\backslash G$, where $G = \operatorname{PSl}(2,\mathbb{R})$ and Γ is a discrete, cocompact subgroup. As in Example 1.1.9, The foliation $\mathcal{H}$ of G by the left cosets of

$$H = \left\{ \begin{bmatrix} a & b \\ 0 & a^{-1} \end{bmatrix} \right\}_{a>0}$$

passes to a foliation $\mathcal{H}_\Gamma$ of M.

Let $\mathfrak{sl}(2,\mathbb{R})$ denote the Lie algebra of $\operatorname{Sl}(2,\mathbb{R})$, hence also of G. This is identified with the vector space of 2×2 real matrices of trace 0, the bracket operation being given by the commutator product $[A,B] = AB - BA$. The Lie subalgebra $\mathfrak{h} \subset \mathfrak{sl}(2,\mathbb{R})$ of H has basis

$$X = \begin{bmatrix} 1 & 0 \\ 0 & -1 \end{bmatrix},$$

$$Y = \begin{bmatrix} 0 & 1 \\ 0 & 0 \end{bmatrix}$$

and this is completed to a basis of $\mathfrak{sl}(2,\mathbb{R})$ by adjoining

$$Z = \begin{bmatrix} 0 & 0 \\ 1 & 0 \end{bmatrix}.$$

The Lie brackets of these fields are

$$\begin{aligned} [X,Y] &= 2Y, \\ [X,Z] &= -2Z, \\ [Y,Z] &= X. \end{aligned}$$

Let X^*, Y^*, Z^* denote the dual basis. Since X, Y, Z can be viewed as left–invariant vector fields on G, the duals can be viewed as left–invariant 1–forms on G. Since X and Y span $\mathfrak{h}$, the form $\omega = Z^*$ defines the foliation of G by the left cosets of H. By Equation (1.3.1),

$$\begin{aligned} d\omega(X \wedge Y) &= 0, \\ d\omega(X \wedge Z) &= 2, \\ d\omega(Y \wedge Z) &= 0. \end{aligned}$$

It follows that

$$d\omega = 2X^* \wedge Z^* = \omega \wedge \eta,$$

where

$$\eta = -2X^*.$$

Similar computations to the above give

$$d\eta = 2Y^* \wedge Z^*,$$

and hence

$$\eta \wedge d\eta = -4X^* \wedge Y^* \wedge Z^*,$$

a nowhere vanishing 3–form on G. Since ω and η are invariant under left translations, they descend to well–defined forms on $M = \Gamma\backslash G$ which (by abuse) we continue to call ω and η. The form $\omega \in A^1(M)$ defines $\mathcal{H}_\Gamma$, $d\omega = \omega \wedge \eta$ and $\eta \wedge d\eta$ is a nowhere vanishing 3–form on M. But M is a compact, orientable 3–manifold without boundary. So

$$\int_M \eta \wedge d\eta \neq 0,$$

implying that

$$\mathrm{gv}(\mathcal{H}_\Gamma) = [\eta \wedge d\eta] \neq 0.$$

Exercise 1.3.15. If $\Gamma \subset \mathrm{PSl}(2, \mathbb{R})$ is a discrete subgroup such that $\mathbf{H}/\Gamma$ is a compact, orientable surface of genus $g \geq 2$, show that

$$\int_M \mathrm{gv}(\mathcal{H}_\Gamma) = \pm 4\pi(1 - g).$$

(Hint. Show that $\pm 4X^* \wedge Y^* \wedge Z^*$ is the Riemannian volume form for a left–invariant metric hyperbolic along the leaves of $\mathcal{H}$.)

Exercise 1.3.16. You are to generalize the construction of $\mathrm{gv}(\mathcal{F})$ to foliations of higher codimension, proceeding as follows.

1. Let $\mathcal{F}$ be a smooth, transversely–oriented (Definition 1.2.14) foliation of codimension q tangent to the $(n-q)$–plane field E. Prove that there is a nonsinglar q–form ω on M such that, for each $x \in M$,

   $$\omega_x(v_1 \wedge \cdots \wedge v_q) = 0 \Leftrightarrow \text{at least one } v_i \in E_x.$$

   Clearly $\omega \in I^q(E)$ and we say that ω defines $\mathcal{F}$.
2. Prove that $d\omega = \omega \wedge \eta$, for some $\eta \in A^1(M)$.
3. Prove that $d\eta \in I^2(E)$ and that $\eta \wedge (d\eta)^q$ is a closed $(2q+1)$–form.
4. Show that $\mathrm{gv}(\mathcal{F}) = [\eta \wedge (d\eta)^q] \in H^{2q+1}(M; \mathbb{R})$ depends only on the foliation $\mathcal{F}$, not on the allowable choices of ω and η.

The class $\mathrm{gv}(\mathcal{F})$ is called the generalized Godbillon–Vey class of $\mathcal{F}$.

Example 1.3.17. Let $\omega \in A^1(M)$ be a closed, nonsingular 1–form on a compact, connected n–manifold M. Since the form is nonsingular, there is an $(n-1)$–plane distribution E such that ω generates the principal ideal $I^*(E)$ (Exercise 1.3.12). Since ω is closed,

$$d(f\omega) = df \wedge \omega;$$

so $d(I^1(E)) \subseteq I^2(E)$ and E is integrable. Let $\mathcal{F}$ be the corresponding foliation. Since ω is nonsingular, there is a vector field $X \in \mathfrak{X}(M)$ such that $\omega(X) > 0$ everywhere. We normalize X so as to assume $\omega(X) \equiv 1$ and let Φ_t denote the flow of X. This flow is transverse to $\mathcal{F}$. If $Y \in \Gamma(E)$, Equation (1.3.1) gives

$$0 = d\omega(X \wedge Y) = X \underbrace{\omega(Y)}_{\equiv 0} - Y \underbrace{\omega(X)}_{\equiv 1} - \omega([X,Y]) = -\omega([X,Y]),$$

implying that $[X, \Gamma(E)] \subseteq \Gamma(E)$. If L is a leaf of $\mathcal{F}$, the fact that L is $(n-1)$–dimensional and the flow is transverse to $\mathcal{F}$ implies that

$$V_L = \bigcup_{t \in \mathbb{R}} \Phi_t(L)$$

is an open subset of M. By Proposition 1.3.21 below, every leaf in V_L is diffeomorphic to L. We have proven that the diffeomorphism classes of leaves form open (necessarily disjoint) subsets of M. So the connectivity of M implies that all of the leaves of $\mathcal{F}$ are diffeomorphic. Finally, since $d\omega = 0$, we can take $\eta = 0$ in the definition of $\mathrm{gv}(\mathcal{F})$ and conclude that $\mathrm{gv}(\mathcal{F}) = 0$. By reasoning similar to that of Exercise 1.2.7, the foliations of T^n by cosets of connected, $(n-1)$–dimensional Lie subgroups are particular cases of this example.

Generally, one cannot expect a compact manifold to admit a closed, nowhere vanishing 1–form ω. For example, the presence of such a form implies that $H^1(M) \neq 0$. Indeed, every exact 1–form df must vanish at a maximum or minimum of $f \in C^\infty(M)$; so ω cannot be exact. In fact, a theorem of D. Tischler [**136**], which we will prove later (Theorem 9.4.2), asserts that M admits such a form ω if and only if M is the total space of a fiber bundle $\pi : M \to S^1$.

Exercise 1.3.18. Let ω, $\mathcal{F}$ and Φ_t be as in Example 1.3.17 and let $P(\omega) \subset \mathbb{R}$ be the group of periods of ω (*cf.* Exercise 1.2.7). Let L be an arbitrary leaf of $\mathcal{F}$ and show that $P(\omega)$ is exactly the set of real numbers t such that $\Phi_t(L) = L$. Proceed as follows.

1. Let L and L' be leaves of $\mathcal{F}$. Show that the set $P(L) \subset \mathbb{R}$ of numbers t such that $\Phi_t(L) = L$ is an additive subgroup of $\mathbb{R}$ and prove that $P(L) = P(L')$. Call this group P.
2. Show that an arbitrary piecewise C^∞ loop σ in M is homotopic to a piecewise C^∞ loop, each C^∞ segment of which lies either in a flow line of Φ_t or in a leaf of $\mathcal{F}$.
3. If $a \in P$, construct a piecewise C^∞ loop s such that $\int_s \omega = a$.

4. If s is a piecewise C^∞ loop and
$$\int_s \omega = a,$$
use step 2 to prove that $\Phi_a(L) = L$, for some leaf L of $\mathfrak{F}$.

Exercise 1.3.19. Let M be a compact, connected n–manifold, $\mathfrak{F}$ a foliation of codimension 1. Prove that $\mathfrak{F}$ is defined by a closed, nonsingular 1–form ω if and only if there is a regular foliated atlas associated to $\mathfrak{F}$ such that the associated holonomy cocycle $\gamma = \{\gamma_{\alpha\beta}\}_{\alpha,\beta\in\mathfrak{A}}$ is of the form
$$\gamma_{\alpha\beta}(y) = y + c_{\alpha\beta},$$
for suitable constants $c_{\alpha\beta} \in \mathbb{R}$. Show that this foliated atlas can be chosen so that the numbers $\{c_{\alpha\beta}\}_{\alpha,\beta\in\mathfrak{A}}$ generate the group $P(\omega)$.

Exercise 1.3.20. Let M be a compact connected n–manifold. A codimension 1 foliation $\mathfrak{F}$ of M is said to be transversely affine if there is a regular foliated atlas associated to $\mathfrak{F}$ whose holonomy cocycle $\gamma = \{\gamma_{\alpha\beta}\}$ is of the form
$$\gamma_{\alpha\beta}(y) = a_{\alpha\beta}y + b_{\alpha\beta},$$
where $a_{\alpha\beta} > 0$ and $b_{\alpha\beta}$ are constants.

1. If $\mathfrak{F}$ is transversely affine, show that there is a 1–form ω defining $\mathfrak{F}$, together with a closed 1–form η, such that $d\omega = \omega \wedge \eta$. Therefore, $\operatorname{gv}(\mathfrak{F}) = 0$.
2. If M admits a transversely–affine foliation, then $H^1(M;\mathbb{R}) \neq 0$.

Remark. The converse of the first part of the above exercise is also true (for a proof of both implications, see [**52**, III-3.11]). These are exactly the foliations having a transverse G–structure (Subsection 1.2.C), where $G = \operatorname{Aff}^+(\mathbb{R})$ is the group of orientation–preserving affine transformations
$$t \mapsto at + b, \quad a > 0,$$
on the real line.

Here is the proposition that was used in Example 1.3.17.

Proposition 1.3.21. *Let Φ_t be a flow on the n–manifold M with infinitesimal generator $X \in \mathfrak{X}(M)$. Let $\mathfrak{F}$ be a foliation (of codimension $q = n - k$) and let $E = T(\mathfrak{F})$. Then, Φ_t carries leaves of $\mathfrak{F}$ diffeomorphically onto leaves of $\mathfrak{F}$ if and only if*
$$[X, \Gamma(E)] \subseteq \Gamma(E).$$

Proof. If Φ_t carries leaves to leaves, it follows that E is also invariant under the flow. Thus, if $Y \in \Gamma(E)$,
$$[X, Y] = \lim_{t\to 0} \frac{\Phi_{-t*}(Y) - Y}{t} \in \Gamma(E).$$

Conversely, if $[X, \Gamma(E)] \subseteq \Gamma(E)$ and if (U, x, y) is a foliated chart for $\mathcal{F}$, we will show that Φ_t carries the plaques $y = \text{const.}$ to other such plaques for all small values of t. The fact that Φ_t carries leaves to leaves for *all* values of t is an easy consequence. Write

$$X|U = \sum_{i=1}^{k} f_i \frac{\partial}{\partial x^i} + \sum_{j=1}^{q} g_j \frac{\partial}{\partial y^j}.$$

By hypothesis,

$$\left[\frac{\partial}{\partial x^i}, X\right] \in \Gamma(E|U), \quad 1 \le i \le k;$$

so $g_j(x, y) = g_j(y)$ is independent of x, $1 \le j \le q$. In U, the flow

$$\Phi_t(a, b) = (x(a, b, t), y(a, b, t))$$

satisfies the system

$$\begin{aligned} \frac{dx^i}{dt} &= f_i(x, y), & 1 \le i \le k, \\ \frac{dy^j}{dt} &= g_j(y), & 1 \le j \le q. \end{aligned}$$

Consequently, given the initial condition $(a, b) \in U$, the last q coordinates $y(a, b, t)$ of the flow line $\Phi_t(a, b)$ depend only on b and t. That is, the plaque $y = y(b, 0) = b$ is carried to the plaque $y = y(b, t)$. □

Chapter 2

Holonomy

Let $(M, \mathcal{F})$ be a foliated manifold. If L is a leaf of $\mathcal{F}$ and s is a path in L, one is interested in the behavior of the foliation in a neighborhood of s in M. Intuitively, an inhabitant of the leaf walks along the path s, keeping an eye on all of the nearby leaves. As he, she or it (hereafter denoted by $s(t)$) proceeds, some of these leaves may "peel away", getting out of visual range, others may suddenly come into range and approach L asymptotically, others may follow along in a more or less parallel fashion or wind around L laterally, *etc.* If s is a loop, then $s(t)$ repeatedly returns to the same point $s(t_0)$ as $t \uparrow \infty$ and each time more and more leaves may have spiraled into view or out of view, *etc.* This behavior, when appropriately formalized, is called the *holonomy* of the foliation.

Example 2.0.1. Let $L = S^{n-1} \times S^1$ be the boundary leaf of the Reeb foliated solid torus (Example 1.1.12). Let s be a longitudinal loop on L, of period $a > 0$ and based at $p = s(0)$, as in Figure 2.1.1 (where the arrow indicates the direction of increasing t). With each complete circuit of s, the walker $s(t)$ sees the nearby planar leaves spiral in closer. Should $s(t)$ walk backwards ($t \downarrow -\infty$), the leaves spiral away. Our observer could quantify this data by carrying a rod J_t, always perpendicular to the home leaf L and having one endpoint at $s(t)$. After one circuit, $J_0 = J_a$ and the points of intersection of this rod with nearby leaves will all have moved closer to the endpoint $s(0) = s(a) = p$. This "first return map" can be viewed as a diffeomorphism $h_s : J_0 \to I$ onto a subinterval $I \subset J_0$ which also has p as an endpoint. This is a contraction mapping to p, called the holonomy of the loop s.

In this chapter, we will discuss various specific ways in which holonomy is implemented. These are: the total holonomy group of foliated bundles, the

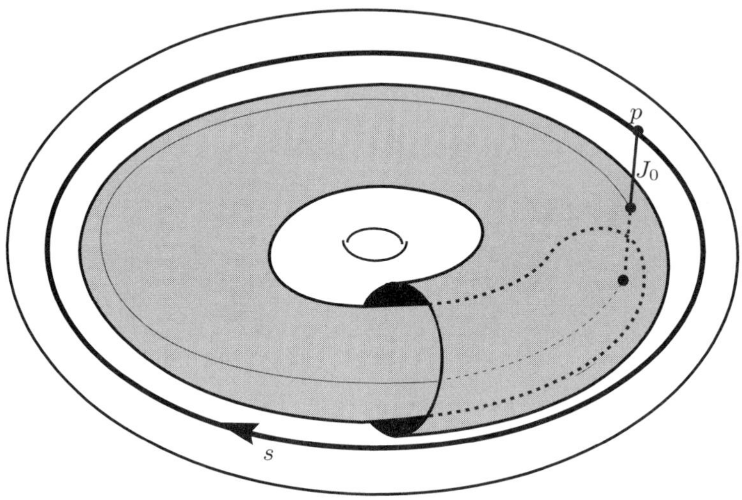

Figure 2.1.1. The holonomy contraction h_s on J_0

holonomy pseudogroup of general foliated manifolds, the germinal holonomy groupoid of general foliated manifolds, the germinal holonomy group of a leaf and the infinitesimal holonomy group of a leaf.

2.1. Foliated Bundles

The easiest case of holonomy to understand is the *total holonomy* of a foliated bundle. This is a generalization of the notion of a *first return map*.

Example 2.1.1. The term "first return map" comes from the theory of dynamical systems. Let Φ_t be a nonsingular C^r flow ($r \geq 1$) on the compact n–manifold M. If $\partial M \neq \emptyset$, we assume that the flow is tangent to the boundary. If we forget its parametrization, the flow is a 1–dimensional foliation $\mathcal{F}$. If we remember the positive direction of flow, but otherwise forget the parametrization, the underlying foliation $\mathcal{F}$ is said to be oriented. Suppose that the flow admits a global cross section N. That is, N is a compact, properly imbedded, C^r submanifold of M, $\dim N = n - 1$, the foliation $\mathcal{F}$ is transverse to N, and every flow line meets N. Because the dimensions of N and of the leaves are complementary, the transversality condition is that

$$T_y(M) = T_y(\mathcal{F}) \oplus T_y(N), \quad \forall\, y \in N.$$

Let $y \in N$ and consider the *ω–limit set* $\omega(y)$ of all accumulation points in M of all sequences $\{\Phi_{t_k}(y)\}_{k=1}^{\infty}$, where $t_k \uparrow \infty$. In Exercise 2.1.2, you will show that $\omega(y)$ is compact, nonempty, and a union of flow lines. If

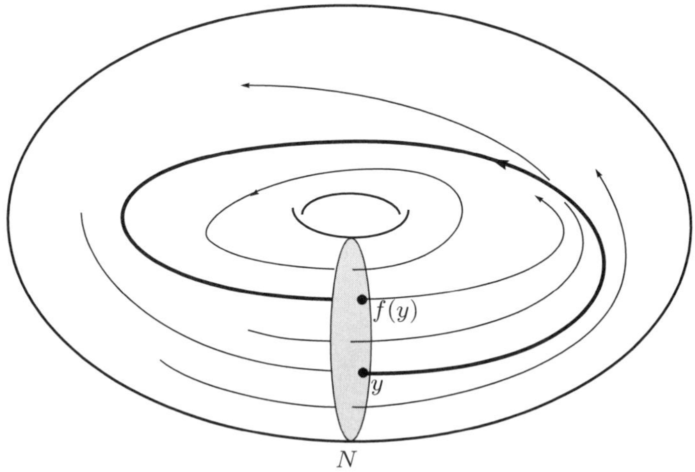

Figure 2.1.2. A cross section N and first return map f

$z = \lim_{k\to\infty} \Phi_{t_k}(y) \in \omega(y)$, there is a value $t^* \in \mathbb{R}$ such that $\Phi_{t^*}(z) \in N$ and it follows that

$$\lim_{k\to\infty} \Phi_{t_k+t^*}(y) = \Phi_{t^*}(z) \in N.$$

Since N is compact and $\mathcal{F} \pitchfork N$, it follows that the set $\{t > 0 \mid \Phi_t(y) \in N\}$ is a monotonically increasing sequence $\{\tau_k(y)\}_{k=1}^{\infty}$ that diverges to infinity. As $y \in N$ varies, let $\tau(y) = \tau_1(y)$, defining in this way a positive function $\tau \in C^r(N)$ (the first return time) such that, for arbitrary $y \in N$, $\Phi_t(y) \notin N$, $0 < t < \tau(y)$, and $\Phi_{\tau(y)}(y) \in N$. These assertions are to be checked in Exercise 2.1.3.

Define $f : N \to N$ by the formula $f(y) = \Phi_{\tau(y)}(y)$ (*cf.* Figure 2.1.2, where $M = S^1 \times D^2$ and $N = D^2$). This is a C^r map. If the flow is reversed, exactly the same construction provides the inverse f^{-1}; so $f \in \mathrm{Diff}^r(N)$. This diffeomorphism is the first return map and τ is called the *first return time*. While the first return time depends on the parametrization of the flow, it should be evident that f depends only on the oriented foliation $\mathcal{F}$. It is possible to reparametrize the flow Φ_t, keeping it nonsingular, of class C^r, and not reversing its direction, so that $\tau \equiv 1$ (Exercise 2.1.4). We assume this in the following.

The assumption that there is a cross section N to the flow is very restrictive, implying that M is the total space of a fiber bundle over S^1. Indeed, on $\mathbb{R} \times N$, define $\sim_f$ to be the equivalence relation generated by

$$(t, y) \sim_f (t - 1, f(y)).$$

Equivalently, this is the orbit equivalence for the action of the additive group $\mathbb{Z}$ on $\mathbb{R} \times N$ defined by

$$k \cdot (t, y) = (t - k, f^k(y)),$$

$\forall\, k \in \mathbb{Z}$, $\forall\,(t, y) \in \mathbb{R} \times N$. The *mapping cylinder* of f is defined to be the C^r manifold

$$M_f = (\mathbb{R} \times N)/{\sim_f} \,.$$

By the definition of the first return map f and the assumption that the first return time is $\tau \equiv 1$, it is immediate that the map

$$\Phi : \mathbb{R} \times N \to M,$$

defined by the flow, induces a canonical C^r diffeomorphism

$$\varphi : M_f \to M.$$

If we make the identification $M_f = M$, then the projection of $\mathbb{R} \times N$ onto $\mathbb{R}$ induces a C^r map

$$\pi : M \to \mathbb{R}/\mathbb{Z} = S^1$$

that makes M into the total space of a fiber bundle over the circle. (In Figure 2.1.2, this is just the projection of $S^1 \times D^2$ onto S^1.) The foliation $\mathcal{F}$ is transverse to the fibers of this bundle and the bundle projection π, restricted to each leaf L, is a covering map $\pi : L \to S^1$. This is an example of what we are going to call a *foliated bundle.*

Take as basepoint $x_0 \in S^1$ the equivalence class $0 + \mathbb{Z}$; so $\pi^{-1}(x_0)$ is the original cross section N. For each loop s on S^1, based at x_0, the homotopy class $[s] \in \pi_1(S^1, x_0)$ is uniquely characterized by $\deg s \in \mathbb{Z}$. The loop s lifts to a path in each flow line and it should be clear that the lift s_y that starts at $y \in N$ ends at $f^k(y) \in N$, where $k = \deg s$. The diffeomorphism $f^k \in \mathrm{Diff}^r(N)$ is also denoted by h_s and is called the *total holonomy* of the loop s. Since this depends only on $[s]$, we have defined a homomorphism

$$h : \pi_1(S^1, x_0) \to \mathrm{Diff}^r(N),$$

called the *total holonomy homomorphism* for the foliated bundle.

Exercise 2.1.2. Let $\Phi : M \times \mathbb{R} \to M$ be a flow (nonsingular or not) on a compact manifold M and, for $x \in M$, define the ω–limit set $\omega(x)$ exactly as in Example 2.1.1. Prove that $\omega(x)$ is a compact, nonempty, Φ–invariant set.

Exercise 2.1.3. Let N be a cross section of a nonsingular C^r flow as in Example 2.1.1. Verify carefully that the construction of the first return time $\tau \in C^r(N)$, as outlined there, really works.

Exercise 2.1.4. Verify the assertion in Example 2.1.1 that a suitable C^r reparametrization of the flow replaces the first return map with the constant $\tau \equiv 1$.

Example 2.1.5. Let $(M, \mathcal{F})$ be a foliated n–manifold of codimension q. Let

$$\pi : M \to B$$

be a fiber bundle with q–dimensional fiber F and connected base space B. Assume that all of these structures are of class C^r, $0 \le r \le \infty$, with the condition that, if $r = 0$, B supports a C^1 structure. Since every maximal C^1 atlas on B contains a C^∞ subatlas [**69**, page 51], no generality is lost in assuming that B is as smooth as desired. Finally, for each $x \in B$, assume that there is a connected, open neighborhood $U \subseteq B$ of x and a local trivialization

$$\begin{array}{ccc} \pi^{-1}(U) & \xrightarrow{\varphi} & U \times F \\ {\scriptstyle \pi}\downarrow & & \downarrow{\scriptstyle p} \\ U & \xrightarrow[\mathrm{id}]{} & U \end{array}$$

where φ is a C^r diffeomorphism (a homeomorphism, if $r = 0$) that carries $\mathcal{F}|\pi^{-1}(U)$ to the product foliation $\{U \times \{y\}\}_{y \in F}$. Here, of course, $\mathcal{F}|\pi^{-1}(U)$ is the foliation with leaves the connected components of $L \cap \pi^{-1}(U)$, where L ranges over the leaves of $\mathcal{F}$. This is the general definition of the term "foliated bundle" $(M, \mathcal{F}, \pi)$ of class C^r.

It should be noted that $\mathcal{F}$ is transverse to the fibers of π (we say that $\mathcal{F}$ is transverse to the fibration) and that the restriction of π to each leaf L of $\mathcal{F}$ is a covering map

$$\pi : L \to B.$$

In particular, each fiber $F_x = \pi^{-1}(x)$ meets every leaf of $\mathcal{F}$. The fiber is a *cross section* of $\mathcal{F}$ in complete analogy with the notion of a cross section of a flow in Example 2.1.1.

Remark that $\mathcal{F}$ being transverse to the fibers does not, of itself, guarantee that the leaves are covering spaces of B. Figure 2.1.3 illustrates a simple version of the problem, showing a foliation of $\mathbb{R}^2$, transverse to the fibration

$$\pi : \mathbb{R}^2 \to \mathbb{R},$$
$$\pi(x, y) = x,$$

but with infinitely many leaves missing the y–axis. (In Figure 2.1.3, it is intended that the "arrowed" leaves, and all above them, are asymptotic to the axis $x = 0$). One calls such a foliation *incomplete* relative to the fibration, meaning that some of the leaves "run off to infinity" as the parameter $x \in B$ approaches some $x_0 \in B$. More precisely, there may be a leaf L and a continuous path $s : [0, a) \to L$ such that $\lim_{t \to a-} \pi(s(t)) = x_0 \in B$, but $\lim_{t \to a-} s(t)$ does not exist in the manifold topology of L. This is analogous to the case of incomplete flows, where some flow lines "go to infinity" in finite time. Although such a leaf L may elsewhere meet $\pi^{-1}(x_0)$, it cannot

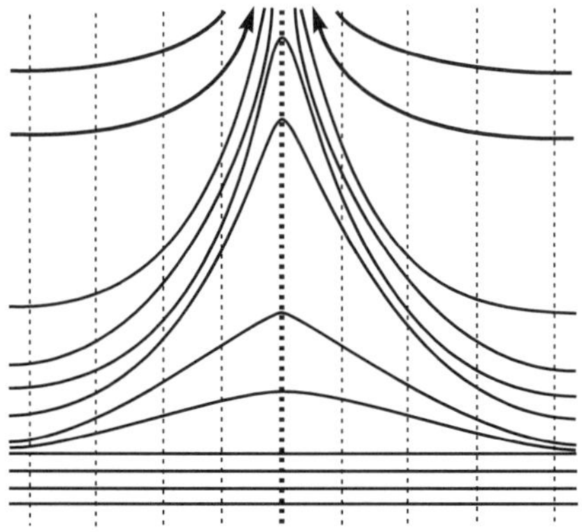

Figure 2.1.3. $\mathcal{F}$ is incomplete relative to the fibration $\pi : \mathbb{R}^2 \to \mathbb{R}$.

evenly cover a neighborhood of x_0, hence cannot be a covering space of B under π. When F is compact, however, it is true that transversality of $\mathcal{F}$ to the fibration does guarantee completeness, hence that $(M, \mathcal{F}, \pi)$ is a foliated bundle (Exercise 2.1.6).

There is an atlas $\mathcal{U} = \{U_\alpha, x_\alpha\}_{\alpha\in\mathfrak{A}}$ on B, consisting of open, connected coordinate charts, together with trivializations

$$\varphi_\alpha : \pi^{-1}(U_\alpha) \to U_\alpha \times F$$

that carry $\mathcal{F}|\pi^{-1}(U_\alpha)$ to the product foliation. Set $W_\alpha = \pi^{-1}(U_\alpha)$ and write $\varphi_\alpha = (x_\alpha, y_\alpha)$ where (by abuse of notation) x_α represents $x_\alpha \circ \pi$ and

$$y_\alpha : \pi^{-1}(U_\alpha) \to F$$

is the submersion obtained by composing φ_α with the canonical projection

$$U_\alpha \times F \to F.$$

We let $\mathcal{W} = \{W_\alpha, x_\alpha, y_\alpha\}_{\alpha\in\mathfrak{A}}$ play a role analogous to that of a foliated atlas. The plaques of W_α are the level sets of y_α and we identify this family of plaques with F via y_α. Since B is assumed to support a C^∞ structure, we appeal to a well–known theorem of J. H. C. Whitehead (*cf.* [**66**, p. 34, Theorem 6.2] and [**29**, Theorem 10.5.4]), fixing a Riemannian metric on B and choosing the atlas $\mathcal{U}$ to be geodesically convex. Thus, $U_\alpha \cap U_\beta$ is always connected. If this intersection is nonempty, each plaque of W_α meets exactly one plaque of W_β. We then define a "holonomy cocycle" $\gamma = \{\gamma_{\alpha\beta}\}_{\alpha,\beta\in\mathfrak{A}}$ by setting

$$\gamma_{\alpha\beta} = y_\alpha \circ y_\beta^{-1} : F \to F.$$

As usual,

1. $\gamma_{\alpha\beta} \circ \gamma_{\beta\delta} = \gamma_{\alpha\delta}$, whenever $W_\alpha \cap W_\beta \cap W_\delta \neq \emptyset$;
2. $\gamma_{\alpha\alpha} = \mathrm{id}_F$, $\forall\, \alpha \in \mathfrak{A}$;

3. $\gamma_{\alpha\beta} = \gamma_{\beta\alpha}^{-1}$, whenever $W_\alpha \cap W_\beta \neq \emptyset$.

Let $\mathcal{P} = \{U_{\alpha_0}, U_{\alpha_1}, \ldots, U_{\alpha_m}\}$ be a chain in $\mathcal{U}$. This will always be taken to mean that $(U_{\alpha_i}, x_{\alpha_i}) \in \mathcal{U}$, $0 \leq i \leq m$, and $U_{\alpha_{i-1}} \cap U_{\alpha_i} \neq \emptyset$, $1 \leq i \leq m$. Each plaque $P_0 \subset W_{\alpha_0}$ initiates a unique plaque chain $\{P_0, P_1, \ldots, P_m\}$ that projects to the chain $\mathcal{P}$ under π. The corresponding *total holonomy transformation* is

$$h_{\mathcal{P}} = \gamma_{\alpha_m \alpha_{m-1}} \circ \gamma_{\alpha_{m-1}\alpha_{m-2}} \circ \cdots \circ \gamma_{\alpha_1 \alpha_0} \in \operatorname{Diff}^r(F).$$

The set $H_{\mathcal{F}}$ of total holonomy transformations contains id_F and is closed under passing to inverses. The subset $H_{\mathcal{F}} \subset \operatorname{Diff}^r(F)$ is not generally closed under composition, however, and so it is not a group. One obtains a group by considering *loops* $\mathcal{P} = \{U_{\alpha_0}, U_{\alpha_1}, \ldots, U_{\alpha_m}\}$, meaning chains in $\mathcal{U}$ such that $U_{\alpha_0} = U_{\alpha_m}$. We say that this loop is based at the chart U_{α_0}. The set $G_{\mathcal{F}}(U_{\alpha_0}) \subset H_{\mathcal{F}}$, corresponding to loops based at a fixed chart U_{α_0}, clearly forms a subgroup of $\operatorname{Diff}^r(F)$, called the *total holonomy group*.

In this situation, it is just as easy (and more natural) to use honest paths and loops on B, rather than chains of charts. Indeed, if $x_0, x_1 \in B$ and if $s : [a, b] \to B$ is a path from x_0 to x_1, the fact that each leaf is a covering space of B under π allows us to lift s uniquely to a path s_z, starting at an arbitrary point $z \in F_{x_0} = \pi^{-1}(x_0)$ and staying in the leaf through z. Setting $h_s(z) = s_z(b)$ defines a C^r map

$$h_s : F_{x_0} \to F_{x_1}.$$

Suitably decomposing $s = s_0 + s_1 + \cdots + s_m$, we can assume that $\operatorname{im}(s_i) \subset U_{\alpha_i}$, for charts $(U_{\alpha_i}, x_{\alpha_i}) \in \mathcal{U}$, $0 \leq i \leq m$. Here, "$s_i + s_{i+1}$" denotes adjunction of a path s_i to a path s_{i+1}, the terminal point of s_i being the initial point of s_{i+1} (with suitable parametrizations). Then $\mathcal{P} = \{U_{\alpha_0}, U_{\alpha_1}, \ldots, U_{\alpha_m}\}$ gives a chain in $\mathcal{U}$ such that

$$y_{\alpha_m} \circ h_s \circ y_{\alpha_0}^{-1} = h_{\mathcal{P}}.$$

Note that

$$h_{\sigma+\tau} = h_\tau \circ h_\sigma.$$

If $-s$ denotes a path s with parametrization reversed, then $h_{-s} = h_s^{-1}$. So

$$h_s : F_{x_0} \to F_{x_1}$$

is a C^r diffeomorphism, called a total holonomy transformation. Those total holonomy transformations defined by loops s, based at a fixed $x_0 \in B$, form a subgroup $G_{\mathcal{F}}(x_0) \subset \operatorname{Diff}^r(F_{x_0})$, called the total holonomy group of the foliated bundle at x_0. Clearly $G_{\mathcal{F}}(x_0) = G_{\mathcal{F}}(U_{\alpha_0})$.

Remark. We want to relate the total holonomy group to the fundamental group $\pi_1(B, x_0)$. There are two possible group composition laws in

$\pi_1(B, x_0)$, the standard one being

$$[\sigma][\tau] = [\sigma + \tau].$$

Alternatively, one could set

$$[\sigma] * [\tau] = [\tau + \sigma]. \tag{2.1.1}$$

That is, one composes by first traversing τ, then σ. These two group structures are canonically isomorphic via the map $[\sigma] \mapsto [-\sigma]$. We will opt for the operation $*$ in this book, guaranteeing that the map

$$h : \pi_1(B, x_0) \to G_{\mathcal{F}}(x_0)$$

which we are about to define is a group homomorphism. Of course, one could use the more standard operation and settle for h being an anti–homomorphism, but equation (2.1.1) will also be consistent with the conventions used in the next chapter in defining the "suspension" construction of foliated bundles.

By standard covering space theory, whenever s_0 and s_1 are paths in B, homotopic via a homotopy that fixes their endpoints, the lifts of s_0 and s_1 to paths in a leaf starting at the same point z must end at the same point w. Thus, $h_{s_0} = h_{s_1}$ and, in particular, we obtain a surjective homomorphism of groups

$$h : \pi_1(B, x_0) \to G_{\mathcal{F}}(x_0),$$

sending each homotopy class $\xi = [s] \in \pi_1(B, x_0)$ to $h_\xi = h_s$.

Exercise 2.1.6. Let $\pi : M \to B$ be a fiber bundle with compact fiber F, and let $\mathcal{F}$ be a foliation of codimension $q = \dim F$. If $\mathcal{F}$ is transverse to the fibration, prove that $(M, \mathcal{F}, \pi)$ is a foliated bundle.

Remark. When $r = 0$, the system of local trivializations

$$\begin{array}{ccc} \pi^{-1}(U_\alpha) & \xrightarrow{\varphi_\alpha} & U_\alpha \times F \\ \pi \downarrow & & \downarrow p \\ U_\alpha & \xrightarrow[\text{id}]{} & U_\alpha \end{array}$$

over the C^∞ atlas $\{U_\alpha, x_\alpha\}_{\alpha \in \mathfrak{A}}$ allows us to define a $C^{\infty,0}$–foliated atlas on M defining $\mathcal{F}$. More generally, for arbitrary values of r, we obtain a $C^{\infty,r}$–foliated manifold $(M, \mathcal{F})$.

The above constructions assemble the holonomy information about $\mathcal{F}$ into a more manageable system. In fact, as the following theorem asserts, the total holonomy homomorphism h contains *all* relevant information about the foliation $\mathcal{F}$.

Theorem 2.1.7. *Let $(M, \mathcal{F}, \pi)$ and $(M', \mathcal{F}', \pi')$ be two foliated bundles of class C^r with the same connected base space B and diffeomorphic fibers F and F'. Assume that the total holonomy homomorphisms*

$$h : \pi_1(B, x_0) \to \mathrm{Diff}^r(F_{x_0}),$$
$$h' : \pi_1(B, x_0) \to \mathrm{Diff}^r(F'_{x_0})$$

are conjugate. That is, there is a C^r diffeomorphism $g : F_{x_0} \to F'_{x_0}$ such that

$$h'_\xi = g \circ h_\xi \circ g^{-1},$$

for all $\xi \in \pi_1(B, x_0)$. Then there is a diagram

$$\begin{array}{ccc} M & \xrightarrow{\varphi} & M' \\ {\scriptstyle\pi}\downarrow & & \downarrow{\scriptstyle\pi'} \\ B & \xrightarrow[\mathrm{id}]{} & B \end{array}$$

where φ is the unique C^r diffeomorphism such that the diagram commutes, $\varphi|F_{x_0} = g$, and φ carries the foliation $\mathcal{F}$ to the foliation $\mathcal{F}'$. We say that φ is an isomorphism of foliated bundles.

Proof. In what follows, we denote by L_z the leaf of $\mathcal{F}$ through $z \in M$ and by L'_w the leaf of $\mathcal{F}'$ through $w \in M'$. We also shorten expressions such as $\sigma + (-\tau)$ to $\sigma - \tau$, where σ and τ are paths with the same terminal point. If σ is a path on a leaf L_y of $\mathcal{F}$ with initial point $y \in F_{x_0}$, we let σ' denote the unique lift of $\pi \circ \sigma$ to a path on $L'_{g(y)}$ starting at $g(y) \in F'_{x_0}$.

We define φ via path lifting as follows. Given $z \in M$, choose $y \in F_{x_0} \cap L_z$. Let $\sigma : [a, b] \to L_z$ be a path from y to z and set $\varphi(z) = \sigma'(b)$. If well defined, this map will have the desired properties. (The mechanical verification is left to the reader.) In turn, these properties force the path lifting in our definition, and so φ is unique.

We show independence of the choice of σ. If $\overline{\sigma} : [a, b] \to L_z$ is another path from y to z, we must show that $\overline{\sigma}'(b) = \sigma'(b)$. Since $\overline{\sigma} - \sigma$ is a loop on L_z, based at y, $s = \pi \circ (\overline{\sigma} - \sigma)$ is a loop on B, based at x_0. The fact that the lift $\overline{\sigma} - \sigma$ of s is a loop at y means that $h_s(y) = y$. Thus,

$$h'_s(g(y)) = g(h_s(g^{-1}(g(y)))) = g(y)$$

and the lift of s to a path on $L'_{g(y)}$ starting at $g(y)$ must also be a loop. That is, σ' and $\overline{\sigma}'$ must have the same terminal point. This shows that $\varphi(z)$ is independent of the choice of path on L_z from y to z.

We show independence of the choice of $y \in L_z \cap F_{x_0}$. Let $\overline{y} \in L_z \cap F_{x_0}$ and let λ be a path from y to $\overline{y}$, ζ a path from $\overline{y}$ to z, and σ a path from y to z, all in L_z. We must show that ζ' and σ' have the same terminal point.

Let $\overline{\sigma} = \lambda + \zeta$ and apply the previous paragraph to conclude that σ' and $\overline{\sigma}'$ have the same terminal point. Also, $\ell = \pi \circ \lambda$ is a loop on B, based at x_0, and $h_\ell(y) = \overline{y}$. It follows that $h'_\ell(g(y)) = g(\overline{y})$, hence that λ' ends at the point $g(\overline{y})$ where ζ' begins. In other words, $\overline{\sigma}' = \lambda' + \zeta'$ and ζ' has the same terminal point as σ'. This completes the proof that φ is well defined. □

Example 2.1.8. We give another characterization of foliated bundles. Let $\pi : M \to B$ be a fiber bundle of class C^r. Let F be the fiber. Recall that a structure cocycle for this bundle is defined by means of a family of local trivializations

$$\begin{array}{ccc} \pi^{-1}(U_i) & \xrightarrow{\varphi_i} & U_i \times F \\ {\scriptstyle\pi}\downarrow & & \downarrow{\scriptstyle p} \\ U_i & \xrightarrow[\text{id}]{} & U_i, \end{array}$$

where $\{U_i\}_{i\in\mathfrak{I}}$ is a cover of B by open, connected subsets. On overlaps, the "change of coordinate" formula is

$$\varphi_i \circ \varphi_j^{-1}(x, y) = (x, g_{ij}(x)(y)),$$

where

$$g_{ij} : U_i \cap U_j \to \operatorname{Diff}^r(F).$$

Basically, the C^r topology on $\operatorname{Diff}^r(F)$ is the topology of uniform convergence of all derivatives of orders $0 \le \ell \le r$ on compact subsets of F. The functions $\{g_{ij}\}_{i,j\in\mathfrak{I}}$ are continuous in this topology and they satisfy the usual cocycle condition. The set $\{g_{ij}(x) \mid x \in U_i \cap U_j\}_{i,j\in\mathfrak{I}}$ generates a subgroup $G \subset \operatorname{Diff}^r(F)$, called the structure group of the bundle relative to this cocycle. We also say that the cocycle is a *reduction* of the structure group to G. If G is a *totally disconnected* subgroup relative to the C^r topology, then each g_{ij} is locally constant on $U_i \cap U_j$. If such a reduction to a totally disconnected group exists, we say that the bundle is *flat* and that the cocycle is a *flat structure*. (*E.g.*, bundles that admit a connection with curvature 0 are flat.) Again, by Whitehead's theorem, we assume that all $U_i \cap U_j$ are connected, hence that all g_{ij}'s are constant on $U_i \cap U_j$.

Theorem 2.1.9. *The C^r fiber bundle $\pi : M \to B$ is flat if and only if there is a C^r foliation $\mathfrak{F}$ of M such that $(M, \mathfrak{F}, \pi)$ is a foliated bundle.*

Proof. The transition functions g_{ij}, being constant on $U_i \cap U_j$, match the product foliation of $U_i \times F$ (with leaves $U_i \times \{y\}$) to the product foliation of $U_j \times F$. Via the diffeomorphisms φ_i, these product foliations define foliations $\mathfrak{F}_i$ of $\pi^{-1}(U_i)$, $i \in \mathfrak{I}$, and each leaf of $\mathfrak{F}_i$ is carried diffeomorphically onto U_i by π. By the above remarks, $\mathfrak{F}_i$ and $\mathfrak{F}_j$ agree on $\pi^{-1}(U_i \cap U_j)$, and so the foliations $\{\mathfrak{F}_i\}_{i\in\mathfrak{I}}$ assemble to form a global foliation $\mathfrak{F}$. This foliation is transverse to the fibers of π and, if L is a leaf, $L \cap \pi^{-1}(U_i)$ is a disjoint union of some of the leaves of $\mathcal{F}_i$; so $(M, \mathcal{F}, \pi)$ is a foliated bundle.

Conversely, assume that the foliation $\mathcal{F}$ exists. Cover B by a system $\{U_i\}_{i=1}^n$ of geodesically convex (hence, simply connected) neighborhoods. These necessarily trivialize $\pi^{-1}(U_i)$ and are evenly covered by $\pi|L : L \to B$, $\forall$ leaves L. Thus, we can select trivializations $\varphi_i : \pi^{-1}(U_i) \to U_i \times F$ that carry each leaf of $\mathcal{F}|\pi^{-1}(U_i)$ onto a leaf $U_i \times \{y\}$ of the product foliation. It follows that the transitions

$$\varphi_i \circ \varphi_j^{-1} : U_i \cap U_j \times F \to U_i \cap U_j \times F$$

preserve the product foliations, hence are of the form

$$(x, y) \mapsto (x, g_{ij}(y)),$$

where $g_{ij} \in \operatorname{Diff}(F)$. □

2.2. The Holonomy Pseudogroup

In the case of general foliations, the total holonomy group of a foliated bundle must be replaced by a local analogue called the *holonomy pseudogroup*. Unfortunately, pseudogroups are awkward things to use. Authors in foliation theory are frequently careless with the holonomy pseudogroup, but our mistakes are usually easy to fix because our intuitions are typically on target. The problems most often have to do with domains and ranges.

Example 2.2.1. Again, let $\mathcal{F}$ be the underlying foliation of a nonsingular C^r flow Φ_t on M, tangent to ∂M. Although Φ_t rarely admits a global cross section, it always admits *local* ones. Indeed, if $w \in M$, let (U, x, y) be a foliated chart for $\mathcal{F}$ about w. Then, the transversal S_w through w, defined as the level set $x = x(w)$, is an $(n-1)$–manifold and $\mathcal{F} \pitchfork S_w$, but not every flow line meets S_w. This is a local cross section to the flow. If $\Phi_a(w) = z$, some $a > 0$, we fix another local cross section S_z through z and define a positive C^r function τ on a suitably small neighborhood $V_w \subset S_w$ of w by requiring that

$$\Phi_{\tau(u)}(u) \in S_z, \qquad \forall\, u \in V_w,$$
$$\tau(w) = a.$$

The path $s(t) = \Phi_t(w)$, $0 \le t \le a$, lies in the leaf of $\mathcal{F}$ through w and connects w and z. The holonomy $h_s : V_w \to S_z$ of $\mathcal{F}$ along s is the map

$$h_s(u) = \Phi_{\tau(u)}(u).$$

Again, reversing the flow provides an inverse h_s^{-1}; so the holonomy is a diffeomorphism of V_w onto a neighborhood $W_z \subset S_z$ of z (*cf.* Figure 2.2.1). If $z = w$, then s is a loop and we choose $S_z = S_w$. In this case, h_s carries

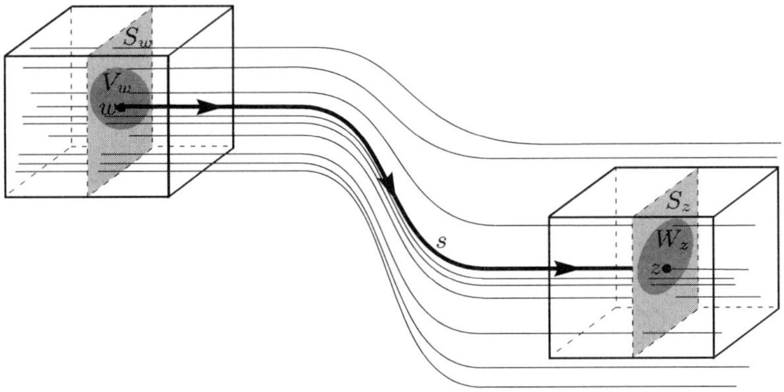

Figure 2.2.1. The holonomy of a nonsingular flow

a neighborhood $V_w \subset S_w$ diffeomorphically onto a neighborhood $W_w \subset S_w$, but generally these neighborhoods do not coincide. In this example, V_w plays a role analogous to that of the "rod" J_0 in Example 2.0.1 and W_w plays the role of I in that example (except that, generally, $W_w \not\subseteq V_w$).

Example 2.2.2. The foliation in Example 2.0.1 has codimension 1 and those in Examples 2.1.1 and 2.2.1 have dimension 1. This made the examples easier to visualize, but the same ideas work well in all dimensions and codimensions. Here is a simple case. Let $(M, \mathcal{F})$ be a C^r–foliated n–manifold, $\operatorname{codim} \mathcal{F} = q$. Fix a regular foliated atlas $\mathfrak{U} = \{U_\alpha, x_\alpha, y_\alpha\}_{\alpha \in \mathfrak{A}}$ for $\mathcal{F}$ and let $\gamma = \{\gamma_{\alpha\beta}\}_{\alpha,\beta \in \mathfrak{A}}$ be the holonomy cocycle, as defined in the remarks beginning on page 27. In each foliated chart U_α, choose a transversal S_α (Definition 1.1.17). The submersion

$$y_\alpha : U_\alpha \longrightarrow \mathbb{H}^q$$

carries S_α diffeomorphically onto $y_\alpha(U_\alpha)$; so we identify these spaces, viewing y_α as a coordinatization of S_α. Thus, if $\alpha, \beta \in \mathfrak{A}$ and $U_\alpha \cap U_\beta \neq \emptyset$, we view $\gamma_{\alpha\beta}$ as a diffeomorphism from an open subset $D(\gamma_{\alpha\beta}) \subset S_\beta$ (the *domain* of $\gamma_{\alpha\beta}$) onto an open subset $R(\gamma_{\alpha\beta}) \subset S_\alpha$ (the *range* of $\gamma_{\alpha\beta}$). In the respective coordinates, the domain and range are

$$\begin{aligned} D(\gamma_{\alpha\beta}) &= y_\beta(U_\alpha \cap U_\beta), \\ R(\gamma_{\alpha\beta}) &= y_\alpha(U_\alpha \cap U_\beta). \end{aligned}$$

Under our identifications, it should be clear that $w \in D(\gamma_{\alpha\beta}) \subseteq S_\beta$ if and only if the plaque $P_w \subset U_\beta$ through w intersects a unique plaque $P_z \subset U_\alpha$. Here, $P_z \cap S_\alpha = \{z\}$ and we write $\gamma_{\alpha\beta}(w) = z$. If s is a path in $P_w \cup P_z$ from

w to z (hence a path from w to z in a leaf L of $\mathcal{F}$), we define the holonomy of this path to be

$$h_s = \gamma_{\alpha\beta} : D(\gamma_{\alpha\beta}) \to R(\gamma_{\alpha\beta}).$$

Intuitively, for each $u \in D(\gamma_{\alpha\beta})$, the local product structure of the foliation allows us to "lift" the path s to a path s_u, starting at $u \in S_\beta$, terminating at the point $v = h_s(u) \in S_\alpha$, and lying in $P_u \cup P_v$. Here, P_u is the U_β–plaque through u, P_v is the U_α–plaque through v, and $P_u \cap P_v \neq \emptyset$.

Remark that h_s does not depend in a very sensitive way on the path s. For *every* path σ in $P_w \cup P_z$ joining w to z, we get $h_\sigma = \gamma_{\alpha\beta}$. One can think of the plaque chain $\mathcal{P} = \{P_w, P_z\}$ as a "fuzzy" version of all such paths and denote the holonomy by $h_{\mathcal{P}}$. In fact, the assignment

$$P \mapsto P \cap S_\delta$$

defines a natural one–to–one correspondence between the set of plaques in U_δ and the q–manifold S_δ, $\forall\, \delta \in \mathfrak{A}$. So we can view the plaques P_w and P_z as fuzzy versions of the respective points $w \in S_\beta$ and $z \in S_\alpha$. The plaque chain $\mathcal{P} = \{P_w, P_z\}$ is a "path" in a leaf with initial point P_w and terminal point P_z. The path is "connected" since $P_w \cap P_z \neq \emptyset$ and, in the notation above, the holonomy of $\mathcal{P}$ is defined by

$$h_{\mathcal{P}}(P_u) = P_v.$$

The plaque chain $\{P_u, P_v\}$ is the "lift" of $\mathcal{P}$ to a path joining P_u to P_v in the leaf through P_v.

The last example introduces the formalism we prefer for careful definitions. Thus, we fix the regular foliated atlas $\mathcal{U}$ and associated holonomy cocycle γ exactly as in Example 2.2.2 and consider *plaque chains*

$$\mathcal{P} = \{P_0, P_1, \dots, P_m\}.$$

As in the proof of Lemma 1.2.16, the term "plaque chain" includes the condition that $P_{i-1} \cap P_i \neq \emptyset$, $1 \le i \le m$. If $U_{\alpha_i} \supset P_i$, $0 \le i \le m$, we see that the composition

$$h_{\mathcal{P}} = \gamma_{\alpha_m\alpha_{m-1}} \circ \gamma_{\alpha_{m-1}\alpha_{m-2}} \circ \cdots \circ \gamma_{\alpha_1\alpha_0}$$

is defined and is a diffeomorphism of a suitable open neighborhood in S_{α_0} onto an open neighborhood in S_{α_m}.

The fact that the regular foliated atlas $\mathcal{U}$ is locally finite, together with the 2nd countability of M, implies that $\mathcal{U}$ is at most countably infinite. Thus, the disjoint union

$$S = \coprod_{\alpha \in \mathfrak{A}} S_\alpha$$

is a q–manifold that imbeds as a submanifold $S \hookrightarrow M$ transverse to $\mathcal{F}$. Remark that, in the coordinates y_α, each diffeomorphism $h_{\mathcal{P}}$ has bounded range in $\mathbb{H}^q$ and has bounded derivatives of all orders $\leq r$. This is because property (1) in Definition 1.2.11 implies that $h_{\mathcal{P}}$ extends C^r–smoothly to the compact closure of its domain.

Definition 2.2.3. Let $h : D(h) \to R(h)$ be a C^r diffeomorphism between open subsets of S. If, for each $w \in D(h)$, there is a neighborhood W of w in $D(h)$ and a plaque chain $\mathcal{P}$ such that $h|W = h_{\mathcal{P}}|W$, then h is called a holonomy transformation defined by the regular foliated atlas $\mathcal{U}$. The set of all such holonomy transformations will be denoted by $\Gamma_{\mathcal{U}}$.

It should be clear from our discussion that, locally, the elements of $\Gamma_{\mathcal{U}}$ can be thought of as transformations between transverse q–manifolds defined by "sliding" along leaves.

Definition 2.2.4. Let N be a q–manifold. A C^r pseudogroup Γ on N is a collection of C^r diffeomorphisms $h : D(h) \to R(h)$ between open subsets of N satisfying the following axioms.

1. If $g, h \in \Gamma$ and $R(h) \subseteq D(g)$, then $g \circ h \in \Gamma$.
2. If $h \in \Gamma$, then $h^{-1} \in \Gamma$.
3. $\mathrm{id}_N \in \Gamma$.
4. If $h \in \Gamma$ and $W \subseteq D(h)$ is an open subset, then $h|W \in \Gamma$.
5. If $h : D(h) \to R(h)$ is a C^r diffeomorphism between open subsets of N and if, for each $w \in D(h)$, there is a neighborhood W of w in $D(h)$ such that $h|W \in \Gamma$, then $h \in \Gamma$.

If $\Gamma' \subseteq \Gamma$ is also a pseudogroup, it is called a subpseudogroup of Γ.

Remarks. By the first three axioms, the pseudogroup Γ is "trying" to be a C^r transformation group on N. The problem is that the transformations are not defined globally on N; so composition may not always be defined. Axioms (3) and (4) imply that, for every open subset $W \subseteq N$, $\mathrm{id}_W \in \Gamma$. By Axioms (4) and (5), a C^r diffeomorphism between open subsets of N belongs to Γ if and only if it belongs to Γ *locally*. In particular, Γ is closed under the operation of *amalgamation*. For example, if $g, h \in \Gamma$ agree on $D(g) \cap D(h)$, then we form a C^r map

$$g \cup h : D(g) \cup D(h) \to R(g) \cup R(h),$$

where

$$(g \cup h)(w) = \begin{cases} g(w), & w \in D(g), \\ h(w), & w \in D(h). \end{cases}$$

This is locally a diffeomorphism. If it is one–to–one, this map is a diffeomorphism (called the amalgamation of g and h) and $g \cup h \in \Gamma$. More generally, one defines the amalgamation of a whole subset $\Theta \subseteq \Gamma$, provided the elements of Θ assemble compatibly into a diffeomorphism. Axiom (5) guarantees that Γ is closed under this operation.

Example 2.2.5. The set $\Gamma^r(N)$ of *all* C^r diffeomorphisms between open subsets of N is a C^r pseudogroup on N. If $\{\Gamma_\lambda\}_{\lambda\in\mathfrak{L}}$ is a nonempty family of subpseudogroups of $\Gamma^r(N)$, then it should be clear that

$$\bigcap_{\lambda\in\mathfrak{L}} \Gamma_\lambda$$

is also a subpseudogroup of $\Gamma^r(N)$. Thus, given $\Theta \subseteq \Gamma^r(N)$, an arbitrary subset, we define the *subpseudogroup generated by* Θ to be the intersection of all the subpseudogroups that contain Θ. There is at least one subpseudogroup containing Θ, namely $\Gamma^r(N)$ itself; so this intersection is a subpseudogroup.

The following is an obvious consequence of our definitions.

Proposition 2.2.6. *Let $\mathfrak{U}$ be a regular foliated atlas of class C^r and let $\gamma = \{\gamma_{\alpha\beta}\}_{\alpha,\beta\in\mathfrak{A}}$ be its holonomy cocycle. Then the set $\Gamma_\mathfrak{U}$ of holonomy transformations is the C^r pseudogroup on S generated by γ, called the* holonomy pseudogroup *of $\mathfrak{U}$.*

Remark. Although $\Gamma_\mathfrak{U}$ depends on the choice of regular foliated atlas, it is frequently and improperly referred to as "the holonomy pseudogroup of the foliation $\mathcal{F}$". This seldom leads to trouble because the pseudogroups of any two regular foliated atlases for the same foliation contain equivalent information. (For an elegant discussion of this, the reader is referred to [**59**]. Here, we content ourselves with a somewhat naive discussion.) Indeed, replace S with the disjoint union $\mathcal{N}$ of all possible imbedded q–manifolds $N \hookrightarrow M$ that are transverse to $\mathcal{F}$. This fails to be 2nd countable, but only because it has uncountably many connected components, each of which is 2nd countable. So $\mathcal{N}$ is a halfway decent q–manifold. One can define diffeomorphisms between open subsets of $\mathcal{N}$ by locally "sliding" along leaves of $\mathcal{F}$, obtaining a pseudogroup $\Gamma_\mathcal{F}$ on $\mathcal{N}$. This pseudogroup, which really does qualify as "the holonomy pseudogroup of $\mathcal{F}$", contains all of our holonomy pseudogroups $\Gamma_\mathfrak{U}$ and restricts to each one of them. Finally, the holonomy pseudogroups of any two regular foliated atlases are, in a fairly obvious sense, "locally conjugate" in $\Gamma_\mathcal{F}$.

2.3. Germinal Holonomy

For general foliations, homotopy–invariant holonomy can be defined along honest paths in leaves only by replacing the holonomy pseudogroup with the associated *groupoid of germs* of holonomy. Recall that the germ $\widehat{g}_x$ of a local homeomorhism g at a point $x \in \operatorname{dom} g$ is the equivalence class of all local homeomorphisms agreeing with g in small neighborhoods of x.

Let $(M, \mathcal{F})$ be a foliated manifold and $\mathcal{U} = \{U_\alpha, x_\alpha, y_\alpha\}_{\alpha \in \mathfrak{A}}$ a regular foliated atlas and S the associated transverse manifold. For $y \in S$, let $s : [a, b] \to L$ be a path in the leaf L through y with $s(a) = y$. Choose a subdivision

$$a = t_0 < t_1 < \cdots < t_{m+1} = b$$

and a chain $\mathcal{P} = \{P_0, P_1, \dots, P_m\}$ of $\mathcal{U}$–plaques with

$$s([t_i, t_{i+1}]) \subset P_i, \quad 0 \leq i \leq m.$$

Definition 2.3.1. A plaque chain $\mathcal{P}$, related to a path s as above, is said to cover s.

If $\mathcal{P}$ covers s, set

$$\widehat{h}_s = \text{germ at } y \text{ of } h_{\mathcal{P}}.$$

Also, let $[s]$ denote the homotopy class of s in L (rel ∂s). That is, the initial point $y = s(a)$ and terminal point $z = s(b)$ are to be fixed throughout the homotopies. It is not obvious that $\widehat{h}_s$ is well defined independently of the choice of partition and chain, let alone the following.

Proposition 2.3.2. *The germ $\widehat{h}_s$ depends only on $[s]$.*

Proof. The initial point y of s determines the component S_{α_0} of S in which it lies, hence determines the initial plaque $P_0 \subset U_{\alpha_0}$ of the plaque chain. Similarly, the terminal plaque P_m is determined by $z = s(b)$. Suppose that the plaque chains

$$\begin{aligned} \mathcal{P} &= \{P_0, P_1, \dots, P_m\}, \\ \mathcal{P}_* &= \{P_0 = P_{0*}, P_{1*}, \dots, P_{\ell *} = P_m\} \end{aligned}$$

both cover s and let

$$\begin{aligned} a &= t_0 < t_1 < \cdots < t_{m+1} = b, \\ a &= t_{0*} < t_{1*} < \cdots < t_{\ell+1 *} = b \end{aligned}$$

be partitions corresponding to these covers. By a small perturbation of either partition, we lose no generality in assuming that the partitions have no common values except a and b. Thus, these partitions can be uniquely amalgamated into a new partition and the two chains into a new associated chain. Suppose that somewhere in this amalgamated data we find strings

$$t_r < \underbrace{t_{k*} < \cdots < t_{p*}}_{*\text{–data}} < t_{r+1},$$

$$P_r, \underbrace{P_{k*}, \dots, P_{p*}}_{*\text{–data}}, P_{r+1},$$

where possibly $p = k$. Since $s(t_{j+1*}) \in P_{j*} \cap P_{j+1*}$ and $s([t_r, t_{r+1}]) \subset P_r$, we see that

$$P_r \cap P_{j*} \cap P_{j+1*} \neq \emptyset, \quad k \leq j < p.$$

Likewise,

$$P_r \cap P_{p*} \cap P_{r+1} \neq \emptyset.$$

Thus, repeated applications of the cocycle property prove that

$$\gamma_{\alpha_{r+1}\alpha_{p*}} \circ \gamma_{\alpha_{p*}\alpha_{p-1*}} \circ \cdots \circ \gamma_{\alpha_{k*}\alpha_r} = \gamma_{\alpha_{r+1}\alpha_r}$$

on a neighborhood of $P_r \in S$. In particular, both sides have the same germ at P_r. Finite repetition of this procedure eliminates all of the $*$–data without changing the germ at y. Similarly, eliminate the other data, leaving only the $*$–data. This proves that $\widehat{h}_s$ depends only on s.

If F_τ is a homotopy of s in L, $F_0 = s$, $F_\tau(a) = s(a)$ and $F_\tau(b) = s(b)$, $0 \leq \tau \leq 1$, we must show that the germ $\widehat{h}_{F_\tau}$ is constant in τ. But for small variations in τ, one need not change the plaque chain $\mathcal{P}$; hence the germ is locally constant in τ. Since $[0, 1]$ is connected, the assertion follows. □

Remark. Assume that $(M, \mathcal{F})$ is a foliated manifold of class at least $C^{1,0+}$ (Definition 1.2.24). Fixing a Riemannian metric on M, we obtain the *normal bundle* $\nu(\mathcal{F}) \subset T(M)$ consisting of vectors perpendicular to $T(\mathcal{F})$. Since $T(\mathcal{F})$ may only be a C^0 subbundle of $T(M)$, the same holds for $\nu(\mathcal{F})$, but by standard methods the normal bundle can be perturbed very slightly to produce a C^∞ subbundle of $T(M)$, transverse to $\mathcal{F}$ and again denoted by $\nu(\mathcal{F})$. Define an *immersed* normal neighborhood $i : N(L) \looparrowright M$ of a leaf L by exponentiating the ϵ–neighborhood of the 0–section of $\nu(\mathcal{F})$ for suitably small $\epsilon > 0$. That is, we have a fiber bundle $\pi : N(L) \to L$ with fibers open q–cells, and a smooth immersion i of the total space into M that carries each fiber $\pi^{-1}(y)$ onto a geodesic disk at $y \in L$. Generally, it is not possible to choose i to be one–to–one but, choosing the normal fibers small enough, we guarantee that they are individually imbedded and that $\mathcal{F}$ is transverse to these fibers. If $L \subset \operatorname{int} M$, the fiber S_y of $N(L)$ at $y \in L$ can be taken to be an imbedded copy of the open unit q-cube $E^q = (-1, 1)^q$ and, if $L \subseteq \partial M$, the fiber is a "half cube" $E^q \cap \mathbb{H}^q$. In either case, the origin $0 \in E^q$ is identified with y. Given points $y, z \in L$, we easily construct a regular foliated atlas $\mathcal{U}$ such that S_y and S_z are connected components of S.

If $s : [a, b] \to L$ is a path from y to z, it is homotopic in L through paths from y to z to an immersed path (in fact, C^1–immersed [**69**, Chapter 2]); so $\widehat{h}_s$ can be computed using such paths. The restriction of $N(L)$ to s is a topologically immersed copy of $[a, b] \times E^q$ and $\mathcal{F}$ induces a foliation of $[a, b] \times E^q$ by curves transverse to the fibers. The curve $[a, b] \times \{0\}$ immerses to s and the other curves immerse to lifts of (parts of) s to the nearby leaves of $\mathcal{F}$. Those lifts that begin at points (a, x) close enough to $(a, 0)$ are complete lifts s_x of s and $\widehat{h}_s$ is the germ of the map taking $s_x(a)$ to $s_x(b)$. Figures 2.1.1 and 2.2.1 illustrate such lifts in respective codimensions 1 and 2. The germinal holonomy $\widehat{h}_s$ depends only on the *germs* $\widehat{S}_y$ and $\widehat{S}_z$ of S_y and S_z, respectively, not otherwise on the choice of $\mathcal{U}$. For these reasons, we tend to suppress reference to the regular foliated atlas when using germinal holonomy.

If $y, z \in S$ lie in a common leaf L, we define

$$\mathcal{G}_y^z = \{\widehat{h}_s \mid s \text{ a path in } L \text{ from } y \text{ to } z\}.$$

If y and $z \in S$ do not lie in a common leaf, we set $\mathcal{G}_y^z = \emptyset$. Remark that there is a natural composition

$$\mathcal{G}_y^z \times \mathcal{G}_x^y \to \mathcal{G}_x^z,$$

defined by

$$(\widehat{h}, \widehat{f}) \mapsto \widehat{h \circ f},$$

$\forall \widehat{h} \in \mathcal{G}_y^z$ and $\forall \widehat{f} \in \mathcal{G}_x^y$. Also, for each $x \in S$, $\mathcal{G}_x^x$ contains an identity. Thus,

$$\mathcal{G}_S = \coprod_{y,z \in S} \mathcal{G}_y^z$$

is a *category* with S as the set of *objects* and $\mathcal{G}_y^z$ the set of *morphisms* from the object y to the object z. This category is special in that each morphism has an inverse that is also a morphism. Thus, for each $x \in S$, $\mathcal{G}_x^x$ is a group.

Definition 2.3.3. A *groupoid* $\mathcal{G}$ *on a set* X is a category with inverses, having X as its set of objects. For $y, z \in X$, the set of morphisms of $\mathcal{G}$ from y to z is denoted by $\mathcal{G}_y^z$.

Example 2.3.4. If $X = \{x\}$ is a singleton, a groupoid on X is simply a group. Every group G is a groupoid on a singleton $X = \{x\}$. Here, we formally define $g(x) = x$, $\forall g \in G$, and define the composition of two morphisms $g_1, g_2 \in G$ to be their group product $g_1 g_2$.

Example 2.3.5. If Γ is a pseudogroup on a manifold N, the set of germs of elements of Γ at points of N forms a groupoid $\mathcal{G}$ on N, composition being induced by the composition in the pseudogroup.

Definition 2.3.6. If $\mathcal{U}$ is a regular foliated atlas on M, associated to the foliation $\mathcal{F}$, and if S is the associated transverse space, then the groupoid $\mathcal{G}_S$ on S of germs of elements of $\Gamma_{\mathcal{U}}$ is called the *holonomy groupoid* of $\mathcal{F}$. If $y \in S$ and L is the leaf of $\mathcal{F}$ through y, the group $\mathcal{G}_y^y$ is called the *holonomy group* of L at y and will be denoted by $\mathcal{H}_y(L)$.

If s is a loop on a leaf L, based at $y \in L \cap S$, then $\widehat{h}_s \in \mathcal{H}_y(L)$. By Proposition 2.3.2, this defines a surjective map

$$\widehat{h} : \pi_1(L, y) \to \mathcal{H}_y(L)$$

and the following is clear.

Lemma 2.3.7. *The surjection $\widehat{h}$ is a group homomorphism, called the holonomy homomorphism of the leaf L.*

Exercise 2.3.8. Fix a positive number r and consider the action of $\mathbb{Z}$ on $\mathbb{R}^n \smallsetminus \{0\}$ generated by the map $v \mapsto rv$. Show that the orbit space of this action is a compact manifold (which we may call M), and describe it. (Hint: find a fundamental domain for the action.) The action preserves a foliation by parallel hyperplanes in $\mathbb{R}^n \smallsetminus \{0\}$ (one of which is punctured) which then induces a foliation of M. Show that there is exactly one compact leaf and describe its holonomy group and holonomy homomorphism.

There is a germinal version of foliated bundle theory that is useful in studying *compact* leaves of general foliations $\mathcal{F}$ of codimension q and class at least $C^{1,0+}$. Let $L \subset M$ be such a leaf and fix a normal neighborhood $\pi : N(L) \to L$. Since L is compact, we can assume that $N(L)$ is imbedded in M. As usual, we arrange that $\mathcal{F}$ be transverse to the fibers S_x, $\forall x \in L$, and that $L \subset N(L)$ be the 0–section. The triple $(N(L), \mathcal{F}|N(L), \pi)$ is not generally a foliated bundle since the completeness property will usually fail. Nevertheless, we call this triple a "foliated normal neighborhood of L". If L is simultaneously a compact leaf in two foliated n–manifolds $(M, \mathcal{F})$ and $(M', \mathcal{F}')$, we say that these foliations have *isomorphic germs* $\widehat{\mathcal{F}}_L$ and $\widehat{\mathcal{F}}'_L$ at L if there exist normal neighborhoods

$$\pi : N(L) \to L \text{ and } \pi' : N'(L) \to L$$

in M and M', respectively, and a commutative diagram

$$\begin{array}{ccc} N(L) & \xrightarrow{\varphi} & N'(L) \\ {\scriptstyle\pi}\downarrow & & \downarrow{\scriptstyle\pi'} \\ L & \xrightarrow[\text{id}]{} & L \end{array}$$

such that φ is a homeomorphism that carries $\mathcal{F}|N(L)$ to $\mathcal{F}'|N'(L)$ and restricts to the identity on L. If the foliations are smooth of class C^r, $r \geq 1$, it is required that φ be a diffeomorphism of class C^r.

Denote the fiber of $N'(L)$ at x by S'_x. Fix $x_0 \in L$ and let $\mathcal{D}_{x_0}$ (respectively, $\mathcal{D}'_{x_0}$) denote the group of all germs at $x_0 \in S_{x_0}$ (respectively, at $x_0 \in S'_{x_0}$) of local homeomorphisms (or diffeomorphisms of suitable smoothness class) of this fiber that fix x_0. The germinal holonomy homomorphisms for L, defined by the respective foliations, will be denoted by

$$\widehat{h} : \pi_1(L, x_0) \to \mathcal{D}_{x_0},$$
$$\widehat{h}' : \pi_1(L, x_0) \to \mathcal{D}'_{x_0},$$

respectively. These holonomy homomorphisms are *conjugate* if there is a homeomorphism g (or diffeomorphism of suitable smoothness class) of a neighborhood of x_0 in S_{x_0} onto a neighborhood of x_0 in S'_{x_0}, carrying x_0 to x_0 and such that

$$\widehat{g}_{x_0}^{-1} \circ \widehat{h}'_s \circ \widehat{g}_{x_0} = \widehat{h}_s,$$

for all loops s at x_0 in L. The following is a germinal analogue of Theorem 2.1.7.

Theorem 2.3.9. *The foliations $\mathcal{F}$ and $\mathcal{F}'$, as above, have isomorphic germs at the compact leaf L if and only if the respective holonomy homomorphisms for L are conjugate.*

Proof. If the germs at L are isomorphic, it is trivial that the respective holonomy homomorphisms are conjugate. For the converse, let $g : W \to W'$ be a diffeomorphism between neighborhoods of x_0 in S_{x_0} and S'_{x_0}, respectively, such that $g(x_0) = x_0$ and

$$\widehat{g}_{x_0}^{-1} \circ \widehat{h}'_s \circ \widehat{g}_{x_0} = \widehat{h}_s,$$

for all loops s at x_0 in L. Fix regular foliated atlases for $\mathcal{F}$ and $\mathcal{F}'$, respectively, having the same plaques on L and such that the transversals for the foliated charts meeting L all lie in the fibers of suitable, imbedded normal neighborhoods $N(L)$ and $N'(L)$. The fact that this can be done is elementary and is relegated to Exercise 2.3.10. Since L is compact and the foliated atlases are regular, there are only finitely many plaques $\{P_i\}_{i=0}^{\ell}$ in L. Suppose that $x_0 \in P_0$ and select plaque chains $\mathcal{P}_j = \{P_0, P_{i_1}, \dots, P_{i_k}, P_j\}$, one for each $j = 0, 1, \dots, \ell$. We take $\mathcal{P}_0 = \{P_0\}$. If $P_j \cap P_r \neq \emptyset$, form the loop

$$\mathcal{P}_j - \mathcal{P}_r = \{P_0, \dots, P_j, P_r, \dots, P_0\}.$$

Since there are only finitely many such loops, we can choose W, W' and a neighborhood $V \subseteq W$ of x_0 so small that all $h_{\mathcal{P}_j - \mathcal{P}_r}$ and $h'_{\mathcal{P}_j - \mathcal{P}_r}$ are defined on W and W', respectively, all $g^{-1} \circ h'_{\mathcal{P}_j - \mathcal{P}_r} \circ g$ are defined on V, and

$$h_{\mathcal{P}_j - \mathcal{P}_r} | V = g^{-1} \circ h'_{\mathcal{P}_j - \mathcal{P}_r} \circ g | V.$$

We can now mimic the proof of Theorem 2.1.7, using only those paths

$$\sigma : [a, b] \to L_z$$

such that $s = \pi \circ \sigma$ is covered (in the sense of Definition 2.3.1) by one of our prechosen plaque chains $\mathcal{P}_i$. □

Exercise 2.3.10. In the proof of Theorem 2.3.9, verify that the regular foliated atlases and normal neighborhoods of the compact leaf L in the respective foliated manifolds $(M, \mathcal{F})$ and $(M', \mathcal{F}')$ can be chosen as specified.

Theorem 2.3.9 has a remarkable consequence, the Reeb stability theorem, which will be discussed in the next section.

Definition 2.3.11. A leaf L has *trivial holonomy* if its germinal holonomy group is the trivial group.

A theorem due to G. Hector [**62**] and to D. B. A. Epstein, K. Millet and D. Tischler [**40**] states that the generic leaf of a foliation has trivial holonomy. More precisely, recall that a subset of a locally compact Hausdorff space is *residual* if it contains a countable intersection of open, dense subsets. The complement of a residual set is contained in the countable union of closed, nowhere dense subsets. Such sets are said to be *meager* and are the topological analogue of sets of measure zero. The two concepts, of course, are logically unrelated, as is illustrated by the existence of Cantor sets in $\mathbb{R}$ of positive Lebesgue measure. Meager sets cannot have interior and may or may not be dense. Residual sets must be everywhere dense and may or may not have interior. We are interested in *$\mathcal{F}$–saturated* sets of these types, a set being $\mathcal{F}$–saturated if it is a union of leaves. One thinks of the leaves in a residual, $\mathcal{F}$–saturated set as "generic".

Theorem 2.3.12. *Let $(M, \mathcal{F})$ be a foliated manifold. The leaves of $\mathcal{F}$ having trivial holonomy group are generic.*

Proof. There are countably many elements $\gamma = \gamma_{\alpha_0\alpha_1} \circ \cdots \circ \gamma_{\alpha_{\ell-1}\alpha_\ell}$ with maximal domain. For each, the set F_γ of fixed points is a closed subset of $D(\gamma)$ and the set–theoretic boundary $X_\gamma = \partial F_\gamma$ is closed and nowhere dense in $D(\gamma)$. By local compactness, X_γ is a countable union of compact, nowhere dense sets. Viewing S as the set of plaques of $\mathcal{U}$, let $Z_\gamma \subset M$ be the union of the compact closures $\overline{P}$ of the plaques $P \in X_\gamma$, itself the union of countably many compact, nowhere dense sets. The countable union of the Z_γ's is meager and is exactly the saturated set of leaves with nontrivial holonomy. □

Exercise 2.3.13. Let $(M, \mathcal{F})$ be a foliated manifold and assume that M is orientable. If L is a nonorientable leaf, $y \in L$, prove that $\mathcal{H}_y(L)$ contains at least one germ of an orientation–reversing holonomy transformation. Use an argument analogous to the proof of Theorem 2.3.12 to prove that, if $\operatorname{codim} \mathcal{F} = 1$, at most countably many leaves are nonorientable. An example in which an infinity of nonorientable leaves exists (and each leaf is dense) is given in [**39**].

Finally, we discuss the *infinitesimal* holonomy group of a leaf. For this, it is essential that everything be smooth of class at least C^1. Given $y \in S_\alpha$, the submersion $y_\alpha : U_\alpha \to \mathbb{R}^q$ induces an isomorphism $\nu_y(\mathcal{F}) \cong \mathbb{R}^q$, where we use the canonical identification of $\mathbb{R}^q$ with $T_y(\mathbb{R}^q)$. If L is the leaf through y and s is a loop on L based at y, the Jacobian $J\widehat{h}_s \in \mathrm{Gl}(q,\mathbb{R}) = \mathrm{Gl}(\nu_y(\mathcal{F}))$ is well defined, giving a group homomorphism

$$J\widehat{h} : \pi_1(L,y) \to \mathrm{Gl}(\nu_y(\mathcal{F})).$$

Definition 2.3.14. The infinitesimal holonomy homomorphism of L at y is $J\widehat{h}$ and the image of this homomorphism is the infinitesimal holonomy group of L at y.

Example 2.3.15. The infinitesimal holonomy has a particularly nice interpretation for transversely orientable foliations $\mathcal{F}$ of codimension 1. In this case, $J\widehat{h}_s = \widehat{h}'_s \in \mathbb{R}^+$ and it is customary to call the composition

$$\pi_1(L,y) \xrightarrow{\widehat{h}'} \mathbb{R}^+ \xrightarrow{\log} \mathbb{R}$$

the infinitesimal holonomy. This is a group homomorphism relative to the additive structure of $\mathbb{R}$. In algebraic topology, there is a standard identification of the first cohomology $H^1(L;\mathbb{R})$ with the set of group homomorphisms

$$\psi : \pi_1(L,y) \to \mathbb{R};$$

so we can view $\log \circ \widehat{h}' \in H^1(L;\mathbb{R})$. In particular, when $H^1(L;\mathbb{R}) = 0$, the infinitesimal holonomy of the leaf is trivial. If $\mathcal{F}$ is of class C^∞, fix a choice of nonsingular form $\omega \in A^1(M)$ defining $\mathcal{F}$. Then, the integrability condition $d\omega = \omega \wedge \eta$ gives a form $\eta \in A^1(M)$ and $d\eta = \omega \wedge \gamma$ (*cf.* Example 1.3.11). It follows that $\eta|L \in A^1(L)$ is a *closed* form. In Exercise 2.3.16, you will show that the de Rham class $[\eta|L] \in H^1(L;\mathbb{R})$ is exactly the infinitesimal holonomy class $\log \circ \widehat{h}'$. In particular, $[\eta|L]$ is independent of the allowable choices of ω and η. This suggests that $\mathrm{gv}(\mathcal{F}) = [\eta \wedge d\eta] \in H^3(M;\mathbb{R})$ should be tied in with the holonomy of the foliation. While this is true, an unpublished theorem of G. Duminy shows that the tie–in is quite subtle and it is not yet completely understood. We will prove Duminy's theorem in Volume II.

Exercise 2.3.16. Verify the assertion in Example 2.3.15 that

$$[\eta|L] = \log \circ \widehat{h}' \in H^1(L;\mathbb{R}).$$

Proceed as follows.

1. Prove that $\eta|L$ is closed and that $[\eta|L]$ is independent of the choice of ω and the subsequent choice of η.
2. Consider a foliated bundle $(S^1 \times [-\epsilon,\epsilon], \mathcal{F}, \pi)$, where

$$\pi : S^1 \times [-\epsilon,\epsilon] \to S^1$$

is the canonical projection, $\mathcal{F}$ is tangent to the boundary $S^1 \times \{\pm\epsilon\}$, and $L = S^1 \times \{0\}$. Show that $[\eta|L]$ is the infinitesimal holonomy class in this case if and only if this holds in general. In the remainder of the exercise, you will verify $\int_\sigma \eta = \log(h'_\sigma(0))$, where σ is the loop on L of degree 1 and h_σ is its total holonomy.

3. Let $\widetilde{M} = \mathbb{R}\times[-\epsilon, \epsilon]$ and consider the action of the infinite cyclic group $\mathbb{Z}$ on $\widetilde{M}$, generated by the diffeomorphism

$$\psi : \mathbb{R} \times [-\epsilon, \epsilon] \longrightarrow \mathbb{R} \times [-\epsilon, \epsilon],$$

$$\psi(s, t) = (s - 1, h_\sigma(t)).$$

Then, $M = \widetilde{M}/\mathbb{Z} \cong S^1 \times [-\epsilon, \epsilon]$ and the foliation $\widetilde{\mathcal{F}}$ of $\widetilde{M}$ by the lines $\mathbb{R} \times \{t\}$, $-\epsilon \le t \le \epsilon$, passes to the foliation $\mathcal{F}$ under this quotient. Show that there is a strictly positive, real–valued function $f(s,t)$ on $\widetilde{M}$ such that $\widetilde{\omega} = f\, dt \in A^1(\widetilde{M})$ is ψ–invariant, hence is a 1–form that passes under the quotient to a 1–form ω on M that defines $\mathcal{F}$.

4. Finally, use the above to find a suitable choice of η for which you can check that $\int_\sigma \eta = \log(h'_\sigma(0))$.

2.4. Reeb Stability

The results in this section are valid for foliations of class $C^{1,0+}$.

Theorem 2.3.9 has the following corollary, the importance of which can hardly be exaggerated.

Theorem 2.4.1 (Reeb Stability Theorem). *If the compact leaf L of a foliated manifold $(M, \mathcal{F})$ has trivial holonomy, then there is a neighborhood of L in M that is a union of leaves that are homeomorphic to L.*

Proof. Indeed, if $\operatorname{codim} \mathcal{F} = q$, the manifold $M' = L \times E^q$ has a foliation $\mathcal{F}' = \{L \times \{y\}\}_{y \in E^q}$ and $L = L \times \{0\}$ has trivial holonomy. (If $L \subseteq \partial M$, use the half cube $E^q \cap \mathbb{H}^q$ in place of the open q–cube E^q.) The projection $\pi : L \times E^q \to L$ makes M' itself a foliated normal neighborhood $N'(L)$ of L with normal fibers $S'_x = \{x\} \times E^q$. Fix a corresponding normal neighborhood $N(L)$ in M and any diffeomorphism $g : S_{x_0} \to S'_{x_0}$ with $g(x_0) = x_0$. Since the holonomy homomorphisms of L relative to both foliations take every loop to the germ of the identity, $\widehat{g}_{x_0}$ is a conjugation of these homomorphisms. Thus, the foliations have isomorphic germs at L and the assertion follows by Theorem 2.3.9. □

Corollary 2.4.2. *If L is a compact, simply connected leaf of a foliated manifold $(M, \mathcal{F})$, then there is a neighborhood of L in M that is a union of leaves that are homeomorphic to L.*

The following generalization of Theorem 2.4.1 is actually a corollary of that result.

Theorem 2.4.3 (Generalized Reeb Stability). *If L is a compact leaf of a foliated manifold $(M, \mathcal{F})$ and if the holonomy group $\mathcal{H}_y(L)$ is finite, then there is a normal neighborhood $p : V \to L$ of L in M such that $(V, \mathcal{F}|V, p)$ is a foliated bundle with all leaves compact. Furthermore, each leaf $L' \subset V$ has finite holonomy group of order at most the order of $\mathcal{H}_y(L)$ and the covering $p|L' : L' \to L$ has k sheets, where $k \leq$ order of $\mathcal{H}_y(L)$.*

Proof. Since $\mathcal{H}_y(L)$ is the quotient of $\pi_1(L, y)$ by a normal subgroup, there is a regular covering $p : F \to L$ with covering group $\mathcal{H}_y(L)$. Since this group is finite, F is a compact manifold. The neighborhood $N(L)$, as a fiber bundle over L, can be pulled back via p to a bundle $N(F)$ over F with the same fiber (an open q–disk). The standard pull–back construction yields a canonical covering $\widetilde{p} : N(F) \to N(L)$ (a C^1 submersion), enabling us to lift the foliation $\mathcal{F}|N(L)$ to a foliation $\widetilde{\mathcal{F}}$ of $N(F)$, transverse to the fibers and having F as a leaf. Let $x \in F$ with $p(x) = y$. Since the covering group is exactly the holonomy group of L,

$$p_* : \pi_1(F, x) \to \pi_1(L, y)$$

injects $\pi_1(F, x)$ onto the subgroup of $\pi_1(L, y)$ with trivial germinal holonomy. It follows that the leaf F of $\widetilde{\mathcal{F}}$ has trivial germinal holonomy. By Theorem 2.4.1, F has a neighborhood in $N(F)$ that is a foliated disk bundle with all leaves homeomorphic to F. The image of this neighborhood by p is a neighborhood W of L in $N(L) \subset M$ consisting of compact leaves. This neighborhood W is a true foliated bundle over L. Since $\pi_1(L, y)$ is a finitely generated group, $\mathcal{H}_y(L)$ is also finitely generated, allowing us to select a possibly smaller normal neighborhood V of L that is a foliated bundle with total holonomy group G isomorphic to $\mathcal{H}_y(L)$ (via passage to germs). Let V_y be the fiber of V over y. It follows that the germinal holonomy group of each leaf L_z through $z \in V_y$ is naturally isomorphic to a quotient of the subgroup of $G \cong \mathcal{H}_y(L)$ that fixes z. All assertions follow. □

Corollary 2.4.4. *If a compact leaf L of a foliated manifold $(M, \mathcal{F})$ has finite fundamental group, then there is a neighborhood V of L in M that is a union of compact leaves of $\mathcal{F}$ that are finite covering spaces of L.*

There is also a Reeb stability theory for *perturbations* of a foliation $\mathcal{F}$. For instance, it is possible to define a topology on the set $\mathcal{F}^r_p(M)$ of p–dimensional C^r foliations of M and to prove (at least for $r \geq 2$) that, if $\mathcal{F}$ has a compact leaf L with finite fundamental group, then there is a neighborhood of $\mathcal{F}$ in $\mathcal{F}^r_p(M)$ consisting of foliations having a compact leaf uniformly near L that is a finite covering space of L. The details are delicate (*cf.* [**52**, pp. 122–128]) and we will be satisfied here with a much more elementary perturbation result.

Definition 2.4.5. Let $(M, \mathcal{F})$ be a foliated manifold of class at least $C^{1,0+}$ and codimension q. Let $k \geq 1$ be an integer and $B^k \subset \mathbb{R}^k$ the open unit ball. A k–parameter perturbation of $\mathcal{F}$ is a $C^{1,0+}$ foliation $\mathcal{K}$ of $B^k \times M$ of codimension $q + k$ such that $\{x\} \times M$ is a union of leaves, $\forall x \in B^k$, and $\mathcal{K}|\{0\} \times M = \mathcal{F}$. The foliations $\mathcal{K}_x = \mathcal{K}|\{x\} \times M$ are called perturbations of $\mathcal{F}$.

The following is an immediate consequence of Corollary 2.4.4, asserting that compact leaves with finite fundamental group are "stable" under small k–parameter perturbations of $\mathcal{F}$.

Theorem 2.4.6 (Reeb Stability for Perturbations). *Let $(M, \mathcal{F})$ be a foliated manifold, L a compact leaf of $\mathcal{F}$ with finite fundamental group, and let $W \subset M$ be an open neighborhood of L. If $\mathcal{K}$ is a k–parameter perturbation of $\mathcal{F}$, then there is $\epsilon > 0$ such that, for every x in the ϵ–neighborhood of 0 in B^k, the perturbed foliation $\mathcal{K}_x$ has a compact leaf $L_x \subset W$ that is a finite covering of L.*

Exercise 2.4.7. Let $M = T^2 \times S^1$ and let $\mathcal{F}$ be the product foliation with leaves $T^2 \times \{z\}$, $z \in S^1$. Exhibit a 3–parameter deformation $(B^3 \times M, \mathcal{K})$ of $\mathcal{F}$ in which there are values of $x \in B^3$ arbitrarily near 0 such that $\mathcal{K}_x$ has no compact leaves and others such that all leaves of $\mathcal{K}_x$ are compact. This shows that the hypothesis in Theorem 2.4.6 that L has finite fundamental group cannot be weakened to finite holonomy group.

Exercise 2.4.8. Produce a compact, foliated 3–manifold $(M, \mathcal{F})$ with all leaves diffeomorphic to T^2 and a 1–parameter perturbation $\mathcal{K}$ of $\mathcal{F}$ such that, for all $s \neq 0$, $\mathcal{K}_s$ has no compact leaves. Proceed as follows.

1. Let $H \in \mathrm{Sl}(2, \mathbb{Z})$ and show that H induces a diffeomorphism h of $T^2 = \mathbb{R}^2/\mathbb{Z}^2$ onto itself. Form the mapping cylinder $M = M_h$ as in Example 2.1.1. As in that example, M fibers over S^1 with fiber T^2 and there is a transverse flow carrying fibers to fibers and having first return time 1. Let t denote the flow parameter.
2. Choose H as above so that it has positive eigenvalues λ and λ^{-1} different from 1 and let v and w be respective nontrivial eigenvectors in $\mathbb{R}^2$. Show that these vectors have irrational slope.
3. Show that the vector fields

$$X = \frac{-1}{\log \lambda} \frac{\partial}{\partial t},$$
$$Y = \lambda^{-t} v,$$
$$Z = \lambda^{t} w$$

on $\mathbb{R} \times \mathbb{R}^2$ pass to well–defined vector fields (again denoted by X, Y and Z) on M, X being tangent to the flow, Y and Z being tangent to the fibration $\mathcal{F}$.

4. Show that the above fields on M satisfy

$$[X, Y] = Y,$$
$$[X, Z] = -Z,$$
$$[Y, Z] = 0.$$

5. For each real number $s \neq 0$, show that the 2–plane distribution spanned by Y and $Z + sX$ is integrable and that each leaf of the resulting foliation $\mathcal{K}_s$ is dense in M. This defines the desired deformation $\mathcal{K}$.

Chapter 3

Basic Constructions

While there are many ways in which foliations are constructed and altered, certain techniques and ideas occur so often that they are "tools of the trade". The ones we discuss here are among the most basic.

3.1. Suspension

We motivate this construction by taking a closer look at Example 1.1.14. The notation from that example is used in the following.

Example 3.1.1. The fundamental group $\pi_1(\Sigma_2, x_0)$ is generated by the loops σ_i, $1 \leq i \leq 4$, pictured in Figure 3.1.1, subject to the single relation

$$\sigma_1 * \sigma_4^{-1} * \sigma_1^{-1} * \sigma_4 = \sigma_3 * \sigma_2^{-1} * \sigma_3^{-1} * \sigma_2. \tag{3.1.1}$$

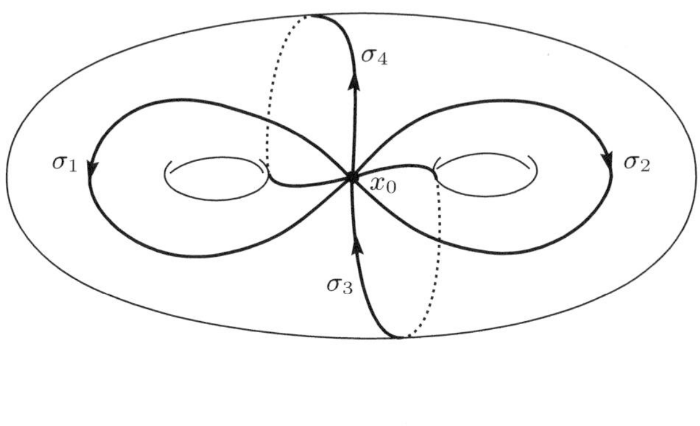

Figure 3.1.1. The generators of the fundamental group of Σ_2

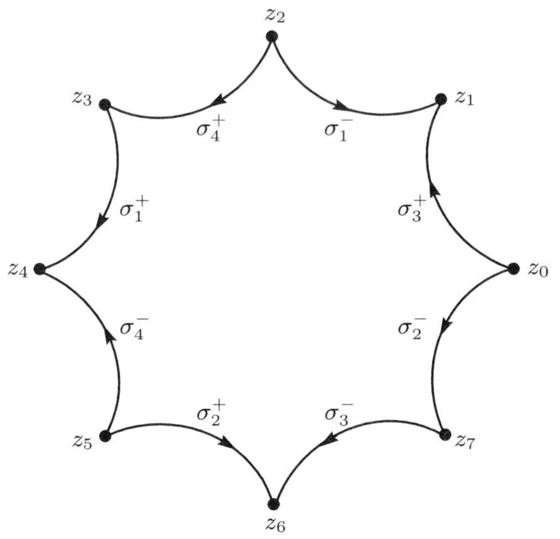

Figure 3.1.2. The fundamental domain Δ of Σ_2

Here, "$*$" denotes the group operation in $\pi_1(\Sigma_2, x_0)$, as defined by equation (2.1.1). If we cut Σ_2 apart along these four loops, the result is a planar octagon Δ (Figure 3.1.2), called the *fundamental domain* of the surface Σ_2. The directed edges of Δ are labeled in Figure 3.1.2 to show how to reglue them to construct Σ_2. This regluing identifies the eight vertices z_i to the single vertex x_0 in Figure 3.1.1. If $z \in \sigma_i \subset \Sigma_2$, the two preimages of z in $\sigma_i^+ \cup \sigma_i^-$ will be denoted by $z^\pm \in \sigma_i^\pm$.

The loop σ_3 (respectively, σ_4), freed from its basepoint, is isotopic in Σ_2 to the circle C_2 (respectively, C_1) in Figure 1.1.8. If $f_1, f_2 \in \mathrm{Diff}_+^r(S^1)$, we define an identification on $\Delta \times S^1$ by setting

$$\begin{aligned}
&(z^-, y) \equiv (z^+, f_1(y)), &&\forall\, z \in \sigma_4,\\
(3.1.2)\qquad &(z^-, y) \equiv (z^+, f_2(y)), &&\forall\, z \in \sigma_3,\\
&(z^-, y) \equiv (z^+, y), &&\forall\, z \in \sigma_1 \cup \sigma_2.
\end{aligned}$$

The reader should compare this with the construction in Example 1.1.14 to see that it seems to be essentially the same, the leaves of the foliation now being assembled by gluing together pieces $\Delta \times \{y\}$ along their edges. The rigorous proof of equivalence relies on Theorems 3.1.4 and 2.1.7. The advantage of our current setup is that it translates rather easily into a construction with universal covering spaces that is quite general. We describe this.

Recall that the universal cover of the 2–holed torus is

$$p : \mathbf{H} \to \Sigma_2,$$

where $\mathbf{H}$ is the hyperbolic plane and the group Γ of covering transformations is a discrete subgroup $\Gamma \subset \mathrm{Sl}(2,\mathbb{R})$ of the isometry group of $\mathbf{H}$.

Remark. A covering space is a principal bundle with discrete structure group (called the covering group or the group of deck transformations). In order to avoid annoying confusions, we must agree on whether this group acts from the right or the left. It is a fairly standard convention that the structure group G of a principal bundle acts on the total space from the *right*; so we agree that deck transformations will do so too. In the case of the hyperbolic plane $\mathbf{H}$ covering the 2–holed torus Σ_2, the right action

$$\mathbf{H} \times \Gamma \to \mathbf{H}$$

is defined by

$$(z, \gamma) \mapsto z \cdot \gamma = \gamma^{-1}(z),$$

the linear fractional transformation defined by $\gamma^{-1} \in \mathrm{PSl}(2,\mathbb{R})$.

If we use the open unit disk model of $\mathbf{H}$, then the fundamental domain pictured in Figure 3.1.2 is a fairly accurate picture of a regular hyperbolic octagon Δ in $\mathbf{H}$, centered at the origin of the unit disk and having all interior angles $\pi/4$. This octagon is carried by p onto Σ_2 with exactly the identifications along the edges indicated in Figure 3.1.2. Each $\gamma \in \Gamma$ carries the fundamental domain to another regular octagon $\Delta \cdot \gamma \subset \mathbf{H}$. If $\gamma \neq \mathrm{id}$, then Δ and $\Delta \cdot \gamma$ are either disjoint, meet along one common edge, or meet at a common vertex. This defines a *tiling* of the hyperbolic plane by regular octagons (most of which do not look all that regular to our Euclidean eyes), each of which is carried by p onto Σ_2 with pairwise edge identifications. The union of the vertices of all these octagons is exactly $p^{-1}(x_0)$.

For a fixed basepoint $z_0 \in p^{-1}(x_0)$, it is standard that the set Γ corresponds one–to–one to $p^{-1}(x_0)$ via

$$\gamma \in \Gamma \leftrightarrow z_0 \cdot \gamma \in p^{-1}(x_0).$$

A path $\widetilde{\sigma}$ in $\mathbf{H}$ from z_0 to $z_0 \cdot \gamma$ projects to a loop $\sigma = p \circ \widetilde{\sigma}$ at x_0 and, since $\mathbf{H}$ is simply connected, the homotopy class of σ in $\pi_1(\Sigma_2, x_0)$ is uniquely determined by γ. Conversely, each loop σ at x_0 lifts to a path $\widetilde{\sigma}$ in $\mathbf{H}$, starting at z_0, whose endpoint $z_\sigma \in p^{-1}(x_0)$ depends only on the class of σ in $\pi_1(\Sigma_2, x_0)$. By abuse of notation, we let σ denote its own homotopy class and we let γ_σ denote the unique covering transformation such that

$$z_0 \cdot \gamma_\sigma = z_\sigma.$$

This sets up a one–to–one correspondence

$$\Gamma \leftrightarrow \pi_1(\Sigma_2, x_0),$$
$$\gamma_\sigma \leftrightarrow \sigma$$

such that

$$(z_0 \cdot \gamma_\sigma) \cdot \gamma_\tau = z_0 \cdot \gamma_{\sigma * \tau} \tag{3.1.3}$$

(Exercise 3.1.2). The point of equation (3.1.3) is that, once the basepoint $z_0 \in p^{-1}(x_0)$ is fixed, the group Γ of covering transformations is canonically isomorphic to the fundamental group $\pi_1(\Sigma_2, x_0)$. Hereafter, we make this identification whenever convenient, denoting γ_σ by σ.

Choose $f_1, f_2 \in \mathrm{Diff}^r_+(S^1)$ and define a group homomorphism

$$h : \pi_1(\Sigma_2, x_0) \to \mathrm{Diff}^r_+(S^1)$$

by setting

$$\begin{aligned} h(\sigma_1) &= f_1, \\ h(\sigma_2) &= f_2, \\ h(\sigma_3) &= \mathrm{id}, \\ h(\sigma_4) &= \mathrm{id}\,. \end{aligned}$$

This extends uniquely to a homomorphism of groups because it respects the relation (3.1.1). On $\mathbf{H} \times S^1$ we define the relation

$$(z, y) \sim (z', y') \Leftrightarrow z' = z \cdot \gamma^{-1} \text{ and } y' = h(\gamma)(y), \text{ for some } \gamma \in \Gamma.$$

Defining an action of Γ on S^1 by

$$\gamma \cdot y = h(\gamma)(y), \quad \forall \gamma \in \Gamma \text{ and } \forall y \in S^1,$$

we write this relation as

$$(z, y) \sim (z \cdot \gamma^{-1}, \gamma \cdot y)$$

and remark that the map

$$\begin{gathered} \Gamma \times (\mathbf{H} \times S^1) \to \mathbf{H} \times S^1, \\ (\gamma, (z, y)) \mapsto (z \cdot \gamma^{-1}, \gamma \cdot y) \end{gathered}$$

is a left group action. Thus, $\sim$ is an equivalence relation whose equivalence classes are the Γ–orbits. The action is free and properly discontinuous. So we obtain a manifold

$$M = \Gamma \backslash (\mathbf{H} \times S^1).$$

If $(z, y) \in \mathbf{H} \times S^1$, let $[z, y]$ denote the corresponding point of M. Remark that, for $\gamma \in \Gamma$, $z \in \mathbf{H}$ and $y \in S^1$,

$$[z \cdot \gamma, y] = [z, \gamma \cdot y].$$

Because of this relation, it is popular to think of M as a kind of "tensor product" over Γ, writing

$$M = \mathbf{H} \times_\Gamma S^1.$$

Every element of M has a preimage in $\Delta \times S^1$ and we claim that the quotient map, restricted to $\Delta \times S^1$, gives exactly the identifications (3.1.2). Indeed, let $\gamma, \lambda \in \Gamma$ be determined by

$$z_3 \cdot \gamma = z_4,$$
$$z_0 \cdot \lambda = z_7.$$

Thus, γ carries σ_4^+ onto σ_4^- and λ carries σ_3^+ onto σ_3^-. Since z_0 is the basepoint, we see that $\lambda = \sigma_2$ and $h(\lambda) = f_2$. But z_3 is not the basepoint, and so $\gamma \neq \sigma_1$. Nevertheless, an elementary computation shows that

$$\gamma = \sigma_3^{-1} * \sigma_1 * \sigma_4^{-1} * \sigma_1 * \sigma_4 * \sigma_1^{-1} * \sigma_3$$

and it follows that $h(\gamma) = f_1$. This gives the identities

$$[z^-, y] = [z^+ \cdot \gamma, y] = [z^+, f_1(y)], \qquad \forall\, z \in \sigma_4 \text{ and } \forall\, y \in S^1,$$
$$[z^-, y] = [z^+ \cdot \lambda, y] = [z^+, f_2(y)], \qquad \forall\, z \in \sigma_3 \text{ and } \forall\, y \in S^1.$$

Similar computations show that the elements of Γ that carry σ_i^+ onto σ_i^-, $i = 1, 2$, are taken to $\mathrm{id} \in \mathrm{Diff}^r_+(S^1)$ by h; so

$$[z^-, y] = [z^+, y], \qquad \forall\, z \in \sigma_1 \cup \sigma_2 \text{ and } \forall\, y \in S^1.$$

(You are asked to verify all of these assertions in Exercise 3.1.3.) This proves that M is obtained from $\Delta \times S^1$ by the identifications (3.1.2). Note that the product foliation of $\mathbf{H} \times S^1$ by leaves $\mathbf{H} \times \{y\}$ is invariant under the action of Γ, hence passes to a foliation of $M = \Sigma_2 \times S^1$, transverse to the circle fibers. This is exactly the foliation produced in Example 1.1.14. So the cut and paste construction in that example has been expressed in an elegant form that will generalise readily.

Exercise 3.1.2. Prove equation (3.1.3) for the universal covering space of an arbitrary manifold B, concluding that, subject to the choice of basepoints, the fundamental group of B (with group composition $*$ as defined in equation (2.1.1)) identifies naturally with the group of covering transformations. Take care, since it is very easy to get things backwards!

Exercise 3.1.3. Verify the assertions in Example 3.1.1 about the values of h on the covering transformations that identify edges of Δ.

The reader familiar with principal bundles will notice that the construction in Example 3.1.1 and the generalization we are about to give is a special case of a standard construction. Given a (right) principal G–bundle $p : P \to B$ and a (left) G–space F, one forms an associated, locally trivial F–bundle

$$\pi : P \times_G F \to B,$$

the total space being the set of equivalence classes $[z,y]$ of points (z,y) of $P \times F$ under the relation

$$(z,y) \sim (z',y') \Leftrightarrow z' = zg^{-1} \text{ and } y' = gy, \quad g \in G.$$

The projection is then well defined by

$$\pi[z,y] = p(y).$$

For details, the reader can refer to any number of standard sources such as [**80, 6, 126**].

In our case, let B be a connected k–manifold, $x_0 \in B$, and let

$$p : \widetilde{B} \to B$$

be the universal cover of B. As above, we adopt the convention that the group Γ of covering transformations acts from the right:

$$\widetilde{B} \times \Gamma \to \widetilde{B},$$
$$(z,\gamma) \mapsto z \cdot \gamma,$$

making $\widetilde{B}$ the total space of a principal bundle with *discrete* structure group Γ. We will assume that B is smooth, hence $\Gamma \subset \mathrm{Diff}^\infty(\widetilde{B})$. Fix a q–manifold F of class C^r, $0 \le r \le \infty$, together with a choice of basepoint $z_0 \in p^{-1}(x_0)$, and let

$$h : \Gamma \to \mathrm{Diff}^r(F)$$

be a group homomorphism ($\mathrm{Diff}^0(F) = \mathrm{Homeo}(F)$). By Exercise 3.1.2, the choice of basepoint fixes an identification $\Gamma = \pi_1(B, x_0)$. We want to show that, under this identification, h is the total holonomy homomorphism of a foliated bundle of class C^r with fiber F and base B. For this, we define the "diagonal" action of Γ on $\widetilde{B} \times F$ to be the left action

$$\Gamma \times (\widetilde{B} \times F) \to \widetilde{B} \times F,$$
$$(\gamma,(z,y)) \mapsto (z \cdot \gamma^{-1}, \gamma \cdot y),$$

where $\gamma \cdot y = h(\gamma)(y)$. Since the action of Γ on $\widetilde{B}$ is free and properly discontinuous, the same follows immediately for the above diagonal action, and the quotient space

$$M = \Gamma \backslash (\widetilde{B} \times F) = \widetilde{B} \times_\Gamma F$$

is a C^r–manifold. The Γ–orbit of (z,y) is a point of M denoted by $[z,y]$. This construction is called the *suspension* of the homomorphism h.

Remark. The mapping cylinder, constructed in Example 2.1.1, is a case of this suspension construction.

Theorem 3.1.4. *The suspension of h yields a foliated bundle $(M, \mathcal{F}, \pi)$ of class C^r, having total space M, fiber F, base space B and total holonomy homomorphism conjugate to h. If $r = 0$, $(M, \mathcal{F})$ is of class $C^{\infty,0}$.*

Proof. The projection $\widetilde{B} \times F \to B$, defined by

$$(z, y) \mapsto p(z),$$

respects the group action, hence passes to a well–defined submersion

$$\pi : M \to B$$

of class C^r. Let $x \in B$ and choose a neighborhood U of x that is evenly covered by p. That is,

$$p^{-1}(U) = \coprod_{i \in \mathfrak{I}} U_i,$$

and

$$p|U_i : U_i \to U$$

is a diffeomorphism, $\forall\, i \in \mathfrak{I}$. The family $\{U_i \times F\}_{i \in \mathfrak{I}}$ consists of disjoint, open subsets of $\widetilde{B} \times F$ that are permuted simply transitively by the diagonal action of Γ. Thus, the equivalence relation defined by the action of Γ makes no internal identifications in any of the sets $U_i \times F$ and the quotient projection carries each one C^r–diffeomorphically onto $\pi^{-1}(U)$. (If $r = 0$, "C^r–diffeomorphically" means "homeomorphically".) The inverse φ_i of this diffeomorphism gives a commutative diagram

$$\begin{array}{ccccc} \pi^{-1}(U) & \xrightarrow{\varphi_i} & U_i \times F & \xrightarrow{p \times \mathrm{id}} & U \times F \\ {\scriptstyle \pi}\downarrow & & {\scriptstyle p}\downarrow & & \downarrow{\scriptstyle p_1} \\ U & \xrightarrow[\mathrm{id}]{} & U & \xrightarrow[\mathrm{id}]{} & U \end{array}$$

where p_1 is projection onto the first factor. Since $x \in B$ is arbitrary, this proves that $\pi : M \to B$ is a C^r bundle with fiber F. The product foliation of $\widetilde{B} \times F$ by leaves $\widetilde{B} \times \{y\}$, $\forall\, y \in F$, is preserved by the action of Γ, hence passes to a C^r foliation $\mathfrak{F}$ of M, transverse to the fibers of π. If $W = \pi^{-1}(U)$ is as above, the restriction $\mathfrak{F}|W$ has leaves that are carried by $(p \times \mathrm{id}) \circ \varphi_i$ to the leaves $U \times \{y\}$ of the product foliation of $U \times F$. This proves that $(M, \mathfrak{F}, \pi)$ is a C^r–foliated bundle. Finally, we compute the total holonomy homomorphism, provisionally denoted by $\widehat{h}$. Using the basepoint $z_0 \in p^{-1}(x_0)$, we fix an identification $F = \pi^{-1}(x_0)$ by

$$y \leftrightarrow [z_0, y].$$

Let $\sigma \in \pi_1(B, x_0)$ and let $w = [z_0, y] \in \pi^{-1}(x_0)$. The lift $\widehat{\sigma}$ of σ to a path on a leaf, starting at w, can be obtained by letting $\widetilde{\sigma}$ denote the lift to a path on $\widetilde{B}$, starting at z_0, and setting

$$\widehat{\sigma}(t) = [\widetilde{\sigma}(t), y].$$

If we parametrize the paths on $[0, 1]$, this gives

$$\widehat{h}_\sigma([z_0, y]) = [\widetilde{\sigma}(1), y] = [z_0 \cdot \sigma, y] = [z_0, h(\sigma)(y)].$$

That is, $\widehat{h} = h$. Of course, this identification of $\widehat{h}$ with h depends on the choice of identification of F with $\pi^{-1}(x_0)$, but the conjugacy class of $\widehat{h}$ is independent of choices. The final assertion is the remark on page 52. $\square$

In particular, the total holonomy homomorphism in Example 3.1.1 is h. This is clearly the same as the total holonomy homomorphism for the foliated bundle constructed in Example 1.1.14. So Theorem 2.1.7 implies that these foliated S^1–bundles are isomorphic. Theorem 3.1.4 and Theorem 2.1.7 give the following classification theorem.

Theorem 3.1.5. *Let F be a q–manifold of class at least C^r and B a connected k–manifold of class C^∞. Given $x_0 \in B$, there is a one-to-one correspondence between the set of conjugacy classes $[h]$ of group homomorphisms*

$$h : \pi_1(B, x_0) \to \mathrm{Diff}^r(F)$$

and the set of isomorphism classes $[M, \mathcal{F}, \pi]$ of C^r–foliated bundles $(M, \mathcal{F}, \pi)$ with fiber F and base B. This correspondence is defined by setting

$$[M, \mathcal{F}, \pi] \leftrightarrow [h]$$

if and only if the total holonomy homomorphism of $(M, \mathcal{F}, \pi)$ is conjugate to h.

There is a germinal version of the suspension construction. For this, let L be a compact, connected k–manifold, $k = n - q$, and let $\mathcal{D}^r_q$ denote the group of all germs at 0 of C^r diffeomorphisms of open neighborhoods of $0 \in \mathbb{R}^q$ that fix 0. Fix a basepoint $x_0 \in L$ and let

$$\widehat{h} : \pi_1(L, x_0) \to \mathcal{D}^r_q$$

be a group homomorphism. We will construct a C^r–foliated n–manifold $(W, \mathcal{F})$ having L as a compact leaf with germinal holonomy homomorphism conjugate to $\widehat{h}$.

Let $p : \widetilde{L} \to L$ be the universal cover and let $\Delta \subset \widetilde{L}$ be a fundamental domain for the group Γ of covering transformations. Thus, Δ is compact, $\widetilde{L}$ is covered by the Γ–translates of Δ, and

$$\Delta \cap \Delta \cdot \gamma \subseteq \partial\Delta,$$

$\forall \gamma \in \Gamma \smallsetminus \{\mathrm{id}\}$. By the compactness of Δ, there are only finitely many $\gamma \in \Gamma$ such that $\Delta \cap \Delta \cdot \gamma \neq \emptyset$. Call this finite subset Γ_0 and remark that it is a symmetric set of generators of Γ containing the identity. Take a small enough open neighborhood U of Δ in $\widetilde{L}$ so that

$$U \cap U \cdot \gamma \neq \emptyset \Leftrightarrow \gamma \in \Gamma_0.$$

We set $x \sim x \cdot \gamma$ whenever $\gamma \in \Gamma_0$ and $x, x \cdot \gamma \in U$, noting that $U/{\sim} = L$ canonically. (The reader can check that $\sim$ is an equivalence relation on U.)

For each $\gamma \in \Gamma_0$, choose a representative $f_\gamma \in \widehat{h}(\gamma)$. We can assume that all $f_\gamma : D \to \mathbb{R}^q$ have common domain an open ball $D \subset \mathbb{R}^q$ centered at 0, that

$$f_{\gamma^{-1}} = f_\gamma^{-1} \quad (\text{on } D \cap f_\gamma(D)),$$
$$f_{\text{id}} = \text{id}$$

and that, whenever $\gamma, \lambda, \eta \in \Gamma_0$ satisfy $\gamma\lambda = \eta$, then

$$f_\gamma \circ f_\lambda = f_\eta,$$

wherever both sides are defined. On $U \times D$, define the relation

$$(x, v) \sim (x \cdot \gamma^{-1}, f_\gamma(v)),$$

whenever $\gamma \in \Gamma_0$, $\{x, x \cdot \gamma^{-1}\} \subset U$, and $\{v, f_\gamma(v)\} \subset D$. By the choices of f_γ, $\gamma \in \Gamma_0$, it follows that this is an equivalence relation and that it extends the relation on $U = U \times \{0\}$ defined above. Let $W = (U \times D)/\!\sim$ with the quotient topology, a second countable Hausdorff space. It is clear that the projection of $U \times D$ onto the first factor passes to a well–defined, continuous surjection $\pi : W \to L = U/\!\sim$. If $x \in L$ and $V \subset L$ is a sufficiently small neighborhood of x, the open subset $\pi^{-1}(V) \subset W$ is homeomorphic to $V \times D$. These sets form a C^r atlas of dimension n on W. Finally, the product foliation on $U \times D$ is respected by $\sim$; so it passes to a C^r foliation $\mathcal{F}$ on W with L as a leaf and $\widehat{h}$ as the holonomy homomorphism. The details are left to the reader. Combining this with Theorem 2.3.9, we obtain the following germinal analogue of Theorem 3.1.5.

Theorem 3.1.6. *Let L be a compact, connected manifold and fix a basepoint $x_0 \in L$. Then there is a one–to–one correspondence between the set of conjugacy classes $[\widehat{h}]$ of homomorphisms*

$$\widehat{h} : \pi_1(L, x_0) \to \mathcal{D}_q^r$$

and the set of isomorphism classes $[\widehat{\mathcal{F}}_L]$ of germs $\widehat{\mathcal{F}}_L$ at L of C^r foliations of codimension q that have L as a leaf. This correspondence is defined by setting

$$[\widehat{\mathcal{F}}_L] \leftrightarrow [\widehat{h}]$$

if and only if the germinal holonomy homomorphism of $\widehat{\mathcal{F}}_L$ is conjugate to $\widehat{h}$.

3.2. Pullbacks

If $(M, \mathcal{F})$ is a foliated manifold of codimension q and $N \subset M$ is a smoothly immersed submanifold transverse to $\mathcal{F}$, then $\mathcal{F}$ induces a foliation on N, also of codimension q. The leaves of the induced foliation will be the connected components of the sets $L \cap N$ as L ranges over the leaves of $\mathcal{F}$. If

$$i : N \to M$$

is the inclusion immersion, we denote the induced foliation of N by $i^*(\mathcal{F})$ and call it the "pullback" of $\mathcal{F}$ by i. In fact, we will establish something much more general.

Definition 3.2.1. Let $(M, \mathcal{F})$ be a foliated manifold of class at least C^1 and let $f : N \to M$ be a C^1 map. If

$$T_{f(z)}(M) = T_{f(z)}(\mathcal{F}) + f_{*z}(T_z(N)), \quad \forall\, z \in N,$$

we say that f is transverse to $\mathcal{F}$ and write $f \pitchfork \mathcal{F}$.

Remark that the above sum of vector spaces is not required to be direct. The definition actually makes sense if $\mathcal{F}$ is only required to be of class $C^{1,0+}$, but the following would be harder to prove in that case.

Theorem 3.2.2. *If $(M, \mathcal{F})$ is a C^r-foliated manifold of codimension q and a C^s map $f : N \to M$ is transverse to $\mathcal{F}$, where $r, s > 0$, then there is a unique foliation $f^*(\mathcal{F})$ of N, also of codimension q and of smoothness class C^k, $k = \min(r, s)$, the leaves of which are the connected components of $f^{-1}(L)$ as L ranges over the leaves of $\mathcal{F}$.*

Proof. Let L be a leaf of $\mathcal{F}$ and let $z \in f^{-1}(L)$. Fix a foliated chart (U, x, y) for $\mathcal{F}$ about $f(z)$. Then $y : U \to \mathbb{H}^q$ is a submersion and the condition $f \pitchfork \mathcal{F}$ guarantees that

$$y \circ f : f^{-1}(U) \to \mathbb{H}^q$$

is also a submersion. The submersion theorem provides a chart (W, u, v) about z in $f^{-1}(U)$ such that

$$y(f(u, v)) = v.$$

Consequently, the level set $v = c$ is an imbedded submanifold of codimension q and is exactly the subset of W that maps into the level set $y = c$ in U. Varying z over $f^{-1}(L)$, we see that $f^{-1}(L)$ is an immersed submanifold of codimension q. Thus, as L ranges over the leaves of $\mathcal{F}$, the connected components of the manifolds $f^{-1}(L)$ give a decomposition $f^*(\mathcal{F})$ of N into connected, immersed submanifolds of codimension q. Also, by appropriately shrinking the chart (W, u, v) about z, we obtain a foliated chart compatible with this decomposition. Since $z \in N$ is arbitrary, this gives a foliated atlas for $f^*(\mathcal{F})$. This atlas is at least as smooth as the minimum of the smoothness classes of f and $\mathcal{F}$ and all assertions follow. $\square$

In particular, submersions $f : N \to M$ are automatically transverse to any foliation $\mathfrak{F}$ of M, yielding pullbacks $f^*(\mathfrak{F})$. This observation is of particular importance for covering maps.

Corollary 3.2.3. *Let $(M, \mathfrak{F})$ be a C^r–foliated manifold, $r \geq 1$, and let*

$$\pi : \widetilde{M} \to M$$

be a covering map. Then there is a unique lifted foliation, $\widetilde{\mathfrak{F}} = \pi^*(\mathfrak{F})$, of class C^r.

Exercise 3.2.4. Prove Corollary 3.2.3 for foliations of class $C^{1,0}$ and $C^{1,0+}$.

Corollary 3.2.5. *If $(M, \mathfrak{F})$ is of class C^∞ and $f \pitchfork \mathfrak{F}$, where $f : N \to M$ is also C^∞, then*

$$f^*(I^*(T(\mathfrak{F}))) = I^*(T(f^*(\mathfrak{F}))).$$

In particular, if $\mathfrak{F}$ is defined by $\omega \in A^1(M)$, then $f^*(\omega) \in A^1(N)$ defines $f^*(\mathfrak{F})$.

The proof of this corollary is straightforward and will be left to the reader. Since

$$d(f^*(\omega)) = f^*(d\omega) = f^*(\omega \wedge \eta) = f^*(\omega) \wedge f^*(\eta)$$

and

$$f^*(\eta \wedge d\eta) = f^*(\eta) \wedge d(f^*(\eta)),$$

we obtain the following *naturality* property of the Godbillon–Vey class (Example 1.3.11 on page 37).

Corollary 3.2.6. *If the C^∞ map $f : N \to M$ is transverse to a transversely orientable, C^∞ foliation $\mathfrak{F}$ of codimension one, then $f^*(\mathrm{gv}(\mathfrak{F})) = \mathrm{gv}(f^*(\mathfrak{F}))$.*

3.3. Transverse Modifications

There are various basic constructions that involve certain kinds of submanifolds that are transverse to a foliation. One can *glue* foliated manifolds along transverse boundary components; one can *spin* a foliation of codimension one along a transverse boundary component; and one can modify a foliation in a foliated neighborhood of a *closed transversal.*

3.3.A. Transverse gluing. Let M_1 and M_2 be smooth n–manifolds with boundary. If there are no corners, suppose that $S_i \subset \partial M_i$ is a union of boundary components, $i = 1, 2$, and that $\varphi : S_2 \to S_1$ is a diffeomorphism. One forms a C^0 manifold

$$M = M_1 \cup_\varphi M_2$$

from the disjoint union $M_1 \amalg M_2$ by identifying $x \equiv \varphi(x)$, $\forall x \in S_2$. If the manifolds have corners, those corners divide the boundaries into smooth submanifolds with boundaries of their own and S_i should be a union of such submanifolds with disjoint boundaries, $i = 1, 2$. One again forms a C^0 manifold M by gluing M_2 to M_1 via φ.

There is a standard way to put a smooth structure on M. For this, it is best to modify the above construction without really changing the topological manifold M that results. In a neighborhood of S_i, choose a smooth, nonsingular vector field v_i that is transverse to S_i and, at every point of S_i points into M_i, $i = 1, 2$. If $\partial S_i \neq \emptyset$, then ∂S_i is a corner on M separating $S_i \subset \partial M_i$ from another smooth portion $T_i \subset \partial M_i$ and we require that v_i be tangent to T_i, $i = 1, 2$. Such fields are easily constructed locally and then assembled by a smooth partition of unity. Replacing v_i by $\lambda_i v_i$, for a suitable positive function λ_i, $i = 1, 2$, one obtains local flows

$$\Phi^i : S_i \times [0, 2) \to M_i, \quad i = 1, 2.$$

These define smooth normal neighborhoods

$$N_i = \Phi^i(S_i \times [0, 1)) \subset \Phi^i(S_i \times [0, 2)) \subseteq M_i$$

and we use the flows to coordinatize N_i as a copy of $S_i \times [0, 1)$. That is, if $x \in S_i$, the point $\Phi^i_t(x) \in N_i$ has coordinates (x, t), $0 \leq t < 1$. It is standard that $M_i \cong M_i \setminus N_i$. We view M as the manifold obtained from the disjoint union $(M_1 \setminus S_1) \amalg (M_2 \setminus S_2)$ by identifying $(x, t) \equiv (\varphi(x), 1-t)$, $\forall\, x \in S_2$ and $0 < t < 1$. (See Figure 3.3.1.) Since this identification is by a diffeomorphism between the open manifolds $N_i \setminus S_i$, it is easy to produce the differentiable structure. One should think of $M_1' = M_1 \setminus N_1$ as the original manifold to which a collar N_1 is attached to form (the new) M_1. The gluing makes this collar overlap the collar neighborhood N_2 of S_2 in M_2 and gives topologically the same manifold as the more naive gluing operation in the previous paragraph. If the manifolds M_i have corners, M may or may not have corners, but it will surely have boundary.

Example 3.3.1. If $(M_1, \mathcal{F}_1)$ and $(M_2, \mathcal{F}_2)$ are both C^{r+1}–foliated of codimension q, $1 \leq r \leq \infty$, and if S_i is a union of components of $\partial_\pitchfork M_i$, the foliations can sometimes be assembled into a C^r foliation of $M = M_1 \cup_\varphi M_2$. For this, we need a C^{r+1} diffeomorphism $\varphi : S_2 \to S_1$ that matches the foliation $\mathcal{F}_2|S_2$ (*cf.* Example 1.2.21) with the foliation $\mathcal{F}_1|S_1$. It is possible to choose the vector fields v_i above to be tangent to the leaves of $\mathcal{F}_i$ and to be smooth of class C^r. Indeed, the foliations being of class C^{r+1}, their tangent bundles are of class C^r and one easily chooses such vector fields in foliated neighborhoods. These local choices are then assembled by a smooth partition of unity into the desired fields. The collar neighborhoods $N_i = S_i \times [0, 1)$ are called *C^r–foliated collars* and it is clear that the identification

$$(x, t) \equiv (\varphi(x), 1 - t)$$

respects the foliations.

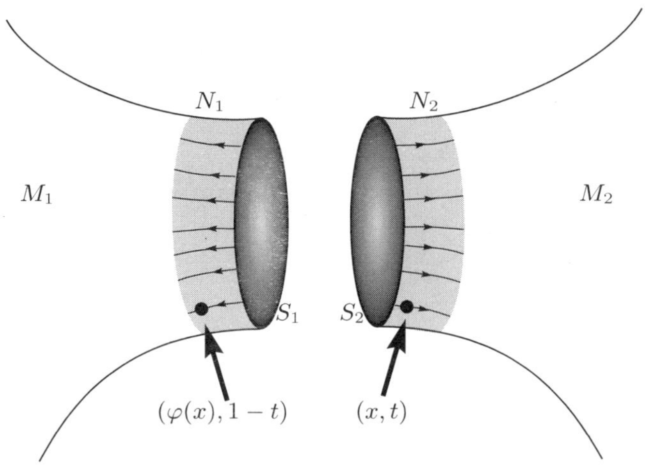

Figure 3.3.1. Smoothing $M_1 \cup_\varphi M_2$

Example 3.3.2. The method of Example 3.3.1 has the disadvantage that, unless $r = \infty$, the resulting foliated manifold has one less degree of smoothness than the initial ones. It also fails for C^1 foliations. Here is a method that preserves the degree of differentiability at the cost of altering $(M_i, \mathfrak{F}_i)$ in a neighborhood of S_i in a topologically harmless way, $i = 1, 2$. Let all notation and conventions be as above, the foliated manifolds being of class C^r, $r \geq 1$. Choose the C^∞ local flow

$$\Phi^i : S_i \times [0, 2) \to M_i, \quad i = 1, 2,$$

as in the discussion preceding Example 3.3.1, with no particular relationship to the foliations. We use Φ^i to identify $S_i \times [0, 2)$ as an open neighborhood of S_i in M_i, $i = 1, 2$. Let $g : [0, 2) \to [0, 2)$ be of class C^∞ and such that

$$\begin{aligned} g(t) &= 0, & 0 \leq t \leq 1, \\ g'(t) &> 0, & 1 < t < 2, \\ g(t) &= t, & 3/2 \leq t < 2. \end{aligned}$$

Then the map $(x, t) \mapsto (x, g(t))$ carries $S_i \times [0, 2)$ smoothly onto itself. This map extends by the identity outside of the collar $S_i \times [0, 2)$, defining a C^∞ map

$$f_i : M_i \to M_i.$$

It is a submersion of $S_i \times [0, 1]$ onto S_i and is a diffeomorphism outside of that collar. Since $\mathfrak{F}_i \pitchfork S_i$, it follows that f_i is transverse to $\mathfrak{F}_i$ and f_i pulls $\mathfrak{F}_i$ back to a foliation $\mathfrak{F}_i'$ *of the same smoothness class* (Theorem 3.2.2) which coincides with $\mathfrak{F}_i$ outside of a collar neighborhood of S_i, $i = 1, 2$.

Furthermore, on the collar $N_i = S_i \times [0,1)$, the leaves of $\mathcal{F}'_i|N_i$ are of the form $L \times [0,1)$, where L ranges over the leaves of $\mathcal{F}_i|S_i$. Now, if $\varphi : S_2 \to S_1$, as in the previous example, is a C^r diffeomorphism matching the foliations, then the gluing identification $(x,t) \equiv (\varphi(x), 1-t)$, $\forall\, x \in S_2$ and $0 < t < 1$, is of class C^r, matching the foliation $\mathcal{F}'_2|N_2$ to the foliation $\mathcal{F}'_1|N_1$.

Remark. By essentially the same constructions, we can glue M to itself along separate portions of its transverse boundary. Here, we assume that S_1 and S_2 are unions of components of $\partial_{\pitchfork} M$ such that $S_1 \cap S_2 = \emptyset$ and we use a gluing map $\varphi : S_2 \to S_1$ that matches the foliations and has a suitable degree of smoothness. The suspension of two diffeomorphisms of S^1, as described in Example 1.1.14, was a case of this construction.

3.3.B. Spinning. Suppose that $(M, \mathcal{F})$ is a C^∞–foliated manifold of codimension one and that S is a component of $\partial_{\pitchfork} M$ such that $\partial S = \emptyset$. Assume also that the foliation $\mathcal{F}|S$ (Example 1.2.21) can be defined by a closed, nonsingular 1–form $\omega \in A^1(S)$. In this situation, we will show how to *spin* $\mathcal{F}$ smoothly along S, obtaining a C^∞ foliation $\mathcal{F}_S$ having S as a tangential boundary component and agreeing with $\mathcal{F}$ outside a collar neighborhood of S. As usual, coordinatize the foliated collar $N = S \times [0,1)$ so that the leaves of $\mathcal{F}|N$ are of the form $L \times [0,1)$, where L is a leaf of $\mathcal{F}|S$. The decomposition

$$T_{(x,t)}(N) = T_x(S) \oplus T_t([0,1))$$

allows us to write a general vector field $\zeta \in \mathfrak{X}(N)$ as

$$\zeta = fv + g\partial_t,$$

where $v \in \mathfrak{X}(S)$, $\partial_t = \partial/\partial t$, and $f, g \in C^\infty(N)$. The closed, nonsingular form $\omega \in A^1(S)$ extends to a closed, nonsingular form (again denoted by ω) on N by

$$\omega(fv + g\partial_t) = f\omega(v).$$

Let $\rho \in C^\infty[0,1)$ vanish identically on $[1/2, 1)$, decrease strictly monotonically from $1 = \rho(0)$ to $0 = \rho(1/2)$ on $[0, 1/2]$, and have the property that all derivatives of all orders ≥ 1 vanish identically at $t = 0$. Then, $\theta \in A^1(S \times [0,1))$, defined by

$$\theta = (1 - \rho(t))\omega + \rho(t)\, dt,$$

agrees with ω on $S \times [1/2, 1)$ and with dt on $S = S \times \{0\}$. Furthermore, θ is integrable. Indeed, since ω is closed,

$$d\theta = -\rho'(t)\, dt \wedge \omega = \begin{cases} \theta \wedge \left(\frac{-\rho'(t)\omega}{\rho(t)}\right) & \text{on } S \times [0, 1/2), \\ \theta \wedge \left(\frac{\rho'(t)\, dt}{1-\rho(t)}\right) & \text{on } S \times (0,1). \end{cases}$$

That is, the Frobenius integrability condition is satisfied locally, hence globally, on N. Since $\theta = dt$ along $S \times \{0\}$, the transverse boundary component

S is now a leaf of the new foliation of the collar $S \times [0, 1)$, while the new foliation coincides with the old outside of the collar $S \times [0, 1/2)$. Thus, these foliations assemble to give a new C^∞ foliation $\mathcal{F}_S$ of M with the announced properties. Intuitively, we have smoothly perturbed the foliation near S so that, as the leaves head for S, the form θ approximates dt and the leaves are deflected more and more into directions parallel to S. This is referred to as "spinning" $\mathcal{F}$ along the transverse boundary component S.

Exercise 3.3.3. Remark that $\mathcal{F}_S$ is transverse to the fibers of $S \times [0, 1/2)$ and show that the holonomy of S on a fiber $\{x_0\} \times [0, 1/2)$ imbeds canonically in the (semi)flow Φ_u on $[0, 1/2)$ generated by the vector field

$$X = \left(\frac{1 - \rho(t)}{\rho(t)}\right) \frac{d}{dt}.$$

More precisely, remark that this field is C^∞–trivial at $t = 0$ and prove that the holonomy about a loop σ based at x_0 is $h_\sigma = \Phi_{-\omega(\sigma)}$, where $\omega(\sigma)$ denotes the integral of ω around σ. (Hint. Observe that, on $S \times (0, 1/2)$, $\mathcal{F}_S$ is defined by the closed form $\omega + (\frac{\rho}{1-\rho})\, dt$.) It follows that this leaf has abelian holonomy.

Exercise 3.3.4. Using the previous exercise, show that $\mathcal{F}_S|S \times [0, 1/2)$ is independent, up to a topological conjugacy that is smooth on $S \times (0, 1/2)$, of the choice of ρ. If two choices of ρ coincide near 0, show that this conjugacy is smooth. If they also coincide near 1/2, show that the corresponding foliations of M are diffeomorphic.

Example 3.3.5. A common situation in which this spinning construction arises is in a C^∞–foliated 3–manifold with a transverse boundary component $S \cong T^2$ such that $\mathcal{F}|S$ is diffeomorphic to a *linear* foliation of T^2 (Example 1.1.5). We have seen (Exercise 1.2.7) that such a foliation is defined by a closed, nonsingular form $\omega \in A^1(T^2)$, and so it can be spun along S. An elementary but important case of this is the Reeb foliation of the solid torus. One begins with the product foliation by disks that induces the linear foliation of slope 0 on the boundary. Spinning the product foliation produces the Reeb foliation.

This construction is useful when gluing two foliated 3–manifolds together along transverse boundary tori. If the foliations do not match via any diffeomorphism, transverse gluing is impossible. In this case, we can spin so that the tori become leaves and then glue. A subtle smoothness problem arises (to be discussed in § 3.4) and our stipulation that all derivatives $\rho^{(m)}(0) = 0$, $m \geq 1$, will be required to make the new foliation C^∞.

3.3.C. Closed transversals. Fix the assumptions that $(M, \mathcal{F})$ is a foliated manifold of codimension q, that M itself is smooth and that $\mathcal{F}$ is of class at least $C^{1,0+}$.

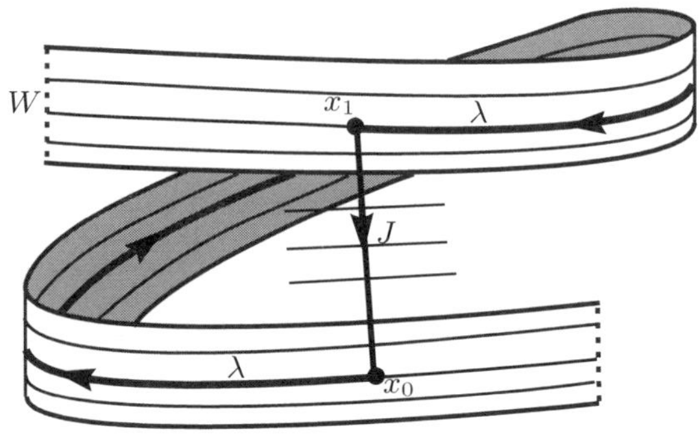

Figure 3.3.2. The "waterfall" construction

Definition 3.3.6. A *closed transversal* to $\mathcal{F}$ is a smoothly imbedded, compact, connected, q–dimensional submanifold $S \hookrightarrow M$ without boundary and everywhere transverse to $\mathcal{F}$.

Remark. No generality is lost in requiring S to be C^∞–imbedded. An arbitrary C^r imbedding, $r \geq 1$, can always be smoothed to a C^∞ imbedding by a C^r homotopy that is arbitrarily small in the C^r topology [**69**, Theorem 2.2.7]. If the homotopy is small enough, the transversality will not be compromised. Note that, because the dimensions of S and $\mathcal{F}$ are complementary, transversality means simply that

$$T_y(S) \cap T_y(\mathcal{F}) = \{0\}, \quad \forall\, y \in S.$$

Closed transversals are only common in foliated manifolds $(M, \mathcal{F})$ of codimension one. In that case, a closed transversal is a smoothly imbedded circle S and the transversality condition only requires that S be nowhere tangent to $\mathcal{F}$.

Lemma 3.3.7. *Suppose that $\mathcal{F}$ is transversely oriented of codimension* 1. *Let $J \subset M$ be a smoothly imbedded, compact arc, everywhere transverse to $\mathcal{F}$ and such that $\partial J \subset L$, where L is a leaf of $\mathcal{F}$. Then there is a closed transversal S to $\mathcal{F}$ meeting exactly the same leaves as J.*

Proof. Let $\partial J = \{x_0, x_1\}$, so indexed that, in the order on J induced by the transverse orientation of $\mathcal{F}$, $x_1 < x_0$. Let $\lambda : [0, 1] \to L$ be a C^1 imbedding with $\lambda(i) = x_i$, $i = 0, 1$. Together, λ and J form a closed, piecewise C^1 loop in M, transverse to $\mathcal{F}$ along J and tangent to $\mathcal{F}$ along λ. This is illustrated in Figure 3.3.2, which bears an abstract resemblance to Escher's *Waterfall* (J is the waterfall, λ the stream). Find a simply connected neighborhood P_λ of $\lambda([0, 1])$ in L and a closed, foliated coordinate chart $W = P \times [-2, 2]$

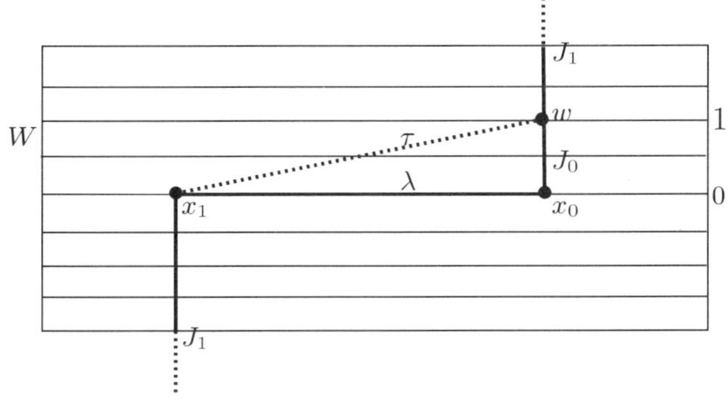

Figure 3.3.3. Deforming the "waterfall" to a closed transversal

such that P_λ is the plaque $P \times \{0\}$. Let J be decomposed into two compact subarcs, $J = J_1 \cup J_0$, where $J_0 \subset W$, the initial point of J_1 is x_1, the terminal point of J_0 is x_0, and $J_1 \cap J_0 = \{w\}$, an interior point of J. We can choose the coordinates (x, y) in $W = P \times [-2, 2]$ so that $J_0 = \{(x_0, y) \mid 0 \leq y \leq 1\}$. So $w = (x_0, 1)$ and x_0 is identified as $(x_0, 0)$. Let $\ell : [0, 1] \to P$ be such that $\lambda(t) = (\ell(t), 0)$. Then $\tau(t) = (\ell(t), 1 - t)$, $0 \leq t \leq 1$, defines an $\mathcal{F}$–transverse arc that stays in W and goes from w to x_1. All of this is illustrated in Figure 3.3.3. Then $\sigma = J_1 \cup \tau$ is a closed transversal with two corners (w and x_1 respectively). Notice that rounding the two corners of σ and slightly perturbing the resulting C^1 loop produces a smoothly imbedded circle S that is transverse to $\mathcal{F}$. From the construction, it is clear that S meets exactly the same leaves of $\mathcal{F}$ as J. □

Corollary 3.3.8. *Suppose that $\mathcal{F}$ is transversely orientable of codimension* 1 *and that M is compact. If the leaf L of $\mathcal{F}$ does not meet a closed transversal, then L is compact.*

Proof. We suppose that L is a noncompact leaf of $\mathcal{F}$ and prove that L meets a closed transversal. Since M can be covered by a finite union of foliated charts, some such chart U has at least two distinct plaques $P_0, P_1 \subset L \cap U$. Then a suitable subarc $J \subset U$ is transverse to $\mathcal{F}$ and has one endpoint in P_1 and one in P_2; so we apply Lemma 3.3.7. □

Remark. If M is simply connected, every line field on M will be orientable. In particular, all foliations of M of codimension one will be transversely orientable. A deep theorem of S. P. Novikov [**99**] asserts that every C^2 foliation of S^3 by surfaces contains a leaf that does not meet a closed transversal. Since S^3 is simply connected, Corollary 3.3.8 applies and this leaf must be compact. A little algebraic topology, together with the Gauss–Bonnet theorem, shows that a compact leaf in S^3 must be diffeomorphic to T^2; so every C^2 foliation of S^3 by surfaces has a toroidal leaf. Novikov goes further, showing that at least one of these toroidal leaves bounds a solid torus in S^3 that is foliated as in Example 1.1.12 (a so–called "Reeb component"). We will shortly construct a famous example of such a foliation, due to G. Reeb [**110**]. A proof of Novikov's theorem will be given in Volume II.

There are certain nice neighborhoods of closed transversals, called "foliated tubular neighborhoods", which are useful in a number of constructions. There are various ways to obtain foliated tubular neighborhoods, two of which we now describe.

Example 3.3.9. Let $(M, \mathcal{F})$ be a foliated n–manifold of codimension q, let $S \subset M$ be a closed transversal, and assume that the foliation is of class at least C^r, $r \geq 3$. Fix a Riemannian metric g on M and let $\pi : E \to S$ be the k–plane subbundle of $T(M)|S$ with fibers perpendicular to the fibers of $T(S)$. If $\epsilon > 0$, the subset

$$E_\epsilon = \{v \in E \mid \|v\| \leq \epsilon\}$$

is a compact neighborhood of the 0–section of E and

$$\pi : E_\epsilon \to S$$

is a fiber bundle with fibers isomorphic to the compact disk of dimension $n - q$ and radius ϵ. The exponential map for the Riemannian metric defines

$$\exp : E_\epsilon \to M,$$

for sufficiently small ϵ, and is a smooth imbedding for possibly smaller choices of ϵ. This is the "orthogonal geodesic spray" along S and its image is a "tubular neighborhood" of S. Unfortunately, this tubular neighborhood construction has little to do with the foliation $\mathcal{F}$. In order to remedy this, replace E with the bundle $T(\mathcal{F})|S$. Each $v \in T(\mathcal{F})$ can be written uniquely as $v = v_S + v_E$, where v_S is tangent to S and $v_E \in E$. The map $v \mapsto v_E$ defines a bundle isomorphism of $T(\mathcal{F})$ onto E. Instead of the geodesic spray for M, consider the induced Riemannian metric g_{y_0} on the leaf L_{y_0} through $y_0 \in S$ and let $\exp_{y_0}$ be the exponential map at y_0 in the Riemannian manifold (L_{y_0}, g_{y_0}). In order to study this map as y_0 varies, choose a foliated chart (U, x, y) about y_0 so that $y = (y^1, \dots, y^q)$ are coordinates about y_0 in S (this is easy). That is, we view y as a point varying in a neighborhood of y_0 in S. Then the system of differential equations for the geodesics out of y in the leaf L_y is

$$\frac{d^2x^k}{dt^2} = \sum_{i,j=1}^{n-q} \Gamma^k_{ij}(x, y) \frac{dx^i}{dt} \frac{dx^j}{dt}, \quad 1 \leq k \leq n - q.$$

The Christoffel symbols are C^{r-2} in the leaf coordinates $x = (x^1, \dots, x^{n-q})$ and C^r in the transverse coordinates $y = (y^1, \dots, y^q)$ of S. Because of our assumption that $r \geq 3$, the Christoffel symbols are at least C^1. The coordinates y are parameters for the system as well as part of the initial conditions. By smooth dependence on initial conditions and parameters, this leafwise geodesic spray defines a C^{r-2}–imbedded tubular neighborhood, imbedding each disk fiber into a leaf. This is the promised foliated tubular neighborhood. If $r = \infty$, the foliated tubular neighborhood is C^∞–imbedded.

Example 3.3.10. We will be interested in foliated tubular neighborhoods mainly in the case that $\mathcal{F}$ is transversely orientable *and* leafwise orientable of codimension one. (Leafwise orientability means that $T(\mathcal{F})$ is an orientable vector bundle.) In that case, E will be isomorphic to the trivial bundle and the standard tubular neighborhood construction (not respecting $\mathcal{F}$) gives a C^∞ imbedding

$$i : S \times D^{n-1} \hookrightarrow M.$$

The image $N(S)$ of i will be recoordinatized as follows. The projection

$$p : S \times D^{n-1} \to D^{n-1}$$

defines a smooth fiber bundle

$$p : N(S) \to D^{n-1}$$

with circle fibers. Choosing the radius of $N(S)$ smaller, if necessary, we assume that the foliation $\mathcal{F}' = \mathcal{F}|N(S)$ is transverse to these fibers. Since the fibers are compact, we conclude that $(N(S), \mathcal{F}', p)$ is a foliated bundle (Exercise 2.1.6 on page 52). Since the base space D^{n-1} is simply connected, total holonomy defines a canonical homeomorphism $h_x : S_x \to S$ from the fiber S_x over $x \in D^{n-1}$ to the fiber S over $0 \in D^{n-1}$. These fit together to give a continuous projection $h : N(S) \to S$ with fibers the leaves of $\mathcal{F}'$. For this, we only need the foliation to be of class $C^{1,0+}$. If $\mathcal{F}$ is of class C^r, $r \geq 1$, then h is of class C^r, while the projection p makes each leaf of $\mathcal{F}'$ a covering space of the simply connected space D^{n-1}, hence carries each of these leaves C^r–diffeomorphically onto D^{n-1}. Thus, $z \mapsto (h(z), p(z))$ recoordinatizes $N(S)$ as $S \times D^{n-1}$, the normal fibers $\{y\} \times D^{n-1}$ now lie in leaves L_y of $\mathcal{F}$, and the new coordinates are exactly as smooth as the foliation $\mathcal{F}$. Notice that, for each concentric sphere $S^{n-2}(c) \subset D^{n-1}$ of radius $c \leq 1$ and centered at the origin,

$$p^{-1}(S^{n-2}(c)) = i(S \times S^{n-2}(c))$$

is C^∞–imbedded in M, although the new S–coordinate on this submanifold is not generally smooth. Of course, everything in sight is just as smooth as $\mathcal{F}$.

Example 3.3.11. Let $(M, \mathcal{F})$ be a foliated n–manifold of codimension one and smoothness class C^k, $k \geq 1$, which is both leafwise and transversely orientable. Let $S \subset M$ be a closed transversal to $\mathcal{F}$ and fix a foliated tubular neighborhood $N(S)$ as in Example 3.3.10. Let the foliated coordinates of $N(S) = D^{n-1} \times S^1$ be the "cylindrical coordinates" (r, z, t). Here, we take t modulo 1 and the level sets $t =$ const. are the leaves of $\mathcal{F}|N(S)$. Intrinsically, these coordinates are smooth (except for their singularity at $r = 0$) but, in M, they are only C^k (as smooth as the foliated diffeomorphism (h, p)). Let $\omega \in A^1(D^{n-1} \times S^1)$ be given by the formula

$$\omega = \cos\lambda(r)\, dr + \sin\lambda(r)\, dt,$$

where

$$\lambda : [0,1] \to [-\pi/2, \pi/2]$$

is smooth, strictly increasing on $[0, 3/4]$, and satisfies

$$\lambda(r) = \begin{cases} -\pi/2, & r = 0, \\ 0, & r = 2/3, \\ \pi/2, & 3/4 \leq r \leq 1. \end{cases}$$

Because cylindrical coordinates are singular at $r = 0$, we require that the derivatives $\lambda^{(m)}(0) = 0$, $\forall m \geq 1$, guaranteeing that ω is of class C^∞ in the intrinsic smooth structure of $D^{n-1} \times S^1$. For the sake of subsequent applications, we will also require that the derivatives $\lambda^{(m)}(2/3) = 0, \forall m \geq 1$. The smooth form ω is integrable, since

$$d\omega = \lambda'(r) \cos\lambda(r)\, dr \wedge dt = \omega \wedge \lambda'(r)\, dt.$$

As r increases from 0 to 2/3, ω turns from $-dt$ to dr. As r continues to increase to 1, ω turns until it becomes $\equiv dt$ for $3/4 \leq r \leq 1$. Since the coordinate $z \in S^{n-2}$ does not appear in the formula for ω, the corresponding foliation of $D^{n-1} \times S^1$ is rotationally symmetric. This foliation is depicted in Figure 3.3.4, where sample values of the covector field ω are also indicated. Notice that the foliation is unchanged near $\partial N(S)$. So it fits with the rest of the foliation of M, yielding a new C^k foliation of M in which a Reeb component $\mathcal{R}$ has been inserted along the original closed transversal S. The leaves in $M \smallsetminus \mathcal{R}$, as they approach the boundary of the Reeb component, grow cylindrical "tails" (actually, these are what topologists call *ends*) which spin along this boundary asymptotically. This modification of $\mathcal{F}$ goes by the unlikely name of "turbulization" (from the French *tourbillonnement*).

3.4. Tangential Gluing

We assume that $(M_i, \mathcal{F}_i)$ is of class C^r, $r \geq 1$, and of codimension one, $i = 1, 2$. Let $S_i \subset \partial_\tau M_i$ be a union of tangential boundary components,

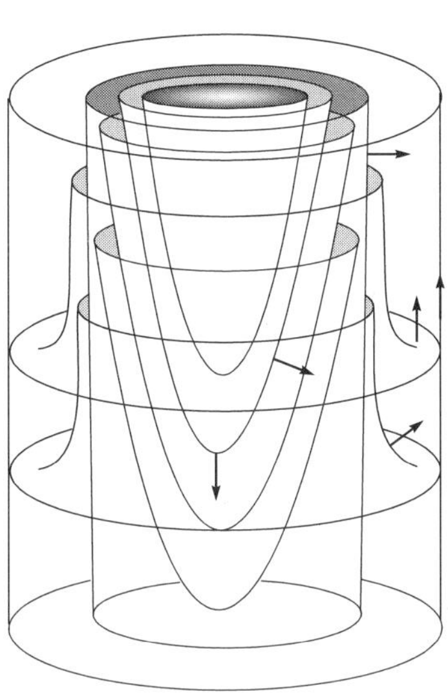

Figure 3.3.4. Turbulizing along a closed transversal

$i = 1, 2$ and let $\varphi : S_2 \to S_1$ be a diffeomorphism. Identify S_2 to S_1 by $x \equiv \varphi(x)$, forming the quotient manifold

$$M = M_1 \cup_\varphi M_2.$$

The C^r structure on M restricts to the original C^r structures on M_1 and M_2, respectively. Thus, the foliations will fit together to give a $C^{r,0+}$ foliation $\mathcal{F} = \mathcal{F}_1 \cup_\varphi \mathcal{F}_2$ of M. We investigate conditions under which $\mathcal{F}$ can be made smooth of class C^r.

Definition 3.4.1. Let $(M, \mathcal{F})$ be a foliated manifold of class C^r and codimension one, $r \geq 1$. Let L be a component of $\partial_\tau M$, $x \in L$, and let $\mathcal{H}_x(L)$ be the germinal holonomy group of the boundary leaf L. Suitably parametrizing the transversal out of x, we realize all $\widehat{h}_\sigma \in \mathcal{H}_x(L)$ as germs at $0 \in [0, \infty)$ of local C^r diffeomorphisms h_σ. If, for each $\widehat{h}_\sigma \in \mathcal{H}_x(L)$, $h'_\sigma(0) = 1$ and $h_\sigma^{(m)}(0) = 0$, $2 \leq m \leq r$, we say that the foliation $\mathcal{F}$ is *infinitesimally C^r-trivial at the boundary leaf L.*

In other words, h_σ is C^r-tangent to the identity at 0, $\forall\, \sigma \in \pi_1(L, x)$. We could just as well parametrize the transversal out of x as $(-\infty, 0]$. Rotating the graph of $h_\sigma : [0, \epsilon) \to [0, \infty)$ through π radians produces a corresponding formula $h_\sigma : (-\epsilon, 0] \to (-\infty, 0]$ which is also C^r-tangent to the identity at 0.

Proposition 3.4.2. *If $(M_i, \mathcal{F}_i)$ is as above, $i = 1, 2$, and if $\mathcal{F}_i$ is infinitesimally C^r-trivial along each component of S_i, $i = 1, 2$, then there is a C^r*

structure on $M = M_1 \cup_\varphi M_2$, agreeing on M_1 and M_2 with the structures there, such that the resulting foliation $\mathcal{F} = \mathcal{F}_1 \cup_\varphi \mathcal{F}_2$ of M is of class C^r.

Proof. Recall that $M = M_1 \cup_\varphi M_2$ where $\varphi : S_2 \to S_1$ is a diffeomorphism between unions of tangential boundary components of M_1 and M_2. We can assume that S_i is a single leaf of $\mathcal{F}_i$, $i = 1, 2$. We are given that $\mathcal{F}_i$ is infinitesimally C^r–trivial along S_i and we need to show that $\mathcal{F} = \mathcal{F}_1 \cup_\varphi \mathcal{F}_2$ is smooth (of class C^r) at S, the leaf that is the result of identifying S_2 to S_1 by φ. Restricting our attention to suitable collars, we may as well assume that $M_1 = S \times [0, \epsilon)$ and $M_2 = S \times (-\epsilon, 0]$, and that these are foliated normal neighborhoods relative to $\mathcal{F}_1$ and $\mathcal{F}_2$, respectively.

Let $\mathcal{D}_1$ be the group of germs at 0 of local C^r diffeomorphisms of $[0, \infty)$, $\mathcal{D}_2$ the group of germs at 0 of the local C^r diffeomorphisms of $(-\infty, 0]$, and $\mathcal{D}$ the group of germs at 0 of the local C^r diffeomorphisms of $\mathbb{R}$ that fix 0. Let $\widehat{h}_i : \pi_1(S) \to \mathcal{D}_i$, $i = 1, 2$, be the corresponding holonomy homomorphisms. For $[\sigma] \in \pi_1(S)$, pick representatives $h^i_\sigma \in \widehat{h}_i([\sigma])$, $i = 1, 2$. These both have one–sided first derivatives equal to 1 at 0 and all relevant higher–order derivatives equal to 0 at 0. Thus,

$$h_\sigma(t) = \begin{cases} h^1_\sigma(t), & t \geq 0, \\ h^2_\sigma(t), & t \leq 0 \end{cases}$$

defines a C^r diffeomorphism in a neighborhood of 0. This defines a homomorphism

$$\widehat{h} : \pi_1(S) \longrightarrow \mathcal{D}$$

and, by Theorem 3.1.6, we obtain a C^r–foliated manifold $(S \times (-\epsilon, \epsilon), \mathcal{F}')$. By Theorem 2.3.9, we conclude that the germ at S of $\mathcal{F}$ is homeomorphic to the germ at S of $\mathcal{F}'$. From the construction, it is clear that this homeomorphism is a C^r diffeomorphism on the complement of S. So it can be used to define a C^r structure in a neighborhood of S in M agreeing with those of $M_1 \smallsetminus S$ and $M_2 \smallsetminus S$. □

Remark. Let $(M, \mathcal{F})$ have a tangential boundary component S and consider the trivial foliation of $S \times (-\epsilon, 0]$ by leaves $S \times \{t\}$, $-\epsilon < t \leq 0$. If $\mathcal{F}$ is infinitesimally C^r–trivial along S, we can glue on this trivially foliated $S \times (-\epsilon, 0]$ by the identity map on $S = S \times \{0\}$, obtaining a C^r foliation. This is an immediate consequence of Proposition 3.4.2. Conversely, if gluing on a trivially foliated $S \times (-\epsilon, 0]$ produces a C^r foliation, $\mathcal{F}$ must have been infinitesimally C^r–trivial along S. Indeed, the holonomy of S in the new foliation is represented by local C^r diffeomorphisms on $(-\epsilon, \epsilon)$, all of which are the identity on $(-\epsilon, 0]$. It follows that all derivatives of holonomy of order $\leq r$ at 0 must agree with the derivatives of the identity.

Example 3.4.3. An important case of the above remark is implicit in Example 3.3.11, where the Reeb component $0 \le r \le 2/3$ is infinitesimally C^∞-trivial along its toroidal boundary $r = 2/3$. To see this, recall that we required the derivatives $\lambda^{(m)}(2/3)$ to vanish, $\forall\, m \ge 1$. Thus, we can replace λ in that construction by a C^∞ function

$$\rho(r) = \begin{cases} \lambda(r), & 0 \le r \le 2/3, \\ 0, & 2/3 \le r \le 1. \end{cases}$$

The corresponding foliation of $D^{n-1} \times S^1$, defined by the form

$$\omega = \cos \rho(r)\, dr + \sin \rho(r)\, dt,$$

consists of a Reeb component for $0 \le r \le 2/3$ and, on $2/3 \le r \le 1$, is a product foliation with leaves parallel copies of $S^{n-2} \times S^1$. Hereafter, our standard Reeb component will be the restriction of this foliation to $0 \le r \le 2/3$. Of course, we can rescale this if we wish so that $0 \le r \le 1$. Two Reeb components constructed in this way can be glued together by any C^∞ diffeomorphism

$$\varphi : S^{n-2} \times S^1 \to S^{n-2} \times S^1$$

to obtain a compact, C^∞–foliated manifold without boundary.

For example, one can decompose S^3 into the union of two solid tori (Exercise 3.4.4), glued together along their boundary, thereby producing a C^∞ foliation called the Reeb foliation of S^3.

Exercise 3.4.4. Realize the 3–sphere as

$$S^3 = \{(z, w) \in \mathbb{C}^2 \mid |z|^2 + |w|^2 = 1\},$$

the set of unit vectors in $\mathbb{C}^2$. Let

$$K_1 = \{(z, w) \in S^3 \mid |z| \ge |w|\},$$
$$K_2 = \{(z, w) \in S^3 \mid |z| \le |w|\}.$$

Show that this decomposes S^3 into a union of solid tori and that

$$T = K_1 \cap K_2$$

is the boundary torus of K_i, $i = 1, 2$. Show also that the circles on T that are meridians relative to one of the solid tori will be longitudes relative to the other.

The infinitesimal C^∞ triviality of the Reeb foliation at the toroidal leaf is a special case of the following.

Exercise 3.4.5. Show that the foliation, produced by spinning a suitable foliation $\mathcal{F}$ along a transverse boundary component S (page 84), is infinitesimally C^∞–trivial at S.

Finally, the following exercise illustrates the need for care in constructing Reeb foliated solid tori.

Exercise 3.4.6. Show that the Reeb component in Example 1.1.12 is not infinitesimally C^r–trivial at the boundary for any $r \geq 1$.

3.5. Orientation Covers

Recall that, if a connected n–manifold M is not orientable, it has a canonical, connected, 2–fold cover

$$\pi : \widetilde{M} \to M$$

such that $\widetilde{M}$ is orientable. This is called the orientation cover and can be formed as follows. At each point $x \in M$, there are two choices of orientation $\{\pm\mu_x\}$ of $T_x(M)$. As a set, $\widetilde{M}$ is the collection of all these orientations as $x \in M$ varies freely. The covering map π is defined by

$$\pi(\pm\mu_x) = x, \quad \forall x \in M.$$

By local triviality of the tangent bundle, there is a unique way to extend a choice of orientation μ_x to a continuous choice of orientation at all points of a small, connected neighborhood U of x in M. Such a local extension is viewed as a neighborhood $\widetilde{U}$ of μ_x in $\widetilde{M}$ and these neighborhoods generate an n–manifold topology on $\widetilde{M}$ relative to which π is a 2–fold covering. The differentiable structure of M is lifted to a differentiable structure of $\widetilde{M}$, relative to which the covering projection is locally a diffeomorphism. This makes sense, even if M is orientable, in which case $\widetilde{M}$ falls into two connected components, each of which is carried diffeomorphically onto M by π. Connectedness of $\widetilde{M}$ is exactly the condition that M be nonorientable. In a tautological manner, the local orientations of M lift to a canonical orientation of $\widetilde{M}$. If there is a foliation $\mathcal{F}$ on M (of arbitrary codimension), π lifts $\mathcal{F}$ to a well–defined foliation $\widetilde{\mathcal{F}} = \pi^*(\mathcal{F})$.

Given a connected, foliated manifold $(M, \mathcal{F})$, there is a corresponding construction of a 2–fold cover

$$\pi : \widetilde{M} \to M$$

such that $\pi^*(\mathcal{F})$ is transversely orientable. One proceeds exactly as above, using local orientations of the normal bundle $\nu(\mathcal{F})$ in place of local orientations of $T(M)$. Again, $\widetilde{M}$ will be connected if and only if $\nu(\mathcal{F})$ is a nonorientable vector bundle. Combining these observations, we have the following.

Proposition 3.5.1. *Let $(M, \mathcal{F})$ be a foliated manifold, M connected. If M is nonorientable and/or $\mathcal{F}$ is not transversely orientable, there is a connected* *covering space (either 2–fold or 4–fold)*

$$\pi : \widetilde{M} \to M$$

such that $\widetilde{M}$ is orientable and the lifted foliation $\widetilde{\mathfrak{F}}$ is transversely orientable. In this case, the leaves of $\widetilde{\mathfrak{F}}$ are all orientable.

The final assertion is elementary. A loop on a leaf L must preserve both the orientation of $T(\widetilde{M})$ and of $\nu(\widetilde{\mathfrak{F}})$, hence preserves the orientation of $T(L)$.

This proposition often allows us to reduce the proof of a theorem to the special case in which everything in sight is orientable. This is especially useful for foliated manifolds of codimension one.

3.6. Deformations

Topologists are interested in properties of geometric structures that are invariant under various sorts of deformations. In this section we define and discuss certain types of deformations and related equivalences of foliations that have proven to be useful.

For simplicity of exposition, we assume that $\partial_\tau M = \emptyset = \partial_\pitchfork M$ throughout, thereby avoiding corners and "corners with corners". The reader will see that, with a little care, one can accomodate the more general cases.

3.6.A. Isotopy. Let $\mathfrak{F}_0$ and $\mathfrak{F}_1$ be C^r foliations of M, both of codimension q, and suppose that $r \geq 1$. A C^r diffeomorphism $f : M \to M$ is automatically transverse to any foliation of M. If $\mathfrak{F}_1 = f^*(\mathfrak{F}_0)$, we can say that $\mathfrak{F}_1$ is C^r–diffeomorphic to $\mathfrak{F}_0$. While diffeomorphism is obviously a useful equivalence relation between foliations, the following is much stronger, preserving everything of significance about the foliated manifold.

Definition 3.6.1. Let $\mathfrak{F}_0$ and $\mathfrak{F}_1$ be C^r foliations of M, $r \geq 1$. These foliations are C^r*–isotopic* if there is a C^r map $f : M \times [0,1] \to M$ such that $f_t = f|M \times \{t\}$ is a C^r diffeomorphism of M onto itself, $0 \leq t \leq 1$, $f_0 = \mathrm{id}$, and $f_1^*(\mathfrak{F}_0) = \mathfrak{F}_1$. The map f is the isotopy and we write $\mathfrak{F}_t = f_t^*(\mathfrak{F}_0)$, $0 \leq t \leq 1$.

Intuitively, $\mathfrak{F}_0$ is smoothly deformed to $\mathfrak{F}_1$ through the family $\{\mathfrak{F}_t\}_{0\leq t\leq 1}$ of mutually C^r–diffeomorphic foliations. It is common to denote the isotopy by f_t or $\mathfrak{F}_t$.

Setting $\varphi_t = f_t^{-1}$, $0 \leq t \leq 1$, we define

$$\varphi : M \times [0,1] \to M,$$
$$\varphi(x,t) = \varphi_t(x).$$

Lemma 3.6.2. *The map $\varphi : M \times [0,1] \to M$ is smooth of class C^r.*

Proof. This is a simple application of the inverse function theorem. Indeed, set

$$F : M \times [0,1] \to M \times [0,1],$$
$$F(x,t) = (f(x,t),t).$$

This map is of class C^r and has nonsingular Jacobian matrix at each point (x,t); hence it has a C^r inverse. Since

$$\Phi : M \times [0,1] \to M \times [0,1],$$
$$\Phi(x,t) = (\varphi(x,t),t)$$

is the inverse of F, it follows that φ is of class C^r. □

We set $\mathcal{F}_t = \varphi_t(\mathcal{F}_0)$, $0 \le t \le 1$. This is the foliation having as leaves the C^r–immersed submanifolds $\varphi_t(L) \subset M$, where L ranges over the leaves of $\mathcal{F}_0$. It is clear that $\{\mathcal{F}_t\}_{0\le t\le 1}$ is the same 1–parameter family of foliations as in Definition 3.6.1, so, this "push–forward" isotopy is implied by the "pull–back" version in that definition. Conversely, the second version implies the first by the same argument, but has the advantage that it continues to make sense in the C^0 case as well.

Definition 3.6.3. Let $\mathcal{F}_0$ and $\mathcal{F}_1$ be C^r foliations of M, $r \ge 0$. These foliations are C^r*–isotopic* if there is a C^r map $\varphi : M \times [0,1] \to M$ such that $\varphi_t = \varphi|M \times \{t\}$ is a C^r diffeomorphism (a homeomorphism if $r = 0$) of M onto itself, $0 \le t \le 1$, $\varphi_0 = \text{id}$, and $\varphi_1(\mathcal{F}_0) = \mathcal{F}_1$. The map φ is the isotopy and we write $\mathcal{F}_t = \varphi_t(\mathcal{F}_0)$, $0 \le t \le 1$.

Again, it is common to denote the isotopy by φ_t or $\mathcal{F}_t$. An interesting C^0 isotopy of foliations will be constructed in Section 9.5.

3.6.B. Integrable homotopy.

We turn to an equivalence relation which is closely related to isotopy.

Definition 3.6.4. Let $\mathcal{F}_0$ and $\mathcal{F}_1$ be C^r foliations of M, $r \ge 1$. An *integrable* C^r *homotopy* between these foliations is a C^r foliation $\mathcal{H}$ of $M \times [0,1]$ of codimension q that is transverse to $M \times \{t\}$, $0 \le t \le 1$, inducing $\mathcal{F}_0$ on $M \times \{0\}$ and $\mathcal{F}_1$ on $M \times \{1\}$.

Here again one thinks of this as defining a deformation of $\mathcal{F}_0 = \mathcal{H}_0$ to $\mathcal{F}_1 = \mathcal{H}_1$ through the intermediate C^r foliations $\mathcal{H}_t$ induced by $\mathcal{H}$ on $M \times \{t\}$.

Example 3.6.5. Let ω_0 and ω_1 be closed, nowhere vanishing 1–forms on M, both transverse to a given oriented, 1–dimensional foliation $\mathcal{L}$. Here, transversality means that, if w is a nowhere vanishing vector field tangent to $\mathcal{L}$ and determining the orientation of $\mathcal{L}$, then $\omega_i(w) > 0$ everywhere,

$i = 0, 1$. By the Frobenius theorem, ω_0 and ω_1 define respective smooth, transversely oriented foliations $\mathcal{F}_0$ and $\mathcal{F}_1$ of codimension 1, both transverse to $\mathcal{L}$ and having transverse orientations that agree with the orientation of $\mathcal{L}$. Assume that the forms are cohomologous and write $\omega_1 - \omega_0 = df$, for a suitable function $f \in C^\infty(M)$. Define $\Omega \in A^1(M \times [0,1])$ by

$$\Omega_{(x,t)} = t\omega_{1x} + (1-t)\omega_{0x} + f(x)\, dt_{(x,t)}, \quad \forall\, (x,t) \in M \times [0,1].$$

Then,

$$d\Omega = dt \wedge (\omega_1 - \omega_0) + df \wedge dt = dt \wedge df - dt \wedge df = 0.$$

The vector field w, as above, can be viewed as a vector field on $M \times [0,1]$, independent of $t \in [0,1]$. It is clear that $\Omega(w) > 0$ everywhere. So Ω is a closed, nowhere vanishing 1–form, defining a foliation $\mathcal{H}$ that is transverse to the factors $M \times \{t\}$, $0 \le t \le 1$. Since $\Omega | M \times \{i\} = \omega_i$, $i = 0, 1$, $\mathcal{H}$ is an integrable homotopy between $\mathcal{F}_0$ and $\mathcal{F}_1$.

While isotopy implies integrable homotopy, the converse is generally false unless M is compact.

Proposition 3.6.6. *If $\mathcal{F}_0$ and $\mathcal{F}_1$ are C^r–isotopic, $r \ge 1$, they are integrably C^r–homotopic. Conversely, if M is compact and $r \ge 0$, then integrably C^{r+1}–homotopic foliations are C^r–isotopic.*

Proof. First assume that there is a C^r isotopy $f : M \times [0,1] \to M$ between $\mathcal{F}_0$ and $\mathcal{F}_1$. Here, $r \ge 1$ and we take f as in Definition 3.6.1. Since f_t is transverse to $\mathcal{F}_0$, $0 \le t \le 1$, f itself is transverse to $\mathcal{F}_0$. Define the C^r foliations $\mathcal{H} = f^*(\mathcal{F}_0)$ and $\mathcal{F}_t = f_t^*(\mathcal{F}_0)$, $0 \le t \le 1$. Evidently,

$$T_{(x,t)}(\mathcal{F}_t) = T_{(x,t)}(\mathcal{H}) \cap T_{(x,t)}(M \times \{t\}), \quad \forall\, (x,t) \in M \times [0,1].$$

Since $T_{(x,t)}(\mathcal{F}_t)$ has codimension q in $T_{(x,t)}(M \times \{t\})$ and $T_{(x,t)}(\mathcal{H})$ has the same codimension q in $T_{(x,t)}(M \times [0,1])$, it follows easily that $\mathcal{H}$ is transverse to $M \times \{t\}$ and induces $\mathcal{H}_t = \mathcal{F}_t$ there, $0 \le t \le 1$. That is, $\mathcal{H}$ is the desired integrable homotopy of class C^r.

Assume that M is compact and that $\mathcal{H}$ is an integrable homotopy of class C^{r+1} between $\mathcal{F}_0$ and $\mathcal{F}_1$, $r \ge 0$. We will construct a C^r isotopy $\varphi : M \times [0,1] \to M$ as in Definition 3.6.3. Let

$$p : M \times [0,1] \to [0,1]$$

be the canonical projection. By the transversality of $\mathcal{H}$ to the fibers of p,

$$p_{*(x,t)} : T_{(x,t)}(\mathcal{H}) \to T_t([0,1])$$

is surjective with kernel exactly $T_{(x,t)}(\mathcal{H}_t)$, $\forall\, (x,t) \in M \times [0,1]$. Since $\mathcal{H}$ is smooth of class C^{r+1}, $T(\mathcal{H})$ is of class C^r and we can choose a C^r section $v \in \Gamma(T(\mathcal{H}))$ that is carried by p_* exactly to $\partial/\partial t \in \mathfrak{X}([0,1])$. Indeed, by the local triviality of the bundles, one chooses v locally, piecing these choices together with a smooth partition of unity. By the compactness of M, solution curves $\Phi_t(x,0)$ of the local flow of v are defined for $0 \le t \le 1$ and every $x \in M$ (*cf.* [**69**, page 179]). Each such curve lies entirely in the leaf of $\mathcal{H}$ through $(x,0)$ and, at each time $t \in [0,1]$, meets $M \times \{t\}$. In particular, Φ_t carries $\mathcal{F}_0 = \mathcal{H}_0$ onto $\mathcal{H}_t$, $0 \le t \le 1$. If

$$\pi : M \times [0,1] \to M$$

is the canonical projection, the desired isotopy of $\mathcal{F}_0$ and $\mathcal{F}_1$ is defined by $\varphi(x,t) = \pi(\Phi_t(x,0))$. □

By Example 3.6.5, we obtain the following result, due essentially to J. Moser [**94**].

Corollary 3.6.7. *Let M be compact. Then closed, cohomologous, nowhere vanishing 1–forms on M, transverse to a common, oriented 1–dimensional foliation $\mathcal{L}$, define foliations that are smoothly isotopic through foliations transverse to $\mathcal{L}$.*

3.6.C. Homotopy. There are two quite different ways in which the notion of an integrable homotopy can be generalized usefully. The first generalization drops the "integrable" condition.

Definition 3.6.8. The foliations $\mathcal{F}_0$ and $\mathcal{F}_1$ of M, both of codimension q, are *C^r–homotopic* if there is a C^r foliation $\mathcal{F}$ of $M \times [0,1]$ of codimension $q+1$ such that $M \times \{t\}$ is $\mathcal{F}$–saturated, $0 \le t \le 1$, with $\mathcal{F}_i = \mathcal{F}|M \times \{i\}$, $i = 0,1$. If $\mathcal{F}$ is only required to be of class $C^{r,0+}$, $r \ge 1$, with $\mathcal{F}_t = \mathcal{F}|M \times \{t\}$ of class C^r, $0 \le t \le 1$, the foliations are $C^{r,0+}$–homotopic.

Both C^r isotopy and integrable C^r homotopy are equivalence relations on the class of C^r–foliated manifolds (Exercise 3.6.16). It is trivial to check that C^r homotopy is reflexive and symmetric, but transitivity seems to be a problem since it is hard to glue C^r–foliated manifolds together along tangential boundary components so as to preserve the smoothness class. This problem disappears for $C^{r,0+}$ homotopy. Another easy way to sidestep the problem is to pass to the equivalence relation generated by C^r homotopy. The result is called *piecewise C^r homotopy* and can be defined in a more direct way as follows.

Definition 3.6.9. The foliations $\mathcal{F}_0$ and $\mathcal{F}_1$ of M, both of codimension q, are piecewise C^r–homotopic if they admit a $C^{r,0+}$ homotopy $\mathcal{F}$ and a partition $0 = t_0 < t_1 < \cdots < t_m = 1$ such that $\mathcal{F}|M \times [t_{j-1}, t_j]$ is of class C^r, $1 \le j \le m$.

Again, in all of these cases, we can view $\mathcal{F}_t = \mathcal{F}|M \times \{t\}$ as defining a continuously varying family of C^r foliations of M, $0 \le t \le 1$, but this family cannot necessarily be "integrated" to give a foliation $\mathcal{H}$ of $M \times [0,1]$

of codimension q. While isotopy (hence, in the compact case, integrable homotopy) preserves all qualitative features of foliated manifolds, homotopy does not.

Example 3.6.10. On $\mathbb{R}^2 \times [0,1]$, define a foliation by lines such that each factor $\mathbb{R}^2 \times \{t\}$ is saturated, being foliated by the lines of constant slope t. By quotienting out the action of $\mathbb{Z}^2$, we obtain a 1–dimensional foliation $\mathcal{F}$ of $T^2 \times [0,1]$, restricting to the linear foliation $\mathcal{F}_t$ of $T^2 \times \{t\}$ of slope t, $0 \le t \le 1$. This is a smooth (in fact, C^ω) homotopy of $\mathcal{F}_0$ to $\mathcal{F}_1$ with qualitative features totally discontinuous in t. Indeed, for rational values of t, $\mathcal{F}_t$ has all leaves compact while, for irrational values, each leaf of $\mathcal{F}_t$ is dense in $T \times \{t\}$. This homotopy is not integrable.

Exercise 3.6.11. It turns out that $\mathrm{gv}(\mathcal{F})$ is not an invariant of smooth homotopy. This may seem surprising since De Rham cohomology is homotopy invariant. Try to use this property of cohomology to prove that the Godbillon–Vey class is homotopy invariant, identifying the (in fact) insurmountable difficulty.

3.6.D. Concordance and cobordism. For smoothly foliated manifolds of codimension 1 (not necessarily compact), integrably homotopic foliations have the same Godbillon–Vey class (Exercise 3.6.17), but there are difficult examples showing that $\mathrm{gv}(\mathcal{F}_t)$ can vary continuously and nontrivially with t for a general smooth homotopy $\mathcal{F}_t$. In the following generalization of integrable homotopy, called *concordance*, $\mathrm{gv}(\mathcal{F})$ is preserved (Exercise 3.6.17), although we lose the intermediate foliations $\mathcal{F}_t$. One can introduce a generalized foliation, called a *Haefliger structure* (*cf.* [**12**, Section 7]), and view concordance as a homotopy through Haefliger structures.

Definition 3.6.12. The foliations $\mathcal{F}_0$ and $\mathcal{F}_1$ of codimension q on M are C^r–concordant if there is a C^r foliation $\mathcal{H}$ of codimension q on $M \times [0,1]$ that is transverse to $M \times \{i\}$ and induces $\mathcal{F}_i$ there, $i = 0, 1$.

The qualitative features of concordant foliations $\mathcal{F}_0$ and $\mathcal{F}_1$ can be quite different, as the following example shows.

Example 3.6.13. Suppose that $\mathcal{F}_1$ has codimension 1 on M and that $\mathcal{F}_0$ is obtained from $\mathcal{F}_1$ by turbulization along a closed transversal σ (Example 3.3.11). We claim that these foliations are concordant. Indeed, consider the foliation on $M \times [-1,1]$ with leaves $L \times [-1,1]$, where L ranges over the leaves of $\mathcal{F}_1$. Turbulize this foliation along the closed transversal $\sigma \times \{0\}$, obtaining a foliation $\mathcal{F}'$ of $M \times [-1,1]$ that is transverse to $M \times \{-1,0,1\}$ and restricts to $\mathcal{F}_0$ on $M \times \{0\}$ and to $\mathcal{F}_1$ on $M \times \{\pm 1\}$. This is illustrated in Figure 3.6.1. The foliation $\mathcal{F} = \mathcal{F}'|M \times [0,1]$ is the desired concordance. Remark that, for some value of $t \in (0,1)$, $M \times \{t\}$ is tangent to the compact

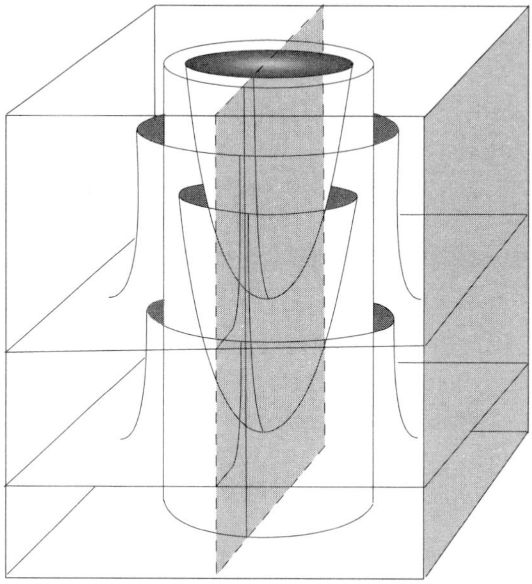

Figure 3.6.1. Turbulizing yields a concordant foliation.

leaf of $\mathcal{F}$ bordering the Reeb component. This is the essential obstruction to making the concordance an integrable homotopy.

Together with Exercise 3.6.17, this example proves the following interesting fact.

Proposition 3.6.14. *If* $(M, \mathcal{F})$ *is a smoothly foliated manifold of codimension* 1 *and* $\mathcal{F}'$ *is obtained from* $\mathcal{F}$ *by turbulizing along a closed transversal, then* $\mathrm{gv}(\mathcal{F}') = \mathrm{gv}(\mathcal{F})$.

Finally, concordance generalizes to an interesting equivalence relation (cobordism) for compact, oriented, foliated manifolds. In the following definition, we use the convention that $-M$ denotes the oppositely oriented version of an oriented manifold M.

Definition 3.6.15. The compact, oriented, C^r–foliated manifolds $(M_0, \mathcal{F}_0)$ and $(M_1, \mathcal{F}_1)$ of codimension q are C^r*–cobordant* if there is an oriented manifold W such that $\partial W = M_1 \cup -M_2$, together with a C^r foliation $\mathcal{H}$ of codimension q on W, transverse to ∂W and inducing $\mathcal{F}_i$ on M_i, $i = 0, 1$.

Exercise 3.6.16. Using the construction in Example 3.3.2, prove that (C^r) isotopy, integrable homotopy, concordance and cobordism are equivalence relations on the class of C^r–foliated manifolds. Prove that the relation of piecewise C^r homotopy, as formulated in Definition 3.6.9, is the equivalence relation on this class generated by C^r homotopy.

Exercise 3.6.17. The Godbillon–Vey class is invariant under smooth isotopy, smooth integrable homotopy and smooth concordance on the class of smooth foliations of codimension one. Prove this by using the *naturality* of $\mathrm{gv}(\mathcal{F})$ (Corollary 3.2.6).

Exercise 3.6.18. If $(M, \mathcal{F})$ is a compact, oriented and transversely oriented, C^∞–foliated 3–manifold of codimension 1, define the *Godbillon–Vey number* to be $\mathrm{gv}_\#(\mathcal{F}) = \int_M \eta \wedge d\eta$, where $\mathrm{gv}(\mathcal{F}) = [\eta \wedge d\eta]$. Prove that the Godbillon–Vey number is an invariant of C^∞ cobordism.

Example 3.6.19. It is a somewhat startling result of W. Thurston [**132**] that $\mathrm{gv}_\#(\mathcal{F})$ takes all nonnegative real values as $\mathcal{F}$ varies over all smooth, codimension–one foliations of S^3. Thurston's exposition is a bit brief, but the interested reader will find a detailed account in the book of I. Tamura [**130**, pp. 169–179]. In particular, S^3 supports an uncountable infinity of distinct isotopy classes, concordance classes and cobordism classes of smooth foliations of codimension 1.

Chapter 4

Asymptotic Properties

It is useful to treat the leaves of a foliation somewhat as if they were orbits of a dynamical system. Dynamical concepts such as *minimal set*, *α– and ω–limit set*, *saturated set*, *etc.*, carry over to foliations quite naturally.

4.1. Minimal Sets

The notion of a "minimal set" in dynamical systems generalizes easily to foliation theory. This notion is especially useful in analyzing the structure of compact foliated manifolds of codimension one.

Definition 4.1.1. Let $(M, \mathcal{F})$ be a foliated manifold. A subset $X \subseteq M$ is $\mathcal{F}$–saturated if it is a union of leaves of $\mathcal{F}$. A minimal set $X \subseteq M$ is a closed, nonempty, $\mathcal{F}$–saturated subset that has no proper subset with these properties.

Example 4.1.2. A closed leaf L is clearly a minimal set. In the Reeb foliation of S^3, the toroidal leaf is the only minimal set. In a fiber bundle $\pi : M \to B$, each leaf is a minimal set.

Example 4.1.3. Slightly more interesting examples are provided by *minimal foliations*, those in which every leaf is dense in M. It should be evident that, in such a foliated manifold, M itself is the only minimal set. Examples of this include the linear foliations of tori with noncompact leaves and the foliation of $T^1(\Sigma_g)$ constructed in Example 1.1.9. More generally, a minimal set X consisting of locally dense leaves will be a connected component of M. Indeed, by definition, X will be the closure of any one of its leaves, hence will be closed and connected. Since each leaf of X is locally dense in M, X will also be open in M, hence a connected component.

Definition 4.1.4. Let $(M, \mathcal{F})$ be a foliated manifold. A minimal set $X \subset M$ of $\mathcal{F}$ is *exceptional* if it is neither a single closed leaf of $\mathcal{F}$ nor a connected component of M.

Exercise 4.1.5. Let $\operatorname{codim} \mathcal{F} = 1$ and let $X \subset M$ be an exceptional minimal set. Prove that, through every point $z \in X$, there passes an arc J transverse to $\mathcal{F}$ and such that $J \cap X$ is a Cantor set. By contrast, give an example of a foliation of codimension 2 with an exceptional minimal set that is a smoothly imbedded submanifold.

Example 4.1.6. We construct an example of an exceptional minimal set. This will be the unique minimal set in a real analytic foliation of

$$M = \Sigma_2 \times S^1,$$

constructed by suspending a pair of diffeomorphisms $f, g \in \operatorname{Diff}^{\omega}_{+}(S^1)$ (*cf.* Example 1.1.14).

It will be convenient to realize the hyperbolic plane $\mathbf{H}$ as the open unit disk in $\mathbb{C}$ (the Poincaré disk model). Indeed, the linear fractional transformation

$$z \mapsto \frac{z - i}{z + i}$$

carries the open upper half plane diffeomorphically onto the open unit disk. We use this to transfer hyperbolic geometry to the disk, noting that hyperbolic geodesics are carried to circular arcs in the disk that are orthogonal to the boundary S^1 (the "circle at infinity"). These are the geodesics in the new model of $\mathbf{H}$ and include the straight line segments through the origin 0. The orientation–preserving hyperbolic isometries now form a subgroup of the linear fractional transformations that leave the unit disk invariant. As such, they extend by the same formula to diffeomorphisms of S^1, the circle at infinity. Let

$$f : \mathbf{H} \to \mathbf{H}$$

be the hyperbolic rotation about 0 through $2\pi/3$ and also denote by

$$f : S^1 \to S^1$$

the induced map on the circle. Since 0 is the origin, this also happens to be a Euclidean rotation through the same angle. Let s be a geodesic which, as indicated in Figure 4.1.1, does not pass through 0 and and suppose that there is a compact arc $I \subset S^1$ of length strictly less than $2\pi/3$ subtended by s. Let Q be the Euclidean midpoint of s and let

$$g : \mathbf{H} \to \mathbf{H}$$

be the hyperbolic rotation about Q through the angle π, an isometry carrying s to $-s$. The extension

$$g : S^1 \to S^1$$

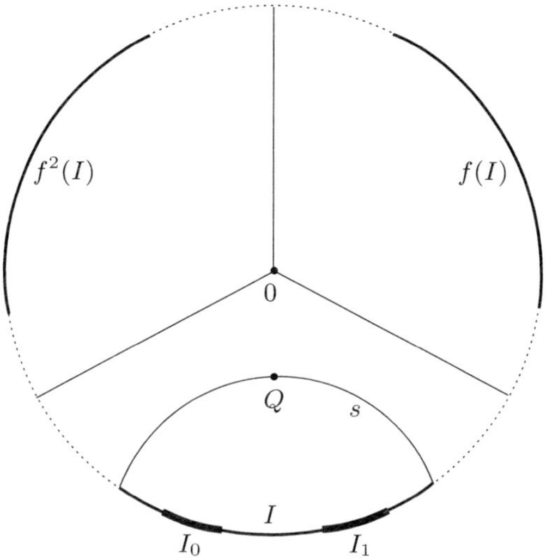

Figure 4.1.1. The action of f and g on S^1

of this rotation interchanges the complementary subarcs of S^1 separated by the endpoints of s. Thus, g carries $f(I)$ to a compact arc $I_0 \subset \operatorname{int} I$ and it carries $f^2(I)$ to another such arc $I_1 \subset \operatorname{int} I$ disjoint from I_0 (Figure 4.1.1). Let G be the subgroup of $\operatorname{Diff}_+^\omega(S^1)$ generated by f and g. We set

$$h_0 = g \circ f,$$
$$h_1 = g \circ f^2,$$

remarking that

$$I_0 = h_0(I),$$
$$I_1 = h_1(I).$$

Similarly, form disjoint, compact subarcs

$$h_0(I_0) = I_{00} \subset \operatorname{int} I_0, \qquad h_1(I_0) = I_{10} \subset \operatorname{int} I_1,$$
$$h_0(I_1) = I_{01} \subset \operatorname{int} I_0, \qquad h_1(I_1) = I_{11} \subset \operatorname{int} I_1.$$

Inductively, one defines, for each finite sequence $(i_0, i_1, \dots, i_r) \in \{0,1\}^{r+1}$, a compact arc

$$h_{i_0}(I_{i_1 i_2 \dots i_r}) = I_{i_0 i_1 \dots i_r} \subset \operatorname{int} I_{i_0 i_1 \dots i_{r-1}},$$

the two choices of the index i_r giving a pair of disjoint, compact subarcs of $\operatorname{int} I_{i_0 i_1 \dots i_{r-1}}$. As $r \uparrow \infty$, the lengths of these arcs shrink to 0 (h_0 and h_1 are contraction mappings on the metric space I). So each infinite sequence

$$\iota = (i_0, i_1, \dots, i_r, \dots) \in \{0,1\}^{\mathbb{N}} = \mathfrak{S}$$

determines a unique nesting

$$I_{i_0} \supset I_{i_0 i_1} \supset \cdots \supset I_{i_0 i_1 \dots i_r} \supset \cdots$$

of compact intervals shrinking to a unique point

$$z_\iota = \bigcap_{j=0}^{\infty} I_{i_0 i_1 \dots i_j} \in I.$$

It is evident from this construction that

$$C_0 = \{z_\iota\}_{\iota \in \mathfrak{S}}$$

is a Cantor set. In Exercise 4.1.7 you will check that the Cantor set

$$C = C_0 \cup f(C_0) \cup f^2(C_0)$$

is invariant under the group G and that the G–orbit of each point of S^1 clusters exactly on C. The set of G–orbits on S^1 corresponds one–to–one to the set of leaves of the suspended foliation $\mathcal{F} = \mathcal{F}(f, g)$; hence the G–invariant set C corresponds to an $\mathcal{F}$–saturated set $X \subset M$. The compactness of C implies that X is compact and the fact that the G–orbit of each point in C is dense in C implies that each leaf in X is dense in X. In fact, each G–orbit in S^1 limits on C. So X is the unique minimal set for the real analytic foliation $\mathcal{F}$. This minimal set is neither a closed leaf nor is it anywhere dense in M.

One can alter this construction slightly so as to produce a minimal set that is all of M. For this, one simply chooses the geodesic s to subtend an arc on S^1 of length exactly $2\pi/3$. Details are left for Exercise 4.1.9.

Exercise 4.1.7. Check the assertion in Example 4.1.6 that the Cantor set

$$C = C_0 \cup f(C_0) \cup f^2(C_0)$$

is G–invariant and that $G \cdot z$ clusters exactly on C, $\forall\, z \in S^1$.

Exercise 4.1.8. In Example 4.1.6, define $\tau : C_0 \to C_0$ by

$$\tau(z_\iota) = \begin{cases} h_0^{-1}(z_\iota), & z_\iota \in C_0 \cap I_0, \\ h_1^{-1}(z_\iota), & z_\iota \in C_0 \cap I_1. \end{cases}$$

1. Prove that each $z_\iota \in C_0$ has a neighborhood in C_0 on which τ is a homeomorphism.
2. Show that the entire pseudogroup on C_0, induced by the holonomy pseudogroup of the foliation, is generated by the restrictions of τ to local homeomorphisms.
3. Define the (one–ended) shift map $\sigma : \mathfrak{S} \to \mathfrak{S}$ by

   $$\sigma(i_0, i_1, \dots, i_r, \dots) = (i_1, i_2, \dots, i_r, \dots)$$

   and prove that

   $$\tau(z_\iota) = z_{\sigma(\iota)}.$$

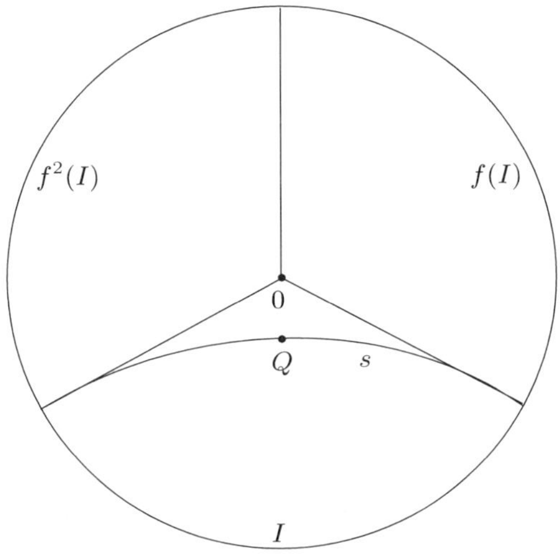

Figure 4.1.2. Modifying Example 4.1.6 to have dense leaves

Thus, the holonomy pseudogroup on C_0 is perfectly modeled by the symbolic dynamical system $(\mathfrak{S}, \sigma)$, called a *topological Markov chain.* This exemplifies a class of exceptional minimal sets called *Markov minimal sets.* Many open questions about exceptional minimal sets have been resolved for the Markov ones [**23, 25, 143**].

4. Show that the points $x_0 = z_{(0,0,\dots,0,\dots)}$ and $x_1 = z_{(1,1,\dots,1,\dots)} \in C_0$ are endpoints of gaps of the Cantor set C and that each endpoint of every gap of C is in the G–orbit of one of these two points. The corresponding two leaves, L_0 and L_1 are called the "border leaves" of X.
5. Show that, on a suitable open arc J_i through x_i, the element $h_i|J_i$ of the holonomy pseudogroup is a contraction to x_i, $i = 0, 1$. That is, each of the border leaves L_0 and L_1 in the previous exercise supports a 2–sided holonomy contraction.

Exercise 4.1.9. Modify Example 4.1.6 by choosing the geodesic s to subtend an arc I of length $2\pi/3$ as in Figure 4.1.2. In this case, prove that every leaf is dense in M.

Example 4.1.10. There is a delicate construction of an orientation–preserving, C^1 diffeomorphism

$$f : S^1 \to S^1$$

that leaves invariant a Cantor set $C \subset S^1$ and has the property that, for each $z \in S^1$, the orbit $\{f^n(z)\}_{n\in\mathbb{Z}}$ clusters exactly on C. This construction, due to Denjoy [**35**], can be suspended over S^1 to give a C^1 foliation of $S^1 \times S^1 = T^2$ by lines. This foliation has a minimal set X, corresponding to C, which is neither a closed leaf (there are none) nor is it anywhere dense in T^2. Every leaf clusters on X; so this is the unique minimal set.

Roughly speaking, f is constructed by modifying a rotation

$$R_\rho : S^1 \to S^1$$

through ρ radians. Here, ρ is to be an irrational multiple of 2π. One selects an orbit $\{R_\rho^n(z)\}_{n\in\mathbb{Z}}$ of this rotation, necessarily a dense subset of S^1, and inserts at each point $R_\rho^n(z)$ a tiny interval I_n. We require that the lengths $|I_n|$ sum to a finite quantity, say

$$\sum_{n=-\infty}^{\infty} |I_n| = \epsilon;$$

so we just get a new circle with length $2\pi + \epsilon$. The map f is chosen to agree with R_ρ outside of the intervals I_n and to send I_n diffeomorphically onto I_{n+1}, $\forall n \in \mathbb{Z}$. It is possible to carry this out so that $f \in \mathrm{Diff}^1_+(S^1)$. Intuitively, the foliation is obtained from an irrational slope foliation by cutting T^2 along one leaf and sewing in an infinite strip whose width diminishes very rapidly. The minimal set X consists of all the other leaves of the original foliation plus the two boundary components of the strip. The strip itself is foliated as a product $I \times \mathbb{R}$ with leaves $\{t\} \times \mathbb{R}$. Each of these leaves is a *smoothly imbedded* submanifold $L \subset T^2$ with $\overline{L} = L \cup X$. For more details, see [**15**, pp. 55–59]. A theorem of R. Sacksteder [**115**], to be discussed in Chapter 8, implies that this C^1 foliation is not even homeomorphic to a C^2 foliation. This fact was proven directly by Denjoy, Sacksteder's theorem being a profound generalization of Denjoy's.

Proposition 4.1.11. *Let $(M, \mathcal{F})$ be a foliated manifold and let $X \subset M$ be $\mathcal{F}$-saturated. Then the closure of this set in M is $\mathcal{F}$-saturated. In particular, the closure of each leaf is $\mathcal{F}$-saturated.*

Proof. Let L' be any leaf of $\mathcal{F}$ and suppose that $L' \cap \overline{X} \neq \emptyset$. We must show that $L' \subseteq \overline{X}$. Let $z \in L' \cap \overline{X}$ and choose a sequence $\{z_k \in X\}_{k=1}^\infty$ that converges to z in M. Let (U, x, y) be a foliated chart for $\mathcal{F}$ with $z \in U$ and assume *wlog* that all $z_k \in U$. If $P \subseteq L' \cap U$ is the plaque through z, either $P \subset X$ or $z_k \notin P$, $\forall k \geq 1$. In this latter case, the fact that $z = \lim_{k\to\infty}(z_k)$ and that X is $\mathcal{F}$-saturated implies that an infinite family of plaques contained in $X \cap U$ clusters on the plaque P. In either case, therefore, $P \subset \overline{X}$. Since $z \in L' \cap \overline{X}$ is arbitrary, this proves that the closed, nonempty subset $L' \cap \overline{X} \subseteq L'$ is also open in the manifold topology of L'. Since L' is connected, we conclude that $L' \cap \overline{X} = L'$. □

Theorem 4.1.12. *Let $(M, \mathcal{F})$ be a foliated manifold, $X \subseteq M$ a nonempty, $\mathcal{F}$–saturated set. If there is an associated regular foliated atlas $\mathfrak{U}$ and a finite set of charts $\{(U_\alpha, x_\alpha, y_\alpha)\}_{\alpha=1}^{\ell}$ in $\mathfrak{U}$ such that every leaf of $\overline{X}$ meets at least one of the sets U_α, then the closure $\overline{X}$ contains a minimal set.*

Proof. Choose transversals $S_\alpha \subset U_\alpha$, $1 \le \alpha \le \ell$. By the definition of a regular foliated atlas, we can choose these transversals so that their closures are pairwise disjoint, compact, imbedded, q–dimensional boxes in M. Let

$$T = \overline{S}_1 \cup \cdots \cup \overline{S}_\ell,$$

a compact space, and let Γ_T be the pseudogroup on T induced by the holonomy pseudogroup of the regular atlas. Then $\overline{X} \cap T$ is a compact, nonempty, Γ_T–invariant set. An elementary application of Zorn's lemma implies that there is a minimal, nonempty, closed, Γ_T–invariant set $Y \subseteq \overline{X} \cap T$. The saturation $Y_{\mathcal{F}}$ of Y is the union of leaves of $\mathcal{F}$ that meet Y. It is clear from our construction that $Y_{\mathcal{F}} \subseteq \overline{X}$. Furthermore, if $Z \subset Y_{\mathcal{F}}$ were a proper, closed, nonempty $\mathcal{F}$–saturated subset, then $Z \cap T \subset Y$ would be a proper, closed, nonempty, Γ_T–invariant subset, contradicting the minimality of Y. It remains to be shown that $Y_{\mathcal{F}}$ is closed in M. By Proposition 4.1.11, any leaf L' meeting $\overline{Y}_{\mathcal{F}}$ is contained in $\overline{Y}_{\mathcal{F}} \subseteq \overline{X}$. By hypothesis, each leaf of $\overline{X}$ meets $\operatorname{int} T$; so we consider $z \in L' \cap \operatorname{int} T$. That is, $z \in U_\alpha$, for some α, and there is a sequence

$$\{z_k\}_{k=1}^{\infty} \subset Y_{\mathcal{F}} \cap U_\alpha$$

converging to z. Viewing y_α as a submersion of U_α onto $S_\alpha \subset T$, we see that

$$\{y_\alpha(z_k)\}_{k=1}^{\infty} \subset Y$$

converges to $y_\alpha(z)$. By the compactness of Y, $y_\alpha(z) \in Y$ and $L' \subseteq Y_{\mathcal{F}}$. That is, $Y_{\mathcal{F}} = \overline{Y}_{\mathcal{F}}$. □

Corollary 4.1.13. *If $(M, \mathcal{F})$ is a compact foliated manifold, then the closure of each $\mathcal{F}$–saturated set contains a minimal set.*

Example 4.1.14. If the manifold M is not compact, it is possible that the foliation $\mathcal{F}$ of M has no minimal sets. Let L be the 2–ended ladder and $\{C_k\}_{k=0}^{\infty}$ a sequence of essential circles on L, as depicted in Figure 4.1.3. We make $M = L \times J$ into a foliated product bundle, where the fiber is the open interval $J = (-1, 1)$. Fix a sequence $\{g_k\}_{k=0}^{\infty} \subset \operatorname{Diff}_c^\infty(J)$ of compactly supported diffeomorphisms and cut M along the open annulus $C_k \times J$, regluing by $(z^-, t) \equiv (z^+, g_k(t))$, $0 \le k < \infty$. We show how to pick these diffeomorphisms so that the resulting C^∞ foliation $\mathcal{F}$ has no minimal sets.

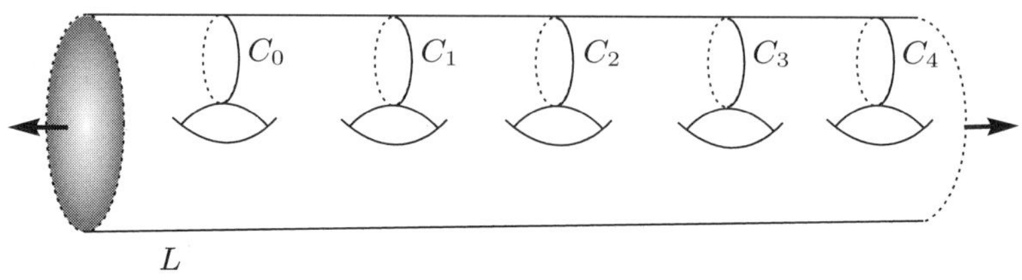

Figure 4.1.3. The 2–ended ladder L

Let $\{a_k\}_{k=0}^{\infty} \subset (0,1)$, $a_k \uparrow 1$. For each $k \geq 0$, choose g_k to have support $[-a_k, a_k]$ and to be a contraction of $(-a_k, a_k]$ to a_k. Also, for $k \geq 1$, require that $g_k(-a_{k-1}) = a_{k-1}$. Let L_k denote the leaf of $\mathfrak{F}$ through a_k, $k \geq 0$. In Exercise 4.1.15, you will be asked to show that, for each $k \geq 0$,

$$\overline{L}_k = \bigcup_{\ell=k}^{\infty} L_\ell.$$

If $Y \subseteq \overline{L}_k$ were a minimal set, it would have to contain a leaf L_ℓ, some $\ell \geq k$; hence it would *properly* contain $\overline{L}_{\ell+1}$, a contradiction. More generally, you will show that, for an arbitrary leaf F, either $F = L_k$, some $k \geq 0$, or

$$\overline{F} = F \cup L_0 \cup L_1 \cup \cdots \cup L_k \cup \cdots,$$

concluding likewise that $\overline{F}$ contains no minimal sets. The point is that, as the closure of a leaf approaches the missing boundary $L \times \{\pm 1\}$, it keeps acquiring new leaves that stay arbitrarily "close" to $L \times \{\pm 1\}$.

Exercise 4.1.15. In Example 4.1.14, verify the assertions about $\overline{L}_k$ and $\overline{F}$.

4.2. Ends of Manifolds

In order to describe the ways that leaves approach other leaves, the topological concept of *ends* of a space is essential. Although this makes sense for all connected, finite–dimensional cell complexes, we will limit our attention to connected manifolds.

Roughly, the "ends" of a noncompact, connected manifold L are the distinct ways to go to infinity on L. These ends will be ideal points at infinity and will compactify L. Thus, the plane has *one* ideal point at infinity that compactifies it to S^2. There is only one "way to infinity" in the sense that the complement of any compact subset of $\mathbb{R}^2$ has only one unbounded

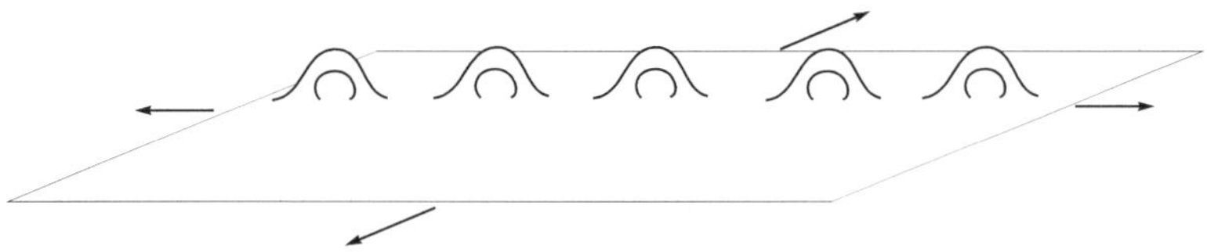

Figure 4.2.1. The infinite Loch Ness monster

component. The cylinder $S^1 \times \mathbb{R}$ has two ends, $\pm\infty$, again compactifying it to S^2. Here, arbitrarily large compact subsets can be found whose complements have two unbounded components, but none can be found with more than two such components. The line $\mathbb{R}$ also has ends $\pm\infty$, compactifying it to a closed, bounded interval $[-\infty, \infty]$. The "infinite Loch Ness monster" (Figure 4.2.1) has one end. The compactification is a sphere with an infinite sequence of handles clustering at ∞ (thought of as the "north pole"). The two–ended ladder (Figure 4.1.3) compactifies to a sphere with a sequence of handles clustering at the north and south poles (which are the ends $\pm\infty$). These last two compactifications are not manifolds. The Alexander horned sphere (Figure 4.2.2) has a Cantor set of ends and compactifies to S^2. We make this idea more precise.

Let L be a noncompact, connected manifold. We agree that an "unbounded" subset of L is one whose closure in L is noncompact. Denote by $\{K_\alpha\}_{\alpha \in \mathfrak{A}}$ the family of all compact subsets of L. We consider descending chains

$$U_{\alpha_1} \supsetneq U_{\alpha_2} \supsetneq \cdots \supsetneq U_{\alpha_n} \supsetneq \cdots,$$

where each U_{α_k} is an *unbounded* connected component of $L \smallsetminus K_{\alpha_k}$ and

$$\bigcap_{k=1}^{\infty} U_{\alpha_k} = \emptyset.$$

Two such descending chains $\mathcal{U} = \{U_{\alpha_k}\}_{k=1}^{\infty}$ and $\mathcal{V} = \{V_{\beta_k}\}_{k=1}^{\infty}$ are "equivalent" ($\mathcal{U} \sim \mathcal{V}$) if, for each $k \geq 1$, there is $n > k$ such that $U_{\alpha_k} \supset V_{\beta_n}$ and $V_{\beta_k} \supset U_{\alpha_n}$. It is clear that "$\sim$" is an equivalence relation.

Definition 4.2.1. If

$$\mathcal{U} = \{U_{\alpha_1} \supsetneq U_{\alpha_2} \supsetneq \cdots \supsetneq U_{\alpha_k} \supsetneq \cdots\}$$

is as above, the equivalence class of this descending chain (under $\sim$) is called an end e of L. We say that $\mathcal{U} \in e$ is a fundamental open neighborhood system

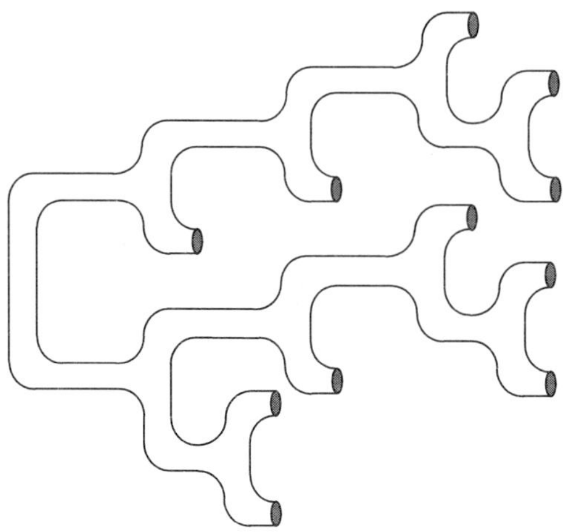

Figure 4.2.2. A Cantor set of ends

of e in L. The set of all ends of L is denoted by $\mathcal{E}(L)$. If an open subset $U \subseteq L$ contains some fundamental neighborhood system of an end e of L, then U is said to be a neighborhood of e in L.

As indicated above, the ends of L should be thought of as ideal points at infinity. For instance, it is an elementary application of the definition to check that $\mathcal{E}(\mathbb{R})$ is a set with two elements, call them $\pm\infty$ and, if $n \geq 2$, $\mathcal{E}(\mathbb{R}^n)$ is a singleton ∞.

Exercise 4.2.2. If M is a compact, connected manifold and $X \subset M$ is a compact, totally disconnected subset, set $L = M \smallsetminus X$ and exhibit a canonical one–to–one correspondence $X \leftrightarrow \mathcal{E}(L)$. For example, the surface indicated in Figure 4.2.2 is homeomorphic to $S^2 \smallsetminus X$, where X is an imbedded Cantor set.

In the above exercise, remark that, under the identification $X = \mathcal{E}(L)$, $L \cup \mathcal{E}(L)$ is identified with M, inheriting a natural compact topology.

In the general case, we want to topologize $L \cup \mathcal{E}(L)$ so as to obtain a compactification of L. The main thing needed is a description of how sequences in this set converge to the ideal points. A sequence $\{x_k\}_{k=1}^{\infty}$ of points in L converges to an end $e \in \mathcal{E}(L)$ if every neighborhood U of e in L contains all but finitely many x_k. A sequence $\{e_k\}_{k=1}^{\infty}$ of ends converges to the end e if every neighborhood U of e in L is a neighborhood in L of all but finitely many e_k.

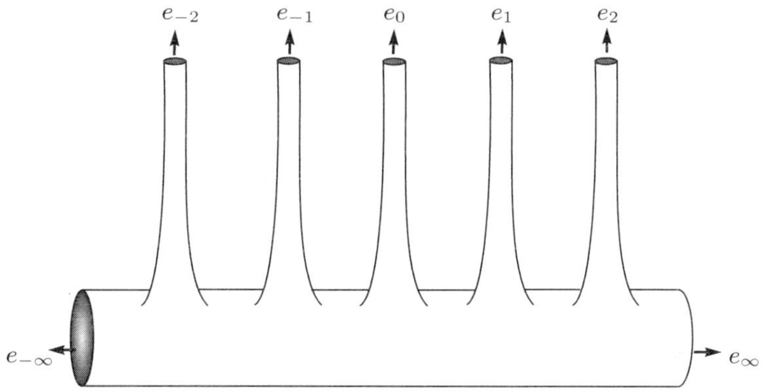

Figure 4.2.3. Isolated ends clustering at two limit ends

Example 4.2.3. Consider the surface in Figure 4.2.3. There are infinitely many ends. The bi–infinite sequence $\{e_k\}_{-\infty<k<\infty}$ consists of isolated ends. Indeed, each has a neighborhood (the corresponding vertical tube) that is not a neighborhood of any other end. On the other hand, an arbitrary neighborhood of e_∞ is a neighborhood of e_k, $k \geq N$, and an arbitrary neighborhood of $e_{-\infty}$ is a neighborhood of e_k, $k \leq -N$, for sufficiently large values of N. That is,

$$\lim_{k\to\infty} e_k = e_\infty,$$
$$\lim_{k\to-\infty} e_k = e_{-\infty}.$$

More precisely, we return to the general situation and put a topology on $L^* = L \cup \mathcal{E}(L)$ as follows. All open subsets $U \subset L$ are to be open in L^*. If such U is connected and unbounded, with compact boundary, then it belongs to the neighborhood system of at least one end e. If $U^* \subset L^*$ denotes the union of U and all ends of which it is a neighborhood, then U^* is also to be an open subset of L^*. The topology of L^* is the smallest containing all of the above open sets.

Proposition 4.2.4. *The space L^* is compact, connected and second countable. The endset $\mathcal{E}(L) \subset L^*$ is a compact, totally disconnected, separable, metrizable subspace and $L \subset L^*$ is an open subset with its usual manifold topology.*

In the following exercise, the proof of this proposition is broken down into manageable steps. For readers not already familiar with ends, this exercise is highly recommended.

Exercise 4.2.5. In order to establish Proposition 4.2.4, proceed as follows.

1. Fix an exhaustion $K_1 \subset K_2 \subset \cdots \subset K_r \subset \cdots$ of L by compact, connected submanifolds. If e and e' are distinct ends of L, let $r(e,e')$ denote the smallest positive integer such that distinct components of $L \smallsetminus K_{r(e,e')}$ are neighborhoods of e and e', respectively. If $e = e'$, set $r(e,e') = \infty$ and, in all cases, define
$$d(e,e') = e^{-r(e,e')}.$$
Show that this defines a metric on $\mathcal{E}(L)$, making the endset a compact, totally disconnected, separable metric space.
2. Show that the topology on $\mathcal{E}(X)$ defined by the metric d coincides with its relative topology as a subspace of L^*.
3. It is obvious that L, with its manifold topology, is an open, dense subset of L^*. Using this observation and the above, conclude that L^* is compact, connected and second countable.

Given a compact, totally disconnected, separable, metrizable space X, one defines the derived set X' to be the set of cluster points of X, again compact, totally disconnected, separable and metrizable. Inductively, the kth derived set $X^{(k)}$ is the set of cluster points in $X^{(k-1)}$. If ω denotes the first infinite ordinal, one defines $X^{(\omega)} = \bigcap_{k=1}^{\infty} X^{(k)}$. We define *derived sets*, indexed by all ordinals, as follows.

Definition 4.2.6. Let X be a compact, totally disconnected, and metrizable space. The 0th derived set of X is $X^{(0)} = X$. If the γth derived set $X^{(\gamma)}$ has been defined for an ordinal γ, then $X^{(\gamma+1)}$ is the set of cluster points of $X^{(\gamma)}$. If γ is a limit ordinal and $X^{(\beta)}$ has been defined for all ordinals $\beta < \gamma$, then the γth derived set is $X^{(\gamma)} = \bigcap_{\beta<\gamma} X^{(\beta)}$.

It can be shown that, for some *countable* ordinal γ, either $X^{(\gamma)}$ is finite and nonempty or $X^{(\gamma)} = X^{(\Omega)}$ is a Cantor set, where Ω is the first uncountable ordinal [**102**]. One often writes X' for the first derived set $X^{(1)}$ and X'' for $X^{(2)}$.

For $X = \mathcal{E}(L)$, we set $\mathcal{E}^{(\gamma)}(L) = X^{(\gamma)}$. For example, if L is the surface pictured in Figure 4.2.3, $\mathcal{E}'(L) = \{e_\infty, e_{-\infty}\}$. On the other hand, if L is the surface pictured in Figure 4.2.2, $\mathcal{E}^{(\Omega)}(L) = \mathcal{E}(L)$.

Definition 4.2.7. An end $e \in \mathcal{E}(L)$ is said to be of type $\gamma < \Omega$ if e is an isolated point in $\mathcal{E}^{(\gamma)}(L)$. If $e \in \mathcal{E}^{(\Omega)}(L)$, then e has type Ω. The manifold L itself has (topological) type $\gamma \le \Omega$ if γ is the least upper bound to the types of ends of L. If L is compact (equivalently, $\mathcal{E}(L) = \emptyset$), then the type of L is taken to be -1.

In other words, if there is a countable ordinal γ such that $\mathcal{E}^{(\gamma)}(L)$ is finite and nonempty, then L has type γ. If $\mathcal{E}^{(\Omega)}(L)$ is a Cantor set, then L has type Ω.

4.3. Ends of Leaves and Limit Sets

Let L be a leaf of a compact, foliated manifold $(M, \mathcal{F})$. Recall that $\{K_\alpha\}_{\alpha \in \mathfrak{A}}$ denotes the family of all compact subsets of L. Let $W_\alpha = L \smallsetminus K_\alpha$ and let $\overline{W}_\alpha$ denote the closure of W_α in M.

Definition 4.3.1. The *limit set* of L is the set $\lim L = \bigcap_{\alpha \in \mathfrak{A}} \overline{W}_\alpha$. This is also called the *asymptote* of L.

The proof of the following is elementary and will be left to the reader.

Lemma 4.3.2. *For each noncompact leaf L of $\mathcal{F}$, $\lim L$ is a compact, nonempty, $\mathcal{F}$–saturated set.*

For example, if $M = T^3$ is the 3–torus and $\mathcal{F}$ is a linear foliation by dense planes or cylinders, then each leaf has all of T^3 as its limit set. In the case of the Reeb foliation of the solid torus, however, each of the planar leaves has limit set the toral boundary leaf. The toral leaf, of course, has empty asymptote.

Definition 4.3.3. The leaf L is asymptotic to a leaf F if $F \subseteq \lim L$. The leaf L is *proper* if it is not asymptotic to itself.

More intuitively, in a foliation $\mathcal{F}$ of codimension q, a leaf L is asymptotic to a leaf L' if and only if, for each $x \in L'$ and each q–manifold N through x transverse to $\mathcal{F}$, $N \cap L$ clusters at x. Thus, L' is proper if N can be chosen so that $N \cap L' = \{x\}$.

Let $e \in \mathcal{E}(L)$ have a fundamental neighborhood system $\{U_{\alpha_k}\}_{k=1}^{\infty}$ in L. Let $\overline{U}_{\alpha_k}$ denote the closure of U_{α_k} in M.

Definition 4.3.4. The e–limit set of L is $\lim_e L = \bigcap_{k=1}^{\infty} \overline{U}_{\alpha_k}$. This is also called the asymptote of e.

For example, the α–limit set and the ω–limit set of a (noncompact) flowline L are $\lim_{-\infty} L$ and $\lim_{\infty} L$, respectively, where $\pm\infty$ are the two ends of $L \cong \mathbb{R}$. In the case of the dense leaved linear foliations of T^3, each end of each leaf has limit set all of T^3. In general, we have:

Lemma 4.3.5. *For each leaf L of $\mathcal{F}$ and each end $e \in \mathcal{E}(L)$, $\lim_e L$ is a compact, nonempty $\mathcal{F}$–saturated set, not depending on the choice of fundamental neighborhood system of e.*

The proof of Lemma 4.3.5 is similar to that of Lemma 4.3.2 and is left to the reader.

Definition 4.3.6. The end e is asymptotic to each leaf of $\lim_e L$.

More intuitively, if $\operatorname{codim} \mathcal{F} = q$ and $e \in \mathcal{E}(L)$ is asymptotic to a leaf L', then for each $x \in L'$, each q–manifold N through x that is transverse to $\mathcal{F}$, and each neighborhood $U \subseteq L$ of e, $N \cap U$ clusters at x.

Lemma 4.3.7. *For each leaf L of $\mathcal{F}$ and each $e \in \mathcal{E}(L)$, $\lim_e L \subseteq \lim L$. Furthermore,*

$$\lim L = \bigcup_{e \in \mathcal{E}(L)} \lim_e L.$$

Proof. Choose an exhaustion

$$K_{\alpha_1} \subset K_{\alpha_2} \subset \cdots \subset K_{\alpha_k} \subset \cdots \subset L$$

of L by compact subsets. For $e \in \mathcal{E}(L)$, there is a unique unbounded component U_{α_k} of $W_{\alpha_k} = L \smallsetminus K_{\alpha_k}$ that is a neighborhood of e in L. It is elementary that

$$\lim_e L = \bigcap_{k=1}^{\infty} \overline{U}_{\alpha_k} \subseteq \bigcap_{k=1}^{\infty} \overline{W}_{\alpha_k} = \lim L,$$

giving the first assertion. For the second, if $x \in \lim L$, there will be a subsequence $\{x_k\}_{k=1}^{\infty} \subset L$ that diverges to infinity in L (*i.e.*, is not contained in any compact subset of L), but converges to x in M. By passing to a subsequence, we can assume that

$$e = \lim_{k \to \infty} x_k$$

in $L \cup \mathcal{E}(L)$. Evidently, $e \in \mathcal{E}(L)$ and $x \in \lim_e L$. □

Definition 4.3.8. An end e of a leaf L is proper if $L \not\subseteq \lim_e L$. It is *totally proper* if $\lim_e L$ is a union of proper leaves.

Intuitively, an end $e \in \mathcal{E}(L)$ of a foliation of codimension q is proper if, given any $x \in L$ and any neighborhood $U \subset L$ of e such that $x \notin U$, there exists a small transverse q–manifold through x that does not intersect U.

Lemma 4.3.9. *Every totally proper end of L is also a proper end.*

Proof. Let e be a totally proper end of L. If $L \subset \lim_e L$, then L must be a proper leaf. But Lemma 4.3.7 implies that $L \subset \lim L$, a contradiction. □

Example 4.3.10. Consider a linear foliation of T^3 by dense cylinders. Each leaf L has two ends $e_{\pm\infty}$ and $\lim_{e_{\pm\infty}} L = T^3$; so neither end is proper. Turbulize (Example 3.3.11) the foliation along a closed transversal. The locally dense leaves of the new foliation $\mathcal{F}'$ unite to form an open set U and

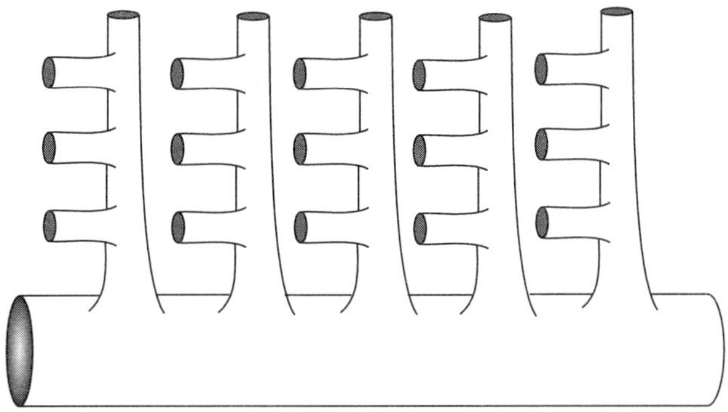

Figure 4.3.1. A leaf L'' of type 2

each leaf L' of $\mathcal{F}'|U$ has a doubly infinite sequence $\{e_k\}_{k=-\infty}^{\infty}$ of isolated ends such that $\lim_{k\to\pm\infty} e_k = e_{\pm\infty}$. Indeed, L' is diffeomorphic to the manifold pictured in Figure 4.2.3. Furthermore, for each $k \in \mathbb{Z}$, $\lim_{e_k}(L') = T$, the toral boundary of the Reeb component. These ends, therefore, are totally proper. But $\lim_{e_{\pm\infty}}(L') = U \cup T$ and, in particular, $L' \subset \lim_{e_{\pm\infty}}(L')$, showing that the ends $e_{\pm\infty}$ are not proper.

If, in this example, we had turbulized along two disjoint closed transversals, the topological type of the leaves L' of $\mathcal{F}'|U$ would be the same, but the totally proper ends would fall into two classes, some asymptotic to the toral boundary leaf T_1 of one of the Reeb components, the others asymptotic to the boundary T_2 of the other Reeb component. If the "directions of spin" along T_1 and T_2 are suitably chosen, there will be an arc τ across U from T_1 to T_2 that is transverse to $\mathcal{F}'$. If we excise the interiors of the two Reeb components and glue T_1 to T_2 by a suitable diffeomorphism (Section 3.4) that identifies the endpoints of τ, this transverse arc becomes a closed transversal σ in the resulting (smoothly) foliated manifold. Turbulizing along σ converts each of the leaves L' of $\mathcal{F}'|U$ to a leaf L'' diffeomorphic to the one depicted in Figure 4.3.1. This surface is of type 2. The isolated ends in $\mathcal{E}(L'')$ are asymptotic to the boundary leaf T of the new Reeb component. The leaf $T_1 \equiv T_2$ has "grown" one end, which is also asymptotic to T. Call this new proper leaf F. The isolated ends in $\mathcal{E}^{(1)}(L'')$ are also totally proper, being asymptotic exactly to $T \cup F$. Finally, $\mathcal{E}^{(2)}(L'')$ consists of two ends that are not proper.

Remark. In a foliation of codimenson 1, one can define the notion of a *semiproper* leaf L. A transverse arc N through $x \in L$ has a linear order. Hence $N \cap L$ might fail to cluster on x from one side, although it might cluster on x from the other side. In this case, the leaf L is said to be semiproper, although it might not be proper. A transverse arc N meets an exceptional minimal set in an open subset X of a Cantor set. So endpoints of gaps of X lie on nonproper but semiproper leaves of the exceptional minimal set.

Example 4.3.11. Let $\mathcal{F}$ have an exceptional minimal set Z and let $L \subset Z$ be a semiproper leaf. If $e \in \mathcal{E}(L)$, then $\lim_e L = Z$ (minimality of Z); so no end of L is proper. The type of L is given by the following very deep and unpublished theorem of G. Duminy (1977). (A proof is available in preprint form [**22**].)

Theorem 4.3.12 (Duminy). *Let L be a semiproper leaf in an exceptional minimal set. Then $\mathcal{E}(L)$ is homeomorphic to the Cantor set and L has type Ω.*

Pierce the semiproper leaf L (hence, every leaf in Z) by a closed transversal σ and turbulize along σ, replacing Z with a saturated set Z'. Let $L' \subset Z'$ be the semiproper leaf corresponding to L. The original ends of L form a Cantor subset $C \subset \mathcal{E}(L')$ and $\mathcal{E}(L') \smallsetminus C$ consists of isolated ends e, each having $\lim_e L'$ equal to the toral boundary leaf of the Reeb component. The set of isolated ends of L' clusters exactly on C. These isolated ends are totally proper and the ends in C are nonproper. Since $\mathcal{E}^{(1)}(L') = C$ is a Cantor set, the leaf L' is also of type Ω.

Part 2

Codimension One

Foreword to Part 2

This part of the book is intended as a short course on the theory of foliations of codimension one. The choice of topics is subjective, representing only a fraction of the significant results. The theory of 3–manifolds foliated by surfaces is largely ignored here. Its special methods merit separate treatment in Volume II.

The structure theory of compact, foliated manifolds $(M, \mathfrak{F})$ of codimension one is fairly well understood. These foliations admit a transverse, 1–dimensional foliation $\mathfrak{L}$ of class C^∞, allowing us to find a *biregular* foliated atlas (one that is a regular foliated atlas for $\mathfrak{F}$ and $\mathfrak{L}$ simultaneously). If the holonomy is of class C^r, $r \geq 1$, the differential calculus of functions of one variable has many profound applications. This is especially true if $r \geq 2$, in which case the mean value theorem from freshman calculus can sometimes be applied with almost magical results. These facts are the principal reasons for the rich structure theory in codimension one.

Unless otherwise stipulated, $(M, \mathfrak{F})$ will be a compact, connected, transversely oriented, foliated n–manifold of codimension one. At the very minimum, if no higher differentiability is assumed, $\mathfrak{F}$ will be integral to an $(n-1)$–plane field $E \subset T(M)$ of class C^0.

Chapter 5

Basic Structures

5.1. Biregular Covers

In Example 1.2.25, we outlined the consequences of the assumption that $\mathcal{F}$ is integral to a continuous hyperplane field E. Here we supply the details. For the moment, we need not assume that $\mathcal{F}$ is transversely oriented nor that M is compact.

Lemma 5.1.1. *There is a C^∞ line bundle $L \subset T(M)$, everywhere transverse to E and tangent to $\partial_\pitchfork M$.*

Proof. Let $N \cong (0,1] \times \partial_\pitchfork M$ be smoothly imbedded in M as a normal neighborhood of $\partial_\pitchfork M$. Fix the identification of N with $(0,1] \times \partial_\pitchfork M$. A Riemannian metric on M defines a C^0 line field L_0 on $M_0 = M \setminus \partial_\pitchfork M$ everywhere transverse to $E|M_0$. Viewing L_0 as a continuous section of the smooth bundle $P(M_0)$ of projective spaces associated to $T(M_0)$ and appealing to standard approximation results in differential topology, we can perturb L_0 slightly to a smooth line field, again denoted by L_0. Since the perturbation can be arbitrarily small, the smooth line field continues to be transverse to $E|M_0$. Likewise, the metric relativized to $\partial_\pitchfork M$ defines a C^0 line field $L_1' \subset T(\partial_\pitchfork M)$ everywhere transverse to $V = E \cap T(\partial_\pitchfork M)$ and we can perturb this to a smooth line field L_1'' in $T(\partial_\pitchfork M)$ transverse to V. This extends to a C^∞ line field L_1 on $M_1 = (1-\epsilon, 1] \times \partial_\pitchfork M$. Choosing $\epsilon > 0$ smaller, if necessary, guarantees that L_1 is transverse to $E|M_1$. Let $\{\lambda_0, \lambda_1\}$ be a smooth partition of unity, subordinate to the open cover $\{M_0, M_1\}$ of M. If $\mathcal{F}$ is transversely oriented, let v_i be the smooth, unit length sections of L_i that agree with the transverse orientation of $\mathcal{F}|M_i$, $i = 0, 1$, and remark that $v = \lambda_0 v_0 + \lambda_1 v_1$ defines a smooth vector field everywhere transverse

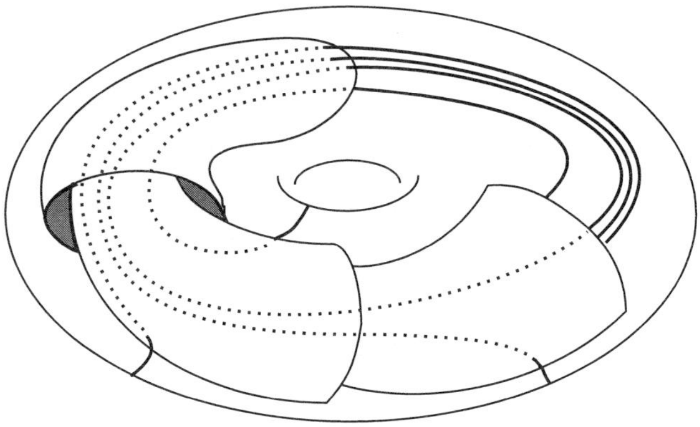

Figure 5.1.1. A foliation by curves transverse to a Reeb component

to E and tangent to $\partial_{\pitchfork} M$. This field spans the desired line field. If $\mathcal{F}$ is not transversely orientable, the vector fields v_i are not globally defined, but they can be locally defined up to sign. Let

$$\pi : \widetilde{M} \to M$$

be the 2–fold covering space such that the lifted foliation $\widetilde{\mathcal{F}}$ is transversely orientable (Section 3.5). There are vector fields $\widetilde{v}_i$ on $\widetilde{M}_i = \pi^{-1}(M_i)$, $i = 0, 1$, transverse to $\widetilde{\mathcal{F}}$, tangent to $\partial_{\pitchfork} \widetilde{M}$, each carried to its negative by the non-trivial covering transformation, and each agreeing with a common transverse orientation of $\widetilde{\mathcal{F}}$. These are defined by coherently piecing together the lifts of the local fields on M, which were only defined up to sign. If $\{\widetilde{\lambda}_0, \widetilde{\lambda}_1\}$ is the lifted partition of unity, the line field $\widetilde{L}$ spanned by

$$\widetilde{v} = \widetilde{\lambda}_0 \widetilde{v}_0 + \widetilde{\lambda}_1 \widetilde{v}_1$$

is invariant under the covering group and descends to a line field on M with all the desired properties. □

Corollary 5.1.2. *There is a 1–dimensional C^∞ foliation $\mathcal{L}$, transverse to $\mathcal{F}$ and tangent to $\partial_{\pitchfork} M$.*

Indeed, the Frobenius integrability condition is automatic for smooth line fields. For example, Figure 5.1.1 illustrates such a transverse foliation $\mathcal{L}$ in a Reeb foliated solid torus. This can be chosen so that $\mathcal{L}$ has exactly one compact leaf, the core circle $\{0\} \times S^1 \subset D^{n-1} \times S^1$, and the other leaves meet the boundary transversely. From now on, given $\mathcal{F}$, we fix a choice of the transverse one–dimensional foliation $\mathcal{L}$. Our regular atlases will be *biregular* in the following sense.

Definition 5.1.3. A biregular foliated atlas for $(M, \mathcal{F}, \mathcal{L})$ is an atlas

$$\mathcal{U} = \{(U_\alpha, x_\alpha, y_\alpha)\}_{\alpha \in \mathfrak{A}}$$

on M that is simultaneously a regular foliated atlas for $\mathcal{F}$ and $\mathcal{L}$. Here, the $\mathcal{F}$–plaques are the level sets of y_α and the $\mathcal{L}$–plaques are the level sets of x_α. In the local change of coordinate formulas,

$$\begin{aligned} x_\alpha &= x_\alpha(x_\beta), \\ y_\alpha &= y_\alpha(y_\beta), \end{aligned}$$

$x_\alpha(x_\beta)$ is of class C^∞ and, if $\mathcal{F}$ is of class C^r, the formula $y_\alpha(y_\beta)$ is of class C^r.

Such an atlas will usually be referred to simply as a *biregular cover.*

Proposition 5.1.4. *Biregular covers exist.*

Proof. Let W be an open neighborhood of an arbitrary point $w \in M$. We claim that W contains a foliated chart (U, x, y) about w in which $y =$ const. defines the plaques of $\mathcal{F}$ and the transversals $x =$ const. are plaques of $\mathcal{L}$. Indeed, let $(U, \widetilde{x}, y)$ be a foliated chart about w for $\mathcal{F}$, $(V, x, \widetilde{y})$ a foliated chart for $\mathcal{L}$ (with the usual roles of x and $\widetilde{y}$ reversed) such that $U \subseteq V \subseteq W$. Choosing U smaller, if necessary, we can assure that (U, x, y) is a rectangular neighborhood. This is clearly a foliated chart with the required properties. As usual, it follows that, on the overlaps of two such charts, the change of y coordinates is locally independent of x and the change of x coordinates is locally independent of y. Since the charts $(V, x, \widetilde{y})$ can be chosen to be compatible with the C^∞ structure of $\mathcal{L}$, the change of x coordinates will be C^∞. Similarly, if $\mathcal{F}$ is of class C^r, the change of y coordinates will be C^r. One completes the proof in analogy with the proof of Lemma 1.2.17. □

Corollary 5.1.5. *The codimension–one foliation $\mathcal{F}$, integral to a C^0 hyperplane field $E \subset T(M)$, is of class $C^{\infty,0+}$. The foliated manifold $(M, \mathcal{F})$ is of class C^r, $1 \le r \le \infty$, if and only if the holonomy pseudogroup $\Gamma_{\mathcal{U}}$ of a suitable regular foliated atlas $\mathcal{U}$ is of class C^r.*

Proof. Let $\{U_\alpha, x_\alpha, y_\alpha\}_{\alpha \in \mathfrak{A}}$ be a biregular cover. Since $\mathcal{L}$ is of class C^∞, the change of coordinate formulas $x_\alpha = x_\alpha(x_\beta)$ are of class C^∞, proving that $(M, \mathcal{F})$ is of class $C^{\infty,0}$. In fact, the leaves of $\mathcal{F}$ are C^1–immersed in M relative to a C^∞ regular atlas for the C^∞ foliation $\mathcal{L}$ and the subbundle $E \subset T(M)$ is exactly the image of $T(\mathcal{F})$ under these immersions; so $(M, \mathcal{F})$

is of class $C^{\infty,0+}$. If a regular atlas $\mathcal{U}$ for $\mathcal{F}$ has $\Gamma_{\mathcal{U}}$ of class C^r, the argument in the proof of Proposition 5.1.4 allows us to choose a biregular atlas

$$\mathcal{U}' = \{U_\alpha, x_\alpha, y_\alpha\}_{\alpha\in\mathfrak{A}}$$

for $(M, \mathcal{F}, \mathcal{L})$ such that $\Gamma_{\mathcal{U}} = \Gamma_{\mathcal{U}'}$. Thus, the coordinate changes $x_\alpha(x_\beta)$ are C^∞ and $y_\alpha(y_\beta)$ are C^r, proving that $(M, \mathcal{F})$ is of class C^r. The converse is immediate. □

5.2. Open, Saturated Sets

In the following, we assume that $(M, \mathcal{F}, \mathcal{L})$ is as above, *with M compact.* The integer $r \geq 0$ measures the differentiability class of $(M, \mathcal{F})$, this foliated manifold being of class C^r if $r \geq 0$ and of class $C^{\infty,0+}$ if $r = 0$. We will simplify the discussion considerably by assuming that *the foliation is transversely oriented and that $\partial_{\pitchfork} M = \emptyset$.* The latter condition guarantees that each leaf has empty boundary. As discussed in Section 3.5, if $\mathcal{F}$ is not transversely orientable, it can be lifted to a transversely–oriented foliation on a double covering of M. Also, if $\partial_{\pitchfork} M \neq \emptyset$, one can *double* M along $\partial_{\pitchfork} M$ to produce a foliated manifold with no transverse boundary. This means that we glue two copies of $(M, \mathcal{F})$ along $\partial_{\pitchfork} M$ via the identity map, smoothing everything as in Subsection 3.3.A. These tricks often make it possible to extend theorems that have been proven only for transversely orientable foliations, without transverse boundary, to the general case.

Definition 5.2.1. The set of all open, $\mathcal{F}$–saturated subsets of M will be denoted by $\mathcal{O}(\mathcal{F})$.

Definition 5.2.2. A set $W \in \mathcal{O}(\mathcal{F})$ is called a *foliated product* if it is connected and $\mathcal{L}|W$ fibers W by open intervals over some $(n-1)$–manifold L.

This is a foliated interval bundle in our general sense of foliated bundles. (The completeness property discussed in Example 2.1.5 will be apparent shortly.) Our assumption that $\mathcal{F}$ is transversely oriented guarantees that W is an orientable interval bundle, hence is a trivial bundle. This is why we use the term *foliated product.* In this case, $W \cong L \times (0, 1)$, but the foliation $\mathcal{F}|W$ generally is not diffeomorphic to the product foliation.

Example 5.2.3. Consider a Reeb component $(M, \mathcal{F})$ and let L be one of the noncompact leaves, F the compact boundary leaf. Then $L \cup F$ is closed and, evidently, $\mathcal{F}$–saturated. The complement of $L \cup F$ is an element of $\mathcal{O}(\mathcal{F})$ which is easily seen to be a foliated product $W \cong L \times (0, 1)$. In this case, $\mathcal{F}|W$ is the product foliation. Remark that an inhabitant of $W \cong L \times (0, 1)$

might complete its universe to $\widehat{W} \cong L \times [0, 1]$ by adding the "floor" $L \times \{0\}$ and the "ceiling" $L \times \{1\}$. The inclusion map

$$i : W \hookrightarrow M$$

extends to a smooth immersion

$$\widehat{\imath} : \widehat{W} \to M$$

that identifies the floor and the ceiling each to the leaf L. We call L the *border* of W and denote it by δW. Thus,

$$\widehat{\imath}(\widehat{W}) = W \cup \delta W = \operatorname{int} M.$$

If L_1 and L_2 are distinct noncompact leaves of the Reeb component, then the complement $W \in \mathcal{O}(\mathcal{F})$ of $L_1 \cup L_2 \cup F$ has two components, W_1 and W_2, each of which is a foliated product. Each component of W has one of L_1 or L_2 as floor and the other as ceiling. This time, $\widehat{\imath}$ smoothly imbeds $\widehat{W}_j$ in M and $\delta W_j = L_1 \cup L_2$, $j = 1, 2$.

Since we assume M to be compact, any biregular cover $\mathcal{U}$ will be finite, say $\mathcal{U} = \{U_i, x_i, y_i\}_{i=1}^m$. We choose the coordinates so that

$$\overline{U}_i = \overline{E}^{n-1} \times [-1, 1],$$

where E^{n-1} is the open cube $(-1, 1)^{n-1}$ and the last coordinate $-1 < y < 1$ of U_i orients the $\mathcal{L}$–plaques $\{x\} \times (-1, 1)$ in conformity with the transverse orientation of $\mathcal{F}$. If $W \in \mathcal{O}(\mathcal{F})$, each component of $W \cap \overline{U}_i$ will be of one of the following forms:

- $\overline{E}^{n-1} \times [-1, 1]$,
- $\overline{E}^{n-1} \times [-1, a_i)$,
- $\overline{E}^{n-1} \times (b_i, 1]$,
- $\overline{E}^{n-1} \times (a_{\alpha i}, b_{\alpha i})$, where α ranges over a possibly infinite index set $\mathcal{A}_i$.

As a consequence of the remark on page 28, every $h \in \Gamma_{\mathcal{U}}$ that is a restriction of a pure word $\gamma_{i_1 i_2} \circ \gamma_{i_2 i_3} \circ \cdots \circ \gamma_{i_{n-1} i_n}$ has a C^r extension $\overline{h}$ to the closure in $\overline{S}$ of its domain.

Definition 5.2.4. The $\mathcal{F}$–plaques of the form $E^{n-1} \times \{t_i\}$, where $t_i = a_i$, b_i, $a_{\alpha i}$, or $b_{\alpha i}$, $1 \leq i \leq m$, $\alpha \in \mathcal{A}_i$, are called *border plaques* of W. If $\partial M \neq \emptyset$, plaques $E^{n-1} \times \{\pm 1\} \subseteq W \cap \partial M$ will also be called border plaques of W. The border plaques such that $t_i = a_i$ or b_i, together with the plaques $E^{n-1} \times \{\pm 1\} \subseteq W \cap \partial M$, are called *special.* The union of all border plaques of W is denoted by δW and is called the border of W.

In Figure 5.2.1, the shaded regions are portions of W and the part of ∂M shown is also part of W. We illustrate the various possibilities for $W \cap \overline{U}_i$ and

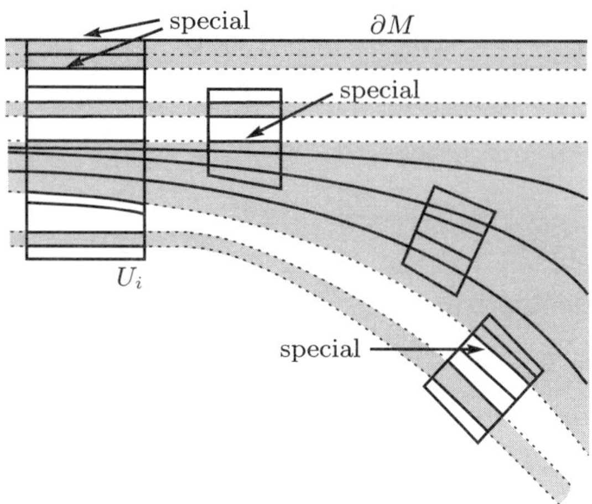

Figure 5.2.1. The possibilities for $W \cap \overline{U}_i$

point out the special border plaques. If W is replaced by $W' = W \smallsetminus \partial M$ in Figure 5.2.1, then the plaques of $\overline{U}_i$ marked "special" remain border plaques but are no longer special.

It is easy to give a crude description of $W \in \mathcal{O}(\mathcal{F})$ and of its border.

Lemma 5.2.5. *If $W \in \mathcal{O}(\mathcal{F})$, the border δW is a union of leaves. If W is connected, then δW is a finite union of leaves. Finally, if W is connected and at least one leaf in δW contains no special border plaque, then $W \cong L \times (a, b)$ is a foliated product and δW consists of one or two leaves diffeomorphic to L.*

Proof. Where convenient, we will view plaques as the points on which the holonomy pseudogroup $\Gamma_{\mathcal{U}}$ acts. It is clear that, if P is a border plaque of W, then its orbit $\Gamma_{\mathcal{U}}(P)$ is a set of border plaques. The plaques in a $\Gamma_{\mathcal{U}}$–orbit unite to form a leaf; so δW is a union of leaves. Now assume that W is connected and consider two cases.

Case 1. Every leaf $L \subseteq \delta W$ contains a special border plaque. There are only finitely many such plaques; so δW contains only finitely many leaves.

Case 2. Some leaf $L \subseteq \delta W$ contains no special border plaque. Choose a component $\overline{E}^{n-1} \times (a_{\alpha i}, b_{\alpha i})$ of $W \cap \overline{U}_i$ and suppose that one of the two border plaques, say $P = E^{n-1} \times \{a_{\alpha i}\}$, is in L. Let $\overline{\gamma}_{ji}$ be defined at P and let $P_t = E^{n-1} \times \{t\}$, $a_{\alpha i} \leq t \leq b_{\alpha i}$. We claim that $\overline{\gamma}_{ji}$ is also defined at P_t. If not, then $\overline{\gamma}_{ji}(P) \subset L$ would be a special border plaque, contradicting the hypothesis. By induction on $N \geq 1$, it follows that, whenever

$$\overline{\gamma} = \overline{\gamma}_{i_N i_{N-1}} \circ \cdots \circ \overline{\gamma}_{i_1 i}$$

is defined at P, then $\overline{\gamma}$ is also defined at P_t. Let

$$\Gamma_P = \{\overline{\gamma} \mid \overline{\gamma}(P) \text{ is defined}\}.$$

Then $\bigcup_{\overline{\gamma}\in\Gamma_P} \overline{\gamma}(P_t)$ is an open subset of the leaf L_t containing P_t. In Exercise 5.2.6, you will show that this subset is also closed in L_t, hence is the whole leaf L_t. It follows that the $\mathcal{F}$–saturation W' of the open arc $\{0\} \times (a_{\alpha i}, b_{\alpha i})$ is a foliated product of the form $L \times (a, b)$, where L is the leaf through P. But W' is open in W and, in Exercise 5.2.7, you will show that it is also closed in W. The connectivity of W and the fact that $W' \neq \emptyset$ imply that $W = W'$. As in Example 5.2.3, δW may consist of one or two leaves of $\mathcal{F}$. $\square$

Exercise 5.2.6. In the proof of Lemma 5.2.5, it is asserted that

$$L'_t = \bigcup_{\overline{\gamma}\in\Gamma_P} \overline{\gamma}(P_t)$$

is a closed subset of the leaf L_t containing P_t. In order to prove this, show first that the closure in M of each plaque $\overline{\gamma}(P_t)$ is a subset of L'_t. If a sequence $\{y_k\}_{k=1}^{\infty} \subset L'_t$ converges in L_t to a point y, show that a subsequence lies in a single plaque $\overline{\gamma}(P_t)$, some $\overline{\gamma} \in \Gamma_P$. Thus, y is in the closure of that plaque, hence $y \in L'_t$. (Hint. By the definition of a regular foliated atlas, a plaque through y meets only finitely many other plaques.)

Exercise 5.2.7. It is asserted in the proof of Lemma 5.2.5 that the open, saturated subset $W' \subseteq W$ is also closed in W. For this, assume that a sequence $\{y_k\}_{k=1}^{\infty} \subset W'$ converges in W to a point y. If $y \notin W'$, prove that a sequence of points of δW also converges to y. Since $W \in \mathcal{O}(\mathcal{F})$, this is a contradiction.

Proposition 5.2.8. *Let $W_1 \supseteq W_2 \supseteq \cdots \supseteq W_k \supseteq \cdots$ be a descending sequence of elements of $\mathcal{O}(\mathcal{F})$ such that $\bigcap_{k=1}^{\infty} W_k$ has empty interior. Then, for k sufficiently large, each component of W_k is a foliated product.*

Proof. Set $X = \bigcap_{k=1}^{\infty} W_k$. Without serious loss of generality, we replace each W_k with $W_k \smallsetminus \partial M$, hence X with $X \smallsetminus \partial M$. Then, since $\operatorname{int}(X) = \emptyset$, one can slightly alter each U_i so as to assure that $\{U_i\}_{i=1}^{m}$ is still a biregular cover, but

$$X \cap \partial_\tau \overline{U}_i = \emptyset, \quad 1 \leq i \leq m.$$

Thus, for k large enough,

$$W_1 \cap W_2 \cap \cdots \cap W_k \cap \delta U_i = W_k \cap \partial_\tau \overline{U}_i = \emptyset, \quad 1 \leq i \leq m.$$

For these W_k, no border plaque can be special. So each component is a foliated product by Lemma 5.2.5. $\square$

Corollary 5.2.9. *If $V \in \mathcal{O}(\mathcal{F})$, then all but finitely many components of V are foliated products.*

Proof. If V has only finitely many components, there is nothing to prove. By 2nd countability, the components of V can be enumerated as $\{V_j\}_{j=1}^{\infty}$. Let $W_k = V \smallsetminus V_1 \cdots \smallsetminus V_k$ and apply Proposition 5.2.8. □

Let $W \in \mathcal{O}(\mathcal{F})$ be connected. Fix any Riemannian metric g on M and restrict that metric to W. Let $d : W \times W \to [0, \infty)$ be the topological metric induced by $g|W$. Finally, let $\widehat{W}$ be the completion of W in the metric d. The following propositions are fairly intuitive and details are left to the reader.

Proposition 5.2.10. *The space $\widehat{W}$ has the structure of a connected manifold (generally noncompact) of class C^r with finitely many boundary components and $W = \operatorname{int}(\widehat{W})$.*

Proposition 5.2.11. *The inclusion $i : W \hookrightarrow M$ extends to a C^r immersion $\widehat{\imath} : \widehat{W} \to M$ carrying $\partial\widehat{W}$ onto δW. If $L \subseteq \delta W$ is a border leaf, then $\widehat{\imath}^{-1}(L)$ consists of one or two components of $\partial\widehat{W}$, each of them being carried one–to–one onto L by $\widehat{\imath}$.*

Proposition 5.2.12. *The immersion $\widehat{\imath} : \widehat{W} \to M$ is transverse both to $\mathcal{F}$ and to $\mathcal{L}$. Thus, $\widehat{\mathcal{F}} = \widehat{\imath}^*(\mathcal{F})$ is a well–defined foliation on $\widehat{W}$ of codimension one and the same smoothness class as $\mathcal{F}$, while $\widehat{\mathcal{L}} = \widehat{\imath}^*(\mathcal{L})$ is a smooth, one–dimensional foliation transverse to $\widehat{\mathcal{F}}$. Finally, relative to the foliation $\widehat{\mathcal{F}}$, $\partial\widehat{W} = \partial_\tau\widehat{W}$.*

If W is a foliated product, then $\widehat{W}$ is also a foliated bundle which is trivial as a bundle, the interval fibers being compact. We will call $\widehat{W}$ (or, more properly, $(\widehat{W}, \widehat{\mathcal{F}})$) a foliated product also.

Remark. Since the fiber of $(\widehat{W}, \widehat{\mathcal{F}})$ is compact, the completeness property discussed in Example 2.1.5 is evident. The completeness of $(W, \mathcal{F}|W)$ is an immediate consequence.

For many purposes, analysis of the foliation $\mathcal{F}|W$, for $W \in \mathcal{O}(\mathcal{F})$, is more easily carried out for the foliation $\widehat{\mathcal{F}}$ in the manifold $\widehat{W}$. If $\widehat{W}$ is a foliated product, holonomy questions can be pursued in the total holonomy *group* and life becomes particularly easy. For general $W \in \mathcal{O}(\mathcal{F})$, all but finitely many components are foliated products; so we need a good analysis for those connected elements $W \in \mathcal{O}(\mathcal{F})$ that are not foliated products.

When $W \in \mathcal{O}(\mathcal{F})$ is connected, but is not a foliated product, we will show that $\widehat{W}$ has a kind of *octopus* decomposition

$$\widehat{W} = K \cup V_1 \cup \cdots \cup V_\ell,$$

where K is a compact, connected manifold with boundary and corners and $\widehat{\mathcal{F}}|K$ foliates this manifold so that the corners divide $\partial_\pitchfork K$ from $\partial_\tau K$. Furthermore, for $1 \le i \le \ell$, each V_i is a closed, noncompact submanifold with boundary and corners and the foliations $\widehat{\mathcal{F}}_i = \widehat{\mathcal{F}}|V_i$ and $\widehat{\mathcal{L}}_i = \widehat{\mathcal{L}}|V_i$ make V_i into a foliated interval bundle (a product bundle) such that $K \cap V_i = \partial_{\pitchfork} V_i$ is exactly one of the components of $\partial_{\pitchfork} K$. Although generally $\ell \neq 8$, we adopt "octopus decomposition" as a technical term.

Definition 5.2.13. If $W \in \mathcal{O}(\mathcal{F})$ is connected, a decomposition

$$\widehat{W} = K \cup V_1 \cup \cdots \cup V_\ell,$$

as above, will be called an octopus decomposition. The foliated manifolds $(V_i, \widehat{\mathcal{F}}_i)$, $1 \leq i \leq \ell$, are called the *arms* of the decomposition and $(K, \widehat{\mathcal{F}}|K)$ is called the *nucleus*.

Of course, when $\widehat{W}$ is a foliated product, it has an octopus decomposition, but this is uninteresting. Also, if $\widehat{W}$ is compact, there is an uninteresting octopus decomposition with no arms. The point of this decomposition is to be able to use total holonomy arguments in the arms, relegating the harder structural problems to the compact, foliated manifold $(K, \widehat{\mathcal{F}}|K)$.

Proposition 5.2.14. *If $W \in \mathcal{O}(\mathcal{F})$ is connected, then $\widehat{W}$ has an octopus decomposition.*

Proof. The foliated manifold $(\widehat{W}, \widehat{\mathcal{F}}, \widehat{\mathcal{L}})$ has a biregular cover, induced by $\mathcal{U}$, as follows. For $(U_i, x_i, y_i) \in \mathcal{U}$, the possible components of $U_i \cap W$ (the interiors of the components of $\overline{U}_i \cap W$ listed on page 127), are completed to open subsets of $\widehat{W}$ by adjoining the appropriate border plaques described in Definition 5.2.4. These open subsets form a (generally infinite) biregular cover, denoted by $\widehat{\mathcal{U}}$. All but finitely many elements of $\widehat{\mathcal{U}}$ will be of the form

$$E^{n-1} \times [a_{\alpha i}, b_{\alpha i}] \quad \text{(two plaques having been added).}$$

The union of the remaining elements of $\widehat{\mathcal{U}}$ is a relatively compact, relatively open submanifold K_0 of $\widehat{W}$. By adding finitely many more elements of $\widehat{\mathcal{U}}$, we can assume also that K_0 is connected. The compact closure $\overline{K}_0$ is clearly $\widehat{\mathcal{L}}$–saturated, as is $\widehat{W} \smallsetminus K_0$. This latter falls into finitely many, connected, foliated products. One would like to take these as the arms and $\overline{K}_0$ as the nucleus, but there are some problems.

A minor problem is that some components of $\widehat{W} \smallsetminus K_0$ may be compact. Simply adjoin these to $\overline{K}_0$.

A slightly more serious problem is that the interface between $\overline{K}_0$ and $\widehat{W} \smallsetminus K_0$ is not smooth in general. But the jagged features are due to the rectangular closures of finitely many elements of $\widehat{\mathcal{U}}$ and can rather easily be smoothed, as we now show. If V_i is a component of $\widehat{W} \smallsetminus K_0$, we can identify $V = B \times [-1, 1]$, where $B \times \{\pm 1\}$ are connected submanifolds of $\partial \widehat{W}$ with compact, piecewise smooth boundary. There is a small, compact neighborhood N of ∂B in B with $\partial N \smallsetminus \partial B$ smooth and we can adjoin the collar $N \times [-1,1]$ to $\overline{K}_0$. Carrying out this procedure for the finitely many components of $\widehat{W} \smallsetminus K_0$, we slightly enlarge $\overline{K}_0$ to a compact, connected manifold K with boundary and corners, foliated by $\widehat{\mathcal{F}}|K$. The corners divide ∂K into $\partial_\tau K = K \cap \partial\widehat{W}$ and $\partial_\pitchfork K = C_1 \cup \cdots \cup C_\ell$, where C_j is the smooth interface between K and the arm $V_j = B_j \times [-1,1]$, $1 \le j \le \ell$.

The remaining problem is that, while we have arranged that the components $V_1, \dots, V_\ell$ of $\widehat{W} \smallsetminus \operatorname{int} K$ be noncompact foliated products with corners and that

$$\partial_\pitchfork K = \bigcup_{j=1}^{\ell} \partial_\pitchfork V_j,$$

a disjoint union, we have not guaranteed that $\partial_\pitchfork V_j$ is connected, $1 \le j \le \ell$. For this, write $V_j = B_j \times [-1,1]$. The connected, noncompact manifold B_j has compact boundary. If ∂B_j is connected, so is $\partial_\pitchfork V_j = \partial B_j \times [-1,1]$ and there is nothing to do. Otherwise, find a properly imbedded arc $\sigma \subset B_j$ starting at one component of ∂B_j and ending at another. A closed normal neighborhood $N(\sigma)$ of this arc is a compact tube in B_j joining the two boundary components. Adjoining the $\mathcal{L}$–saturation $N(\sigma) \times [-1,1]$ to K reduces by 1 the number of components of $\partial_\pitchfork V_j$. Finite repetition of this "tunneling" procedure completes the proof. □

Remarks. The boundary leaves $L_1, \dots, L_p$ of $\widehat{W}$ have decompositions

$$L_i = A_i \cup B_i^1 \cup \cdots \cup B_i^{r_i}, \quad 1 \le i \le p, \tag{5.2.1}$$

where $A_i = K \cap L_i$ is a compact, connected submanifold with boundary and each B_i^k is the intersection of L_i with one of the arms. Clearly, B_i^k is a noncompact, connected submanifold with just the one boundary component $\partial B_i^k = B_i^k \cap A_i$. It is *not* generally true that a given boundary leaf L_i will meet every arm. That is, $r_i \le \ell$, where strict inequality is possible. These assertions are implicit in the proof of Proposition 5.2.14.

There is a version of the octopus decomposition result that holds when $\mathcal{F}$ is not assumed to be transversely orientable. The arms may be nonorientable interval bundles in this case. If we drop the assumption that $\partial_\pitchfork M = \emptyset$, the leaves of $\mathcal{F}$ may have boundary. Then the nucleus K and the arms V_j may have higher–order corners (the ordinary corners may have corners). In this case, if $\dim M = 3$, the argument we gave to make $\partial_\pitchfork V_j$ connected does not work. Indeed, $B_j \cap A_j$ may be two or more disjoint intervals and the tunneling process in the above proof will not connect them. There is no way around this difficulty which sometimes complicates the study of foliated 3–manifolds with transverse boundary.

In the following example, we show that $\widehat{W}$ may not be a foliated product. Octopus decompositions are generally unavoidable.

Example 5.2.15. In Example 4.1.6, we created a foliated manifold $(M, \mathcal{F})$ having an exceptional minimal set X. Let

$$W = M \smallsetminus X \in \mathcal{O}(\mathcal{F}).$$

In Exercise 4.1.8, part 4 implies that W is connected and, in fact, is a foliated product and that $\partial\widehat{W}$ consists of the two leaves $L_0, L_1 \subset X$ corresponding to the endpoints of gaps in the transverse Cantor set C. By part 5 of the same exercise, a leaf $L \subset W$ meets each gap of the Cantor set $C \subset S^1$ in infinitely many points. Thus, by the construction in Lemma 3.3.7, there is a closed transversal $\sigma \subset W$ to $\mathcal{F}$. By turbulizing along σ (Example 3.3.11), we obtain a new foliated manifold $(M, \mathcal{F}')$ which is no longer a foliated bundle. Indeed, W now contains a Reeb component. Since $X \cap \sigma = \emptyset$, the exceptional minimal set has not been affected, but $W \in \mathcal{O}(\mathcal{F}')$ is no longer a foliated product. Remark that $\partial\widehat{W}$ still consists of the two leaves $L_0, L_1 \subset X$ bounding the gaps in the transverse Cantor set. If we ignore the foliation, $W \cong L_1 \times (a, b)$ is topologically a Cartesian product.

Finally, form $W' \in \mathcal{O}(\mathcal{F}')$ by taking the complement in W of the Reeb component. Then $\partial\widehat{W'} = L_0 \cup L_1 \cup T$, where T is the toral boundary of the Reeb component and $\widehat{W'}$ is not even topologically a Cartesian product.

5.3. Proper and Semiproper Leaves

Recall the notions of "proper leaf" (Definition 4.3.3) and "semiproper leaf" (*cf.* the remark on page 117). Basically, a proper leaf L does not cluster on itself, while a semiproper leaf is allowed to cluster on itself from one side only. This condition only makes sense for transversely–orientable foliations of codimension 1 and is not meant to exclude the possibility that the leaf is proper.

Exercise 5.3.1. If L is a semiproper leaf that is not proper, prove that $\overline{L}$ is transversely a Cantor set. Give an example to show that $\overline{L}$ need not be an exceptional minimal set.

Lemma 5.3.2. *A leaf L is semiproper if and only if it is a border leaf of some $W \in \mathcal{O}(\mathcal{F})$.*

Proof. Clearly, a leaf $L \subseteq \delta W$ cannot enter W, hence cannot cluster on itself on the side bordering W. Conversely, if L is semiproper, let J be a transverse arc out of a point $x_0 \in L$ that meets L only in $\{x_0\}$. The $\mathcal{F}$–saturation W of $\operatorname{int} J$ is an element of $\mathcal{O}(\mathcal{F})$ and $L \subseteq \delta W$. □

Definition 5.3.3. Let L be a semiproper leaf, let $x_0 \in L$ and J be as in the above proof and let $x \in J \smallsetminus \{x_0\}$. Suppose that every holonomy transformation $h_\sigma \in \Gamma_{\mathcal{U}}$ defined via loops σ on L based at x_0 either is not defined on x or has $h_\sigma(x) = x$. Then x will be called a fixed point for the holonomy of L. If there are such fixed points arbitrarily close to x_0 in J, we will say that L is *semistable* on the proper side defined by J.

The following elementary result, known as the Dippolito semistability theorem, is very useful.

Theorem 5.3.4 (Dippolito [**36**]). *Allow the possibility that $\partial_\pitchfork M \neq \emptyset$ and let L be a semiproper leaf that is semistable on the proper side defined by $J = [x_0, y_0)$. Then there is a point $y_1 \in J \smallsetminus \{x_0\}$ such that the $\mathfrak{F}$–saturation W of (x_0, y_1) is a foliated product with border leaves the (distinct) leaves through x_0 and y_1. Furthermore, there is a sequence $\{y_k\}_{k=1}^\infty \subset (x_0, y_1]$, converging monotonically to x_0, such that the leaf L_k through y_k is carried by the $\widehat{\mathfrak{L}}$–fibration $\pi : \widehat{W} \to L$ homeomorphically onto L, $\forall\, k \geq 1$.*

Proof. We will give the proof under the assumption that $\partial_\pitchfork M = \emptyset$ and invite the reader, in Exercise 5.3.5, to remove that restriction. Let $U \in \mathcal{O}(\mathfrak{F})$ be the $\mathfrak{F}$–saturation of the open, transverse arc $J \smallsetminus \{x_0\}$. Pull the picture back to $\widehat{U}$, where we identify L with the component of $\widehat{\imath}^{-1}(L)$ that borders U on the side determined by J. That is, we view $L \subset \partial\widehat{U}$, $J \subset \widehat{U}$, and $J \cap L = \{x_0\}$. The decomposition

$$L = A \cup B^1 \cup \cdots \cup B^r$$

(*cf.* (5.2.1)) can be chosen so that $x_0 \in \mathrm{int}(A)$. Choose $x_j \in \partial B_j$ and fix paths σ_j in A from x_0 to x_j, $1 \leq j \leq r$.

The classification of foliation germs at compact leaves (Theorem 2.3.9) works equally well for germs of $\widehat{\mathfrak{F}}$ at the compact submanifold A of the leaf L and allows us to choose J small enough that, through each of the holonomy fixed points $z \in J$, there passes a leaf A_z of the foliated normal neighborhood $N(A)$ that projects diffeomorphically onto A along the leaves of $\widehat{\mathfrak{L}}$. Choosing J even smaller, if necessary, also ensures that h_{σ_j} is defined on all of J, $1 \leq j \leq r$.

Let $z \in \mathrm{int}\, J$ be one of the holonomy fixed points and let $z_j = h_{\sigma_j}(z)$. Since B_j is one of the two boundary components of an arm V of $\widehat{W}$, $h_\sigma(z_j)$ is defined for all loops σ on B_j that are based at x_j. Since

$$h_{\sigma_j + \sigma - \sigma_j} = h_{\sigma_j}^{-1} \circ h_\sigma \circ h_{\sigma_j}$$

is defined at z and fixes that point, it follows that h_σ fixes z_j. That is, the total holonomy group of the foliated bundle V fixes z_j; so the whole leaf of $\widehat{\mathfrak{F}}|V$ through z_j must project diffeomorphically onto B_j. This argument works for each $j = 1, 2, \ldots, r$. So we assemble these finitely many pictures to produce an open submanifold L_z' of the leaf L_z of $\widehat{\mathfrak{F}}$ through z that projects diffeomorphically along the leaves of $\widehat{\mathfrak{L}}$ onto L. Call this projection π. If

$\{w_\ell\}_{\ell=1}^\infty \subset L'_z$ converges in L_z to a point w, we claim that $w \in L'_z$. You are asked to prove this in Exercise 5.3.6. It follows that L'_z is open and closed in L_z, hence is that entire leaf.

Since the holonomy fixed points cluster at x_0, we choose the sequence $\{y_k\}_{k=1}^\infty$ from those fixed points and all assertions follow. $\square$

Exercise 5.3.5. If $\partial_\pitchfork M \neq \emptyset$, double M along $\partial_\pitchfork M$, fitting the foliations together as usual to obtain a foliated manifold $(M', \mathcal{F}')$ without transverse boundary. Show that the truth of Theorem 5.3.4 for $(M', \mathcal{F}')$ implies its truth for $(M, \mathcal{F})$. (Hint. If $L' \subset M'$ is the double of L, work at $x_0 \in \partial L \subset L'$ and show that a loop on L', based at x_0 and properly crossing ∂L only once, fixes the fixed points of the holonomy of L. Now do an induction on the number of crossings.)

Exercise 5.3.6. In the proof Theorem 5.3.4, it is asserted that the open subset $L'_z \subset L_z$ is also a closed subset. Prove this by an argument analogous to that in Exercise 5.2.6.

Chapter 6

Compact Leaves

Our ongoing hypothesis remains fixed: M is a compact, connected n–manifold and, unless otherwise stipulated, the codimension–1 foliation $\mathcal{F}$ is transversely oriented. We allow $\partial_{\pitchfork} M \neq \emptyset$. We orient the leaves of $\mathcal{L}$ by the transverse orientation of $\mathcal{F}$. If J is a subarc of a leaf T of $\mathcal{L}$, we use interval notation such as $J = (x, y)$, $J = [x, y)$, *etc.*, where $x, y \in T$ and $x < y$ in the orientation of T.

6.1. The Set of Compact Leaves

Our first goal is to prove the following theorem of A. Haefliger [**58**]. The proof will only use smoothness of class $C^{1,0+}$.

Theorem 6.1.1 (Haefliger). *Whether or not $\mathcal{F}$ is transversely orientable, the union of the compact leaves of $\mathcal{F}$ is a compact subset of M. If $\mathcal{F}$ is transversely orientable, the union of the compact leaves of a given homeomorphism type is also compact and there are only finitely many homeomorphism types of compact leaves.*

Example 6.1.2. The condition that $\mathcal{F}$ be transversely orientable is essential for the conclusion that the union of compact leaves in a given homeomorphism class be compact. Here is a simple example. Let P^2 denote the real projective plane. Then, as an additive group,

$$\pi_1(P^2) = \{0, 1\} \pmod 2.$$

Define the homomorphism

$$h : \pi_1(P^2) \to \mathrm{Diff}^{\omega}(S^1)$$

so that $h(1)$ is complex conjugation. By suspension, obtain a foliated circle bundle

$$\pi : M \to P^2.$$

Since conjugation is an orientation–reversing transformation of S^1, the foliation is not transversely orientable. Each leaf is a covering space of P^2, hence is either a sphere or a projective plane. The nontrivial loop on P^2 lifts to a closed loop on the leaves corresponding to the two fixed points ± 1 of $h(1)$ in S^1; so these leaves are copies of P^2. The lift of this loop to every other leaf opens up because $z \neq \overline{z}$ whenever $z \in S^1 \setminus \{\pm 1\}$. Thus, these leaves are homeomorphic to S^2 and form a proper, open subset of M.

Temporary Hypothesis. M is oriented, $\partial_\tau M = \emptyset$ and $\mathcal{F}$ is transversely oriented.

Under the temporary hypothesis, each compact leaf L is oriented by the orientation of M and the transverse orientation of $\mathcal{F}$. Furthermore, $\partial M = \partial_\pitchfork M$; so we obtain a homology class $[L] \in H_{n-1}(M, \partial M; \mathbb{Z})$. Indeed, an oriented triangulation of this leaf can be viewed as a singular $(n-1)$–chain c and, since $\partial L \subset \partial M$, the oriented character of the triangulation implies that $\partial c \in C_{n-2}(\partial M; \mathbb{Z})$ (all faces of the $(n-1)$–simplices cancel except those in ∂M). That is, $c \in Z_{n-1}(M, \partial M; \mathbb{Z})$, a cycle modulo ∂M, and we set $[L] = [c]$.

We are going to use the *homology intersection pairing*. Since M is a compact, connected, oriented n–manifold, there are canonical isomorphisms

$$H_p(M; \mathbb{Z}) = H^{n-p}(M, \partial M; \mathbb{Z}),$$
$$H_q(M, \partial M; \mathbb{Z}) = H^{n-q}(M; \mathbb{Z}).$$

This is *Poincaré duality*, arguably the most important theorem in the algebraic topology of manifolds. (A modern proof, due to Milnor, will be found in [**56**, Part III].) Coupled with the cup product in cohomology,

$$H^{n-p}(M, \partial M; \mathbb{Z}) \times H^{n-q}(M; \mathbb{Z}) \to H^{2n-p-q}(M, \partial M; \mathbb{Z}),$$

Poincaré duality produces the *intersection product* in homology

$$H_p(M; \mathbb{Z}) \times H_q(M, \partial M; \mathbb{Z}) \to H_{p+q-n}(M; \mathbb{Z}),$$

which we will write

$$(\alpha, \beta) \mapsto \alpha \cdot \beta.$$

In all of the above, we can use homology and cohomology over $\mathbb{R}$ equally well.

Remark. The name "intersection product" comes from a remarkable geometric fact. If $\alpha = [P]$ and $\beta = [Q]$, where $P \subset \operatorname{int} M$ is a compact, oriented, p–manifold without boundary and $Q \subset M$ is a compact, oriented q–manifold with $\partial Q \subset \partial M$, one can move P by an isotopy so that $P \pitchfork Q$. (If $p+q < n$, this means that $P \cap Q = \emptyset$.) Then $P \cap Q \subset \operatorname{int} M$ is a compact, oriented submanifold without boundary, having dimension $p + q - n$. The claim is that

$$[P] \cdot [Q] = [P \cap Q].$$

Of course, not every homology class in M is representable by a submanifold.

Our use of homology intersection theory will be restricted to the following situation. Given the compact leaf L and a closed transversal σ to $\mathcal{F}$, we orient σ by the transverse orientation of $\mathcal{F}$ and view $[\sigma] \in H_1(M; \mathbb{Z})$. The intersection product

$$H_1(M; \mathbb{Z}) \times H_{n-1}(M, \partial M; \mathbb{Z}) \to H_0(M; \mathbb{Z}) = \mathbb{Z}$$

allows us to interpret $[\sigma]\cdot[L]$ as a nonnegative integer. In fact, our hypotheses imply that $[\sigma] \cdot [L]$ is the cardinality of the finite set of points $\sigma \cap L$. If this is nonzero, both σ and L will be homologically nontrivial.

Lemma 6.1.3. *If a compact leaf L of $\mathcal{F}$ meets a closed transversal, then the class $[L] \in H_{n-1}(M, \partial M; \mathbb{Z})$ is nontrivial.*

Let $\{L_k\}_{k=1}^{\infty}$ be a sequence of distinct compact leaves of $\mathcal{F}$. Let $x_k \in L_k$, $1 \le k < \infty$, be a sequence such that $\lim_{k\to\infty} x_k = x$ in M. Let L be the leaf of $\mathcal{F}$ through x. If U is a biregular chart and $x \in U$, we can assume that all $x_k \in U$. Evidently, the $\mathcal{F}$–plaques $P_k \subset U$ with $x_k \in P_k$ are distinct. Let $J \subset U$ be the $\mathcal{L}$–plaque through x. Clearly, we can replace x_k by the point $P_k \cap J$, if necessary, so as to assume that $\{x_k\}_{k=1}^{\infty} \subset J$. We can also assume that, in the orientation of J, $x_k \downarrow x$ and $L_k \cap [x, x_k) = \emptyset$.

Lemma 6.1.4. *The leaf L is compact and, for k sufficiently large, every compact leaf meeting $[x, x_k]$ is homeomorphic to L.*

Proof. First we claim that $L \cap (x, x_k] = \emptyset$, k sufficiently large. If not, pass to a subsequence and choose $y_k \in L \cap (x_{k+1}, x_k)$, $\forall k \ge 1$. Corresponding to the subarc $[y_k, y_{k-1}]$, Theorem 3.3.7 provides a closed transversal σ_k that meets L_k, but does not meet L_j, $j < k$. Consider the sequence

$$\{[L_k]\}_{k=1}^{\infty} \subset H_{n-1}(M, \partial M; \mathbb{R})$$

of nontrivial homology classes. If $\alpha = \sum_{j=1}^{r} a_j[L_j] = 0$, then $0 = [\sigma_r]\cdot\alpha = a_r$. Iterating this for $r-1, \dots, 2, 1$, in turn, shows that $a_j = 0$, $1 \le j \le r$. Since this holds for all $r \ge 1$, it follows that $H_{n-1}(M, \partial M; \mathbb{R})$ is infinite dimensional, contradicting compactness of the manifold M. Thus, L is semiproper on the side defined by $[x, x_1]$.

Since x_k is the closest point of $L_k \cap J$ to x, it must be a fixed point for the holonomy of L. Thus, the Dippolito semistability theorem applies (Theorem 5.3.4) and shows that $L \cong L_k$, all large values of k. Indeed, every compact leaf that meets $(x, x_k]$, k sufficiently large, is also homeomorphic to L. □

We deduce Theorem 6.1.1 when the temporary hypothesis holds. Indeed, the above shows that the union X of all compact leaves is closed in M, hence compact, and that the union X_L of compact leaves of a given homeomorphism type L is both open and closed in X. These form an open, disjoint cover of the compact set X. Hence there are only finitely many distinct homeomorphism types L. Since X_L is also a closed subset of X, it is compact.

Proof of Theorem 6.1.1. Whether or not M is orientable or $\mathfrak{F}$ transversely orientable, we can find a finite sheeted covering $p : \widetilde{M} \to M$ such that $\widetilde{M}$ is oriented and $\widetilde{\mathfrak{F}} = p^*(\mathfrak{F})$ is transversely oriented (Proposition 3.5.1). Evidently, the union $\widetilde{X}$ of the compact leaves of $\widetilde{\mathfrak{F}}$ is carried by p onto the union X of the compact leaves of $\mathfrak{F}$. Since $\widetilde{X}$ is compact, so is X.

Assume that $\mathfrak{F}$ is transversely oriented. If M is nonorientable with $\partial_\tau M = \emptyset$, lift to the orientation cover

$$p : \widetilde{M} \to M$$

to produce a foliated manifold $(\widetilde{M}, \widetilde{\mathfrak{F}})$ in which the assertions are all true. Let

$$\gamma : \widetilde{M} \to \widetilde{M}$$

be the nontrivial covering transformation. Note that, for an arbitrary leaf L of $\mathfrak{F}$, $p^{-1}(L)$ is either a pair of leaves of $\widetilde{\mathfrak{F}}$, each carried diffeomorphically onto L by p and interchanged by γ, or it is a single leaf, invariant under γ and doubly covering L. If $\{L_k\}_{k=1}^{\infty}$ is a sequence of distinct compact leaves of $\mathfrak{F}$, choose $x_k \in L_k$ and $x_k \to x \in M$, and let L be the leaf through x. We consider two cases.

Case 1. For infinitely many values of k, $p^{-1}(L_k) = \{F_k, \gamma(\mathfrak{F}_k)\}$, where F_k and $\gamma(F_k)$ are distinct leaves carried by p homeomorphically onto L_k. We can pass to a subsequence so that this holds for all $k \geq 1$. Thus, in the above proof, we obtain two converging sequences of leaves,

$$F_k \to F,$$
$$\gamma(F_k) \to F',$$

implying that $\gamma(F) = F'$. Since γ is of order 2, the only way that $F = F'$ is that F_k and $\gamma(F_k)$ lie on opposite sides of F, in which case, γ would reverse the transverse orientation at F. This would contradict the assumption that $\mathfrak{F}$ is transversely orientable. Thus, $F \neq F'$ and, applying the covering map p, we see that

$$L = p(F) \cong F \cong F_k \cong L_k,$$

for all $k \geq 1$.

Case 2. For infinitely many values of k, $p^{-1}(L_k) = F_k$, where F_k is a single leaf of $\widetilde{\mathcal{F}}$. Again, we can assume that this holds for all $k \geq 1$, and that $\{F_k\}_{k=1}^{\infty}$ clusters on F. By the Dippolito semistability theorem, there is a one–sided neighborhood of F of the form $F \times [x, x_1]$ and a sequence $x_k \downarrow x$ such that $F_k = F \times \{x_k\}$, for all $k \geq 1$. Since this neighborhood is γ–invariant, it projects into M as a one–sided neighborhood $L \times [p(x), p(x_1)]$ and $L \cong L_k$, $k \geq 1$.

It only remains to remove the restriction that $\partial_\tau M = \emptyset$. For this, we simply double along the tangential boundary. The resulting foliated manifold is transversely oriented by keeping the same transverse orientation in one copy of $(M, \mathcal{F})$ and the opposite one in the other copy. Since we have used only smoothness class $C^{1,0+}$, no smoothness problems arise. It is trivial that the truth of the assertions in the doubled foliated manifold implies their truth in $(M, \mathcal{F})$. □

Theorem 6.1.1 yields a nice globalization of the Reeb stability theorem.

Theorem 6.1.5. *Let $(M, \mathcal{F})$ be a compact, connected, transversely orientable, foliated manifold of codimension one. If there is a compact leaf L having finite fundamental group, then every leaf is homeomorphic to L and M is homeomorphic either to $L \times [0, 1]$, foliated as a product, or to the total space of a fiber bundle $p : M \to S^1$ having the leaves of $\mathcal{F}$ as fibers.*

Proof. Consider the holonomy group $\mathcal{H}_y(L)$. This is the homomorphic image of the finite group $\pi_1(L, y)$, hence is a finite group. Let $\widehat{f} \in \mathcal{H}_y(L)$ be represented by an orientation–preserving homeomorphism f of $(-1, 1)$ into $\mathbb{R}$ with $f(0) = 0$. Here, y corresponds to 0. Since $\widehat{f}$ is of finite order, we can choose f so that $f^k = \text{id}$, for some integer $k \geq 1$. If $f \neq \text{id}$, we can assume (replacing f by f^{-1}, if necessary) that, for some value of $t \in (0, 1)$, $0 < f(t) < t$. This uses the fact that f preserves orientation and implies that $f^k(t) < f^{k-1}(t) < \cdots < t$, contradicting $f^k = \text{id}$. Thus, $\mathcal{H}_y(L)$ is the trivial group and Theorem 2.4.1 implies that the union X_L of leaves homeomorphic to L is open in M. By Theorem 6.1.1, X_L is also closed in M, hence $X_L = M$. Since each leaf has an $\mathcal{F}$–saturated neighborhood of the form $U \cong L \times (-\epsilon, \epsilon)$, foliated as a product, it is clear that the quotient space $M/\mathcal{F}$ is a compact, connected 1–manifold B and that the quotient map $p : M \to B$ is a fiber bundle. By the classification of 1–manifolds, B is homeomorphic either to S^1 or $[0, 1]$. Bundles over $[0, 1]$ are automatically trivial. □

6.2. The Thurston Stability Theorem

Reeb stability has been greatly improved by W. Thurston [**133**]. His proof was simplified by G. Reeb and P. Schweitzer [**111**], who used nonstandard analysis. As usual, nonstandard proofs have standard analogues. The one given here was formulated independently by J. Cantwell (unpublished) and W. Schachermayer [**116**]. For another direct proof, see J. P. Jouanolou [**78**].

Theorem 6.2.1 (Thurston Stability Theorem). *Let $(M, \mathcal{F})$ be a compact, connected, transversely–orientable, foliated manifold of codimension one and smoothness class C^1. If there is a compact leaf L with $H^1(L;\mathbb{R}) = 0$, then every leaf is homeomorphic to L and M is homeomorphic either to $L\times[0,1]$, foliated as a product, or to the total space of a fiber bundle $p : M \to S^1$ having the leaves of $\mathcal{F}$ as fibers.*

As in the proof of Theorem 6.1.5, we only need to show that the germinal holonomy group $\mathcal{H}(L)$ is trivial (for simplicity of notation, we omit reference to a basepoint). The key is an elementary technical lemma. Let

$$f, g, h : (-\epsilon, \epsilon) \to (-\delta, \delta)$$

be orientation–preserving C^1 imbeddings that fix 0 and satisfy

$$f'(0) = g'(0) = h'(0) = 1.$$

If we write

$$\begin{aligned} f(x) &= x + \widetilde{f}(x), \\ g(x) &= x + \widetilde{g}(x), \\ h(x) &= x + \widetilde{h}(x), \end{aligned}$$

then $\widetilde{f}(0) = 0$ and $\widetilde{f}'(0) = 0$, and similarly for $\widetilde{g}$ and $\widetilde{h}$.

Lemma 6.2.2. *Suppose there is a sequence $\{x_m\}_{m\geq 1} \subset (-\epsilon, \epsilon)$, converging to 0 and such that the following limits exist:*

$$\begin{aligned} \lim_{m\to\infty} \frac{\widetilde{f}(x_m)}{\widetilde{h}(x_m)} &= A, \\ \lim_{m\to\infty} \frac{\widetilde{g}(x_m)}{\widetilde{h}(x_m)} &= B. \end{aligned}$$

Then

$$\begin{aligned} \lim_{m\to\infty} \frac{\widetilde{f \circ g}(x_m)}{\widetilde{h}(x_m)} &= A + B, \\ \lim_{m\to\infty} \frac{\widetilde{f^{-1}}(x_m)}{\widetilde{h}(x_m)} &= -A. \end{aligned}$$

Proof. First, write

$$(f \circ g)(x) = g(x) + \widetilde{f}(g(x)) = x + \widetilde{g}(x) + \widetilde{f}(x) + (\widetilde{f}(g(x)) - \widetilde{f}(x)).$$

Then, by the mean value theorem and the fact that $g(x) - x = \widetilde{g}(x)$, there is ξ between x and $g(x)$ such that

$$\widetilde{f \circ g}(x) = \widetilde{f}(x) + \widetilde{g}(x) + \widetilde{g}(x)\widetilde{f}'(\xi).$$

As $x \to 0$, so does $\xi \to 0$; hence $\widetilde{f}'(\xi) \to \widetilde{f}'(0) = 0$. Therefore,

$$\begin{aligned} \lim_{m\to\infty} \frac{\widetilde{f \circ g}(x_m)}{\widetilde{h}(x_m)} &= \lim_{m\to\infty} \frac{\widetilde{f}(x_m)}{\widetilde{h}(x_m)} + \lim_{m\to\infty} \frac{\widetilde{g}(x_m)}{\widetilde{h}(x_m)} + \lim_{m\to\infty} \frac{\widetilde{g}(x_m)}{\widetilde{h}(x_m)} \widetilde{f}'(\xi_m) \\ &= A + B. \end{aligned}$$

Similarly,

$$\begin{aligned} x &= f^{-1}(f(x)) \\ &= x + \widetilde{f}(x) + \widetilde{f^{-1}}(f(x)); \end{aligned}$$

so

$$\begin{aligned} -\widetilde{f}(x) &= \widetilde{f^{-1}}(f(x)) + \widetilde{f^{-1}}(x) - \widetilde{f^{-1}}(x) \\ &= \widetilde{f^{-1}}(x) + \widetilde{f}(x)(\widetilde{f^{-1}})'(\xi). \end{aligned}$$

Therefore,

$$\begin{aligned} \lim_{m\to\infty} \frac{\widetilde{f^{-1}}(x_m)}{\widetilde{h}(x_m)} &= \lim_{m\to\infty} \left(-\frac{\widetilde{f}(x_m)}{\widetilde{h}(x_m)} - \frac{\widetilde{f}(x_m)}{\widetilde{h}(x_m)} (\widetilde{f^{-1}})'(\xi_m)\right) \\ &= -A. \end{aligned}$$

□

Since $H^1(L;\mathbb{R}) = \mathrm{Hom}(\pi_1(L), \mathbb{R})$, the following is sufficient to prove Theorem 6.2.1.

Proposition 6.2.3. *If L is a compact leaf of $\mathcal{F}$ with orientable normal bundle $\nu(L)$ and with holonomy $\mathcal{H}(L) \neq \{\mathrm{id}\}$, then there is a nontrivial homomorphism $\varphi : \pi_1(L) \to \mathbb{R}$.*

Proof. As remarked in Example 2.3.15, the nonvanishing of the infinitesimal holonomy of L implies the existence of φ. Assume, therefore, that the infinitesimal holonomy is trivial. Let $\alpha, \beta, \gamma, \dots, \lambda$ be C^1 imbeddings into $\mathbb{R}$ of neighborhoods of 0 that fix 0 and whose germs define a system of generators $\widehat{\alpha}, \widehat{\beta}, \widehat{\gamma}, \dots, \widehat{\lambda} \in \mathcal{H}(L)$. Let $\widetilde{\alpha}, \widetilde{\beta}, \widetilde{\gamma}, \dots, \widetilde{\lambda}$ be as in Lemma 6.2.2. Since $\mathcal{H}(L) \neq \{\mathrm{id}\}$, there is a sequence $\{x_n\}_{n=1}^{\infty} \subset (0, \epsilon)$, converging monotonically to 0, such that, for each $n > 0$, at least one of

$$\widetilde{\alpha}(x_n), \widetilde{\beta}(x_n), \widetilde{\gamma}(x_n), \dots, \widetilde{\lambda}(x_n) \neq 0.$$

Let μ_n denote the maximum absolute value of these numbers and set

$$\begin{aligned}N_\alpha &= \{n \geq 1 \mid |\widetilde{\alpha}(x_n)| = \mu_n\},\\ &\vdots\\ N_\lambda &= \{n \geq 1 \mid |\widetilde{\lambda}(x_n)| = \mu_n\}.\end{aligned}$$

Then $\mathbb{Z}^+ = N_\alpha \cup N_\beta \cup N_\gamma \cup \cdots \cup N_\lambda$. Reordering the generators, if necessary, we assume that N_α is infinite. Then, by passing to a subsequence, if necessary, we assume that $0 \neq |\widetilde{\alpha}(x_n)| = \mu_n$, for all n. But

$$\begin{gathered}\left|\frac{\widetilde{\beta}(x_n)}{\widetilde{\alpha}(x_n)}\right| \leq 1,\\ \vdots\\ \left|\frac{\widetilde{\lambda}(x_n)}{\widetilde{\alpha}(x_n)}\right| \leq 1;\end{gathered}$$

so we can assume (again passing to a subsequence, if necessary) that the following limits exist:

$$\begin{aligned}a &= \lim_{n\to\infty} \frac{\widetilde{\alpha}(x_n)}{\widetilde{\alpha}(x_n)},\\ b &= \lim_{n\to\infty} \frac{\widetilde{\beta}(x_n)}{\widetilde{\alpha}(x_n)},\\ &\vdots\\ l &= \lim_{n\to\infty} \frac{\widetilde{\beta}(x_n)}{\widetilde{\alpha}(x_n)}.\end{aligned}$$

These are not all 0 since $a = 1$. By Lemma 6.2.2, the assignments

$$\begin{gathered}\widehat{\alpha} \mapsto a,\\ \widehat{\beta} \mapsto b,\\ \vdots\\ \widehat{\lambda} \mapsto l\end{gathered}$$

extend uniquely to an additive homomorphism (nontrivial since $a = 1$)

$$\rho : \mathcal{H}(L) \to \mathbb{R}.$$

The composition

$$\pi_1(L) \xrightarrow{\widehat{h}} \mathcal{H}(L) \xrightarrow{\rho} \mathbb{R}$$

defines the desired homomorphism, nontrivial because $\widehat{h}$ is surjective and ρ is nontrivial. □

Exercise 6.2.4. Some smoothness is essential in Theorem 6.2.1. Let Γ' be the group of isometries of the hyperbolic plane $\mathbf{H}$ generated by reflections on the sides of a geodesic triangle with angles $\pi/2$, $\pi/3$ and $\pi/7$. Let Γ be the subgroup of Γ' that preserves orientation. Let $M = \widetilde{\mathrm{Sl}}(2,\mathbb{R})/\Gamma$.

(1) Show that M is a compact homology 3–sphere over $\mathbb{R}$ with fundamental group having presentation $\{a, b, c \mid a^2 = b^3 = c^7 = abc\}$.

(2) The action of Γ extends to an action on the "circle at infinity" fixing one point. By suspension, construct a foliation having M as a leaf with nontrivial holonomy.

Exercise 6.2.5. In higher codimension q, the infinitesimal holonomy of a leaf L is an element $J\widehat{h} \in \mathrm{Hom}(\pi_1(L), \mathrm{Gl}(q,\mathbb{R}))$; so the vanishing of $H^1(L;\mathbb{R})$ does not imply the triviality of $J\widehat{h}$. If L is a compact leaf with vanishing first real cohomology *and* with $\mathrm{Hom}(\pi_1(L), \mathrm{Gl}(q,\mathbb{R})) = 1$, then L has a saturated neighborhood of compact leaves homeomorphic to L. Using multivariable calculus and Theorem 2.4.1, argue as above to prove this.

Exercise 6.2.6. Assuming that $(M, \mathcal{F})$ is compact, connected, of codimension one, but *not* assuming transverse orientability, show that, if some leaf L is compact with finite fundamental group, then all leaves are compact with finite fundamental group and that only finitely many homeomorphism classes of leaves occur. If the foliation is of class C^1 and the compact leaf has $H^1(L;\mathbb{R}) = 0$, draw the analogous conclusion. You may use the fact that, if $H^1(L;\mathbb{R}) = 0$ and $p : \widetilde{L} \to L$ is a finite sheeted covering, then $H^1(\widetilde{L};\mathbb{R}) = 0$ also.

6.3. Compact Leaves and Closed Transversals

Let $(M, \mathcal{F})$ be a compact, connected, transversely–oriented, foliated manifold of codimension one and let L be a leaf. Recall that, if L does not meet a closed transversal, then L is compact (Corollary 3.3.8). On the other hand, given that the leaf L is compact, whether it does or does not meet a closed transversal has interesting topological consequences, especially if $\dim M = 3$.

In this section, we will assume that $\partial_\tau M = \emptyset$. We fix a transverse orientation and use it to determine the positive direction of transverse curves to $\mathcal{F}$ and the positive and negative sides of proper leaves.

Definition 6.3.1. The *accessible set* $A(L)$ of a leaf L is the set of $x \in M$ for which there is a nontrivial, positively directed, $\mathcal{F}$–transverse curve from L to x.

In this definition, it should be emphasized that the transverse curve is not required to be a part of a leaf of the transverse 1–dimensional foliation $\mathcal{L}$. That would put a positive lower bound to the angle the curve can make with any leaf, but we often need curves which, while transverse to $\mathcal{F}$, make sufficiently small angles with $\mathcal{F}$. The following properties are evident.

1. $A(L)$ is open, $\mathcal{F}$–saturated, and connected.
2. If L' is a leaf, $L' \subset A(L)$, then $A(L') \subseteq A(L)$.
3. A leaf L meets a closed transversal if and only if $L \subset A(L)$.
4. If L and L' meet a common closed transversal, then $A(L) = A(L')$.

Lemma 6.3.2. *If $A(L) \neq M$, then the closure of this accessible set is a compact, connected manifold with boundary $L_1 \cup L_2 \cup \cdots \cup L_k$, a finite union of compact leaves. Furthermore, $A(L)$ lies on the positive side of L_i, $1 \leq i \leq k$.*

Proof. For short, write $U = A(L)$. Since $U \in \mathcal{O}(\mathcal{F})$, it is clear that the set–theoretic boundary $\overline{U} \setminus U$ is a union of leaves.

Let $L' \subset \overline{U} \setminus U$ be a leaf. We claim that $L' \subset \delta U$, but that U cannot accumulate on L' from the negative side. Indeed, let $\{x_n\}_{n=1}^{\infty} \subset U$ be a sequence that converges to $x \in L'$. Such a sequence exists since $L' \subset \overline{U}$. We may assume that the sequence converges from either the positive or negative side of L'. If the convergence is from the negative side, then there is a positively directed transverse arc from x_m to x when m is sufficiently large. But $x_m \in U$; so there is a positively directed transversal from L to x_m. Combining these produces such a transverse curve from L to x, contradicting the fact that $x \notin U$. The convergence is, therefore, from the positive side of L', and we may suppose that all of the points involved are on an $\mathcal{L}$–plaque S through x. Clearly, S is positively directed from x_{m+1} to x_m, $\forall m$; so every point of S between x_m and x_{m+1} is in U. Since the sequence converges to $x \in L'$, it follows that $L' \subset \delta U$ and that U lies on the positive side of L'.

Since $\overline{U}$ is compact, the above analysis shows that $\overline{U}$ is a compact 3–manifold with boundary $\partial \overline{U} = \delta U$, necessarily a finite union of compact leaves, and that U lies on the positive side of each boundary leaf. □

Definition 6.3.3. The foliation $\mathcal{F}$ is *taut* if every leaf meets a closed transversal. In the case that $\partial_\tau M \neq \emptyset$, the foliation is taut if each leaf either meets a closed transversal or a transverse arc from one component of $\partial_\tau M$ to another.

Although currently we are assuming that $\partial_\tau M = \emptyset$, later in the book we will need the concept of tautness in the general case. For instance, a theorem of Dennis Sullivan, to be proven in Chapter 10, asserts that there exists a Riemannian metric on M relative to which every leaf of $\mathcal{F}$ is a minimal submanifold if and only if $\mathcal{F}$ is taut. Other critical uses of this concept will occur in our study of foliated 3–manifolds in Volume II.

Corollary 6.3.4. *The foliation $\mathcal{F}$ is taut if and only if no finite family of compact leaves $\{L_1, \dots, L_q\}$ cobounds a submanifold $Q \subset M$ in such a way that Q lies on the positive side of L_i, $1 \le i \le q$.*

Proof. If $\mathcal{F}$ is taut and L is a compact leaf, the existence of a closed transversal σ through L implies that L cannot be a boundary leaf of a submanifold Q as above. Conversely, if no such submanifold Q exists, Lemma 6.3.2 implies that $A(L) = M$, for every compact leaf L. In particular, $L \subset A(L)$, which implies that there is a closed transversal through L. □

In compact, foliated 3–manifolds without boundary, we will prove another interesting consequence [**54**] of Lemma 6.3.2.

Theorem 6.3.5 (S. Goodman)**.** *Let $(M, \mathcal{F})$ be a compact, oriented, transversely–oriented, C^∞–foliated 3–manifold without boundary. Let L be a leaf of $\mathcal{F}$ that meets no closed transversal. Then L is a torus.*

Thus, in closed, foliated 3–manifolds, the absence of toral leaves guarantees tautness. The C^∞ hypothesis is for convenience, enabling us to define the Euler class $e(\mathcal{F}) \in H^2(M;\mathbb{R})$ via differential forms. This is a characteristic class of the vector bundle $T(\mathcal{F})$ with the remarkable property that, if L is a compact leaf with Euler characteristic $\chi(L)$, then the value of $e(\mathcal{F})$ on $[L] \in H_2(M;\mathbb{R})$ is exactly $\chi(L)$ (Proposition 6.3.7). The following exercise leads you through the main technical details.

Exercise 6.3.6. Let $(M, \mathcal{F}, g)$ be a smoothly foliated, Riemannian 3–manifold, with $\mathcal{F}$ transversely oriented. Let ∇ be the connection on $T(\mathcal{F})$ that is the Levi–Civita connection induced by the Riemannian metric g along the leaves of $\mathcal{F}$. Let R be the curvature tensor of ∇. Let $\{(U_\alpha, x_\alpha, y_\alpha)\}_{\alpha\in\mathfrak{A}}$ be a biregular atlas for $(M, \mathcal{F}, \mathcal{L})$ and let (s^1_α, s^2_α) be a smooth, orthonormal frame field for $T(\mathcal{F})|U_\alpha$, $\alpha \in \mathfrak{A}$. Then,

$$R(X,Y)s^i_\alpha = \sum_{j=1}^{2} \Omega^\alpha_{ij}(X,Y)s^j_\alpha$$

defines a matrix

$$K^\alpha = \begin{bmatrix} \Omega^\alpha_{11} & \Omega^\alpha_{12} \\ \Omega^\alpha_{21} & \Omega^\alpha_{22} \end{bmatrix}$$

of 2–forms on U_α.

1. Show that this matrix has the form

$$K^\alpha = \begin{bmatrix} 0 & \Omega^\alpha \\ -\Omega^\alpha & 0 \end{bmatrix}.$$

2. Show that Ω^α is independent of the choice of orthonormal frame field (s^1_α, s^2_α). Conclude, in particular, that Ω^α and Ω^β agree on overlaps $U_\alpha \cap U_\beta$. Thus, they fit together to define $\Omega \in A^2(M)$.
3. Show that Ω is closed. For this, work in U_α and choose the frame field (s^1_α, s^2_α) first along one $\mathcal{F}$–plaque and then extend it by parallel translation along $\mathcal{L}|U_\alpha$. If θ_{ij} are the connection forms for $\nabla|U_\alpha$ relative to this frame, then each θ_{ij} vanishes on fields tangent to $\mathcal{L}|U_\alpha$, $1 \le i, j \le 2$. Use this fact and one of the Cartan structure equations to conclude that $d\Omega = 0$.
4. Show that $[\Omega] \in H^2(M;\mathbb{R})$ is independent of the choice of Riemannian metric. For this, let $g = g_0$ and let g_1 be another choice of metric. Then $g_t = tg_1 + (1-t)g_0$ is a Riemannian metric on M, $0 \le t \le 1$. Use the corresponding forms Ω_t to define a closed 2–form Ω_* on $M \times [0,1]$ that restricts to Ω_i on $M \times \{i\}$, $i = 0, 1$. (Hint. Ω_* should vanish on any pair of fields, one of which is tangent to the $[0,1]$–fibers. Test $d\Omega_*$ on fields d/dt and X, Y, where X and Y are tangent to the M–fibers and independent of t.) Set $2\pi e(\mathcal{F}) = [\Omega]$.
5. The sectional curvature of a leaf L at a point $z \in L \cap U_\alpha$ is
$$\kappa(z) = \left\langle R(s^1_\alpha, s^2_\alpha)s^2_\alpha, s^1_\alpha \right\rangle_z .$$
Prove that $\Omega|L = \kappa\omega_L$ where ω_L is the volume form of $g|L$.

Proposition 6.3.7. *If, under the above hypotheses, $L \subset M$ is a compact leaf of $\mathcal{F}$, then the value of the Euler class $e(\mathcal{F})$ on $[L] \in H_2(M;\mathbb{R})$ is the Euler characteristic $\chi(L)$. In particular, if L separates M, then L is a torus.*

Proof. The Gauss–Bonnet theorem [**90**, p. 305] gives
$$2\pi e(\mathcal{F})[L] = \int_L \kappa\omega_L = 2\pi\chi(L).$$
If L separates M, then $[L] = 0$ in $H_2(M;\mathbb{R})$; so $0 = e(\mathcal{F})[L] = \chi(L)$. By the classification of surfaces and the fact that L is orientable, L must be a torus. □

A rational homology 3–sphere M has the property that $H_*(M;\mathbb{Q})$ is the same as $H_*(S^3;\mathbb{Q})$. In addition to the honest 3–sphere, the lens spaces are rational homology spheres, as are all closed, connected 3–manifolds with $\pi_1(M)$ a perfect group, and many others. If $L \subset M$ is a compact leaf of $\mathcal{F}$, the vanishing of $H^2(M;\mathbb{R})$ implies that $[L] = 0$, hence the following.

Corollary 6.3.8. *If M is a rational homology 3–sphere, then every compact leaf of $\mathcal{F}$ is a torus.*

Lemma 6.3.9. *Continue to assume that* $\dim M = 3$ *and let F be a leaf of $\mathcal{F}$. If L is a leaf in $\partial\overline{A(F)}$, then L is a torus.*

Proof. Let $B = \overline{A(F)}$. If $L_1, \ldots . L_r$ are the components of ∂B,

$$0 = \int_B d\Omega = \int_{\partial B} \Omega = 2\pi \sum_{i=1}^{r} \pm\chi(L_i).$$

Here, the sign $\pm$ in $\pm\chi(L_i)$ is positive if and only if the orientation of L_i *as a leaf of a transversely–oriented foliation of an oriented manifold* is the same as its orientation *as a boundary leaf of* B. By Lemma 6.3.2, it follows that the sign is the same for all summands, hence that $0 = \sum_{i=1}^{r} \chi(L_i)$. By Reeb stability, if some $L_i \cong S^2$, then all leaves are 2–spheres and we get the contradiction that $0 = 2r$. Thus, $\chi(L_i) \leq 0$, $1 \leq i \leq r$; so we must have $\chi(L_i) = 0$ and every boundary leaf of B is a torus. Since $L \subset \partial B$, the proof is complete. □

Proof of Theorem 6.3.5. Let L be a (necessarily compact) leaf of $\mathcal{F}$ that meets no closed transversal. Thus $L \not\subset A(L)$; so $A(L) \neq M$ and L is a boundary leaf of $B = \overline{A(L)}$. The assertion follows from Lemma 6.3.9. □

Exercise 6.3.10. Let $(M, \mathcal{F})$ be as in Theorem 6.3.5. Prove that $\mathcal{F}$ is concordant (Definition 3.6.12) to a foliation in which the only compact leaves (if any) are tori.

Exercise 6.3.11. Let $(M, \mathcal{F})$ be as in Theorem 6.3.5. Define a partial order $\succeq$ on the set of compact leaves of $\mathcal{F}$ (the "Novikov order") by setting $F \succeq T$ iff every closed transversal meeting F also meets T. The following steps outline a proof that $F \succeq T$ implies $\chi(F) \geq \chi(T)$.

1. Define $A \subseteq M$ to be the set of points x such that x lies in the interior of a transverse arc from F to T that does not properly cross either F or T. Prove that $B = \overline{A}$ is a compact submanifold with ∂B a finite union of compact leaves, including F and T. Remark that the positive side of F and the negative side of T meet A.
2. Let L be a component of ∂B other than T or F and prove that either L is a torus or the negative side of L does not meet A. (Hint. If L is not a torus, there is a closed transversal σ through L that can be chosen to meet L only once. Show that σ must meet F or T and consider four cases.)
3. Using the Euler class, prove that $\chi(F) - \chi(T) \geq 0$.

It follows that, if F and T are compact leaves with $F \succeq T \succeq F$, then $F \cong T$.

Exercise 6.3.12. If $M = T^3$, show that any two compact, nontoral leaves must be homologous, hence diffeomorphic. (Hint. The intersection product is Poincaré dual to the cup product.)

Chapter 7

General Position

Let $(M, \mathcal{F})$ be a C^2–foliated manifold of codimension one, N a compact k–manifold, $0 < k < n$, and consider maps $f : N \to \text{int}(M)$. If $\partial N \neq \emptyset$, we assume that $f|\partial N$ is transverse to $\mathcal{F}$.

Since continuous maps can be homotoped to smooth ones, the homotopy being arbitrarily small, we lose no generality in assuming that f is smooth. Similarly, assume that f is an immersion. Our goal is to deform this map by an arbitrarily small homotopy f_t, $0 \leq t \leq 1$, to an immersion that is transverse to $\mathcal{F}$ at all but a finite number of points. We will also arrange that these points of tangency be of an especially nice type, called *Morse singularities*, and say that f has been homotoped to a map in "general position" with respect to the foliation. Since the deformation is arbitrarily small, if one starts with an immersion (respectively, an imbedding), it can be arranged that the result is a smooth immersion (respectively, a smooth imbedding) and, indeed, that the homotopy f_t gives an immersion (respectively, imbedding) at each parameter value t. This result has remarkable applications, one of which is a theorem of A. Haefliger on nonexistence of real analytic foliations of codimension one on manifolds with finite fundamental group (Theorem 7.3.1). This will be proven in the last section of this chapter. Other important applications of the general position result are mostly to foliated 3–manifolds and will be discussed in Volume II.

7.1. The Morse Lemma and Homotopies

In this section, we discuss the technical details of the general position theory. It will not be necessary to assume that M is compact.

Definition 7.1.1. If $f_i : N \to M$ is a smooth immersion, $i = 0, 1$, then an *isotopy* of f_0 to f_1 is a smooth homotopy f_t, $0 \le t \le 1$, of f_0 to f_1 such that each f_t is an immersion. If f_t is a smooth imbedding, $0 \le t \le 1$, then the homotopy is called an isotopy of imbeddings.

Let $p \in N$ and let $y = f(p)$. Let $\mathfrak{U} = \{U_\beta, x_\beta, y_\beta\}_{\beta=1}^m$ be a finite subset of a biregular, foliated atlas so chosen that $f(N) \subset \bigcup_{\beta=1}^m U_\beta$. Let $U_\alpha \in \mathfrak{U}$ be chosen so that $y \in U_\alpha$. Let $W_\alpha = f^{-1}(U_\alpha)$ and consider the composition $\varphi_\alpha = y_\alpha \circ f : W_\alpha \to \mathbb{R}$. Evidently, f fails to be transverse to $\mathcal{F}$ at $y = f(p)$ if and only if p is a critical point of φ_α. The "nicest" types of critical points are *Morse singularities*. In order to define this notion, we need some preliminary considerations.

Lemma 7.1.2. *Let $\varphi : N \to \mathbb{R}$ be smooth (of class at least C^2) and let $p \in N$ be a critical point of φ. Let X and Y be vector fields defined on a neighborhood of p. Then $Y_p(X(\varphi)) = X_p(Y(\varphi))$ and, in particular, this number depends only on the vectors X_p and Y_p, not on their extensions to fields.*

Proof. Indeed, $X_p(Y(\varphi)) - Y_p(X(\varphi)) = [X, Y]_p(\varphi) = 0$ since p is a critical point. □

Definition 7.1.3. The *Hessian* $h_p\varphi : T_pN \otimes T_pN \to \mathbb{R}$ of φ at the critical point p is the symmetric bilinear form defined by $h_p\varphi(X_p, Y_p) = X_p(Y(\varphi))$, where Y is any smooth extension of Y_p to a vector field on a neighborhood of p.

Definition 7.1.4. The critical point p of φ is a *nondegenerate* critical point (also called a Morse singularity) if the Hessian $h_p\varphi$ is a nondegenerate bilinear form. The (Morse) *index* of the critical point p is the number of negative eigenvalues of $h_p\varphi$. If φ has only nondegenerate critical points, it is called a Morse function.

Definition 7.1.5. We say that the immersion $f : N \to M$ is in *general position* with respect to the foliation $\mathcal{F}$ if the local representation φ_β of f is a Morse function, $\beta = 1, 2, \dots, m$.

Lemma 7.1.6 (Morse Lemma). *If $p \in N$ is a nondegenerate critical point of φ with index λ, then there is a local coordinate system $x_1, x_2, \dots, x_k$, defined on a neighborhood of p, in which $p = (0, 0, \dots, 0)$ and*

$$f(x_1, x_2, \dots, x_k) = f(0) - x_1^2 - \cdots - x_\lambda^2 + x_{\lambda+1}^2 + \cdots + x_k^2.$$

For the elementary proof, consult [**88**, pp. 6–8].

Corollary 7.1.7. *The critical points of a Morse function are isolated. In particular, the points of N at which f as in* Definition 7.1.5 *fails to be transverse to $\mathcal{F}$ are isolated, hence finite in number.*

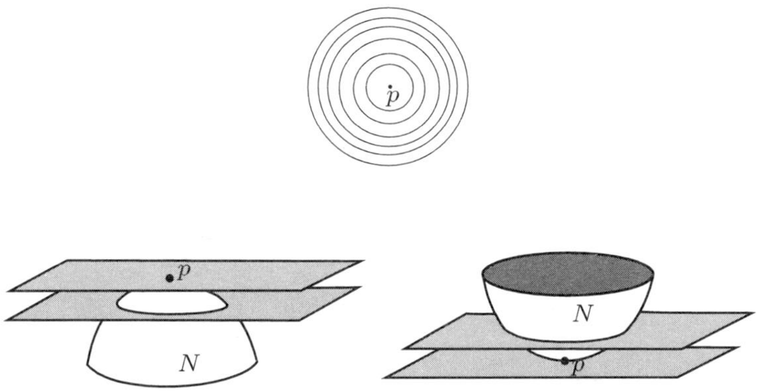

Figure 7.1.1. $\lambda = 0$ or 2

Remark. In the case that $\dim(N) = 2$, let $p \in N$ be a singularity of $\varphi = \varphi_\alpha = y_\alpha \circ f$. By Lemma 7.1.6, we have

1. $\varphi(x, y) = \varphi(0, 0) + x^2 + y^2$ ($\lambda = 0$),
2. $\varphi(x, y) = \varphi(0, 0) - x^2 + y^2$ ($\lambda = 1$),
3. $\varphi(x, y) = \varphi(0, 0) - x^2 - y^2$ ($\lambda = 2$).

If we picture the image of f in M so that the leaves of $\mathcal{F}$ near $f(p)$ look like horizontal planes, then, near $f(p)$, the surface $f(N)$ looks like a "peak" when $\lambda = 2$ and like a "pit" when $\lambda = 0$. These cases, called *centers*, are pictured in Figure 7.1.1, along with the level curves $\varphi =$ constant in a neighborhood of p on N. When $\lambda = 1$, the picture relative to the foliation looks like a "mountain pass". This case, called a *saddle*, is illustrated in Figure 7.1.2, where the level curves near p on N are also shown.

Let $C^r(N, \mathbb{R})$, $1 \leq r < \infty$, denote the space of real–valued C^r functions on N with the topology of uniform convergence in all derivatives of order $j \leq r$. In the space $C^\infty(N, \mathbb{R})$, convergence means C^r–convergence, $\forall\, r \geq 1$. The following is also proven in [**88**].

Theorem 7.1.8. *Let N be a compact manifold (possibly with boundary). If $2 \leq r \leq \infty$, then the set of Morse functions on N is open and dense in $C^r(N, \mathbb{R})$.*

Remark. Since the set of Morse functions is C^r–open, $r \geq 2$, it follows that all perturbations of a Morse function that are small enough in the C^2 topology are also Morse functions.

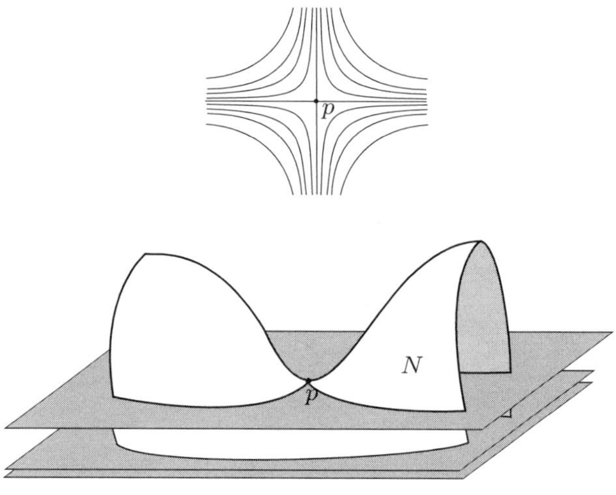

Figure 7.1.2. $\lambda = 1$

Corollary 7.1.9. *Let $W \subset \mathbb{R}^n$ be an open ball and let $O, V \subset W$ be open sets such that $\overline{O} \subset V$ and $\overline{V} \subset W$. Let $\varphi : \overline{W} \to \mathbb{R}$ be a C^k map, $2 \leq k \leq \infty$. Then, given $\epsilon > 0$ and an integer $2 \leq r < k+1$, there is a C^k map $\psi : \overline{W} \to \mathbb{R}$, which is ϵ–close to φ in the C^r topology, such that $\psi|O$ is a Morse function and $\psi|(\overline{W} \smallsetminus V) = \varphi|(\overline{W} \smallsetminus V)$.*

Proof. Consider the open cover $\{U_0 = V, U_1 = \overline{W} \smallsetminus \overline{O}\}$ of $\overline{W}$. Let $\{\lambda_0, \lambda_1\}$ be a smooth partition of unity subordinate to $\{U_0, U_1\}$. Remark that, even if $k = \infty$, our hypothesis guarantees that r is finite. Thus, there is a constant $c > 0$, depending only on λ_0 and r, such that, for any $g \in C^r(\overline{W}, \mathbb{R})$, the C^r norm satisfies $||\lambda_0 g||_r \leq c||g||_r$. Let $\epsilon > 0$ be given. By an application of Theorem 7.1.8 to the compact manifold $\overline{W}$, we can choose a Morse function $g \in C^r(\overline{W}, \mathbb{R})$ such that $||\varphi - g||_r < \epsilon/c$. Set $\psi = \lambda_0 g + \lambda_1 \varphi$, remarking that ψ agrees with the Morse function g on O and with φ on $\overline{W} \smallsetminus V$. Then, $\varphi - \psi = (\varphi - \lambda_1\varphi) - \lambda_0 g = \lambda_0(\varphi - g)$ and

$$||\varphi - \psi||_r = ||\lambda_0(\varphi - g)||_r < c(\epsilon/c) = \epsilon.$$

□

Return to the map $f : N \to M$ as at the beginning of this section.

Theorem 7.1.10. *Let $f : N \to \mathrm{int}(M)$ be a C^k map, $2 \leq k \leq \infty$, such that $f|\partial N$ is transverse to $\mathfrak{F}$. Then, given $\epsilon > 0$ and an integer $2 \leq r < k+1$, there is a C^k immersion $g : N \to \mathrm{int}(M)$, ϵ–near f in the C^r topology and satisfying the following:*

1. *$g|\partial N$ is transverse to $\mathcal{F}$;*
2. *g is in general position with respect to $\mathcal{F}$;*
3. *If $\{p_1, \dots, p_\ell\}$ is the set of points in N such that g is tangent to $\mathcal{F}$ at $g(p_i)$, $1 \le i \le \ell$, then $g(p_i)$ and $g(p_j)$ lie in distinct leaves of $\mathcal{F}$, whenever $i \ne j$.*

Proof. Fix a biregular cover $\{(U_\alpha, x_\alpha, y_\alpha\}_{\alpha\in\mathfrak{A}}$ for $(M, \mathcal{F})$. Pass to a finite subcover $\{W_\alpha = f^{-1}(U_\alpha)\}_{\alpha=1}^m$ of N and let $\{O_\alpha\}_{\alpha=1}^m$ and $\{V_\alpha\}_{\alpha=1}^m$ be open covers of N such that $\overline{O}_\alpha \subset V_\alpha \subset \overline{V}_\alpha \subset W_\alpha$, $\alpha = 1, 2, \dots, m$.

Let d be a metric on M that gives the manifold topology and let

$$\delta = \min\{\delta_\alpha\}_{\alpha=1}^m,$$

where

$$\delta_\alpha = \inf\{d(f(x), y) \mid y \in \partial\overline{U}_\alpha, x \in \overline{V}_\alpha\}.$$

For $g : N \to M$, let $d_0(f, g) = \sup\{d(f(x), g(x)) \mid x \in N\}$. If $d_0(f, g) < \delta$, then $g(\overline{V}_\alpha) \subset U_\alpha$, $\alpha = 1, 2, \dots, m$. So all of the maps $y_\alpha \circ g$ are well defined on an open neighborhood of $\overline{V}_\alpha$. All maps g that we consider are to satisfy $d_0(f, g) < \delta$.

By finite induction on $j = 1, 2, \dots, m$, we construct g_j to be $(j\epsilon/m)$–close to f in the C^r topology, to agree with f outside of $\bigcup_{\alpha=1}^j W_\alpha$, and to be such that $y_\alpha \circ g_j|O_\alpha$ is a Morse function, $1 \le \alpha \le j$. Then g_m will be in general position with respect to $\mathcal{F}$ and ϵ–close to f. With ϵ sufficiently small, it will follow that $g_m|\partial N$ is also transverse to $\mathcal{F}$. It will then be necessary to modify g_m further so as not to lose these properties, but to guarantee that the points of tangency lie on different leaves.

First consider the case $j = 1$. By Corollary 7.1.9, choose $\psi_1 : W_1 \to \mathbb{R}$, as C^r–close to $\varphi_1 = y_1 \circ f$ as desired, equal to φ_1 on $W_1 \setminus \overline{V}_1$, and Morse on O_1. Then $g_1|W_1$ should have its first $n-1$ U_1–coordinate functions the same as those of $f|W_1$ and its last coordinate function equal to ψ_1. Since this agrees with f outside a compact subset of W_1, it extends smoothly to agree with f on the rest of N. Since the C^r approximation is as close as desired, we make $||g_1 - f||_r < \epsilon/m$.

Inductively, suppose that $g_{j-1} : N \to M$ has been constructed according to the above specifications, some $j \in \{1, 2, \dots, m\}$. Again use Corollary 7.1.9 to produce smooth $g_j : N \to M$, agreeing with g_{j-1} outside $\overline{V}_j$, Morse on O_j, and with $||g_j - g_{j-1}||_r < \epsilon/m$. Making this norm even smaller, if necessary, and using the fact that the set of Morse functions on $\bigcup_{\alpha=1}^{j-1} O_\alpha$ is open in the C^r topology, we see that g_j is also a Morse function on $\bigcup_{\alpha=1}^{j-1} O_\alpha$, hence on $\bigcup_{\alpha=1}^{j} O_\alpha$. Finally, $||g_j - f||_r \le ||g_j - g_{j-1}||_r + ||g_{j-1} - f||_r < j\epsilon/m$.

We have constructed g_m as desired. Set $\theta_\alpha = y_\alpha \circ g_m$ on W_α, $1 \le \alpha \le m$. By choosing the biregular cover finer, if necessary, guarantee that θ_α has at most one Morse singularity, $\forall \alpha$, and that the singularity is in O_α. Let $p \in W_\alpha$ be such a singularity and let $\lambda : W_\alpha \to \mathbb{R}$ be a smooth function such that $\lambda | O_\alpha \equiv 1$ and $\lambda|(W_\alpha \smallsetminus \overline{V}_\alpha) \equiv 0$. For $\delta > 0$ sufficiently small, $\tilde{\theta}_\alpha = \theta_\alpha + \delta\lambda$ is still a Morse function on W_α and has p as its only singularity there. Modify g_m on W_α by replacing the coordinate function θ_α with $\tilde{\theta}_\alpha$. This modification agrees with g_m outside of $\overline{V}_\alpha$. So we obtain a map

$$\tilde{g}_m : N \to M,$$

still in general position with respect to $\mathcal{F}$ and having the same tangencies as g_m, except that $\tilde{g}_m(p)$ is shifted to an $\mathcal{F}$–plaque arbitrarily near the one containing $g_m(p)$. It is clear that systematic applications of this construction produce $g : N \to M$, in general position with respect to $\mathcal{F}$ and so close to g_m in the C^r topology that $g|\partial N$ is still transverse to $\mathcal{F}$, but with all points of tangency on distinct leaves of $\mathcal{F}$. □

Exercise 7.1.11. In Corollary 7.1.9, show that φ and ψ are homotopic by a homotopy φ_t such that $\varphi_t|(\overline{W} \smallsetminus V) = \varphi|(\overline{W} \smallsetminus V)$, $0 \le t \le 1$. Use this to show that g and f are homotopic in Theorem 7.1.10.

Corollary 7.1.12. *Let $\sigma : S^1 \to M$ be a smooth, closed transversal to $\mathcal{F}$. Then, if σ is homotopic to 0 in M, there is a smooth immersion $g : D^2 \to M$ such that $g^{-1}(\mathcal{F}) = \mathcal{F}^*$ is a foliation of D^2 with isolated singularities, all of which are centers and saddles, and $\mathcal{F}^*$ is transverse to ∂D^2. Furthermore, no two saddle points are connected by a leaf of $\mathcal{F}^*$.*

Proof. The map σ extends to a map $f : D^2 \to M$. Apply Theorem 7.1.10 to f. □

Exercise 7.1.13. Let $f : D^2 \to M$ be in general position with respect to $\mathcal{F}$, $\mathcal{F}^* = f^{-1}(\mathcal{F})$. Show that there is a smooth vector field v on D^2, directed inwardly along ∂D^2, whose associated flow has flow lines the leaves of $\mathcal{F}^*$ and fixed points the singularities of $\mathcal{F}^*$.

As a result of this exercise, the classical Poincaré–Bendixson theorem allows us to describe the possibilities for $\mathcal{F}^*$ in considerable detail.

7.2. Poincaré–Bendixson Theory

Let Φ_t be a flow on the disk D^2 generated by a vector field v transverse to ∂D^2 and pointing inward along the boundary. We assume that the singularities are centers and saddles, that no trajectory limits on two distinct saddles, and we let $\mathcal{F}^*$ denote the underlying singular foliation of D^2. If $p \in D^2$, $\Phi_t(p)$ is defined for all $t \ge 0$. It is possible that $\Phi_t(p)$ is also defined

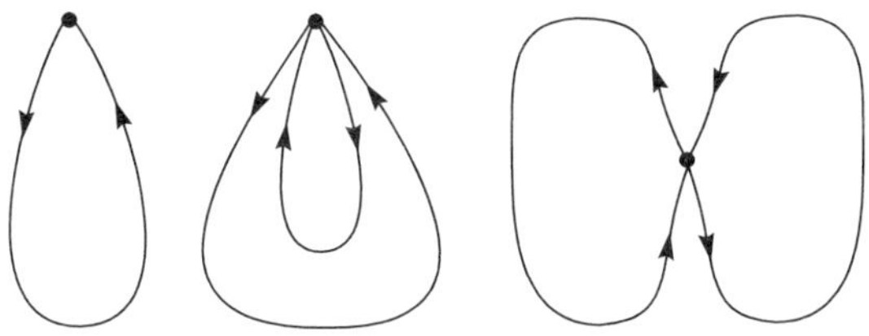

Figure 7.2.1. Closed graphs

for all $t \leq 0$, but the alternative possibility is that this trajectory meets ∂D^2 in finite time $t(p) \leq 0$. We allow both cases by allowing $t(p) = -\infty$ and we denote the trajectory in int D^2 by $\gamma_p(t) = \Phi_t(p)$, $t(p) < t < \infty$.

Suppose that $p \in D^2$ is a saddle singularity of $\mathcal{F}^*$. Locally, there are two trajectories $\gamma(t)$ such that $\lim_{t\to\infty} \gamma(t) = p$ and two such that $\lim_{t\to-\infty} \gamma(t) = p$. These trajectories are called *separatrices* at p, the first type being *stable* and the second type *unstable*. Globally, it is possible for a single separatrix to approach p in both forward and backward time, hence to be both stable and unstable. By our assumption, no trajectory can be a separatrix at more than one saddle.

Definition 7.2.1. A *graph* $\mathcal{G}$ is a connected set formed of saddles and some of the separatrices at these saddles such that, if $p \in \mathcal{G}$ is a saddle, then $\mathcal{G}$ contains at least one stable and one unstable separatrix of p.

Because of our hypotheses, the only compact graphs that can lie in the interior of D^2 are as pictured in Figure 7.2.1. We will call these *closed* graphs. Each loop in a closed graph has well–defined $\mathcal{F}^*$–holonomy on one side and a full figure–eight graph has $\mathcal{F}^*$–holonomy on one side. This is illustrated in Figure 7.2.2. The picture is completely general if we allow ∂D^2 to either encircle the whole figure eight or to lie "within" one of the simple loops. A careful description of the holonomy uses the parametrization of $\mathcal{F}^*$ as a flow Φ_t. On the arc $J_+ \cong [0,1]$, the point 0 lies on the figure–eight graph and the flow defines a smooth first return map $h : (0, \epsilon] \to (0, 1]$ for small enough $\epsilon > 0$. We extend this continuously to a topological imbedding $h : [0, \epsilon] \to [0, 1]$ by setting $h(0) = 0$. This is the one–sided holonomy of the loop that goes "once around" the figure eight. A similar construction defines one–sided holonomy on J_- for the loop that goes once around the closed separatrix.

Definition 7.2.2. Let $\gamma = \gamma_p$ be a trajectory. The ω–limit set is the set

$$\omega(\gamma) = \{\lim_{n\to\infty} \gamma(t_n) \mid t_n \uparrow \infty\}.$$

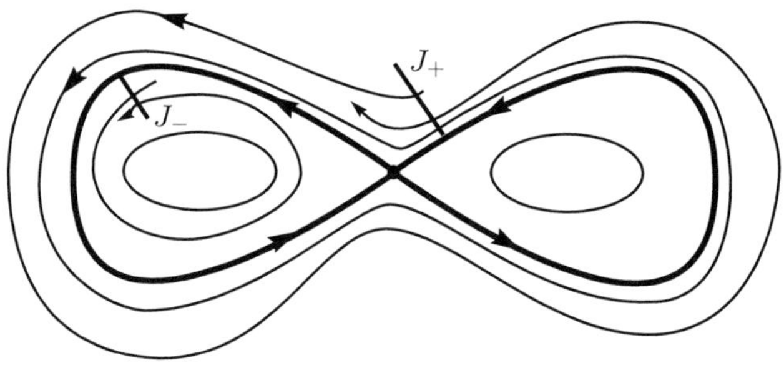

Figure 7.2.2. One–sided holonomy for a closed graph

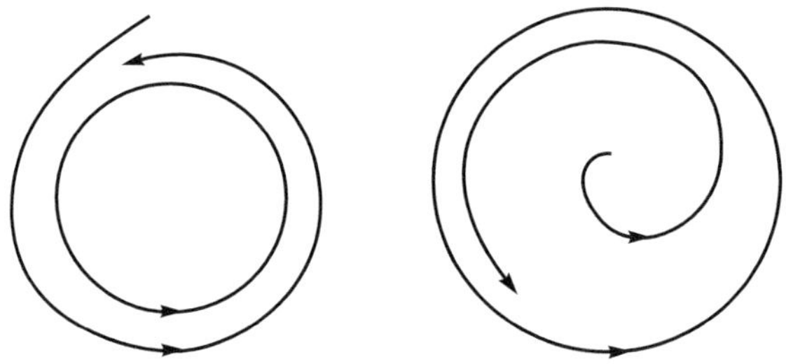

Figure 7.2.3. Nonsingular limit cycles

The α–limit set is

$$\alpha(\gamma) = \{\lim_{n\to\infty} \gamma(t_n) \mid t_n \downarrow -\infty\}.$$

Theorem 7.2.3 (Poincaré–Bendixson). *If $\gamma = \gamma_p$ is a trajectory, then $\omega(\gamma)$ is either a closed trajectory, a saddle singularity, or a closed graph. Similarly, $\alpha(\gamma)$ is either empty (the case in which $t(p) > -\infty$), a closed trajectory, a saddle singularity, or a closed graph.*

If the trajectory γ is not a separatrix, then neither limit set can be a saddle point. The case of Theorem 7.2.3 in which the limit set contains no singularity is proven in Hirsch and Smale [**70**] and the general case is given as an exercise there (p. 249, Exercise 1).

Definition 7.2.4. A *limit cycle* $\mathcal{G}$ is a nonempty limit set $\alpha(\gamma)$ or $\omega(\gamma)$ that does not consist of a saddle point and satisfies $\mathcal{G} \cap \gamma = \emptyset$.

Examples of limit cycles that are closed trajectories are given in Figure 7.2.3. Limit cycles that are graphs are pictured in Figures 7.2.4 and 7.2.5.

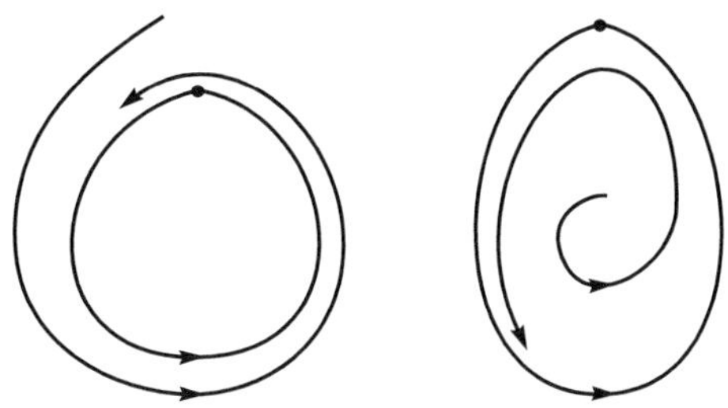

Figure 7.2.4. Limit cycles containing one separatrix

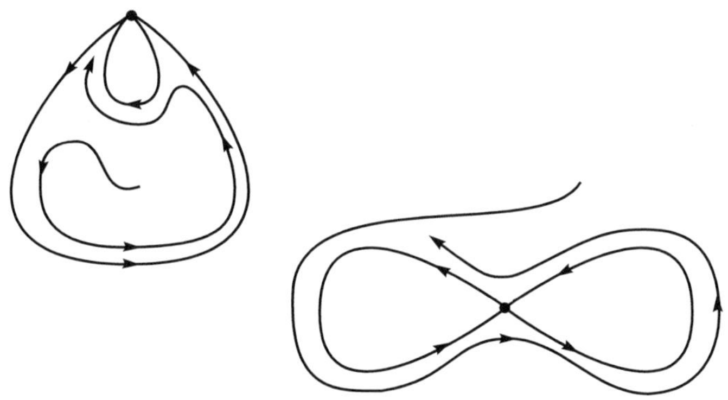

Figure 7.2.5. Figure–eight limit cycles

Because of our hypotheses, these (and the corresponding pictures with all arrows reversed) are the only possibilities.

Consider the set $\mathcal{S}$ of all compact limit cycles of $\mathcal{F}^*$. We define a partial ordering on $\mathcal{S}$ as follows. If $\mathcal{G} \in \mathcal{S}$, let $O(\mathcal{G})$ be the union of the (one or two) components of $D^2 \smallsetminus \mathcal{G}$ that do not contain ∂D^2 and let $D(\mathcal{G})$ denote the closure of $O(\mathcal{G})$.

Definition 7.2.5. If $\mathcal{G}_1, \mathcal{G}_2 \in \mathcal{S}$, then $\mathcal{G}_1 \preceq \mathcal{G}_2$ if $D(\mathcal{G}_1) \subseteq D(\mathcal{G}_2)$.

Lemma 7.2.6. *Let $\{\mathcal{G}_\alpha\}_{\alpha \in \mathcal{A}}$ be a linearly ordered, decreasing set of limit cycles and let $\mathcal{G}_\infty$ be the boundary of the set $\bigcap_{\alpha\in\mathcal{A}} D(\mathcal{G}_\alpha)$. Then $\mathcal{G}_\infty$ is either a closed orbit or a compact graph.*

Proof. If some cofinal subset $\{\mathcal{G}_\beta\}_{\beta\in\mathcal{B}}$ is constant, then $\mathcal{G}_\infty = \partial D(\mathcal{G}_\beta)$ and we are done. Assume, therefore, that all terms are distinct. Only a finite number of limit cycles can be graphs; so we assume that every $\mathcal{G}_\alpha$ is a

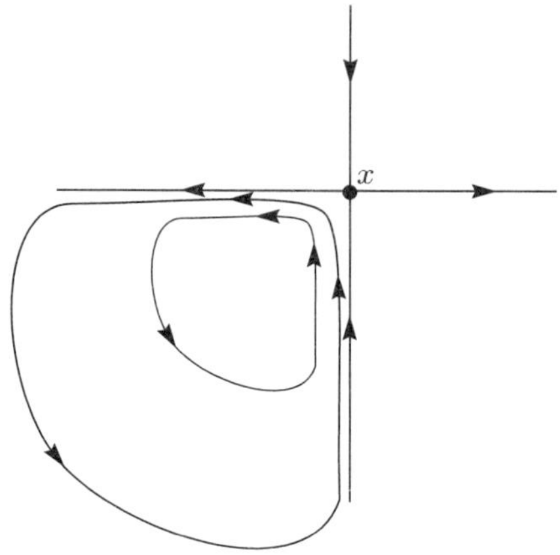

Figure 7.2.6. The limit cycles near a saddle $x \in \mathcal{G}_\infty$

closed orbit. Every $D(\mathcal{G}_\alpha)$ is a closed disk bounded by the closed orbit $\mathcal{G}_\alpha$ and $X = \bigcap_{\alpha \in \mathcal{A}} D(\mathcal{G}_\alpha)$ is a nonempty, compact, $\mathcal{F}^*$–invariant set, as is its boundary $\mathcal{G}_\infty$.

Let $x \in \mathcal{G}_\infty$ and let $J_x(\epsilon)$ be an arc of length at most $\epsilon > 0$, having x as one endpoint and transverse to $\mathcal{F}^*$ (except at x, should that be a singularity) such that $J_x(\epsilon) \cap (D^2 \smallsetminus X)$ clusters at x. Then $J_x(\epsilon)$ meets $\mathcal{G}_{\alpha_n}$ and $D^2 \smallsetminus \mathcal{G}_{\alpha_n}$ for a cofinal subsequence $\{\mathcal{G}_{\alpha_n}\}_{n=1}^\infty$ of $\{\mathcal{G}_\alpha\}_{\alpha \in \mathcal{A}}$. Using the compactness of $\mathcal{G}_\infty$, we find $N > 0$ such that every point of $\mathcal{G}_\infty$ is ϵ–close to $\mathcal{G}_{\alpha_n}$, $\forall n \geq N$.

If $x \in \mathcal{G}_\infty$ is a singular point, it must be a saddle (centers cannot be approached by limit cycles). Thus, whether x is singular or not, it is evident from the local structure of $\mathcal{F}^*$ about x (Figure 7.2.6) that there is a sequence $\{I_n \subset \mathcal{G}_{\alpha_n}\}_{n=1}^\infty$ of arcs of lengths bounded away from 0, converging uniformly either to a nondegenerate arc through x in $\mathcal{G}_\infty$ or to the union of $\{x\}$ and two nondegenerate arcs in $\mathcal{G}_\infty$. In particular, $\mathcal{G}_\infty$ contains a nonsingular point p.

If $\mathcal{G}_\infty$ contains a separatrix, then it contains a closed graph. Suppose that $p \in \mathcal{G}_\infty$ is a nonsingular point on an orbit γ_p that is not a separatrix. Consider $\omega(p) = \omega(\gamma_p) \subseteq \mathcal{G}_\infty$. By the Poincaré–Bendixson theorem, $\omega(p)$ is either a closed orbit, a saddle singularity, or a compact graph. Since we assume that γ_p is not a separatrix, $\omega(p)$ is not a saddle singularity. We conclude that, in all cases, $\mathcal{G}_\infty$ contains either a closed orbit or a closed graph.

Suppose that $\mathcal{G}_\infty$ contains a closed orbit γ and let J_ϵ be a transverse arc through γ of length $\epsilon > 0$. For $\delta > \epsilon > 0$ sufficiently small, the holonomy $h_\gamma : J_\epsilon \to J_\delta$ is defined. Here γ is identified as a loop that goes once around itself. If a trajectory meets the domain of h_γ in at least one point x, then $h_\gamma(x) \neq x$ implies that the trajectory winds forever in one direction (either in forward or backward time) around γ, hence is noncompact. If $h_\gamma(x) = x$, the trajectory is compact and stays in an η–neighborhood of γ, where $\eta \to 0$ as $\epsilon \to 0$. In particular, for n sufficiently large, the closed trajectories $\mathcal{G}_{\alpha_n}$ meet $\operatorname{dom} h_\gamma$ for all large n, hence lie entirely in an η–neighborhood of γ. Since $\eta > 0$ is arbitrary, it follows that $\{\mathcal{G}_{\alpha_n}\}_{n=1}^\infty$ converges uniformly to γ. Since every point of $\mathcal{G}_\infty$ is a cluster point of $\{x_n \in \mathcal{G}_{\alpha_n}\}_{n=1}^\infty$, it follows that $\gamma = \mathcal{G}_\infty$ and we are done.

Finally, if $\mathcal{G}_\infty$ contains a closed graph, one uses one–sided holonomy either around one separatrix or around a figure eight, arguing exactly as above to prove that $\mathcal{G}_\infty$ is exactly this closed graph. □

Let $\widetilde{\mathcal{S}}$ denote the union of $\mathcal{S}$ and the set of all $\mathcal{G}_\infty$ as in Lemma 7.2.6. The definition of the order $\preceq$ carries over unchanged to $\widetilde{\mathcal{S}}$.

Lemma 7.2.7. *The partially ordered set $\widetilde{\mathcal{S}}$ is inductive, hence contains a minimal element.*

Proof. Again let $\{\mathcal{G}_\alpha\}_{\alpha\in\mathcal{A}}$ be a linearly ordered subset of $\widetilde{\mathcal{S}}$ and set

$$\mathcal{G}_* = \partial \bigcap_{\alpha\in\mathcal{A}} D(\mathcal{G}_\alpha).$$

As before, pick a subsequence $\{\mathcal{G}_n = \mathcal{G}_{\alpha_n}\}_{n=1}^\infty$ converging uniformly to $\mathcal{G}_*$. Each $\mathcal{G}_n$ is either a limit cycle or a uniform limit of limit cycles. Hence we can assume that every $\mathcal{G}_n$ is a limit cycle. By Lemma 7.2.6, $\mathcal{G}_* \in \widetilde{\mathcal{S}}$. Thus, $\widetilde{\mathcal{S}}$ is inductive and Zorn's lemma gives a minimal element. □

Remark. It was implicitly assumed in the above that there is a limit cycle. Since the trajectories that start at ∂D^2 never return to the boundary, the Poincaré–Bendixson theorem implies that the only alternative is that *all* of these trajectories have saddles as ω–limit sets. But there are infinitely many of these trajectories and only finitely many saddles. We also used the fact that a limit cycle separates D^2 (the Jordan curve theorem). On other compact surfaces, the situation can be quite different, as you are asked to show in the following exercises.

Exercise 7.2.8. Let P be the "pair of pants" surface, obtained by deleting the interiors of a pair of disjoint disks from $\operatorname{int} D^2$. Produce a flow on P, transverse to ∂P, having no limit cycles and having only one singularity, a saddle. With a suitable choice of the direction of the flow, its velocity field will point inward along two boundary components and outward along the third.

Exercise 7.2.9. Let N be the surface obtained by deleting an open disk from the torus. Produce a flow, transverse to ∂N and having only one singularity, a saddle. This flow is to have one limit cycle, which will not separate N.

7.3. Analytic Foliations

As an application of the general position result, we prove the nonexistence theorem for real analytic foliations due to A. Haefliger.

Theorem 7.3.1 (Haefliger). *If M is a compact n–manifold with finite fundamental group and empty boundary, then there does not exist a real analytic foliation of M of codimension one.*

Remark that, by passing to a double cover if necessary, no generality is lost in assuming that $\mathcal{F}$ is transversely orientable.

Let M be an n–manifold and let $\mathcal{F}$ be a codimension–one foliation that admits a closed transversal τ. Assume that the class $[\tau] \in \pi_1(M)$ is of finite order k and choose a closed, imbedded transversal σ, close to τ but homotopic to τ^k. Then σ is nullhomotopic in M. We will prove that this implies that $\mathcal{F}$ cannot be a real analytic foliation. In particular, if M is a compact manifold without boundary, transverse orientability implies that there exists a closed transversal τ to $\mathcal{F}$. If $\pi_1(M)$ is finite, $[\tau]$ has finite order and Theorem 7.3.1 follows.

By Corollary 7.1.12, extend σ to a map $f : D^2 \to M$ in general position relative to $\mathcal{F}$. Let $\mathcal{F}^* = f^{-1}(\mathcal{F})$ be the foliation of D^2 with center and saddle singularities, transverse to ∂D^2. By the condition that no tangencies lie on a common leaf of $\mathcal{F}$, no two saddle points are connected by a leaf of $\mathcal{F}^*$.

Let Φ_t be the flow, pointing inward along ∂D^2, given by Exercise 7.1.13, v the velocity field of this flow pointing inward along ∂D^2.

Proposition 7.3.2. *If there is a closed, nullhomotopic transversal to $\mathcal{F}$, then there is a leaf L of $\mathcal{F}$ and a loop β on L such that the holonomy around β is germinally trivial on one side of L and germinally nontrivial on the other side.*

Proof. As above, we consider the foliation $\mathcal{F}^* = f^{-1}(\mathcal{F})$ of D^2 and apply the theory of the previous section. By Lemma 7.2.7, let $\mathcal{G}$ be a minimal element of $\widetilde{\mathcal{S}}$ and consider the various possibilities for $\mathcal{G}$.

Case 1. The cycle $\mathcal{G} = \gamma$ is a closed orbit. This is the easy case, from which we get the idea for the other cases. Since $\mathcal{G}$ is minimal in $\widetilde{\mathcal{S}}$, it bounds

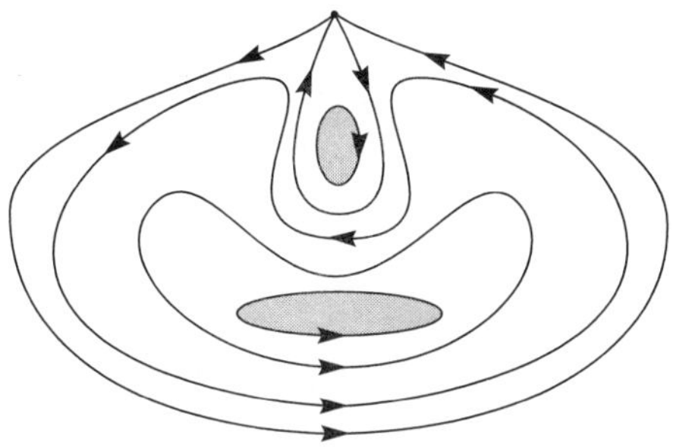

Figure 7.3.1. Case 2

a subdisk D of D^2 in which a neighborhood of $\mathcal{G}$ is a union of closed orbits. Indeed, minimality implies that $\mathcal{F}^*|D$ is made up of closed orbits, singular points, and separatrices that issue from and return to a saddle. Since there are only finitely many singularities, none of which are on $\mathcal{G}$, a neighborhood of $\mathcal{G} = \partial D$ in D must be a union of closed orbits.

Let $[-1,1] \cong J \subset D^2$ be an $\mathcal{F}^*$–transverse arc through $x_0 \in \mathcal{G}$ such that $0 \in J$ represents the point x_0 and $[-1,0)$ represents $J \cap \operatorname{int}(D)$. The holonomy $h_\gamma : (-\epsilon,\epsilon) \to [-1,1]$ fixes $(-\epsilon,0]$ pointwise, but is germinally nontrivial on $[0,\epsilon)$. Indeed, either $\mathcal{G}$ itself is a limit cycle or limit cycles of $\mathcal{F}^*$ intersect $(0,\epsilon)$ in points that cluster at 0. It is clear that the germ of h_γ at 0 is conjugate to the holonomy germ of $\beta = f(\mathcal{G})$ and the assertion is proven in this case.

Case 2. The cycle $\mathcal{G}$ again bounds a subdisk D, foliated as above, but $\mathcal{G}$ is the union of a saddle point p and a separatrix σ_1 that is both stable and unstable for p (see Figure 7.3.1, where the shaded regions contain only saddles, centers, separatrices and compact trajectories). In this case, $\operatorname{int}(D)$ must contain another separatrix σ_2 that is both stable and unstable for p. Otherwise, the other separatrices for p would lie in $D^2 \smallsetminus D$ and prevent $\mathcal{G}$ from being an element of $\widetilde{\mathcal{S}}$. Let $\mathcal{G}'$ denote the graph $\{p\} \cup \sigma_2$. The one–sided holonomy of the figure–eight loop $\mathcal{G} + \mathcal{G}'$ is clearly trivial. Otherwise, $\operatorname{int} D$ would contain a limit cycle, contradicting the minimality of $\mathcal{G}$ in $\widetilde{\mathcal{S}}$. That is, the loop $\beta = f(\mathcal{G} + \mathcal{G}')$ on L has trivial holonomy on one side of L (which we will call the positive side). We must show that the germinal holonomy of β is nontrivial on the other (negative) side of L.

The one–sided holonomy of $\mathcal{G}$ is germinally nontrivial. This is because either $\mathcal{G}$ itself is a limit cycle or limit cycles accumulate on $\mathcal{G}$ from that side.

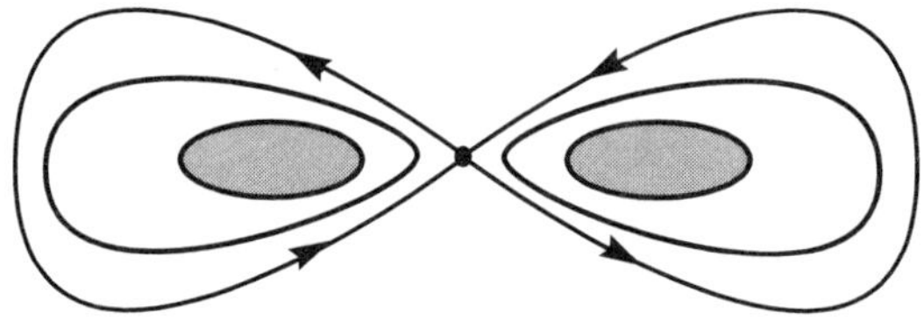

Figure 7.3.2. Case 3

This means that the germinal holonomy of $\sigma = f(\mathcal{G})$ on the negative side of L is nontrivial. On the other hand, the one–sided holonomy of $\mathcal{G}'$ is trivial. Otherwise, the open disk it bounds would contain a limit cycle. This disk lies in int D, contradicting the minimality of $\mathcal{G}$. Therefore, the holonomy of $\sigma' = f(\mathcal{G}')$ is trivial on the negative side of L. It follows that $\beta = \sigma + \sigma'$ has nontrivial holonomy on the negative side of L.

Case 3. The only remaining possibility is that $\mathcal{G}$ is a figure–eight graph as in Figure 7.3.2, where the shaded regions contain only saddles, centers, separatrices and compact trajectories. The one–sided holonomy of $\mathcal{G}$ is germinally nontrivial because, if $\mathcal{G}$ itself is not a limit cycle, limit cycles accumulate on $\mathcal{G}$ from that side. Correspondingly, $\beta = f(\mathcal{G})$ has nontrivial holonomy on, say, the positive side of L. By the minimality of $\mathcal{G}$, the individual loops, each made up of a separatrix and saddle, have trivial one–sided holonomy, from which it follows that β has trivial holonomy on the negative side of L. □

Real analytic foliations have germinal holonomy given by a power series expansion about 0. Such a germ cannot be trivial on one side of 0 without being trivial. We have proven the following.

Corollary 7.3.3. *If the codimension–one foliation $\mathcal{F}$ of M admits a closed transversal that determines an element of finite order in $\pi_1(M)$, then $\mathcal{F}$ cannot be a real analytic foliation.*

Theorem 7.3.1 is an immediate consequence.

Exercise 7.3.4. Let $i_* : \pi_1(L,x) \to \pi_1(M,x)$ be the homomorphism induced by the inclusion of a leaf L and let $\widehat{h} : \pi_1(L,x) \to \mathcal{H}_x(L)$ be the holonomy homomorphism. If $\mathcal{F}$ is real analytic, prove that $\ker i_* \subseteq \ker \widehat{h}$. (The Reeb foliation gives a counterexample for C^∞ foliations.)

Exercise 7.3.5. Does the conclusion of Theorem 7.3.1 hold if we only assume that $H_1(M;\mathbb{Z})$ is finite?

Chapter 8

Generalized Poincaré–Bendixson Theory

We continue to assume that $(M, \mathcal{F})$ is a compact, oriented, transversely–oriented, foliated manifold of codimension one. In order to use the results of Chapter 5, we assume that $\partial_{\pitchfork} M = \emptyset$, but this restriction can usually be removed by doubling along $\partial_{\pitchfork} M$. For most of the results in this chapter, it will be essential that all foliations be of smoothness class C^2. So we fix this assumption, save mention to the contrary.

The Poincaré–Bendixson theorem for flows on a disk (Theorem 7.2.3) works equally well, with minor modifications, for flows on S^2. The theory, which describes the possible limit sets of flow lines, was extended by A. J. Schwartz [**117**] to flows of class C^2 on closed surfaces. Schwartz showed that the limit sets were either the whole manifold, singular points, or limit cycles analogous to those in the classical theory. In a key step, the C^2 hypothesis, together with the mean value theorem, was used to rule out exceptional minimal sets. Subsequently, R. Sacksteder [**115**] extended this result to a profound analysis of holonomy in exceptional minimal sets in compact, C^2–foliated manifolds of codimension one. Other aspects of the Poincaré–Bendixson theorem have been generalized to the codimension–one setting by various authors (*e.g.*, Plante [**106, 105, 104**], Moussu–Pelletier [**95**], Nishimori [**98**], Tsuchiya [**139, 140**], Hector [**61, 64**], Duminy–Hector (unpublished), Cantwell–Conlon [**18**]).

In this chapter, we give a detailed account of that part of the theory that has been called the "architecture" of foliations [**64**] or the "theory of levels" [**24**]. We will describe the limit sets of leaves and their ends by a hierarchy of "local minimal sets". For leaves of finite depth, the analogy with Theorem 7.2.3 will be particularly evident in the way that the ends of such leaves spiral in on leaves at lower depth.

8.1. Foliations and Freshman Calculus

In the theory of foliations of codimension one, there are delicate topological distinctions between degrees of differentiability. The most striking examples distinguish certain C^1 foliations from those of class C^2. This story begins with Denjoy's foliation (Example 4.1.10) and the following lemma of Nancy Kopell [**82**].

Lemma 8.1.1 (Kopell Lemma). *Let* $f \in \operatorname{Diff}^2_+[0,1]$ *and* $h \in \operatorname{Homeo}_+[0,1]$. *Suppose that*

$$f \circ h = h \circ f$$

and that f *is a contraction of* $[0,1)$ *to* 0. *If* $h|[0,1)$ *is a* C^2 *diffeomorphism of* $[0,1)$ *onto itself and fixes some point of* $(0,1)$, *then* $h = \mathrm{id}_{[0,1]}$.

This lemma will be an easy corollary of results to be proven in this section. Denjoy's theorem follows from a theorem of Sacksteder (Theorem 8.2.1) which we will also prove.

Example 8.1.2. Choose a C^∞ diffeomorphism

$$f : [0,1] \to [0,1]$$

such that $f|[0,1)$ is a contraction mapping to 0. Choose $t_0 \in (0,1)$ and set $t_k = f^k(t_0)$, $-\infty < k < \infty$. Let

$$h_0 : [t_1, t_0] \to [t_1, t_0]$$

be a C^∞ diffeomorphism that is a contraction of $[t_1, t_0)$ to t_1 and is C^∞–tangent to the identity at t_1 and t_0. Set

$$h_k = f^k \circ h_0 \circ f^{-k}, \quad \forall\, k \in \mathbb{Z},$$

and define h on $[0,1]$ by

$$h(t) = \begin{cases} h_k(t), & t \in [t_{k+1}, t_k], \\ 0, & t = 0, \\ 1, & t = 1. \end{cases}$$

It is clear that h is a homeomorphism, that $h(t_k) = t_k$, $-\infty < k < \infty$, and that $h \circ f = f \circ h$. Also, each h_k is a C^∞ diffeomorphism, C^∞–tangent to the identity at t_{k+1} and t_k; so $h|(0,1)$ is a C^∞ diffeomorphism. By Lemma 8.1.1, h cannot be a C^2 diffeomorphism at 0. In fact, f^{-1} is a contraction of $(0,1]$

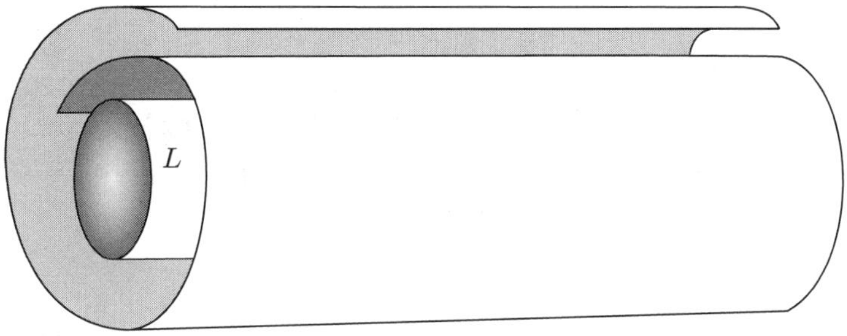

Figure 8.1.1. A planar leaf spirals on a cylindrical leaf.

to 1 and the lemma implies that h fails to be a C^2 diffeomorphism both at 0 and 1. By a construction of Dennis Pixton [**103**], careful choices of the C^∞ contraction f and of h_0 will yield a C^1 diffeomorphism $h : [0,1] \to [0,1]$.

In order to translate this into the context of foliations, note that

$$\pi_1(T^2, x_0) = \mathbb{Z} \oplus \mathbb{Z}$$

and consider the free abelian basis $\{\alpha, \beta\}$ of this group given by the respective generators of the fundamental groups of the factors of $T^2 = S^1 \times S^1$. Define a group homomorphism

$$\varphi : \pi_1(T^2, x_0) \to \text{Homeo}_+[0,1]$$

by setting

$$\begin{aligned} \varphi(\alpha) &= f, \\ \varphi(\beta) &= h. \end{aligned}$$

Since f and h commute, this does extend uniquely to a homomorphism. By suspension, we obtain a $C^{\infty,0}$ foliation $\mathcal{F} = \mathcal{F}(f,h)$ of $T^2 \times [0,1]$, transverse to the interval fibers. There are two compact leaves

$$T_i = T^2 \times \{i\}, \quad i = 0, 1,$$

and there is a cylindrical leaf L, corresponding to the orbit $\{f^k(t_0)\}_{k=-\infty}^{\infty}$, which winds in asymptotically on T_0 and T_1. The remaining leaves are copies of $\mathbb{R}^2$ winding around the cylinder L as indicated in Figure 8.1.1. Any foliation homeomorphic to $\mathcal{F}$ clearly has total holonomy generated by homeomorphisms f and h having the prescribed properties. By Lemma 8.1.1, such a foliation cannot be smooth of class C^2, but Pixton's construction allows us to produce a C^1 foliation homeomorphic to $\mathcal{F}$. This is, perhaps, the simplest example of the way that crude topological features of a foliation can obstruct smoothability.

8.1.A. Basic estimates. Let Γ be an orientation–preserving, C^2 pseudogroup in $\mathbb{R}$, finitely generated by $\Gamma_1 = \{h_1, h_2, \dots, h_m\}$. It will be assumed that the closure of the domain of h_i is a compact interval D_i and that h_i extends to a unique C^2 diffeomorphism

$$\overline{h}_i : D_i \to R_i = \overline{h}_i(D_i) \subset \mathbb{R}, \quad 1 \le i \le m.$$

We may suppose that Γ_1 is a *symmetric* generating set. That is, $h \in \Gamma_1$ implies that $h^{-1} \in \Gamma_1$. Let $\Lambda \subset \mathbb{R}$ be a bounded interval containing the domains D_i (hence, by symmetry, the ranges R_i) of $\overline{h}_i$, $1 \le i \le m$. Remark that all of this holds for the holonomy pseudogroup $\Gamma_{\mathcal{U}}$ for a regular, foliated atlas $\mathcal{U}$ of a compact, transversely–oriented, foliated manifold $(M, \mathcal{F})$ with $\operatorname{codim} \mathcal{F} = 1$. In this case, Γ_0 is the holonomy cocycle $\{\gamma_{ij}\}$.

In what follows, we will use $|J|$ to denote the length of a bounded interval $J \subset \mathbb{R}$. By the above assumptions, we can fix constants $A > 0$, $B > 0$, $\theta > 0$, and $\lambda > 1$ such that

1. $h_i' > A$, $1 \le i \le m$,
2. $|h_i''| \le B$, $1 \le i \le m$,
3. $\theta = B/A$,
4. $\lambda = \exp(6\theta|\Lambda|)$.

If $g = h_{i_n} \circ h_{i_{n-1}} \circ \cdots \circ h_{i_1}$, we set $g_p = h_{i_p} \circ \cdots \circ h_{i_1}$, $p \le n$. Here we allow $p = 0$, in which case $g_0 = \mathrm{id}$. If $u_0 \in \operatorname{dom} g$, set $u_p = g_p(u_0)$, $0 \le p \le n$. The following estimate is fundamental.

Lemma 8.1.3. *If $u_0, v_0 \in \operatorname{dom} g$, then*

$$\frac{g'(u_0)}{g'(v_0)} < \exp\left(\theta \sum_{p=0}^{n-1} |u_p - v_p|\right).$$

Proof. First, we have

$$\begin{aligned}
0 &< \frac{h_i'(u)}{h_i'(v)} \\
&= \left|1 + \frac{h_i'(u) - h_i'(v)}{h_i'(v)}\right| \\
&= \left|1 + \frac{h_i''(\xi)(u-v)}{h_i'(v)}\right| \qquad \text{(by the mean value theorem)} \\
&\leq 1 + \left|\frac{h_i''(\xi)}{h_i'(v)}\right| |u-v| \\
&< 1 + \theta|u-v| \\
&\leq \exp(\theta|u-v|),
\end{aligned}$$

for all $u, v \in \operatorname{dom} h_i$. Now, we use the chain rule and this inequality to obtain

$$\begin{aligned}
\frac{g'(u_0)}{g'(v_0)} &= \frac{h_{i_n}'(u_{n-1})h_{i_{n-1}}'(u_{n-2})\cdots h_{i_1}'(u_0)}{h_{i_n}'(v_{n-1})h_{i_{n-1}}'(v_{n-2})\cdots h_{i_1}'(v_0)} \\
&= \prod_{p=0}^{n-1} \frac{h_{i_{p+1}}'(u_p)}{h_{i_{p+1}}'(v_p)} \\
&< \prod_{p=0}^{n-1} \exp(\theta|u_p - v_p|) \\
&= \exp\left(\theta \sum_{p=0}^{n-1} |u_p - v_p|\right).
\end{aligned}$$

□

Let $C \subset \bigcup_{i=1}^{m} \operatorname{dom} h_i$ be a compact, Γ–invariant set. For example, take $C = X \cap S$ and $\Gamma = \Gamma_{\mathfrak{U}}$, where $X \subset M$ is an exceptional minimal set and the biregular cover $\mathfrak{U} = \{U_i\}_{i=1}^{\ell}$ has been chosen so that X meets no boundary plaque of $\overline{U}_i$, $1 \leq i \leq \ell$. In general, however, we do not assume that C is a Cantor set.

Definition 8.1.4. A *gap* of C is a compact interval $J \subset \Lambda$ with $\partial J = J \cap C$.

Definition 8.1.5. The element $g = h_{i_n} \circ \cdots \circ h_{i_1} \in \Gamma$ is called a *chain* of length n. If $u_0 \in \operatorname{dom} g$, g is a chain at u_0. It is a *simple chain* at u_0 if the points $u_0, u_1, \ldots, u_n$ are distinct. It is a *simple loop* at u_0 if $u_0, \ldots, u_{n-1}$ are distinct and $u_n = u_0$. A chain $g = \ell^{-1} \circ f \circ \ell$ is a *basic loop* at u_0 if ℓ is a simple chain at u_0 and f is a simple loop at $\ell(u_0)$. If the chain is of any one of these three types, it will be said to be a *fundamental chain* at u_0.

Lemma 8.1.6 (Key Lemma). *Let J_0 be a gap of C, K_0 a compact (possibly degenerate) subinterval of Λ such that*

- $J_0 \cap K_0 = \{u_0\}$ *is a single point of* ∂J_0,
- $|K_0|/|J_0| \leq 1/\lambda$.

If $g \in \Gamma$ is a fundamental chain at u_0 and if $J_0 \cup K_0 \subseteq \operatorname{dom} g$, then

1. $|gK_0| < |gJ_0|$,
2. $g'(u)/g'(v) < \lambda$, $\forall u, v \in J_0 \cup K_0$.

Proof. Since $\lambda > 1$, it is clear that $|K_0| < |J_0|$. We set $K_p = g_p K_0$ and $J_p = g_p J_0$ and assume, inductively on p, that $|K_i| < |J_i|$, $0 \leq i \leq p-1$. Then, for arbitrary $u, v \in J_0 \cup K_0$,

$$\begin{aligned}
\frac{g_p'(u)}{g_p'(v)} &< \exp\left(\theta \sum_{i=0}^{p-1} |g_i(u) - g_i(v)|\right) \\
&\leq \exp\left(\theta \sum_{i=0}^{p-1} |K_i \cup J_i|\right) \\
&< \exp\left(\theta \sum_{i=1}^{p-1} 2|J_i|\right) \\
&\leq \exp(6\theta|\Lambda|) \\
&= \lambda,
\end{aligned}$$

since each J_i appears at most three times and, otherwise, $\operatorname{int}(J_j \cap J_k) = \emptyset$, $k \neq j$. By the mean value theorem, there are $u \in K_0$ and $v \in J_0$ such that

$$\frac{|K_p|}{|J_p|} = \frac{g_p'(u)|K_0|}{g_p'(v)|J_0|} < \lambda \frac{1}{\lambda} = 1.$$

By finite induction on p, the assertions follow. □

Definition 8.1.7. If g is a chain of length n and if $J_0 \subseteq \operatorname{dom} g$ is an interval, then g is a chain at J_0. The chain g is a simple chain at J_0 if it is a chain at J_0 and all $J_p = g_p J_0$ have disjoint interiors, $0 \leq p \leq n$. It is a simple loop at J_0 if g_{n-1} is a simple chain at J_0 and $J_n = J_0$. It is a basic loop at J_0 if $g = \ell^{-1} \circ f \circ \ell$, where ℓ is a simple chain at J_0 and f is a simple loop at ℓJ_0. Finally, in all of these cases, g is called a fundamental chain at J_0.

Remark. In Lemma 8.1.6, the fact that J_0 is a gap of C implies that the fundamental chain at $u_0 \in \partial J_0$ is also a fundamental chain of the same type at J_0. It will not be true, in general, that g is also a fundamental chain at K_0. One of the powerful features of Lemma 8.1.6 is the ability to control the dynamics of g on K_0, in spite of the possibility that g is not fundamental there. For the study of exceptional minimal sets, this is critical. In other applications, one is only interested in the gaps (e.g., in Lemma 8.1.1). In those cases, the basic constant λ could be taken to be $\exp(3\theta|\Lambda|)$, the factor 6 in our definition of λ being needed only to keep track of K_p, $1 \le p \le n$.

8.1.B. Local minimal sets. A remarkable property of compact, C^2–foliated manifolds $(M, \mathcal{F})$ is that the foliated manifolds $(U, \mathcal{F}|U)$, $U \in \mathcal{O}(\mathcal{F})$, satisfy the hypotheses of Theorem 4.1.12, hence have minimal sets.

Theorem 8.1.8 (Hector, Cantwell–Conlon). *Let $U \in \mathcal{O}(\mathcal{F})$ be connected and let F be a leaf of $\mathcal{F}|U$. Then $\overline{F} \cap U$ contains a minimal set of $\mathcal{F}|U$.*

In what follows, we keep the hypotheses of the previous subsection. Our goal is to prove Theorem 8.1.8.

Definition 8.1.9. Let $J \subset \Lambda$ be a gap of the compact, Γ–invariant set C. Then the set of restrictions, $g|J$, where g is a basic loop at J, is denoted by $G_0(J)$.

Clearly, $G_0(J) \subset \operatorname{Diff}^2_+(J)$.

Definition 8.1.10. The group $G(J)$ is the subgroup of $\operatorname{Diff}^2_+(J)$ generated by $G_0(J)$.

It is clear that $G_0(J)$ is a symmetric generating set of $G(J)$, but it may well be infinite.

Lemma 8.1.11. *If $g \in G_0(J)$ and if $z \in J$, then $1/\lambda < g'(z) < \lambda$.*

Proof. Since $gJ = J$, the mean value theorem implies the existence of a point $w \in J$ such that $g'(w) = 1$. Then Lemma 8.1.6 implies that $g'(z) < \lambda$ and that $1/g'(z) < \lambda$. □

Theorem 8.1.8 will be deduced as a consequence of the following result.

Lemma 8.1.12. *Let $b \in (x, y) = \operatorname{int} J$. Then there is a point $a \in (x, b)$ such that $a < g(b)$, $\forall g \in G_0(J)$.*

Proof. Choose $a \in (x, b)$ so close to x that $(b-x)/(a-x) > \lambda$. Suppose there is some $f \in G_0(J)$ such that $a \ge f(b)$. We must deduce a contradiction.

Let $g = f^{-1} \in G_0(J)$. Then $g(a) \ge b$. By the mean value theorem, there is a point z between x and a such that

$$g'(z) = \frac{g(a) - g(x)}{a - x} = \frac{g(a) - x}{a - x}.$$

By Lemma 8.1.11 and the fact that $g(a) \ge b$, we obtain

$$\lambda > g'(z) = \frac{g(a) - x}{a - x} \ge \frac{b - x}{a - x} > \lambda,$$

the desired contradiction. □

Example 8.1.13. In Example 4.1.14, we constructed an open, C^∞–foliated 3–manifold with no minimal set. Here, we produce a compact, $C^{\infty,0}$–foliated 3–manifold $(M, \mathcal{F})$ having an open, $\mathcal{F}$–saturated set W such that $\mathcal{F}|W$ has no minimal set. The construction, which is basically the same as in Example 4.1.14, will clearly violate Lemma 8.1.12, hence cannot possibly be smooth of class C^2.

Realize $\mathbb{R}$ as the subarc of $S^1 \subset \mathbb{C}$ complementary to some point of S^1 in the open upper half plane (say $\mathbb{R} = S^1 \smallsetminus \{\sqrt{-1}\}$), coordinatized so that $[-1, 1]$ is the subarc of S^1 from -1 to $+1$ in the counterclockwise direction. We produce a pseudogroup of homeomorphisms on S^1, supported on a compact arc $K \subset \mathbb{R}$ with $[-1, 1] \subset \operatorname{int} K$, violating the conclusion of Lemma 8.1.12. Choose f so that it moves all points in $\operatorname{int} K$ counterclockwise and so that $f(-1) = 1$. Of course, f can be C^∞. Set $J_k = f^k[-1, 1], \forall\, k \in \mathbb{Z}$, noting that the bi–infinite sequence $\{J_k\}_{k\in\mathbb{Z}}$ consists of arcs with disjoint interiors, exactly filling $\operatorname{int} K$. We use the diffeomorphisms $g_k \in \operatorname{Diff}^\infty_c(-1, 1)$ that were constructed in Example 4.1.14 to fashion $g \in \operatorname{Homeo}_+(S^1)$. This homeomorphism will leave each of the arcs J_k invariant and

$$g|J_k = \begin{cases} f^k \circ g_k \circ f^{-k}, & k \geq 0, \\ \mathrm{id}, & k < 0. \end{cases}$$

Also, $g|(S^1 \smallsetminus \operatorname{int} K) = \mathrm{id}$. The pseudogroup Γ on S^1 will be generated by f and g. Since these are supported in $K \subset \mathbb{R}$, the set $\Gamma_0 = \{g, g^{-1}, f, f^{-1}\}$ conforms to our general hypotheses except, perhaps, for C^2 smoothness of g at the upper endpoint of K.

Let C denote the closure of the set $\{f^k(-1)\}_{k=-\infty}^{\infty}$. Remark that this is a compact, Γ–invariant set and that J_k is a gap of C, $-\infty < k < \infty$. The elements f^k are simple chains at $J = J_0$ and the set of nontrivial basic loops at J is

$$G_0(J) = \{g_k^{\pm 1} = f^{-k} \circ g^{\pm 1} \circ f^k\}_{k \geq 0}.$$

By the construction in Example 4.1.14, we have points $\{-a_k\}_{k=0}^\infty \subset \operatorname{int} J$, converging monotonically to -1, with the property that, whenever $b \in (-a_k, a_k)$ and $\ell > k$, the basic loop $g_\ell^{-1} \in G_0(J)$ satisfies $g_\ell^{-1}(b) < -a_k$. By Lemma 8.1.12, g cannot be a C^2 diffeomorphism at the upper endpoint of K.

We can suspend $f, g \in \operatorname{Homeo}_+(S^1)$ to construct a $C^{\infty,0}$ foliation $\mathcal{G}$ of $\Sigma_2 \times S^1$. The interesting part of this foliation is in $\Sigma_2 \times K$. This is a foliated interval bundle with exactly two compact leaves, the components of $\Sigma_2 \times \partial K$. Let $U = \Sigma_2 \times \operatorname{int} K \in \mathcal{O}(\mathcal{G})$ and consider the proper leaf $L \subset U$ corresponding to the orbit $\{f^k(-1)\}_{k\in\mathbb{Z}}$. This leaf is a minimal set of $\mathcal{G}|U$ and is assembled as follows. Cut open the 2–holed torus Σ_2 along the circle in Figure 1.1.8 over which f will be the gluing map. This produces the

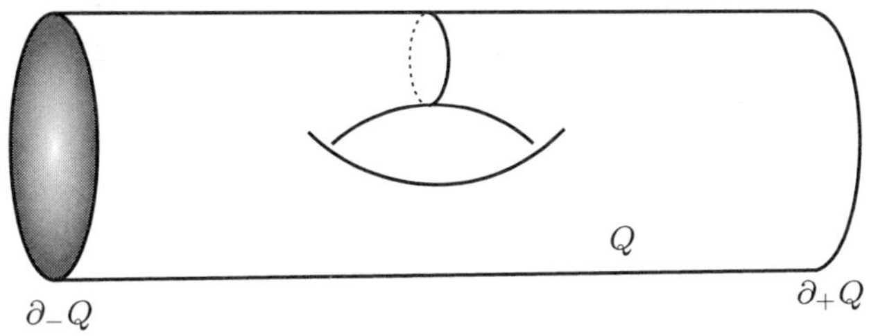

Figure 8.1.2. The basic segment Q for assembling L

surface Q in Figure 8.1.2 in which the other essential circle in Figure 1.1.8 survives. Then glue together a bi–infinite sequence $\{Q_k\}_{k\in\mathbb{Z}}$ of copies of the segment Q so that the right boundary circle $\partial_+ Q_k$ is identified to the left boundary circle $\partial_- Q_{k+1}$, $-\infty < k < \infty$. The resulting leaf L is the 2–ended ladder in Figure 4.1.3. It should be clear that the saturation of $(-1,1)$ is a foliated product $W = U \smallsetminus L \in \mathcal{O}(\mathcal{G})$ and, since g fixes each point $f^k(-1)$, the total holonomy of that foliated product is generated by $\{g_k\}_{k=0}^{\infty}$. Indeed, the foliation $\widehat{\mathcal{G}}$ of $\widehat{W}$ is obtained by cutting $L \times [-1,1]$ along each of the annuli $C_k \times S^1$ (where the C_k's are as in Figure 4.1.3) and regluing via $(z^-, t) \equiv (z^+, g_k(t))$, $k \geq 0$. In other words, $(W, \mathcal{G}|W)$ is homeomorphic to the foliated manifold $(M, \mathcal{F})$ of Example 4.1.14, hence has no minimal set. Lemma 8.1.12 prevents the foliation $\mathcal{G}$ from being C^2–smoothable, the problem being concentrated at the compact leaf through the upper endpoint of K. With some effort, it is possible to C^1–smooth this construction.

We turn to the proof that Lemma 8.1.12 implies Theorem 8.1.8. In the compact, C^2–foliated manifold $(M, \mathcal{F})$, we take $\Gamma = \Gamma_{\mathcal{U}}$ and the symmetric generating system is $\Gamma_1 = \{\gamma_{ij}\}$.

Definition 8.1.14. Let L be a semiproper leaf of $\mathcal{F}$. The holonomy of L is *unbounded* on a given proper side if L is not semistable on that side.

Theorem 8.1.15. *Let $U \in \mathcal{O}(\mathcal{F})$ be connected and let $L \subseteq \partial\widehat{U}$ be a boundary component, $x \in L$. Let $J = [x, y]$ be a nondegenerate, compact subarc of the leaf of $\widehat{\mathcal{L}}$ through x. Suppose that the holonomy of L is unbounded on the side determined by J. Then there is a compact subarc $[a, b] \subset (x, y)$, as near to x as desired, such that every leaf of $\mathcal{F}$ that meets $(x, b]$ also meets (a, b).*

This theorem asserts that new leaves of $\mathcal{F}|U$ do not keep appearing arbitrarily close to L on the side determined by J. The reader should recall that the appearance of new leaves arbitrarily close to an "ideal boundary leaf" was exactly the behavior in Example 4.1.14 that prevented the existence of minimal sets.

The proof of Theorem 8.1.15 requires the following lemma. Let Γ_L denote the pseudogroup on J induced by the holonomy around loops on L based at the point x.

Lemma 8.1.16. *Under the hypotheses of* Theorem 8.1.15, *there is a compact subarc* $[a, b] \subset (x, y)$, *as near to* x *as desired, and a (generally nonsymmetric) generating set* Γ_* *of* $\Gamma_L|[x, b]$, *such that, for every* $f \in \Gamma_*$, $[x, b] = \operatorname{dom} f$ *and* $a < f(b) \leq b$.

Proof of Theorem 8.1.15. By Dippolito's semistability theorem, choose $[a, b]$ so near to x that there are no Γ_L–fixed points in $(x, b]$. Thus, there is some $f \in \Gamma_*$ such that $a < f(b) < b$. It follows that the only way Theorem 8.1.15 can fail is that there exists $z \in (x, a]$ such that

$$z_0 = \sup\{\Gamma_L(z) \cap (x, b]\} \leq a.$$

Let Γ_* be as in Lemma 8.1.16. For each $g \in \Gamma_*$, $g(z_0) < g(b) \leq b$ and so $g(z_0) \leq z_0$. By the absence of Γ_L–fixed points, some $g \in \Gamma_*$ must give $g(z_0) < z_0$. But $g(z_0) < z_0 \leq a < g(b)$; so $g^{-1} \in \Gamma_L$ is defined at z_0 and $z_0 < g^{-1}(z_0) < b$, contradicting the definition of z_0. □

Proof of Lemma 8.1.16. First we give the proof in the case that $\widehat{U}$ is a foliated product. In that case, we can choose the biregular cover so that no border plaque of U is special. Let $X = M \setminus U$, a compact, $\mathcal{F}$–saturated set. Let $C = X \cap T$. As usual, view $T \subset \mathbb{R}$ and choose a bounded interval $\Lambda \subset \mathbb{R}$ that contains T. Note that the gaps of C are really fibers of the foliated product. Choose one of these gaps to be $J = [x, y]$ and choose $b \in (x, y)$ as close to x as desired. Since no border plaque is special, the total holonomy group of the foliated product consists of all words $h = (h_{i_m} \circ \cdots \circ h_{i_1})|J$ such that $hJ = J$ and each $h_k = \gamma_{\alpha_k \beta_k}$. Clearly h can be rewritten as a word in basic loops in these generators by inserting suitable strings of the form

$$h_{i_\ell} \circ \cdots \circ h_{i_1} \circ h_{i_1}^{-1} \circ \cdots \circ h_{i_\ell}^{-1}.$$

Thus, the group $G(J)$ is the total holonomy group of the foliated bundle and Lemma 8.1.12 gives a choice of $a \in (x, b)$ so close to x that $g \in G_0(J)$ implies that $g(b) > a$. Also, either $g(b) \leq b$ or $g^{-1}(b) \leq b$; so Γ_* is formed by choosing the appropriate one of g or g^{-1}, $\forall g \in G_0(J)$.

We turn to the general case. If L is compact, it is easy to produce a finite generating set Γ_* as desired. Therefore, we can assume that L is not compact. Let $L = A \cup B^1 \cup \cdots \cup B^q$, $q \geq 1$, be the decomposition (5.2.1) on page 132. Here, recall that the octopus decomposition of $\widehat{U}$ is chosen so that no plaque meeting B^i is a special border plaque, $1 \le i \le q$. Fix $x \in \operatorname{int} A$, $x_i \in \partial B^i$, and paths $\tau_i \subset A$ from x to x_i, $1 \le i \le q$. Then $J = [x, y]$ is a subarc of the leaf of $\widehat{\mathcal{L}}$ issuing from x, chosen small enough that h_{τ_i} carries it into $J_i = [x_i, y_i]$ in the fiber at x_i of the arm. By the Van Kampen theorem, the generators $\Gamma_* = \Gamma_A \cup \Gamma_1 \cup \cdots \cup \Gamma_q$ will consist of a finite system Γ_A corresponding to generators for $\pi_1(A)$ and (generally) infinite systems Γ_i, corresponding to generators for each $\pi_1(B^i)$. It is easy to select Γ_A. This gives a choice of $[a, b] \subset J$ and we set $[a_i, b_i] = h_{\tau_i}[a, b]$. In defining Γ_i, it will not be necessary to change b_i, but it may be necessary to make a_i smaller. The final choice of a should then be the smallest of $h_{\tau_i}^{-1}(a_i)$, $1 \le i \le q$.

To select Γ_i, work entirely in the arm V^i of $\widehat{U}$ bordered by B^i, replacing J with J_i, etc. Since no border plaques of the arm are special, we have reduced the problem, effectively, to the case of a foliated product, treated above. □

Proof of Theorem 8.1.8. Relative to an octopus decomposition of $\widehat{U}$, let $V_1, \dots, V_r$ be the arms and choose a fiber $[x_j, y_j]$ of $V_j = B_j \times [x_j, y_j]$, $1 \le j \le r$. If $B_j \times \{x_j\}$ lies in a component of $\partial\widehat{U}$ that is not semistable, Theorem 8.1.15 allows us to choose $[a_j, b_j] \subset (x_j, y_j)$, near x_j and such that every leaf meeting $(x_j, b_j]$ actually meets (a_j, b_j). If $B_j \times \{x_j\}$ lies in a semistable boundary leaf L, choose $b_j \in (x_j, y_j)$ such that the leaf through b_j projects homeomorphically onto L along the $\mathcal{L}$–fibers. In this case, choose $a_j = b_j$; so the interval $[a_j, b_j]$ has empty interior. Similarly, choose $[b'_j, a'_j]$ near $B_j \times \{y_j\}$. For $\epsilon > 0$ sufficiently small, the compact subintervals

$$(*) \qquad [a_j, b_j],\ [b'_j, a'_j] \text{ and } [b_j - \epsilon, b'_j + \epsilon] \subset (x_j, y_j), \quad 1 \le j \le r,$$

all lie in U. Remark that a leaf F of $\mathcal{F}|U$ satisfies one of the following:

1. F is trapped in one of the $\mathcal{F}$–saturated collar neighborhoods $L \times [x_j, b_j]$ or $L \times [b'_j, y_j]$ of a semistable boundary leaf L;
2. F meets the interior of one of the intervals $(*)$;
3. there is a compact subset $C \subset K \setminus \partial_\tau K$, independent of the choice of F, such that $F \subset C$.

Indeed, if (1) fails and the leaf F enters an arm V_j, it must either pass arbitrarily near a boundary leaf, hence meet (a_j, b_j) or (b'_j, a'_j), or it meets $(b_j - \epsilon, b'_j + \epsilon)$. If the leaf F does not enter any arm, then it cannot meet the $\widehat{\mathcal{F}}$–saturation of any of the intervals $[x_j, b_j)$ or $(b'_j, y_j]$. The intersections of these saturations with K unite to form an open collar neighborhood of $\partial_\tau K$ and $F \subset K$ lies in the complement C of this collar. By the compactness of C, find a finite set of compact, transverse arcs $J_1, \dots, J_p$ in $\operatorname{int} K$ such that all leaves F of type (3) meet the interior of at least one J_k.

In case (1), assume for definiteness that $F \subset L \times [x_j, b_j]$. Since L is semistable, Theorem 5.3.4 implies that $\overline{F}$ is contained in an $\mathcal{F}$–saturated set $L \times [c_j, b_j]$, where $x_j < c_j < b_j$. Then $\overline{F} \cap [c_j, b_j]$ is a compact, Γ–invariant set and the standard application of Zorn's Lemma implies the existence of a minimal, Γ–invariant subset X. The saturation of X is the desired minimal set in $\overline{F} \cap U$. In the remaining cases, no leaf of $\overline{F} \cap U$ satisfies (1); hence every such leaf meets the interior of at least one of the intervals (∗) or of some J_k. It follows that there is a biregular cover of U such that every leaf of $\overline{F} \cap U$ encounters one of finitely many charts U_i in that cover. By Theorem 4.1.12, $\overline{F} \cap U$ contains a minimal set of $\mathcal{F}|U$. □

Definition 8.1.17. A subset $X \subseteq M$ is called a *local minimal set* of $\mathcal{F}$ if there is $U \in \mathcal{O}(\mathcal{F})$ such that X is a minimal set of $\mathcal{F}|U$.

Example 8.1.18. In Example 4.1.6, the foliation $\mathcal{F}$ of $\Sigma_2 \times S^1$ has one minimal set X. This is exceptional and its complement $W \in \mathcal{O}(\mathcal{F})$ is a foliated product. Each leaf L of $\mathcal{F}|W$ clusters in $\widehat{W}$ only on the components L_0 and L_1 of $\partial\widehat{W}$ (step 5 of Exercise 4.1.8). Thus, $\overline{L} \cap W = L$; so L is a minimal set of $\mathcal{F}|W$. Every leaf of $\mathcal{F}$ not in X is a local minimal set.

Let X be a local minimal set. Then X is either a single leaf, an element $U \in \mathcal{O}(\mathcal{F})$, or $\overline{X}$ is transversely a Cantor set. In this last case, X is called an exceptional local minimal set.

Proposition 8.1.19. *A single leaf of $\mathcal{F}$ is a local minimal set if and only if it is a proper leaf.*

Exercise 8.1.20. Prove Proposition 8.1.19 as follows.

1. If $L \subset M$ is a proper leaf, show that $\overline{L} \smallsetminus L$ is compact. Conclude that L is a minimal set in $M \smallsetminus (\overline{L} \smallsetminus L)$.
2. Let $U \in \mathcal{O}(\mathcal{F})$ and suppose that L is a leaf of $\mathcal{F}|U$ that is closed in U. If L is not proper, show that there is a compact subarc J of a leaf of $\mathcal{L}|U$ such that $L \cap J$ is a nonempty, perfect set. Show that this contradicts the fact that L is 2nd countable.

Exercise 8.1.21. By suspending two C^∞ diffeomorphisms f and g of the compact interval $J = [-1, 1]$ over the 2–holed torus Σ, produce a C^∞–foliated manifold in which there are leaves that do not lie in any local minimal set. Proceed as follows.

1. Let $f \in \mathrm{Diff}^\infty_+ J$ be chosen so that $f(x) < x$, $-1 < x < 1$, and set $I = [0, f(0)]$. Let $g_0 \in \mathrm{Diff}^\infty_+ J$ be supported on I with
$$g_0(x) < x, \quad f(0) < x < 0.$$
Thus, I is a fundamental domain of f. Choose $x_0 \in \operatorname{int} I$, and let $I_0 = [g_0(x_0), x_0]$, a fundamental domain of g_0.

2. Inductively, construct diffeomorphisms $g_1, \dots, g_n$ of I, such that the diffeomorphism $f^k \circ g_k \circ f^{-k}$ is supported on the fundamental domain I_{k-1} of $f^{k-1} \circ g_{k-1} \circ f^{-(k-1)}$ and is strictly decreasing on the interior of I_{k-1}, $1 \le k \le n$.
3. Show how to choose these diffeomorphisms such that the sequence $\{g_k\}_{k=0}^{\infty}$ assembles to form a C^∞ diffeomorphism g of J and such that $\bigcap_{k=0}^{\infty} I_k$ is a singleton.
4. Show that the foliation $\mathcal{F}(f,g)$ of $M = \Sigma \times J$, formed as in Example 1.1.14, has exactly a countable infinity of proper leaves and that the union of these leaves is dense in M.
5. Show that each of the uncountably many nonproper leaves is dense in M and does not lie in any local minimal set.
6. Modify the construction of g so as to produce a smooth foliation $\mathcal{F}(f,g)$ in which the union of the proper leaves is an open dense subset, but each of the remaining leaves again fails to lie in any local minimal set. In this case, show that there are still uncountably many of these nonproper leaves, each being nowhere dense but not semiproper.

8.1.C. The generalized Kopell lemma. We return to the abstract pseudogroup setting. Let J be a gap of C and let

$$\mathcal{J} = \{fJ \mid f \text{ is a simple chain at } J\}.$$

Let $\mathcal{J}_* \subseteq \mathcal{J}$.

Definition 8.1.22. The size $|\mathcal{J}_*|$ of $\mathcal{J}_*$ is the sum of the lengths of the elements of $\mathcal{J}_*$.

Remark that $|\mathcal{J}_*| < \infty$. Indeed, the elements of $\mathcal{J}_*$ are subintervals of Λ with disjoint interiors; so $|\mathcal{J}_*| \le |\Lambda| < \infty$.

Let $f_* = h_{i_1} \circ \cdots \circ h_{i_\ell}$ be a simple chain at J such that $f_*J = J_* \in \mathcal{J}_*$, but $J_p \notin \mathcal{J}_*$, $0 \le p < \ell$. Consider all basic loops $g = h_{j_1} \circ \cdots \circ h_{j_m}$ at J_* such that $g_p J_* \in \mathcal{J}_*$, $0 \le p \le m$. Then $\gamma = f_*^{-1} \circ g \circ f_* \in G = G(J)$ is a basic loop at J and the set of all such chains γ is a symmetric subset (*not* generally a subgroup) $G_* \subseteq G$.

Lemma 8.1.23. *If $0 < r < 1 < s$ and if the size $|\mathcal{J}_*|$ is sufficiently small, then $r < g' < s$, uniformly on J_*, for all $g \in f_* G_* f_*^{-1}$.*

Proof. Set $\lambda_* = \exp(3\theta|\mathcal{J}_*|)$. (For the use of the factor 3 instead of 6, *cf.* the Remark on page 170.) By choosing $|\mathcal{J}_*|$ sufficiently small, we arrange that $\lambda_* < \min\{s, 1/r\}$. Letting $|\mathcal{J}_*|$ play the role of $|\Lambda|$ in the earlier estimates, we obtain, as in Lemma 8.1.11, that

$$r \le \frac{1}{\lambda_*} < g' < \lambda_* \le s,$$

uniformly on J_*. □

Lemma 8.1.24 (Generalized Kopell lemma). *Suppose that $J = [x, y]$ and that $x < a < b < y$. If $|\mathfrak{J}_*|$ is sufficiently small, then $a < \gamma(b)$, $\forall \gamma \in G_*$.*

Proof. Choose $s > 1$ so that

$$s - 1 < \frac{b-a}{\lambda(a-x)}.$$

For $\mathfrak{J}_*$ as in the previous lemma, $g'(t) < s$, for all $t \in J_*$ and for all $g \in f_* G_* f_*^{-1}$. Suppose there is $\gamma \in G_*$ with $a \geq \gamma(b)$. Set $g = f_* \circ \gamma^{-1} \circ f_*^{-1}$, $a_* = f_*(a)$, $b_* = f_*(b)$; so $g(a_*) \geq b_*$. We will show that there is a point $t \in J_*$ such that $g'(t) > s$, a contradiction.

Indeed, set $x_* = f_*(x)$ and obtain

$$\begin{aligned} \frac{(g(a_*) - a_*)}{a_* - x_*} &\geq \frac{b_* - a_*}{a_* - x_*} \\ &= \frac{f_*'(\xi)(b-a)}{f_*'(\zeta)(a-x)} \\ &> \frac{b-a}{\lambda(a-x)} \\ &> s - 1. \end{aligned}$$

Here, the mean value theorem provides $\xi \in (a, b) \subset J$ and $\zeta \in (x, a) \subset J$, while Lemma 8.1.6 implies the second last inequality. Thus,

$$\frac{g(a_*) - x_*}{a_* - x_*} > s.$$

By the mean value theorem, there is $t \in (x_*, a_*) \subset J_*$ such that $g'(t) > s$. □

The standard Kopell lemma is a corollary, as follows.

Proof of Lemma 8.1.1. Suppose that $z_0 \in (0,1)$ is fixed by h and set $z_k = f^k(z_0)$, $\forall k \in \mathbb{Z}$. Commutativity implies that h fixes every z_k. Let Γ be the pseudogroup on $[0,1]$ generated by the symmetric set $\{f, h, f^{-1}, h^{-1}\}$ and let

$$C = \{1, \ldots, z_{-k}, z_{-k+1}, \ldots, z_{-1}, z_0, z_1, \ldots, z_k, z_{k+1}, \ldots, 0\},$$

clearly a closed, Γ–invariant set. Let $\mathfrak{J} = \{J_k = [z_{k+1}, z_k]\}_{k \in \mathbb{Z}}$, the set of all gaps of C. Remark that $h_k = h|J_k$ is a simple loop at J_k and that $f^q : J_k \to J_{k+q}$ is a simple chain at J_k. Let $b \in (z_1, z_0)$ and suppose that $h_0(b) \neq b$. Without loss of generality, suppose that $h_0(b) < b$ and, in Theorem 8.1.24, take $a = h_0(b)$ and $J = J_0$. Let $N \geq 0$ be an integer and set $\mathcal{J}_* = \{J_k\}_{|k|\geq N}$. We can make $|\mathcal{J}_*|$ as small as desired by taking N large enough. In Theorem 8.1.24 let $J_* = J_N$ and take

$$\begin{aligned} f_* &= f^N : J_0 \to J_N, \\ \gamma &= f^{-N-1} \circ h_{N+1} \circ f^{N+1} : J_0 \to J_0, \\ g &= f^{-1} \circ h_{N+1} \circ f : J_N \to J_N. \end{aligned}$$

Then,

$$a < f_*^{-1} \circ g \circ f_*(b) = \gamma(b).$$

Since $f \circ h = h \circ f$, this reduces to the contradiction that

$$a < h_0(b) = a.$$

By this contradiction, h fixes each point of J_0. It fixes each point of

$$(0,1) = \bigcup_{k\in\mathbb{Z}} J_k$$

by the commutativity of f and h. □

8.1.D. Junctures. We take up the geometric interpretation of the generalized Kopell lemma, namely that "in C^2 foliations, all junctures are compact".

Let $(M, \mathcal{F})$ be a compact, C^2–foliated manifold as usual, and let $W \in \mathcal{O}(\mathcal{F})$. Suppose that $L \subset \partial\widehat{W}$ is a leaf, fix a subarc $[x, y)$ of a leaf of $\widehat{\mathcal{L}}$ with $x \in L$, and let a leaf $F \subset W$ be a minimal set of $\mathcal{F}|W$ such that $F \cap [x, y)$ clusters at x. Since F is minimal in W, this set clusters *only* at x; so we write $F \cap [x, y) = \{x_k\}_{k=1}^{\infty}$, a sequence converging monotonically to x. By the Dippolito semistability theorem, x_k is not a fixed point for the holonomy of L, provided k is sufficiently large. Thus, we assume that no x_k is such a fixed point, $k \geq 1$. Thus, for any particular k there is a loop σ on L, based at x, such that $h_\sigma(x_k) \neq x_k$ (and, in particular, is defined). Using the reverse loop, if necessary, we can assume that $h_\sigma(x_k) = x_{k+p}$, for some integer $p > 0$. We can assume that p is the smallest such integer that occurs for any value of $k \geq 1$ and any loop σ on L, based at x.

Lemma 8.1.25. *For every loop τ on L, based at x, there is an integer $\nu(\tau)$ such that $h_\tau(x_m) = x_{m+\nu(\tau)p}$, provided that $m \geq 1$ is so large that $h_\tau(x_m)$ is defined. Finally, the assignment $\tau \mapsto \nu(\tau)$ passes to a well-defined group homomorphism*

$$\nu : \pi_1(L, x) \to \mathbb{Z},$$

defining a nondivisible cohomology class $\nu \in H^1(L; \mathbb{Z})$.

The proof of this lemma is left to the reader. We will say that the leaf F *spirals* on L and that the cohomology class ν is the *juncture* of the spiral. In fact, since it is possible that $p > 1$, there will generally be p distinct ends

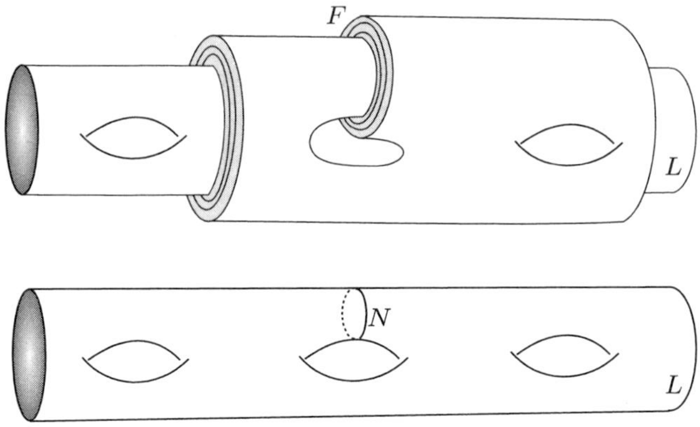

Figure 8.1.3. The leaf F spirals on L

of F that spiral on L and ν is the "juncture" of each these spiraling ends. To clarify this picture and motivate the terminology, we need the following theorem, which will be proven via the generalized Kopell lemma.

Theorem 8.1.26. *The cohomology class ν is compactly supported.*

Before proving this, we consider an example.

Example 8.1.27. Refer to Example 8.1.13. Use the same diffeomorphism f and, for g, use a C^∞ diffeomorphism of S^1, supported on $[-1,1]$ and contracting $[-1,1)$ to -1. This gives a C^∞ foliation $\mathcal{F}$ of $\Sigma_2 \times S^1$. As before, the interesting part of the foliation is in $\Sigma_2 \times K$, where K is a compact interval containing $[-1,1]$ in its interior. This part of the foliation has two compact leaves, the components of $\Sigma_2 \times \partial K$. The open, saturated set

$$U = \Sigma_2 \times \operatorname{int} K$$

contains one minimal set, the proper leaf L that is homeomorphic to the 2–ended ladder (Figure 4.1.3). This is the leaf corresponding to the orbit $\{f^k(-1)\}_{k\in\mathbb{Z}}$. Let $W = U \smallsetminus L \in \mathcal{O}(\mathcal{F})$, also connected. Each leaf $F \subset W$ is a minimal set of $\mathcal{F}|W$. Indeed,

$$\widehat{W} = L \times [-1,1]$$

and the foliation $\widehat{\mathcal{F}}$ is the suspension of g, defined by cutting $L \times [-1,1]$ along the annulus $N \times [-1,1]$, where N is the essential circle in Figure 8.1.3, and regluing by $(z^-,t) \equiv (z^+, g(t))$. Part of the foliation of $\widehat{W}$ near L is also depicted in Figure 8.1.3 and the general leaf F of $\mathcal{F}|W$ is depicted in Figure 8.1.4.

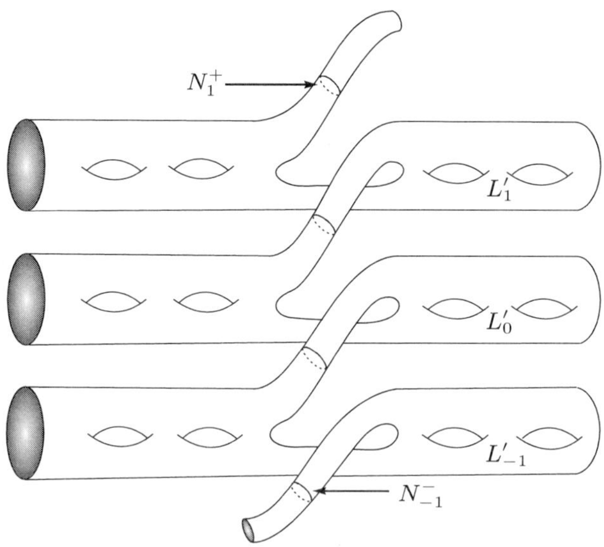

Figure 8.1.4. F is an "infinite repetition" of L', joined along $N^\pm$

It should be clear that, if $F \subset W$ is a leaf and, if

$$F \cap (\{x\} \times [-1,1]) = \{x_k\}_{k\in\mathbb{Z}}$$

is suitably indexed, then $\tau \in \pi_1(L,x)$ has

$$h_\tau(x_k) = x_{k+\nu(\tau)}, \quad \forall\, k \in \mathbb{Z},$$

if and only if the homological intersection number $[\tau]\cdot[N] = \nu(\tau)$. That is, the submanifold N represents the Poincaré dual $[N] \in H_1(L;\mathbb{Z})$ of the cohomology class $\nu \in H^1(L;\mathbb{Z})$ (and, in this case, there are just $p = 1$ spiraling ends of F). Since N is compact, there is a compact neighborhood A of N in L such that, whenever σ is a loop on L that does not meet the interior of A, then $\nu(\sigma) = 0$. This means that, in fact, $\nu \in H^1_c(L;\mathbb{Z})$, the cohomology of L with compact supports. (For noncompact manifolds such as L, Poincaré duality equates ordinary homology with compactly supported cohomology.) The geometric meaning of the words "spiral" and "juncture" are supposed to be clear in this example. Figure 8.1.3 is supposed to illustrate the "spiral" phenomenon and Figure 8.1.4 the "juncture" notion. Indeed, F is formed, topologically, by cutting L along N, producing a manifold L' with boundary two copies $N^\pm$ of N, and then "joining" successively a bi–infinite sequence of copies L'_k of L', attaching L'_k to L'_{k+1} by identifying N_k^+ to N_{k+1}^-. We call F an "infinite repetition" of L', successive copies of L' being joined along the junctures $N^\pm$.

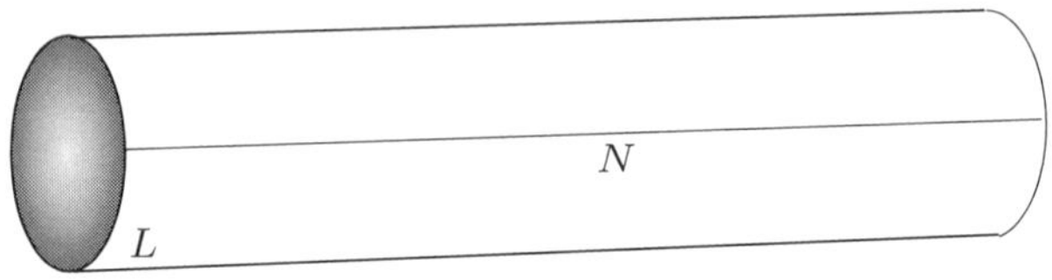

Figure 8.1.5. A noncompact juncture N

Example 8.1.28. Noncompact junctures occur in C^1 foliations. For instance, the juncture N on the cylindrical leaf L in Figure 8.1.1, corresponding to the planar leaf spiraling on L, is the infinite line pictured in Figure 8.1.5. In Figure 4.1.3, the infinite sequence of essential circles C_k can serve as a noncompact juncture as follows. In Example 8.1.27, replace g by letting g_0 be supported on $[-1,1]$, contracting $[-1,1)$ to -1, and defining g_k on $f^k[-1,1]$ to be $f^k \circ g_0 \circ f^{-k}$. Then define $g \in \text{Homeo}^+(K)$ by

$$g(z) = \begin{cases} g_k(z), & z \in f^k[-1,1], \\ z, & z \in \partial K. \end{cases}$$

These noncompact submanifolds $N \subset L$ still have well–defined intersection numbers $[\sigma] \cdot [N]$ with closed loops σ, depending only on $[\sigma] \in H_1(L;\mathbb{Z})$. In fact, $[N]$ can be viewed as a class in the singular homology $H_*^\infty(L;\mathbb{Z})$ defined by chains that are infinite linear combinations of singular simplices. For the noncompact 2–manifold L, Poincaré duality gives $H^1(L;\mathbb{Z}) = H_1^\infty(L;\mathbb{Z})$.

Proof of Theorem 8.1.26. As usual, we fix a decomposition

$$L = A \cup B^1 \cup \cdots \cup B^q,$$

$q \geq 1$, as in (5.2.1) on page 132. As in the proof of Lemma 8.1.16, the family $\mathfrak{J}$ corresponds to interval fibers in an arm $B_i \times J$. Working in this arm, we suppose that $J = [x,y]$ is the fiber containing $F \cap [x,y] = \{x_k\}_{k=1}^\infty$. By Lemma 8.1.24, let $|\mathfrak{J}_*|$ be so small that $x_2 < g(x_1)$, $\forall\, g \in G_*$. It is clear that this implies $g(x_1) = x_1$, hence $g(x_k) = x_k$, $\forall\, k \geq 1$, $\forall\, g \in G_*$. Doing this for each arm, we see that, by choosing the compact nucleus A sufficiently large, the generating set $\Gamma_* = \Gamma_A \cup \Gamma_1 \cup \cdots \cup \Gamma_q$ of Γ_L, produced by the Van Kampen theorem, has the property that $h_\sigma(x_k) = x_k$, $\forall\, h_\sigma \in \Gamma_i$, $1 \leq i \leq q$. It follows that loops τ on L not meeting $\operatorname{int} A$ have $\nu(\tau) = 0$, proving that ν is compactly supported. □

Remark. In order to see that Example 8.1.27 gives a reasonably general picture, we need to know that 1–dimensional cohomology classes $\nu \in H^1(L;\mathbb{Z})$ correspond perfectly to homotopy classes of maps $f : L \to S^1$. Indeed, such a class induces a homomorphism

$$\pi_1(L) \to \pi_1(S^1) = \mathbb{Z}$$

and it can be proven, conversely, that every homomorphism

$$\pi_1(L) \to \pi_1(S^1) = \mathbb{Z}$$

is induced by a continuous map $f : L \to S^1$. By a small homotopy, such a map f is smooth. By Sard's theorem, f has a regular value $z \in S^1$; so $f^{-1}(z)$ is a submanifold $N \subset L$ of codimension one. Compact cohomology classes are represented by *proper* maps f. That is, $N = f^{-1}(z)$ is a compact submanifold. Assume, now, that the leaf L is *oriented* (since $\mathfrak{F}$ is transversely oriented, L will be oriented if M is). When the above remarks are applied to our juncture $\nu \in H^1_c(L;\mathbb{Z})$, it follows that $[\tau] \cdot [N] = \nu(\tau)$, where "$\cdot$" denotes the homological intersection product. We will see later that, near L, the leaf F falls into a finite number of spirals, each of which is an infinite repetition of the manifold L' obtained by cutting L along N.

8.2. Sacksteder's Theorem

We discuss a crucial result on pseudogroups due to R. Sacksteder [**115**]. This will complete the set of technical tools that will be needed in subsequent sections.

Let Γ be a pseudogroup on $\mathbb{R}$ with all the properties in the preceding section. We carry over notation from that section also. Let C be a Γ–invariant Cantor set. We assume that $C \cap \operatorname{int} D_i$ is compact, $\forall h_i \in \Gamma_1$, and that C is covered by the open intervals D_i. We do *not* assume that C is Γ–minimal. Thus, our results will apply to X, an exceptional local minimal set. Indeed, we know that $\overline{X}$ is transversely a Cantor set. Hence we choose the biregular cover $\{U_i\}_{i=1}^r$ so that the top and bottom $\mathfrak{F}$–plaques of $\overline{U}_i$ miss $\overline{X}$, $1 \le i \le r$.

Theorem 8.2.1 (Sacksteder). *Let C be a Γ–invariant Cantor set and let $J = [a,b]$ be a gap of C such that $\Gamma(b)$ clusters at b (from above). Then, $\exists\, x_0 \in C$, as close to b as desired, and $\exists\, \gamma \in \Gamma$ such that $\gamma(x_0) = x_0$ and $\gamma'(x_0) < 1$.*

Remarks. Of course, the symmetric assertion holds with b replaced by a and $\Gamma(a)$ clustering at a from below.

It is clear that γ is a 2–sided contraction of a neighborhood of x_0 to x_0. When a contraction has the property that $\gamma'(x_0) < 1$, it is called a *linear* contraction. The linear contraction γ in Theorem 8.2.1 is often referred to as a "Sacksteder contraction" and x_0 is called a "Sacksteder point".

If X is an exceptional local minimal set, but not a minimal set, let $U \in \mathcal{O}(\mathcal{F})$ be such that $X \subset U$ is minimal for $\mathcal{F}|U$. The gap $[a, b]$ should be chosen so that the leaf through b lies in U. Then the hypothesis that $\Gamma(b)$ clusters at b is automatically satisfied because of minimality in U.

Before proving Theorem 8.2.1, we point out applications.

Corollary 8.2.2. *Let $Y \subseteq M$ be a local minimal set with more than one leaf. If each leaf in Y has trivial linear holonomy group, then $Y \in \mathcal{O}(\mathcal{F})$.*

We emphasize that this corollary, like Theorem 8.2.1 itself, is only true in the presence of smoothness of class at least C^2.

Corollary 8.2.3. *The Denjoy foliation of T^2 is not homeomorphic to a foliation of class C^2.*

Definition 8.2.4. The leaf L of $\mathcal{F}$ is *resilient* if there is a transverse arc $J = [x, y)$, $x \in L$, and a loop σ on L, based at x, such that $h_\sigma : [x, y) \to [x, y)$ is a contraction to x, and $L \cap (x, y) \neq \emptyset$.

Thus, a resilient leaf "captures" itself by a holonomy contraction. The term "resilient" is an unfortunate translation of the French word "ressort", meaning "spring–like". By a deep theorem of Duminy [**38**], which we will prove in the second volume, $\mathrm{gv}(\mathcal{F}) = 0$ for foliations without resilient leaves.

Corollary 8.2.5. *An exceptional local minimal set X contains a resilient leaf.*

Proof. As remarked above, the Sacksteder contraction γ, with $\gamma(x_0) = x_0$ and $\gamma'(x_0) < 1$, is a 2–sided holonomy contraction on the leaf $L_0 \subset \overline{X}$ through x_0. Since this point x_0 can be chosen arbitrarily close to any leaf bordering a gap of C, it can be chosen in the set $U \in \mathcal{O}(\mathcal{F})$ such that $X \subset U$ is $\mathcal{F}|U$–minimal. That is, the leaf L_0 is actually in X itself. Therefore, this leaf approaches itself on at least one side, hence is captured by its own 2–sided holonomy contraction γ. □

Remark. Another unpublished theorem of Duminy asserts that a semiproper leaf L in an exceptional minimal set has a Cantor set of ends. A key step in the proof shows that L has a holonomy contraction (not necessarily linear) on its nonproper side. Thus, L is also resilient. This last assertion extends to semiproper leaves of exceptional local minimal sets.

Exercise 8.2.6. Let $U \in \mathcal{O}(\mathcal{F})$ and suppose that U is a local minimal set. If some leaf F of $\mathcal{F}|U$ has nontrivial germinal holonomy, prove that some (possibly different) leaf L of $\mathcal{F}|U$ is resilient.

Exercise 8.2.7. Let $U \in \mathcal{O}(\mathcal{F})$ be a local minimal set and L a resilient leaf of $\mathcal{F}|U$. Show that some resilient leaf L' of $\mathcal{F}|U$ has a holonomy element that is a *linear* contraction. (Hint. Although U is not exceptional, Sacksteder's theorem can be applied to a suitable subpseudogroup $\Gamma' \subset \Gamma$ of the holonomy pseudogroup of $\mathcal{F}|U$. The generators of Γ' should be a carefully chosen pair of elements $f_1, f_2 \in \Gamma$.)

We turn to the proof of Theorem 8.2.1. Let $J_0 = [u_0, v_0]$ be a gap of C. If $g = h_{i_m} \circ \cdots \circ h_{i_1}$ is a chain (a reduced word in elements of Γ_1), set $|g| = m$. Denote by $\mathcal{C}_{u_0}$ the set of chains at u_0. For each $u \in \Gamma(u_0)$, set

$$n(u) = \min\{|g| \mid g \in \mathcal{C}_{u_0},\ g(u_0) = u\}.$$

Definition 8.2.8. If $g \in \mathcal{C}_{u_0}$, $g(u_0) = u$, and $|g| = n(u)$, then g is called a *shortcut* from u_0 to u.

Remark that a shortcut is necessarily a simple chain at u_0.

Lemma 8.2.9. *There exists a set $\mathcal{S}_{u_0}$ of shortcuts h_u from u_0 to u, one for each $u \in \Gamma(u_0)$, such that $g \in \mathcal{S}_{u_0}$ implies that $g_p \in \mathcal{S}_{u_0}$, $0 \le p \le |g|$.*

Proof. We proceed by induction on the lengths $|g|$. If $|g| = 0$, then $u_0 = u$ and there is a unique choice of shortcut, $g = \text{id}$. If the elements of $\mathcal{S}_{u_0}$ of lengths $\le k$ have been chosen, and if $u \in \Gamma(u_0)$, $n(u) = k+1$, choose $u_k \in \Gamma(u_0)$ such that $n(u_k) = k$ and $u = h_{i_{k+1}}(u_k)$, some $h_{i_{k+1}} \in \Gamma_1$. Let $g_k \in \mathcal{C}_{u_0}$ be the shortcut such that $g_k(u_0) = u_k$. Define $g_{k+1} = h_{i_{k+1}} \circ g_k$, a chain of length $k+1$ from u_0 to u (hence a shortcut), and include it in $\mathcal{S}_{u_0}$. □

There are at most two components of $\Lambda \smallsetminus C$ that are not gaps of C. Here, Λ is the bounded interval containing all the domains D_i. Thus, if we set J_g equal to the component of $\Lambda \smallsetminus C$ having $g(u_0)$ as an endpoint, $\forall g \in \mathcal{S}_{u_0}$, we obtain a gap of C for all but at most one $g \in \mathcal{S}_{u_0}$. The proof of the following uses the full force of Lemma 8.1.6.

Lemma 8.2.10. *There exists an arbitrarily small, nondegenerate interval K_0, with $J_0 \cap K_0 = \{u_0\}$, and a constant $\mu > 0$ such that, $\forall g \in \mathcal{S}_{u_0}$,*

1. $K_0 \subset \operatorname{dom} g$,
2. $|gK_0| < |J_g|$,
3. $g'(x)/g'(y) < \mu$, $\forall x, y \in K_0$.

Proof. Since $C \cap \operatorname{int} D_i$ is compact, $\forall h_i \in \Gamma_1$, we choose $\epsilon > 0$ such that, whenever $x \in C \cap \operatorname{int} D_i$, $h_i \in \Gamma_1$, the entire ϵ–neighborhood of x in $\mathbb{R}$ lies in $\operatorname{int} D_i$. Now, $\sum_{g \in \mathcal{S}_{u_0}} |J_g| \le |\Lambda| < \infty$. Therefore, we can choose $N \ge 1$

such that $g \in \mathcal{S}_{u_0}$ and $|g| \geq N$ implies that $|J_g| < \epsilon$, hence that $J_g \subset \operatorname{int} D_i$, whenever $g(u_0) \in \operatorname{int} D_i$, $\forall h_i \in \Gamma_1$.

Let $\gamma_1, \dots, \gamma_r \in \mathcal{S}_{u_0}$ be the elements with $|\gamma_i| \leq N$, so numbered that $\gamma_1, \dots, \gamma_p$ are those with $|\gamma_i| = N$. Let $\mathcal{S}_N = \{g \in \mathcal{S}_{u_0} \mid |g| \geq N\}$, remarking that every $g \in \mathcal{S}_N$ is of the form $g = \gamma \circ \gamma_i$, $1 \leq i \leq p$, where γ is a simple chain at $\gamma_i(u_0)$.

Choose K_0 so small that $K_0 \subset \operatorname{int} \operatorname{dom} \gamma_i$ and

$$|\gamma_i K_0|/|J_{\gamma_i}| \leq 1/\lambda, \quad 1 \leq i \leq r.$$

We will prove that, for $1 \leq i \leq p$, $\gamma_i K_0 \subset \operatorname{int} \operatorname{dom} \gamma$, for each simple chain γ at $\gamma_i(u_0)$ such that $g = \gamma \circ \gamma_i \in \mathcal{S}_{u_0}$. Thus, $K_0 \subset \operatorname{int} \operatorname{dom} g$, $\forall g \in \mathcal{S}_{u_0}$. We proceed by induction on $|\gamma|$ in $g = \gamma \circ \gamma_i$, $1 \leq i \leq p$.

If $|\gamma| = 0$, the assertion is trivial. Assume that the assertion is true for $|\gamma| \leq k$ and let

$$\gamma = h_{i_{k+1}} \circ h_{i_k} \circ \cdots \circ h_{i_1} = h_{i_{k+1}} \circ \widetilde{\gamma}, \quad |\widetilde{\gamma}| = k.$$

Therefore, $\widetilde{\gamma}\gamma_i \in \mathcal{S}_{u_0}$ and $\widetilde{\gamma}\gamma_i K_0$ is defined. Furthermore, since $|J_{\gamma_i}| < \epsilon$, we see that $h_{i_1} J_{\gamma_i}$ is defined *and* $h_{i_1} \circ \gamma_i \in \mathcal{S}_N$, hence $|h_{i_1} J_{\gamma_i}| < \epsilon$. Therefore, $h_{i_2} h_{i_1} J_{\gamma_i}$ is defined and of length $< \epsilon$, etc. Finite repetition of this argument gives, finally, that $\widetilde{\gamma} J_{\gamma_i}$ is defined and $|\widetilde{\gamma} J_{\gamma_i}| < \epsilon$. By Lemma 8.1.6,

$$\frac{|\widetilde{\gamma}\gamma_i K_0|}{|\widetilde{\gamma} J_{\gamma_i}|} < 1;$$

so $|\widetilde{\gamma}\gamma_i K_0| < \epsilon$ and $\widetilde{\gamma}\gamma_i K_0 \subset \operatorname{int} D_{i_{k+1}}$. That is, $K_0 \subset \operatorname{int} \operatorname{dom} g$ as asserted.

Another application of Lemma 8.1.6 gives

$$|g K_0| < |\gamma J_{\gamma_i}| = |J_g|;$$

so (1) and (2) are proven.

For (3), choose $\nu > 0$ such that

$$\gamma_i'(x)/\gamma_i'(y) < \nu, \quad 1 \leq i \leq r, \quad \forall x, y \in K_0.$$

Then, Lemma 8.1.6 and the chain rule imply that $\mu = \nu\lambda$ is as desired. $\square$

Proof of Theorem 8.2.1. We let $J = [a, b]$ play the role of J_0 and b the role of u_0 in the above lemma and choose $K = K_0$ as in the conclusion of the lemma. The Sacksteder point x_0 will be found in $\operatorname{int} K$.

Since $\Gamma(b) \cap \operatorname{int} K$ clusters at b, this set is nonempty. For $z \in \Gamma(b) \cap \operatorname{int} K$, we let $J_z \subset \operatorname{int} K$ be the gap of C with z as an endpoint. Since $\Gamma(b) \cap \operatorname{int} K$ also clusters at each point of itself, we can choose $z \in \Gamma(b) \cap \operatorname{int} K$ so that, for a given $\delta > 0$,

- $|J_z| < \delta$;
- $\operatorname{dist}(z, \partial K) > |J_z|$;

- z is as close to b as desired.

Choose $\gamma \in \mathcal{S}_b$ so that $\gamma(b) = z$. By the lemma, $|\gamma K| < |J_z| < \delta$. So $\gamma K \subset \operatorname{int} K$ and is uniformly as close to b as desired (choice of δ). In choosing δ small, also require that $\delta \in (0, |K|/\mu)$, where μ is as in the lemma.

By the mean value theorem, $\exists\, y_0 \in K$ such that $\gamma'(y_0)|K| = |\gamma K| < \delta$. By the lemma, $\gamma'(z_0)/\gamma'(y_0) < \mu$, $\forall z_0 \in K$; so

$$\gamma'(z_0) < \mu\gamma'(y_0) < \mu\delta/|K| < 1, \ \forall z_0 \in K.$$

It follows that $\gamma : K \to \operatorname{int} K$ is a contraction mapping, hence that there is a point $x_0 \in \operatorname{int} K$ to which γ contracts K and $\gamma'(x_0) < 1$. Since $x_0 \in \gamma K$, it is as close to b as desired. Since $\bigcap_{n=1}^{\infty} \gamma^n(C \cap K) = \{x_0\}$ and C is a closed, Γ–invariant set, $x_0 \in C$. □

The following corollary plays an important role in Section 8.3.

Corollary 8.2.11. *Let C and $[a, b] = J$ be as in* Theorem 8.2.1. *Let $\{x_n\}_{n=1}^{\infty} \subset \mathbb{R} \smallsetminus \{b\}$ converge to b (not necessarily from above). Then, there is an integer $N \geq 1$ such that, if $n \geq N$, then $\Gamma(x_n)$ clusters at b.*

Proof. Fix K as in the proof of Sacksteder's theorem and choose N such that $n \geq N$ implies that $\Gamma(x_n) \cap \operatorname{int} K \neq \emptyset$. This is possible since $\Gamma(b) \cap \operatorname{int} K \neq \emptyset$. If $\Gamma(x_n)$ does not cluster at b, then there is a closest point $y_n \in \overline{\Gamma(x_n)} \cap K$ to b, $y_n \neq b$. We can choose the Sacksteder point $x_0 \in (b, y_n) \subset K$. Then the Sacksteder contraction $\gamma : K \to K$ to x_0 gives $b < x_0 < \gamma(y_n) < y_n$. Since $\overline{\Gamma(x_n)}$ is Γ–invariant, this contradicts the choice of n. □

8.3. The Theory of Levels

The local minimal sets define a very useful hierarchy by "levels" in compact, C^2–foliated manifolds $(M, \mathcal{F})$ that are transversely oriented and of codimension one. As usual, the assumption that $\partial_{\pitchfork} M = \emptyset$ is made, but all results are extended to the general case by doubling along $\partial_{\pitchfork} M$.

Lemma 8.3.1. *Let X and Y be local minimal sets. Then, either $X \cap Y = \emptyset$ or $X = Y$.*

Proof. Let $U, V \in \mathcal{O}(\mathcal{F})$ be such that $X \subseteq U$ is a minimal set of $\mathcal{F}|U$ and $Y \subseteq V$ is a minimal set of $\mathcal{F}|V$. Since $U \smallsetminus V$ is closed in U, X is either disjoint from this set, or is contained in it. Thus, if $X \cap Y \neq \emptyset$, $X \subseteq U \cap V$ and, similarly, $Y \subseteq U \cap V$. Since X and Y are minimal sets for $\mathcal{F}|U \cap V$ and $X \cap Y \neq \emptyset$, we see that $X = Y$. □

Theorem 8.3.2. *Let $U \in \mathcal{O}(\mathcal{F})$ and let $Y \subseteq U$ be the union of all minimal sets of $\mathcal{F}|U$. Then Y is closed in U. Furthermore, if U is connected, at most finitely many minimal sets of $\mathcal{F}|U$ are exceptional.*

Proof. Let $\{y_n\}_{n=1}^{\infty}$ be a sequence in Y that converges to $y \in U$. Let Y_n be the unique minimal set of $\mathcal{F}|U$ such that $y_n \in Y_n$, $n \geq 1$. If infinitely many Y_n are the same, then y is a point in this minimal set and we are done. Thus, by Lemma 8.3.1, we can assume that $Y_n \cap Y_m = \emptyset$, $n \neq m$. Also, all but finitely many Y_n lie in the component of U containing y; so we can assume that U is connected. Thus, no Y_n is open, since that would imply that $U = Y_n$ is the unique minimal set of $\mathcal{F}|U$.

Let $L \subset U$ be the leaf of $\mathcal{F}$ through y. Let Y_* be a minimal set of $\mathcal{F}|U$ in $\overline{L} \cap U$. Given $z \in Y_*$, there are elements $\gamma_n \in \Gamma_{\mathcal{U}}$ such that $\{\gamma_n(y)\}_{n=1}^{\infty}$ converges to z. By passing to a subsequence of $\{y_n\}_{n=1}^{\infty}$ we can suppose that $z_n = \gamma_n(y_n) \in Y_n$ is defined, $\forall n \geq 1$, and that $\{z_n\}_{n=1}^{\infty}$ converges to z. While it is possible that $\gamma_n(y) = z$, it is not possible that any $z_n = z$.

If Y_* is exceptional, we can choose z so that the leaf through z has a Sacksteder contraction. But then, for n sufficiently large, z_n gets captured by this contraction, implying that $Y_* \subseteq \overline{Y}_n \cap U = Y_n$, for all large values of n. But then $Y_n = Y_*$ for these values of n, contradicting the fact that the Y_n's are distinct. Since also Y_* is not open, it follows that $Y_* = L_*$ is a proper leaf.

Using a biregular neighborhood W of z, we can assume that each z_n lies on the $\mathcal{L}$–plaque $J \subset W$ through z. We can replace each z_n with the closest point of $Y_n \cap J$ to z. Hence we assume that no point of $Y_n \cap J$ lies between z_n and z, $\forall n \geq 1$. Again passing to a subsequence, we assume that the convergence is monotone, both for $z_n \to z$ and $\gamma_n(y) \to z$. Although $\{\gamma_n(y)\}_{n=1}^{\infty}$ might be a constant sequence, the convergence of $\{z_n\}_{n=1}^{\infty}$ is strictly monotone.

Evidently, the points z_n are fixed points for the holonomy of L_*. So Dippolito's semistability theorem (Theorem 5.3.4) implies that L_* is a boundary leaf of a foliated product $\widehat{W}$ and that each Y_n is a proper leaf that is a global section of $\widehat{W}$. These Y_n's cluster on L_* from within $\widehat{W}$ and the leaf L approaches L_* on that same side; so $L = L_*$. In particular, $y \in Y$.

Similarly, if U is connected, we prove that only finitely many of the minimal sets of $\mathcal{F}|U$ are exceptional. Otherwise, choose an infinite sequence of such and argue as above that a subsequence $\{Y_n\}_{n=1}^{\infty}$ will consist of proper leaves. □

The above proof has the following interesting consequence.

Corollary 8.3.3. *If $U \in \mathcal{O}(\mathcal{F})$ is connected, the union of the leaves of $\mathcal{F}|U$ that are closed in U is a closed subset of U and contains only finitely many distinct diffeomorphism types.*

The proof is left to the reader. Remark that, taking $U = M$, we obtain a proof of Haefliger's theorem (Theorem 6.1.1) for C^2–foliated manifolds that does not use homological intersection theory.

Definition 8.3.4. A leaf that belongs to a minimal set of $\mathcal{F}$ is said to be at level 0. The union of all leaves at level 0 is denoted by M_0.

By Theorem 8.3.2, applied to the case $U = M$, we see that M_0 is compact. Set $U_0 = M \setminus M_0 \in \mathcal{O}(\mathcal{F})$. By Theorem 8.1.8, either $U_0 = \emptyset$ or $\mathcal{F}|U_0$ has at least one minimal set. In the first case, $M = M_0$. In the second, the union M_1 of M_0 and all of the minimal sets of $\mathcal{F}|U_0$ is closed in M (Theorem 8.3.2), hence is also compact, and we get $U_1 = M \setminus M_1 \in \mathcal{O}(\mathcal{F})$. Again, either $M = M_1$ or we obtain a compact, $\mathcal{F}$–saturated set M_2 as the union of M_1 and all minimal sets of $\mathcal{F}|U_1$. Inductively, one obtains the following.

Theorem 8.3.5. *There is a unique filtration*

$$\emptyset = M_{-1} \subset M_0 \subseteq M_1 \subseteq \cdots \subseteq M_k \subseteq \cdots \subseteq M$$

of M by compact, $\mathcal{F}$–saturated subsets such that

1. *$M_k \setminus M_{k-1}$ is the union of all minimal sets of $\mathcal{F}|(M \setminus M_{k-1})$, $\forall k \geq 0$;*
2. *if $M_k = M_{k+1}$, some $k \geq 0$, then $M_k = M_{k+p} = M$, $\forall p \geq 0$.*

Definition 8.3.6. The leaves of $\mathcal{F}|(M_k \setminus M_{k-1})$ are said to be at level k, $\forall k \geq 0$. If $M_* = \bigcup_{k=0}^{\infty} M_k$, the leaves of $\mathcal{F}|M_*$ are said to be at finite level. The leaves (if any) in $M_\infty = M \setminus M_*$ are said to be at infinite level.

Examples show that M_∞ may be empty or may contain uncountably many leaves (see Exercise 8.1.21). We will see that no other possibilities exist. In any case, the following shows that M_∞ has no interior, hence we cannot continue finding local minimal sets at infinite levels.

Theorem 8.3.7. *The union M_* of all leaves at finite level is dense in M.*

The proof of this theorem will use some lemmas that are interesting in their own right. In both of the following lemmas, L denotes an arbitrary leaf of $\mathcal{F}$.

Lemma 8.3.8. *Let $X \subset \overline{L}$ be a proper, compact, $\mathcal{F}$–saturated subset. Let U be the component of $M \setminus X$ containing L and let $L_1, \ldots, L_r$ be the components of $X \cap \delta U$. Then $X = \overline{L}_1 \cup \cdots \cup \overline{L}_r$.*

Proof. If $X = \emptyset$, there is nothing to prove. Otherwise, X clearly contains $\overline{L}_1 \cup \cdots \cup \overline{L}_r$ and we must demonstrate the reverse inclusion. Let $x_0 \in X$ and let J be an $\mathcal{L}$–plaque through x_0. Then, $L \cap J$ clusters at x_0, hence so does $U \cap J$. Since $x_0 \notin U$, it follows that either $x_0 \in \delta U$ (and we are done) or $\delta U \cap J$ clusters at x_0. In the second case, $L_i \cap J$ clusters at x_0, for at least one value of $i = 1, 2, \dots, r$. Since $x_0 \in X$ was chosen arbitrarily, $X \subseteq \overline{L}_1 \cup \cdots \cup \overline{L}_r$. □

Lemma 8.3.9. *If $X_1 \subset X_2 \subset \cdots \subset X_k \subset \cdots \subset \overline{L}$ is an infinite sequence of proper inclusions of compact, $\mathcal{F}$–saturated subsets of $\overline{L}$, then $X_* = \bigcup_{k=1}^{\infty} X_k$ is dense in $\overline{L}$.*

Proof. Let $X = \overline{X}_*$. If $X \neq \overline{L}$, let U be the component of $M \smallsetminus X$ that contains L and let $L_1, L_2, \dots, L_r$ be the components of $X \cap \delta U$. By Lemma 8.3.8, $X = \overline{L}_1 \cup \cdots \cup \overline{L}_r$. At least one $L_i \not\subseteq X_*$ since, otherwise, $X = \overline{L}_1 \cup \cdots \cup \overline{L}_r \subseteq X_k$, for some k, contradicting the assumption that the inclusions are proper. Renumbering the components of $X \cap \delta U$, if necessary, we assume that $L_1, L_2, \dots, L_q$ are the ones $\not\subseteq X_*$. Then $X_k \subseteq \overline{L}_1 \cup \cdots \cup \overline{L}_q$ for all large values of k.

In order to show that some L_i is asymptotic to itself, $1 \leq i \leq q$, let $y \in L_1 \cap \operatorname{int}(J)$, where J is a compact $\mathcal{L}$–plaque. Choose $x_k \in X_k \cap J$ such that $\{x_k\}_{k=1}^{\infty}$ clusters at y. There is $i \leq q$ such that infinitely many $x_k \in \overline{L}_i$; so L_i is asymptotic to L_1. If also L_1 is asymptotic to L_i, then L_1 is asymptotic to itself and we are done. Otherwise, renumber so that $i = 2$, noting that $\overline{L}_1 \subset \overline{L}_2$ is a proper inclusion. Repeat the same argument for L_2. Either a finite repetition ends with an L_i asymptotic to itself, or we reach a sequence of proper inclusions

$$\overline{L}_1 \subset \overline{L}_2 \subset \cdots \subset \overline{L}_q.$$

But the same argument shows that some L_i, $1 \leq i \leq q$, is asymptotic to L_q and the only possibility is $i = q$.

By the previous paragraph, we assume that L_1 is asymptotic to itself and choose the compact $\mathcal{L}$–plaque, containing $y \in L_1$ in its interior, so that $\overline{L}_1 \cap J$ is a Cantor set. Since $\Gamma_{\mathcal{U}}(y) \cap J$ clusters at y, Corollary 8.2.11 applies and shows that, for k sufficiently large, $\Gamma_{\mathcal{U}}(x_k)$ clusters at y. But this implies that $L_1 \subseteq X_k \subseteq X_*$, a contradiction. □

Proof of Theorem 8.3.7. We must show that $\overline{M}_* = M$. If not, then

$$\emptyset \neq M \smallsetminus \overline{M}_* = W \in \mathcal{O}(\mathcal{F}).$$

Fix a leaf $L \subset W$ and let $X_k = \overline{L} \cap M_k$, $\forall k \geq 0$. Then, the inclusions $X_k \subset X_{k+1}$ must be proper since, otherwise, Theorem 8.1.8 would imply that $\overline{L} \cap (M \smallsetminus M_k) = \emptyset$, hence that $L \subseteq \overline{L} \subseteq M_k \subseteq \overline{M}_*$. By Lemma 8.3.9, $X_* = \bigcup_{k=0}^{\infty} X_k$ is dense in $\overline{L}$, contradicting $L \subset W$. □

Corollary 8.3.10. *Every local minimal set of $\mathcal{F}$ lies at some finite level.*

Proof. Let $U \in \mathcal{O}(\mathcal{F})$ and let X be a minimal set of $\mathcal{F}|U$, $L \subseteq X$ a leaf. Suppose that X does not lie at finite level. In particular, X is not a minimal set of $\mathcal{F}$, and so $\overline{X} \smallsetminus X = \overline{X} \smallsetminus U$ is a compact, nonempty, $\mathcal{F}$–saturated set. Also, the $\mathcal{F}|U$–minimality of X implies that $\overline{L} = \overline{X}$. Set

$$X_k = \overline{X} \cap M_k = \overline{L} \cap M_k, \quad \forall k \geq 0,$$

a compact $\mathcal{F}$–saturated subset of $\overline{L}$.

Since $\overline{L}$ contains a minimal set of $\mathcal{F}$, we see that $X_0 \neq \emptyset$. Furthermore, the inclusion $X_k \subseteq X_{k+1}$ must be a *proper* inclusion, $k \geq 0$. Indeed, the set $\overline{L} \cap (M \smallsetminus M_k)$ is not empty. So Theorem 8.1.8 implies that $\overline{L} \cap (M \smallsetminus M_k)$ contains a minimal set of $\mathcal{F}|(M \smallsetminus M_k)$, necessarily a local minimal set at level $k+1$. By Lemma 8.3.9, $X_* = \bigcup_{k=0}^{\infty} X_k$ is dense in $\overline{L}$. But X cannot meet M_k, for any $k \geq 0$. Indeed, that would imply that $X \cap Y \neq \emptyset$, for some local minimal set Y at level $\leq k$; hence $X = Y$ by Lemma 8.3.1. That is, X_* lies in the compact set $\overline{X} \smallsetminus X$, contradicting the fact that X_* is dense in $\overline{L} = \overline{X}$. □

Theorem 8.3.11. *If $k \geq 0$ and if L is an arbitrary leaf of $\mathcal{F}$, then $\overline{L} \cap M_k$ is a finite union of local minimal sets.*

Proof. Evidently, every local minimal set that meets $\overline{L}$ is contained in $\overline{L}$; so $\overline{L} \cap M_k$ is a union of local minimal sets. Let the inductive hypothesis be that, for any leaf L of $\mathcal{F}$, $\overline{L} \cap M_{k-1}$ is a finite union of local minimal sets. Since $M_{-1} = \emptyset$, this hypothesis is vacuously true for $k = 0$.

Assume the inductive hypothesis for $k - 1$. There are two cases to consider.

Case 1. If $\overline{L} \cap M_k = \overline{L}$, then $L \subseteq M_k$; so $Y = \overline{L} \smallsetminus (\overline{L} \cap M_{k-1})$ is a local minimal set at level k. Then, by the inductive hypothesis,

$$\overline{L} \cap M_k = \overline{L} = Y \cup (\overline{L} \cap M_{k-1})$$

is a finite union of local minimal sets.

Case 2. If $\overline{L} \cap M_k \neq \overline{L}$, let U be the component of $M \smallsetminus (\overline{L} \cap M_k)$ that contains L. Let $L_1, \dots, L_r$ be the leaves of $\overline{L} \cap M_k \cap \delta U$ and remark that $\overline{L}_i \cap M_k = \overline{L}_i$, $1 \leq i \leq r$. Case 1 then implies that $\overline{L}_i$ is a finite union of local minimal sets. By Lemma 8.3.8, $\overline{L} \cap M_k = \overline{L}_1 \cup \cdots \cup \overline{L}_r$, a finite union of local minimal sets. □

Corollary 8.3.12. *A leaf L belongs to a local minimal set if and only if $\overline{L}$ is a finite union of local minimal sets. In the alternative case, $\overline{L}$ contains at least one local minimal set at every level $k \geq 0$.*

Proof. Since every local minimal set is at finite level (Corollary 8.3.10), L is a leaf in a local minimal set if and only if $\overline{L} = \overline{L} \cap M_k$ for k sufficiently large. Such an equality holds if and only if $\overline{L}$ is a finite union of local minimal sets.

Suppose that L does not lie in a local minimal set and set $X_k = \overline{L} \cap M_k$, $k \geq 0$. As usual, this defines an infinite, proper nest

$$X_0 \subset X_1 \subset \cdots \subset X_k \subset \cdots \subset \overline{L};$$

so $\overline{L}$ contains local minimal sets at each level $k \geq 0$. □

Definition 8.3.13. A leaf L of $\mathcal{F}$ is said to be totally proper if every leaf in $\overline{L}$ is a proper leaf. The union of all totally proper leaves at levels $\leq k$ will be denoted by C_k, $\forall k \geq 0$.

Definition 8.3.14. A leaf L of $\mathcal{F}$ is at *depth* 0 if it is compact. Inductively, a leaf L is at depth $k \geq 1$ if $\overline{L} \smallsetminus L$ is a union of leaves at depths $< k$, at least one of which is at depth $k-1$.

Proposition 8.3.15. *The set C_k is compact, $\forall\, k \geq 0$, and a leaf L of $\mathcal{F}$ lies in C_k, for some $k \geq 0$, if and only if $\overline{L}$ is a finite union of leaves.*

Corollary 8.3.16. *A leaf of $\mathcal{F}$ is totally proper at level k if and only if it is a leaf at depth k.*

Exercise 8.3.17. Prove Proposition 8.3.15 and Corollary 8.3.16.

Remark. Only in foliations of class C^2 can we claim that totally proper leaves are always at finite depth. In Example 8.1.13, there were leaves at depth 0, one leaf at depth 1, and all other leaves were proper, but none were at finite depth.

We turn to the set M_∞ of leaves at infinite level. These are the least accessible leaves, particularly to our intuition, but they play an important role, as the following (sketchy) examples indicate.

Example 8.3.18. Let M be a compact, connected, oriented n–manifold without boundary. As a subset of M, the $\mathcal{F}$–saturated set M_∞ is a Borel set and, in particular, is Lebesgue measurable. For $\mathcal{F}$–saturated, measurable sets $X \subseteq M$, one shows that the Godbillon–Vey class $\mathrm{gv}(\mathcal{F})$ can be *integrated* over X to define a cohomology class $\mathrm{gv}(X, \mathcal{F}) \in H^3(M; \mathbb{R})$ (Exercise 8.3.19). If $\dim M = 3$, the identification $H^3(M; \mathbb{R}) = \mathbb{R}$ makes this fairly intuitive. One sets

$$\mathrm{gv}(X, \mathcal{F}) = \int_X \eta \wedge d\eta.$$

This number does not depend on the choice of $\eta \wedge d\eta \in \mathrm{gv}(\mathcal{F})$ (Exercise 8.3.19). At any rate, we view gv as an $H^3(M; \mathbb{R})$–valued, countably additive measure on the σ–algebra $\mathfrak{B}(\mathcal{F})$ of Lebesgue measurable, $\mathcal{F}$–saturated sets. This measure satisfies $\operatorname{gv}(M, \mathcal{F}) = \operatorname{gv}(\mathcal{F})$. It is used to prove Duminy's theorem that $\operatorname{gv}(\mathcal{F}) = 0$ unless some leaf is resilient. One can write

$$\operatorname{gv}(\mathcal{F}) = \operatorname{gv}(M_\infty, \mathcal{F}) + \sum_{k=0}^{\infty} \operatorname{gv}(M_k, \mathcal{F}).$$

If no leaf is resilient, the local minimal sets are either proper leaves or open without holonomy. This is used to prove that $\operatorname{gv}(M_k, \mathcal{F}) = 0$, $\forall\, k \geq 0$. It follows that $\operatorname{gv}(\mathcal{F}) = \operatorname{gv}(M_\infty, \mathcal{F})$ and the last step is to prove that this class vanishes. Duminy proves that $\operatorname{gv}(M_\infty, \mathcal{F}) = 0$ for all C^2 foliations, whether or not they have resilient leaves. The details will be given in Volume II.

Exercise 8.3.19. Let M be a compact, connected, oriented n–manifold without boundary. One version of Poincaré duality identifies the vector spaces

$$H^q(M; \mathbb{R}) = \operatorname{Hom}_{\mathbb{R}}(H^{n-q}(M; \mathbb{R}), \mathbb{R})$$

as follows. Given $[\varphi] \in H^q(M; \mathbb{R})$, represented by the q–form φ, define

$$[\varphi] : H^{n-q}(M; \mathbb{R}) \to \mathbb{R}$$

by

$$[\varphi]([\psi]) = \int_M \varphi \wedge \psi.$$

Recall also that the integration of an n–form θ over M is defined by a partition of unity $\{\lambda_\alpha\}_{\alpha \in \mathfrak{A}}$ subordinate to a coordinate atlas $\{(U_\alpha, x_\alpha)\}_{\alpha \in \mathfrak{A}}$. Using this same data, one easily defines integration of n–forms over measurable subsets $X \subseteq M$, although the number

$$\int_X \theta$$

will not generally be an invariant of $[\theta] \in H^n(M; \mathbb{R})$.

1. If $\mathcal{F}$ is a transversely–oriented, codimension–one foliation of class C^∞ on M, choose the above data so that $\{(U_\alpha, x_\alpha, y_\alpha)\}_{\alpha \in \mathfrak{A}}$ is a biregular, foliated atlas. Let $I^*(\mathcal{F}) \subset A^*(M)$ be the annihilator ideal for the bundle $T(\mathcal{F})$ and let $X \in \mathcal{B}(\mathcal{F})$ (Example 8.3.18). If $\varphi \in I^{n-1}(\mathcal{F})$, prove that
$$\int_{X \cap U_\alpha} \lambda_\alpha d\varphi = 0, \quad \forall\, \alpha \in \mathfrak{A}.$$
(Hint. In the local coordinates $(x_\alpha^1, \dots, x_\alpha^{n-1}, y_\alpha)$, φ has a factor dy_α. Use the theorems of Stokes and Fubini.)
2. Let $\eta \wedge d\eta$ represent $\operatorname{gv}(\mathcal{F})$ and let $[\psi] \in H^{n-3}(M; \mathbb{R})$. Show that, if $\widetilde{\eta} \wedge d\widetilde{\eta}$ and $\widetilde{\psi}$ also represent the classes $\operatorname{gv}(\mathcal{F})$ and $[\psi]$, respectively, then
$$\widetilde{\eta} \wedge d\widetilde{\eta} \wedge \widetilde{\psi} = \eta \wedge d\eta \wedge \psi + d\varphi,$$
where $\varphi \in I^{n-1}(\mathcal{F})$.

3. Using the above, show that

$$\mathrm{gv}(X, \mathcal{F}) : H^{n-3}(M; \mathbb{R}) \to \mathbb{R}$$

is well defined, $\forall\, X \in \mathcal{B}(\mathcal{F})$, by

$$\mathrm{gv}(X, \mathcal{F})[\psi] = \int_X \eta \wedge d\eta \wedge \psi.$$

By the above remarks, this well defines $\mathrm{gv}(X, \mathcal{F}) \in H^3(M; \mathbb{R})$, $\forall\, X \in \mathcal{B}(\mathcal{F})$, and $\mathrm{gv}(X, \mathcal{F})$ is σ–additive on $\mathcal{B}(\mathcal{F})$ by σ–additivity of the integral. In particular, $\mathrm{gv}(M_*, \mathcal{F})$, $\mathrm{gv}(M_\infty, \mathcal{F})$ and $\mathrm{gv}(M_k, \mathcal{F})$ $(0 \le k < \infty)$ are all well–defined elements of $H^3(M; \mathbb{R})$.

Example 8.3.20. Leaves at infinite level also arise in answering a question posed by J. Sondow [**123**]. Slightly rephrased, the question is "what open n–manifolds occur as leaves of foliations in what closed $(n+1)$–manifolds?" It was shown by E. Ghys [**47**] that, for $n \ge 3$, there is a noncompact n–manifold that cannot be a leaf of a C^0 foliation of any compact $(n+1)$–manifold. A similar construction was found independently by T. Inaba, T. Nishimori, M. Takamura and N. Tsuchiya [**74**]. In these examples, the fundamental group contains the obstructions. Simply connected, noncompact 6–manifolds that cannot be realized as leaves of compact 7–manifolds have been found by O. Attie and S. Hurder [**5**], the obstructions being Pontryagin classes.

By contrast, every noncompact, orientable surface occurs as a leaf of a C^∞ foliation in every compact 3–manifold, orientable or not, and every noncompact, nonorientable surface occurs as such a leaf L in every compact, nonorientable 3–manifold [**21**]. For this, one constructs a sequence $\{\mathcal{F}_k\}_{k=1}^\infty$ of foliations converging to a limit foliation $\mathcal{F}$. The foliation $\mathcal{F}_k$ contains a leaf L_k at level $k+1$ that agrees with L in a compact region B_k. These regions exhaust L and, although the topology of L_k becomes progressively more horrible outside B_k as $k \uparrow \infty$, these pathologies are "pushed out of the universe" in the limit. The leaf L appears at infinite level in $\mathcal{F}$; so it is basically impossible to "see" this leaf. In this construction, it turns out that one can find a Riemannian metric on M relative to which L has constant curvature -1. In particular, the hyperbolic plane can be realized as a leaf at infinite level in a suitable C^∞ foliation in every compact 3–manifold.

In what follows, suppose that L is a leaf of $\mathcal{F}$ at infinite level. Such a leaf cannot lie in $\partial_\tau M$ since boundary leaves are minimal sets. Let $x \in L$ and let J be an $\mathcal{L}$–plaque with $x \in \mathrm{int}(L)$. Let $X_k = \overline{L} \cap M_k$, $\forall k \ge 0$. We have seen that $X_* = \bigcup_{k=0}^\infty X_k$ is dense in $\overline{L}$ but, evidently, it contains no leaf that is dense in $\overline{L}$.

Lemma 8.3.21. *If there is no finite upper bound to the levels of the totally proper leaves in X_*, then all leaves in X_* are totally proper.*

Proof. Let $Y_k = \overline{L} \cap C_k$, a compact, $\mathcal{F}$–saturated set by Proposition 8.3.15. By hypothesis, this set is nonempty for each $k \geq 0$. By now, the reader should be able to check that $Y_k = Y_{k+1}$ implies that $Y_k = Y_{k+p}$, $\forall p \geq 0$, contradicting our hypothesis. Then, Lemma 8.3.9 implies that $Y_* = \bigcup_{k=0}^{\infty} Y_k$ is dense in $\overline{L}$. It is clear that $\overline{L}$ cannot contain an open local minimal set. So the only way the conclusion of the lemma can fail is that there is an exceptional local minimal set $X \subset \overline{L}$. But, then, the dense subset $Y_* \subset \overline{L}$ must meet the domain of a Sacksteder contraction in X, implying that some totally proper leaf approaches X, a clear contradiction. □

Let U_k denote the component of $M \smallsetminus X_k$ that contains L. Thus, $\{U_k\}_{k=0}^{\infty}$ is a decreasing nest.

Lemma 8.3.22. *There is an integer $N \geq 0$ such that $\widehat{U}_k$ is a foliated product, $\forall k \geq N$.*

Proof. By Proposition 5.2.8, we only need to show that $\bigcap_{k=1}^{\infty} U_k$ has empty interior. Suppose, then, that U is a nonempty component of the interior of $\bigcap_{k=1}^{\infty} U_k$. Since X_* is dense in $\overline{L}$, it is clear that $L \not\subset U$. Let $L_1, \dots, L_r$ be the leaves of δU. We will prove that these leaves all lie in some X_k, hence that U is a component of $M \smallsetminus X_k$. Since $U \subseteq U_k$, it will follow that $U = U_k$, contradicting the fact that $L \not\subset U$.

Suppose, on the contrary, that some $L_j \not\subseteq X_*$. Then $L_j \subset \overline{X}_* = \overline{L}$. Also, L_j, although semiproper, cannot be proper since, then, $L_j \subseteq M_k$, some $k \geq 0$, and $L_j \subseteq \overline{L} \cap M_k = X_k$, contradicting our supposition. Let $x \in L$ and choose $x_k \in X_k$ so that $\lim_{k\to\infty} x_k = x$ along an $\mathcal{L}$–plaque through x. By Corollary 8.2.11, there is a large enough value of k such that the leaf of $\mathcal{F}$ through x_k is asymptotic to L_j. That is, $L_j \subseteq \overline{L} \cap M_k = X_k$, again contrary to our supposition. □

Lemma 8.3.23. *If L is at infinite level, then each leaf $L' \subset \overline{L}$ at infinite level is dense in $\overline{L}$ and does not have a proper side.*

Proof. Let $N \geq 0$ be as in Lemma 8.3.22. Let $k \geq N$; hence $\widehat{U}_k$ is a foliated product. Clearly, $L' \subset U_k$. By Lemma 8.3.8, $X_k \cap \delta U_k$ is dense in X_k and it follows that every leaf of $\mathcal{F}$ in U_k approaches every point of $X_k \smallsetminus \delta U_k$. In particular, L' is asymptotic to every leaf in X_k with the possible exceptions of $X_k \cap \delta U_k$. In fact, L' is asymptotic to these leaves also. To see this, fix a point $y \in L'$ and select an $\mathcal{L}$–plaque J containing y in its interior. Let J_i be the component of $U_i \cap J$ containing y, $i \geq k$. Since $\bigcap_{i=k}^{\infty} U_i$ has empty interior, the nest $J_k \supseteq J_{k+1} \supseteq \cdots \supseteq J_{k+p} \supseteq \cdots$ is a fundamental system of neighborhoods for y in J. Thus, there is a value of $i \geq k$ such that $X_k \subseteq X_i \smallsetminus \delta U_i$; hence L' is asymptotic to every leaf of X_k. Since $k \geq N$ is arbitrary, L' is asymptotic to every leaf in X_*; hence L' is dense in

$\overline{X}_* = \overline{L}$. Since L' approaches every leaf of $\bigcup_{k=N}^{\infty} X_k \cap \delta U_k$, it follows that L' is asymptotic to itself on both sides. □

Corollary 8.3.24. *Let L be a leaf at infinite level, $x \in L$, and let J be an $\mathcal{L}$–plaque through x. Then, the pseudogroup Γ_L on J, induced by the holonomy around loops on L based at x, has fixed points clustering at x from both sides.*

Proof. For $k \geq N$, the component J_k of $U_k \cap J$ containing x has boundary points $x_k^{\pm}$. Since $\widehat{U}_k$ is a foliated product, these must be fixed points for Γ_L. The set $\{x_k^+\}_{k=N}^{\infty}$ clusters on x from above and $\{x_k^-\}_{k=N}^{\infty}$ clusters on x from below. □

Corollary 8.3.25. *If the foliation $\mathcal{F}$ is of class C^r, $2 \leq r \leq \infty$, then $\mathcal{F}$ is C^r–flat along any leaf L at infinite level.*

Proof. Use the previous corollary and repeated applications of the mean value theorem. □

Remark. Although a leaf at infinite level must have trivial infinitesimal holonomy, one can construct examples to show that the *germinal* holonomy of such a leaf may be nontrivial.

Lemma 8.3.26. *If L is at infinite level, then $\overline{L}$ contains uncountably many leaves at infinite level.*

Proof. Let Z denote the union of the leaves of $\overline{L}$ that are at infinite level. Let N be as above and let J be an $\widehat{\mathcal{L}}$–fiber of $\widehat{U}_N$. Each leaf in Z meets $\operatorname{int}(J)$ in a countable set of points; so it will be sufficient to show that $Z \cap J$ is uncountable. Let $y \in J \cap Z$ and let J_k denote the $\widehat{\mathcal{L}}$–fiber in $\widehat{U}_k$, $k \geq N$, containing y. Let $y' \in J \cap Z$ lie on the same leaf of $\mathcal{F}$ as y, $y' \neq y$. Then, for k_1 sufficiently large, $y' \notin J_{k_1}$. There is a holonomy transformation h in the foliated product $\widehat{U}_N$ such that $y' \in h(J_{k_1})$. Since $J = D_h = R_h$, the iterates of h move J_{k_1} to a countably infinite set of intervals with disjoint interiors. The interiors of these intervals are components of $U_{k_1} \cap J$; so there is a countable infinity of such components. By the same method, applied inductively, we find integers $\{k_m\}_{m=1}^{\infty}$ such that $U_{k_{m+1}} \subset U_{k_m}$ and $U_{k_{m+1}}$ meets each of the countably many intervals $J \cap U_{k_m}$ in countably many intervals.

Consider the set $\mathcal{I}$ of all sequences $\{I_m\}_{m=1}^{\infty}$, where I_m is the closure of a component of $U_{k_m} \cap J$ and $I_1 \supseteq I_2 \supseteq \cdots \supseteq I_m \supseteq \cdots$. Since $\bigcap_{m=1}^{\infty} U_{k_m}$ has empty interior, it follows that $\bigcap_{m=1}^{\infty} I_m$ is a single point of $Z \cap J$. This sets up a one–one correspondence between the uncountable set $\mathcal{I}$ and the set $Z \cap J$. □

8.4. Leaves of Finite Depth

Finite depth foliations are those in which all leaves are proper with a finite upper bound on their depths. Such foliations play an important role in the topology of 3–manifolds, especially in David Gabai's analysis of knot and link complements [**42, 41, 44, 45, 43**] (although these foliations are not always C^2–smoothable [**26**]).

Here we study leaves of finite depth in compact, oriented, transversely–oriented, C^2–foliated manifolds $(M, \mathcal{F})$ of codimension one and arbitrary dimension. The orientation hypotheses induce a canonical orientation on each leaf. As usual in this chapter, confusing complications are avoided by assuming that $\partial_{\pitchfork} M = \emptyset$ unless otherwise indicated.

Our goal is to describe how a totally proper leaf F approaches leaves at lower depths (Theorem 8.4.6). This approach will be by way of a "spiraling" of neighborhoods B of ends of F onto lower depth leaves L_*, completely analogous to the spiraling of the planar leaves on the torus in a Reeb component. We also saw a case of this phenomenon in Example 8.1.27.

Definition 8.4.1. Let $(x, z]$ be a subarc of a leaf of $\mathcal{L}$, oriented by the transverse orientation of $\mathcal{F}$. Let L be the leaf of $\mathcal{F}$ through x and assume that $(x, z] \cap L = \emptyset$. Then we say that z projects (in the negative direction) to L and write $p(z) = x \in L$.

Similarly, one defines projection to L in the positive direction. Let F and L be leaves of $\mathcal{F}$ and let $B \subset F$ be a complete, connected, noncompact submanifold of dimension $n-1$ (where $\dim(M) = n$). Suppose that $N_0 = \partial B$ is a compact, connected submanifold of F. Finally, suppose that each point of B projects to L (say, in the negative direction) and remark that $p : B \to L$ is locally a diffeomorphism.

Definition 8.4.2. We will say that B spirals on L (on the positive side) if the following conditions are satisfied:

1. $B = \bigcup_{i=0}^{\infty} B_i$, where each B_i is a complete, connected, submanifold of dimension $(n-1)$, $\partial B_i = N_i \cup N_{i+1}$ is a union of two compact connected components, and $\operatorname{int} B_i \cap \operatorname{int} B_j = \emptyset$ if $i \neq j$.
2. There is a compact, connected, submanifold $N \subset L$ of dimension $(n-2)$ (called the juncture of the spiral) such that $p|N_i$ maps N_i diffeomorphically onto N, $0 \leq i < \infty$.
3. For each $y \in L$ and each $i \geq 0$, $p^{-1}(y) \cap (B_i \setminus N_{i+1})$ is a single point y_i.
4. For each $y \in L$, the sequence $p^{-1}(y) = \{y_i\}_{i \geq 0}$ converges monotonically to y in $[y, y_0]$.

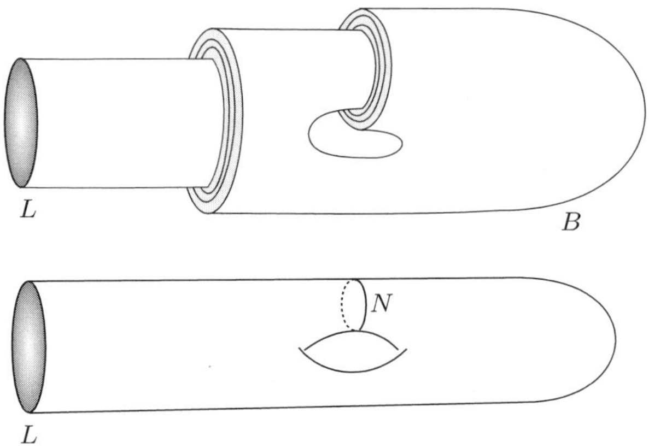

Figure 8.4.1. A periodic end of period L

Similarly, using projection to L in the positive direction, one defines the notion of B spiraling on L on the negative side. In either case,

$$B = B_1 \cup B_2 \cup \cdots \cup B_k \cup \cdots ,$$

where B_k is joined to B_{k+1} along the common boundary component N_{k+1}. We will say that B is an *infinite repetition* of L (*cf.* Example 8.1.27).

Remark. For all $i \geq 0$, the projection p carries int B_i diffeomorphically onto $L \setminus N$. In particular, N does not separate L. We can view this submanifold as a cycle, obtaining $0 \neq [N] \in H_{n-2}(L) = H^1_c(L)$.

Note that, by setting $B(k) = \bigcup_{i \geq k} B_i$, we obtain a fundamental system $\{B(k)\}_{k \geq 0}$ of (closed) neighborhoods of an end $e \in \mathcal{E}(F)$.

Definition 8.4.3. The end $e \in \mathcal{E}(F)$ having fundamental neighborhood system $\{B(k)\}_{k \geq 0}$ is said to be *periodic* of period L (Figure 8.4.1).

Example 8.4.4. If we foliate T^3 with toral leaves T^2 and turbulize along a closed transversal, we produce leaves L at depth 1 that are diffeomorphic to the punctured torus, the one end of L being asymptotic to the boundary of the Reeb component. Let $N \subset L$ be the nonseparating circle depicted in Figure 8.4.1. Consider a foliated product $\widehat{U}$ cobounded by two of these depth 1 leaves. The foliation $\widehat{\mathcal{F}}$ of $\widehat{U} \cong L \times [0,1]$ is the product foliation, but we can refoliate $\widehat{U}$ by suspending $f \in \mathrm{Diff}^\infty_+[0,1]$ which contracts $[0,1)$ to 0. Here, we cut $\widehat{U}$ along the annulus $N \times [0,1]$ and reglue by $(z^-, t) \equiv (z^+, f(t))$.

By requiring f to be C^∞–tangent to the identity at 1 and 0, we guarantee that the new foliation of T^3 is of class C^∞. The typical leaf $F \subset U$ of the new foliation is homeomorphic to the surface in Figure 4.2.3 and has two periodic ends, one with period $L \times \{0\}$ and the other with period $L \times \{1\}$. The spiraling on L of a neighborhood B of one of these periodic ends is illustrated in Figure 8.4.1.

Recall the notion of *type* for leaves and ends of leaves (Definition 4.2.7).

Lemma 8.4.5. *If $e \in \mathcal{E}(F)$ is periodic of period L and if L has finite topological type $k-1$, then e is an end of type k.*

Proof. Indeed, let $e_1, \dots, e_m \in \mathcal{E}(L)$ be the ends of type $k-1$. In the neighborhoods $B(j) = \bigcup_{i\geq j} B_i$ of e, each $\mathcal{E}(B_i) \cong \mathcal{E}(L)$. Let

$$e_1^i, \cdots, e_m^i \in \mathcal{E}(B_i)$$

be the corresponding ends of type $k-1$. Since these cluster exactly on e as $i \to \infty$, e is of type k. □

Remark. Let L be a semiproper leaf in an exceptional minimal set. By an extremely delicate application of Lemma 8.1.6, analyzing the way that L approaches itself on the nonproper side, Duminy has shown that every end e of L is periodic of period L. Exactly as above, it follows that e has a fundamental neighborhood system

$$\left\{ B(j) = \bigcup_{i\geq j} B_i \right\}_{j=1}^{\infty},$$

where $\mathcal{E}(B_i) \cong \mathcal{E}(L)$, $\forall i \geq 1$. Thus, every $e \in \mathcal{E}(L)$ is a cluster point, implying that $\mathcal{E}(L)$ is a Cantor set. For real analytic foliations, the proof is much easier and extends to *every* leaf of the exceptional minimal set. For C^2 foliations, it is an open question whether every leaf of an exceptional minimal set has a Cantor set of ends, although this is true for *Markov* minimal sets [**25**] (*cf.* Exercise 4.1.8). Unfortunately, Duminy has never made available an account of this proof (but, *cf.* [**22**]).

Theorem 8.4.6. *If L is a leaf at depth k, then*

$$L = A \cup B^1 \cup \cdots \cup B^q,$$

where A is a compact, connected, $(n-1)$–dimensional manifold with boundary components $N^1, \dots, N^q$ and

1. $A \cap B^j = N^j$, $1 \leq j \leq q$;
2. $B^i \cap B^j = \emptyset$, $i \neq j$;
3. *B^j spirals on a leaf L^j at level at most $k-1$, $1 \leq j \leq q$;*
4. *for at least one value of j, L^j is at level $k-1$.*

The Poincaré–Bendixson theorem (Theorem 7.2.3) holds for flows in the disk or 2–sphere, describing the way that a nondegenerate flow line spirals in on "limit cycles" in forward and backward time, and Theorem 8.4.6 can be viewed as a generalization of the Poincaré–Bendixson theorem. The proof of Theorem 8.4.6 is the main goal of this section.

Corollary 8.4.7. *If L is a leaf at depth k, then L is of topological type $k-1$.*

Proof. We proceed by induction on k. The case $k = 0$ (L compact) is trivial, since we agree that type -1 means that the endset is empty. If the assertion has been proven for $0 \le k \le p$, assume that L is at level $p+1$ and choose the decomposition

$$L = A \cup B^1 \cup \cdots \cup B^q$$

as in Theorem 8.4.6. If B^i spirals on a leaf L^i of type k_i, then B^i has type $k_i + 1$, by Lemma 8.4.5. The type of L will be the largest of these integers $k_i + 1$, $1 \le i \le q$. Since every leaf in $\overline{L} \smallsetminus L$ is at level $\le p$, the inductive hypothesis implies that $k_i + 1 \le p$. Since at least one of these leaves is at level p, the maximum of $k_i + 1$ is exactly p; so L is of type p. □

Remark. In the proof of Theorem 8.4.6, smoothness of class C^2 will be needed to prove the compactness of the junctures via the generalized Kopell lemma. The fact that this is the only place in which the smoothness assumption is used is crucial for the following smoothability result [**27**].

Theorem 8.4.8. *If all leaves of $(M, \mathcal{F})$ are proper, the following are equivalent:*

1. *$(M, \mathcal{F})$ is homeomorphic to a C^∞–foliated manifold;*
2. *$(M, \mathcal{F})$ is homeomorphic to a C^2–foliated manifold;*
3. *each leaf of $(M, \mathcal{F})$ is at finite depth and all junctures are compact.*

The implication (1) $\Rightarrow$ (2) is trivial. If the foliation is of class C^2, the leaves, being totally proper, are at finite depth and the junctures are compact by Theorem 8.1.26. This proves (2) $\Rightarrow$ (3). The implication (3) $\Rightarrow$ (1) is difficult and will not be treated here.

Exercise 8.4.9. If $(M, \mathcal{F})$ is a C^0–foliated manifold with all leaves at depth ≤ 1, show that all junctures are compact and that $(M, \mathcal{F})$ is homeomorphic to a C^∞–foliated manifold. Do not appeal to Theorem 8.4.8.

Assume that L is a proper leaf. Then, the set $X = \overline{L} \smallsetminus L$ is compact; so the component of $M \smallsetminus X$ containing L is an open, connected, $\mathcal{F}$–saturated set U. Let L_* be a component of $\partial\widehat{U}$. Let $x \in L_*$ and let $J = [x, z]$ be an arc in the leaf of $\widehat{\mathcal{L}}$ that contains x. There are two possibilities.

(1) The set $L \cap (x, z]$ is finite.

(2) The set $L \cap (x, z]$ is a sequence $\{y_m\}_{m=1}^{\infty}$ converging monotonically to the point x.

In case (2), Lemma 8.1.25 asserts there is a smallest integer $\ell \geq 1$ and an integer $K \geq 1$ such that L_* has a holonomy contraction h_σ of J to x with $h_\sigma(y_m) = y_{m+\ell}$, $\forall\, m \geq K$. We then have a cohomology class

$$\nu : \pi_1(L_*, x) \to \mathbb{Z}$$

such that, for τ any loop on L_*, based at x, the holonomy $h_\tau(y_m) = y_{m+\nu(\tau)\ell}$, for all large values of m. By Theorem 8.1.26, $\nu \in H^1_c(L_*; \mathbb{Z})$ is a compactly supported class and, by Poincaré duality, is represented by a compact submanifold $N_* \subset L_*$ (*cf.* the remark on page 183). Since $\nu(\sigma) = 1$, the class ν is *nondivisible* and the manifold N_* representing it can be taken to be connected (Exercise 8.4.10 below). Both ν and N_* are called junctures. Since N_* is compact, we choose a decomposition

$$L_* = A_* \cup B^1_* \cup \cdots \cup B^r_*,$$

as in (5.2.1) on page 132, so that $N_* \subset \operatorname{int} A_*$ and N_* does not separate A_*.

Exercise 8.4.10. Let L be a connected manifold, not necessarily compact but with $\partial L = \emptyset$. Suppose that $N \subset L$ is a compact submanifold of codimension one representing a nondivisible homology class $[N]$. Prove that there is a compact, *connected* manifold $N' \subset L$ such that $[N] = [N']$. If $\partial L \neq \emptyset$, but $\dim L > 2$, prove the same conclusion. (Hint. Try connecting components of N by tubes.)

Remark. In the above, we use the assumption that $\partial L_* = \emptyset$ to get the Poincaré duality isomorphism

$$H^1_c(L_*; \mathbb{Z}) = H_{n-1}(L_*; \mathbb{Z}),$$

hence our requirement that $\partial_\pitchfork M = \emptyset$. If there is boundary, Poincaré duality gives

$$H^1_c(L_*; \mathbb{Z}) = H_{n-1}(L_*, \partial L_*; \mathbb{Z}).$$

The theory generalizes, but $\partial N_* = N_* \cap \partial L_*$ may be nonempty. When $n = 3$, examples show that it may not be possible to have N_* connected (Exercise 8.4.11 below). Consequently, a generalization of Theorem 8.4.6 will have to drop the requirement that $A \cap B^j = N^j$ be connected, at least when $n = 3$. This interferes with some important applications of the theorem, but usually the desired application remains true and is deduced by doubling along $\partial_\pitchfork M$. As a case in point, Theorem 8.4.8 is true when $\partial_\pitchfork M \neq \emptyset$.

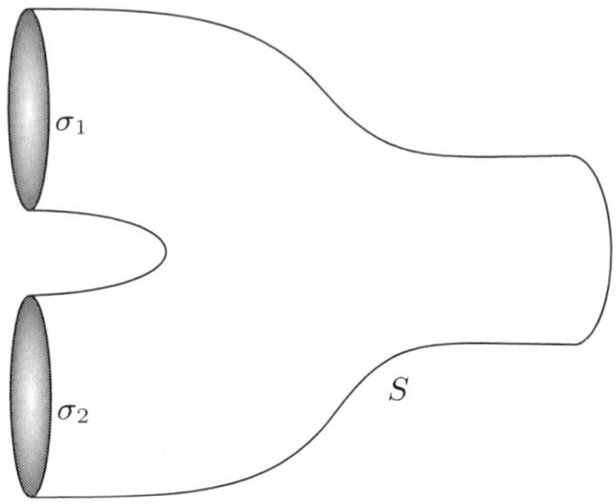

Figure 8.4.2. The pair of pants surface

Exercise 8.4.11. Let S be a "pair of pants", the surface pictured in Figure 8.4.2. It is standard that $\pi_1(S,x)$ is the free group on two generators. The generators are of the form $\sigma_i = \tau_i + \sigma_i^* - \tau_i$, $i = 1, 2$, where σ_1^* and σ_2^* are two of the boundary circles and τ_i is a path from x to a point on σ_i^*, $i = 1, 2$. Define $h : \pi_1(S,x) \to \mathrm{Diff}_+^\infty[0,1]$ to be the homomorphism defined by

$$h(\sigma_1) = f^p,$$
$$h(\sigma_2) = f^q,$$

where $p, q > 1$ are relatively prime integers and f is a contraction of $[0,1)$ to 0. Then $\nu : \pi_1(S,x) \to \mathbb{Z}$, defined by

$$\nu(\sigma_1) = p,$$
$$\nu(\sigma_2) = q,$$

is the juncture for the foliated bundle $(S \times [0,1], \mathcal{F})$ constructed by suspending h. Show that $\nu \in H^1(S;\mathbb{Z})$ is nondivisible, but that no properly imbedded submanifold $N \subset S$, representating the Poincaré dual class $[N] \in H_1(S, \partial S; \mathbb{Z})$, is a connected submanifold.

Exercise 8.4.12. In Exercise 8.4.11, let $p = 2$ and $q = 3$. Show that the typical leaf of $\mathcal{F}$ is made up of a bi–infinite sequence $\{P_i\}_{i=-\infty}^{i=\infty}$ of copies P_i of the connected surface obtained by cutting S along two properly imbedded arcs. We choose the indexing such that, if $J = \{z\} \times [0,1]$, $z \in S$, then the intersection points (z, t_i) of P_i and J satisfy $t_k < t_{k+1}$, $\forall\, k \in \mathbb{Z}$. Show that, for each $k \in \mathbb{Z}$, P_k is attached to P_{k-2}, P_{k-3}, P_{k+2} and P_{k+3}, but not to $P_{k\pm1}$. Thus, the picture of this leaf as a "bi–infinite repetition" of S violates the sequential order implicit in Definition 8.4.2. If we double this foliated manifold along the transverse boundary, we obtain a foliated interval bundle over the 2–holed torus Σ and Exercise 8.4.10 implies that there is a connected manifold representing the juncture. Hence we obtain 2–ended ladders as the noncompact leaves and they are bi–infinite repetitions of Σ conforming to the expected sequential ordering.

Proof of Theorem 8.4.6. The essential feature of N_* is that $\nu(\tau)$ is the homological intersection number of τ with N_*, for every loop τ on L_*. Choose $x \in N_*$. We choose loops $\sigma_0, \sigma_1, \dots, \sigma_s$ generating $\pi_1(A_*, x)$ so that σ_0 is in general position with respect to N_*, meeting that manifold exactly once with $\nu(\sigma_0) = +1$. We require further that σ_k only meet N_* at its endpoints and $\nu(\sigma_k) = 0$, $1 \le k \le s$. Thus, h_{σ_0} is a contraction with $h_{\sigma_0}(y_m) = y_{m+\ell}$ and $h_{\sigma_k}(y_m) = y_m$, $\forall\, m \ge K$, $1 \le k \le s$. If we cut A_* along N_*, we obtain a compact, connected manifold A'_* with fundamental group generated by $\sigma_1, \dots, \sigma_s$. For each $j \ge 0$ and $0 \le i < \ell$, consider the subset $A_{i,j} \subset L$, consisting of points that can be reached from $y_{K+i+j\ell}$ by the lifts of paths on A_*, starting at x and issuing from the same side of N_* as does σ_0, but with interiors not meeting N_*. Adapting the usual Reeb stability argument, we see that this is a submanifold of L diffeomorphic to A'_*, the diffeomorphism being induced by the projection p. Similarly, for each $j \ge 0$ and $0 \le i < \ell$, extend $A_{i,j}$ to $B_{i,j} \subset L$, the set of points that can be reached from $y_{K+i+j\ell}$ by the lifts of paths on L_*, starting at x and issuing from the same side of N_* as does σ_0, but with interiors not meeting N_*. Arguing almost exactly as in the proof of Dippolito's semistability theorem (Theorem 5.3.4), we see that $B_{i,j}$ is a complete, connected, $(n-1)$–dimensional submanifold of L and that $\partial B_{i,j}$ consists of two components, $N_{i,j}$ and $N_{i,j+1}$, each carried diffeomorphically onto N_* by the projection p. It is also evident that

$$p : B_{i,j} \setminus \partial B_{i,j} \to L_* \setminus N_*$$

is a diffeomorphism. Since $\{y_{K+i+m}\}_{m=0}^{\infty}$ converges monotonically to x in $[x, z]$, it follows that

$$C^i = B_{i,0} \cup B_{i,1} \cup \cdots \cup B_{i,m} \cup \cdots \subset L$$

spirals on L_*, $0 \le i < \ell$.

By carrying this construction out for each component L_* of $\partial\widehat{U}$ to which L is asymptotic (from within $\widehat{U}$), we produce finitely many disjoint spirals in L and the complement L_0 in L of the union of the interiors of these spirals is a submanifold bounded away from $\partial\widehat{U}$. Since L is connected and the spirals are disjoint, it is obvious that L_0 is connected and has one boundary component in common with each spiral. If L_0 is compact, we take $A = L_0$ and obtain the required decomposition.

Suppose that L_0 is not compact. Let

$$\widehat{U} = K \cup V_1 \cup \cdots \cup V_p$$

be the nuclear decomposition of $\widehat{U}$ into a compact nucleus K and foliated interval bundles V_i. Since L is proper, relatively closed in U, and L_0 is bounded away from $\partial\widehat{U}$, it follows that L_0 must enter at least one arm V_i. The total holonomy group of that arm (defined on a fiber J) must fix each of the finitely many points $L_0 \cap J$. Thus $L_0 \cap V_i$ falls into finitely many cross sections of V_i.

At this point we assume that L is *totally* proper and, inductively, we assume the truth of Theorem 8.4.6 for leaves at depths $< k =$ depth(L). Since the theorem is vacuously true when the depth is 0, the induction gets off the ground. Since each component of $\partial\widehat{U}$ is (in M) a leaf at depth $< k$, the nuclear decomposition of $\widehat{U}$ can be chosen so that each component of $\partial_\tau V_i$ is a spiral onto a certain leaf $L' \subset \overline{L} \setminus L$ at depth $< k-1$. Then, each of the components of $L_0 \cap V_i$ is forced to be a spiral onto L' also.

Finally, incorporating all of these spirals into our list, we see that the complement A in L of the union of the interiors of these disjoint spirals is contained in K, hence is compact. As remarked above, A is forced to be connected and to have exactly one boundary component in common with each spiral. Each spiral is an infinite repetition of a leaf of $\overline{L} \setminus L$ and every leaf in that set so occurs. By the definition of depth k, these leaves are all at depths $< k$ and at least one is at depth $k-1$. □

The following is a local version [**18**, (6.1)] of Theorem 8.4.6.

Theorem 8.4.13. *Let L be a leaf with an end e such that* $\lim_e(L)$ *is a union of leaves at depths $\le k$, at least one of which, say L_*, is at depth k. Then e is periodic of period L_*.*

In particular, L_* will be the unique leaf in $\lim_e(L)$ at depth k. The proof of Theorem 8.4.13 is surprisingly delicate and will be omitted. A complete account will be found in [**18**, Section 6].

Exercise 8.4.14. Let X be an exceptional minimal set, $U \in \mathcal{O}(\mathcal{F})$ a connected component of $M \setminus X$. If $L \subset U$ is a relatively closed leaf, use the theorem of Duminy discussed in the remark on page 199 to prove that $\mathcal{E}(L)$ is a Cantor set.

Chapter 9

Foliations without Holonomy

We assume that $(M, \mathcal{F})$ is a compact, transversely–oriented, C^2–foliated manifold, $\operatorname{codim} \mathcal{F} = 1$, and we allow $\partial_{\pitchfork} M \neq \emptyset$. (The various uses we will make of the Dippolito semistability theorem are valid when $\partial_{\pitchfork} M \neq \emptyset$ by Exercise 5.3.5 on page 135.) We will not require M to be orientable. Throughout this chapter, $U \in \mathcal{O}(\mathcal{F})$ will be connected.

9.1. Open, Saturated Sets without Holonomy

Assume that $U \in \mathcal{O}(\mathcal{F})$ is connected and has the following property.

Definition 9.1.1. We say that $(U, \mathcal{F}|U)$ is without holonomy if the germinal holonomy group $\mathcal{H}(L)$ is trivial, for every leaf $L \subset U$.

As we will see, either U is a local minimal set or $\mathcal{F}|U$ fibers U over S^1 or $\mathbb{R}$. Of course, the C^2 hypothesis is essential for these assertions, as the Denjoy example shows.

If M were not compact, $(M, \mathcal{F})$ could be C^2–foliated by proper leaves without holonomy, but fail to be a fibration of M over a 1–manifold. Example 1.1.2 gave smooth cases of this on $M = \mathbb{R}^n$, $n \geq 2$. In the case $n = 2$, the *leaf space* $\mathbb{R}^2/\mathcal{F}$ was a locally Euclidean space of dimension one, but was not Hausdorff. When $n \geq 3$, $\mathbb{R}^n/\mathcal{F} \cong \mathbb{R}$, but the leaves fell into two distinct diffeomorphism types. We are going to see that such behavior is impossible when $(U, \mathcal{F}|U)$ satisfies our current hypotheses.

Definition 9.1.2. Let F be a leaf of $\mathcal{F}$. A product neighborhood $N(F)$ of F is an open, $\mathcal{F}$–saturated neighborhood $N(F)$ of F in M that admits an identification $N(F) \cong F \times I$ such that I is a bounded interval, the factor $F \times \{t\}$ is a leaf of $\mathcal{F}|N(F)$, $\forall t \in I$, and the factor $\{z\} \times I$ is a leaf of $\mathcal{L}|N(F)$, $\forall z \in F$.

Remark. If $\partial_\tau M \neq \emptyset$, I may be an open, half open, or closed interval. It is true that $\widehat{N(F)} \cong F \times \overline{I}$ is also a trivially foliated product with fibers the compact intervals $\{z\} \times \overline{I}$ (Exercise 9.1.7), a fact that is not entirely obvious.

Lemma 9.1.3. *If some leaf of $\mathcal{F}|U$ is not dense in U, then every leaf F of $\mathcal{F}|U$ is closed in U and has a product neighborhood $N(F) \subseteq U$.*

Proof. Let $F \subset U$ be an arbitrary leaf and let $Z \subseteq \overline{F} \cap U$ be a minimal set of $\mathcal{F}|U$ (Theorem 8.1.8). Since every leaf in Z has trivial holonomy, Z cannot be exceptional (Theorem 8.2.1). If $Z = U$, every leaf of $\mathcal{F}|U$ is dense in U, contrary to hypothesis. The remaining possibility is that, for each choice of the leaf F, $Z = L$ is a single leaf, closed as a subset of U. In particular, L is a proper leaf of $\mathcal{F}$ (Proposition 8.1.19). Let $x \in L$ and consider a subarc $[x, y) \subset U$ of the leaf of $\mathcal{L}|U$ through x such that $L \cap (x, y) = \emptyset$. We would like to show that every point of $[x, y)$ is a fixed point of the pseudogroup generated by the holonomy of loops on L based at x. This does not follow from $\mathcal{H}(L) = \{1\}$, but it does follow from the assumption that *every* leaf in U has trivial germinal holonomy. Indeed, suppose that $z \in (x, y)$ and that σ is a loop on L, based at x, such that $h_\sigma(z) \neq z$. Replacing σ by $-\sigma$, if necessary, we assume that $x < h_\sigma(z) < z$. Since the germ of h_σ at x is trivial, there is a point $w \in (x, z)$ such that $\lim_{k \to \infty} h_\sigma^k(z) = w$ and, in particular, $h_\sigma(w) = w$. That is, σ lifts to a loop σ' on the leaf L_w through w and $h_{\sigma'}$ is a contraction of $[w, z)$ to w. This contradicts the assumption that $\mathcal{H}(L_w) = \{1\}$; so every point of $[x, y)$ is a fixed point for the holonomy of L. Unless L is a leaf of $\partial_\tau M$, we use the same argument on the other side of L. In any event, by Theorem 5.3.4, we get a product neighborhood $N(L)$ of L. Since $L \subseteq \overline{F} \cap U$, we see that $L = F$ and all assertions follow. □

Theorem 9.1.4. *Either every leaf $F \subset U$ is closed in U and $\mathcal{F}|U$ fibers U over a connected 1–manifold, or each leaf of $\mathcal{F}|U$ is dense in U.*

Proof. By Lemma 9.1.3, either every leaf is dense in U and we are done, or every leaf is closed in U. Under the latter hypothesis, we must prove that the quotient space $U/\mathcal{F}$ is a connected 1–manifold and that the quotient projection $U \to U/\mathcal{F}$ is a fiber bundle. Let F be an arbitrary leaf of $\mathcal{F}|U$ and, by Lemma 9.1.3, let $N(F) \subseteq U$ be a product neighborhood of F. The quotient projection carries each fiber of $\mathcal{L}|N(F)$ homeomorphically onto a neighborhood of the *point* $F \in U/\mathcal{F}$, proving that this quotient space is locally Euclidean of dimension 1. The quotient space is connected (since U is assumed to be connected) and 2nd countable, but Example 1.1.2 shows us that the Hausdorff property might be a problem. In our situation, however,

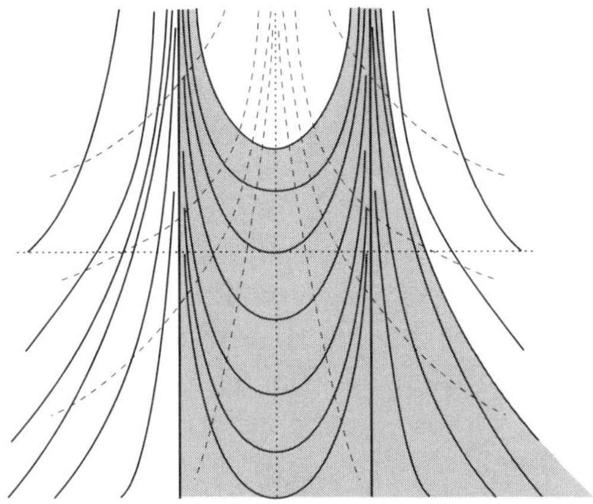

Figure 9.1.1. An $\mathfrak{F}$–saturated neighborhood of F

a *point* $F' \in U/\mathfrak{F}$ can be in the closure of every neighborhood of $F \in U/\mathfrak{F}$ only if the *leaf* F' lies in the closure $\overline{N(F)}$ of every product neighborhood of the leaf $F \subset U$. By shrinking one such neighborhood along the $\mathfrak{L}$–fibers, we produce a sequence $\{N_k(F)\}_{k=1}^{\infty}$ of product neighborhoods such that

$$\overline{N_k(F)} \subset N_{k-1}(F) \subset \widehat{N(F)} = F \times I,$$

where $k \geq 2$ and I is a compact interval. (The equality follows from Exercise 9.1.7, but one can avoid appealing to the exercise by replacing $N(F)$ with a slightly shrunken version.) Evidently, this sequence can be chosen so that

$$\bigcap_{k=1}^{\infty} \overline{N_k(F)} = F,$$

implying that $F' = F$. This analysis proves that $U/\mathfrak{F}$ is a connected 1–manifold and shows simultaneously that the quotient map $U \to U/\mathfrak{F}$ is a fiber bundle. □

Example 9.1.5. It is instructive, in the light of the above proof, to take another look at the case $n = 2$ in Example 1.1.2. In Figure 9.1.1, we redraw Figure 1.1.1, adding dotted curves representing the leaves of a transverse foliation $\mathfrak{L}$. The lines $x = 1$ and $x = -1$ are respective leaves F and F' of $\mathfrak{F}$ and $x = 0$ defines a leaf T of $\mathfrak{L}$. The shaded region is the typical open, $\mathfrak{F}$–saturated neighborhood V of F that does not contain the leaf F'. It is evident, however, that $F' \subset \overline{V}$ for all such neighborhoods V; so the point $F' \in \mathbb{R}^2/\mathfrak{F}$ is in the closure of every neighborhood of $F \in \mathbb{R}^2/\mathfrak{F}$. The reason for this is precisely that no saturated neighborhood of F is a product neighborhood. Indeed, all leaves in such a neighborhood that lie properly to the left of F meet the leaf T of $\mathcal{L}$, while F itself does not meet that leaf. In other words, Dippolito's theorem (Theorem 5.3.4) fails for the leaf F, implying that $(\mathbb{R}^2, \mathcal{F})$ cannot be realized as an open, saturated subset of a compact surface foliated by curves. In the higher–dimensional cases of Example 1.1.2, similar considerations show that the cylindrical leaf F (labeled $c = 0$ in Figure 1.1.2) also violates Dippolito's theorem, but this no longer prevents the leaf space from being Hausdorff. It does, however, prevent the leaf F from having a product neighborhood; so the quotient map fails to be a fibration. In this case, the leaves on one side of F are topological cylinders, those on the other side being topological planes.

Lemma 9.1.6. *If $\mathcal{F}|U$ is a fibration over a 1–manifold B, but $(\widehat{U}, \widehat{\mathcal{F}})$ is not a trivially foliated product, then $B \cong S^1$.*

Exercise 9.1.7. Prove the assertion in the remark on page 206 and use this, together with the well–known classification of 1–manifolds, to prove Lemma 9.1.6. In particular, note that $\widehat{U}$ is a trivially foliated product if $U \cap \partial_\tau M \neq \emptyset$. The trivially foliated case is uninteresting.

Hypothesis. $(\widehat{U}, \widehat{\mathcal{F}})$ is not a trivially foliated product.

Lemma 9.1.8. *Each leaf L of $\partial\widehat{U}$ has a holonomy contraction.*

Proof. Let $x \in L$ and fix a subarc $[x, y)$ of the leaf of $\widehat{\mathcal{L}}$ out of x. Either the leaves of $\mathcal{F}|U$ are each dense in U or $\mathcal{F}|U$ is a fibration over S^1. So Theorem 5.3.4 implies that the holonomy of L does not have fixed points in (x, y) arbitrarily near x. Thus, we find a loop σ on L, based at x, such that $h_\sigma(z) < z$, some $z \in (x, y)$. If $\lim_{k\to\infty} h_\sigma^k(z) = w > x$, we would obtain nontrivial germinal holonomy on the leaf through w. □

9.2. A Leaf-Preserving Flow

In this section we will prove the following generalization of [**115**, Theorem 6].

Theorem 9.2.1 (Sacksteder). *If $(U, \mathcal{F}|U)$ is without holonomy, then there is a C^0 flow*

$$\Phi : \mathbb{R} \times \widehat{U} \to \widehat{U}$$

that fixes the points of $\partial\widehat{U}$, has the leaves of $\mathcal{L}|U$ as nondegenerate flow lines, preserves the foliation $\widehat{\mathcal{F}}$, carrying leaves diffeomorphically to leaves, and is transitive on the set of leaves of $\mathcal{F}|U$.

Sacksteder states and proves this theorem for the case $U = M$. The general case is no harder and will be needed in the proof of Duminy's vanishing theorem for gv($\mathcal{F}$) [**38**] in Volume II. Instead of following Sacksteder's proof, we will use an argument suggested by S. P. Novikov [**99**] and formulated rigorously by H. Imanishi [**75**].

Theorem 9.2.1 is easy in the case that $\mathcal{F}|U$ is a fibration $\pi : U \to S^1$. Just lift any nonsingular vector field on S^1 to a vector field along the leaves of $\mathcal{L}|U$. The field will be complete and its flow can never reach $\partial\widehat{U}$ in forward or backward time; hence it extends to a C^0 flow fixing $\partial\widehat{U}$ pointwise. It is a leaf–preserving flow because its restriction to U projects to a well–defined flow on the leaf space S^1. It is transitive on the set of leaves in U because the flow on S^1, being nonsingular, is transitive on S^1. The case of dense leaves is harder and we restrict our attention to that case.

An application of Lemma 3.3.7 allows us to fix a smoothly imbedded, transverse circle $\Sigma \subset U$. Remark that Σ meets every leaf of $\mathcal{F}|U$.

Exercise 9.2.2. Prove that the transverse foliation $\mathcal{L}$ can be chosen so that Σ is a leaf.

9.2.A. Normal fences.

By Exercise 9.2.2, Σ can be covered by proper subarcs that identify naturally with components of the transverse space S. Thus, if $x, y \in \Sigma$ lie on the same leaf L of $\mathcal{F}$, a path σ on L from x to y defines a holonomy germ $\widehat{h}_\sigma$ at x with target y. This defines a holonomy groupoid $\mathcal{G}_\Sigma$ on Σ and, similarly, one sees that there is induced a holonomy pseudogroup Γ_Σ on Σ, consisting of those local diffeomorphisms that, at each point of their domain, have germ in $\mathcal{G}_\Sigma$.

Definition 9.2.3. Let $(a, c) \subset \Sigma$ be an open subarc, oriented by the transverse orientation of $\mathcal{F}$, with $a \neq c$, let $b \in (a, c)$, let L be the leaf of $\mathcal{F}$ through b, and let $\sigma : [0, 1] \to L$ be a continuous path with $\sigma(0) = b$, $\sigma(1) \in \Sigma$. A normal fence along σ is a continuous map

$$H : [0, 1] \times (a, c) \to M$$

such that

1. $H(0, x) = x$, $\forall\, x \in (a, c)$;
2. $H(t, b) = \sigma(t)$, $0 \le t \le 1$;
3. H immerses $\{t\} \times (a, c)$ into a leaf of $\mathcal{L}$, $0 \le t \le 1$;
4. $H|([0, 1] \times \{x\})$ is a curve in a leaf of $\mathcal{F}$, $\forall\, x \in (a, c)$.

If H is such a normal fence, it is obvious that $h : (a, c) \to \Sigma$, defined by

$$h(x) = H(1, x),$$

is an immersion that has local restrictions to elements of Γ_Σ.

Lemma 9.2.4. *Every normal fence $H : [0, 1] \times (a, c) \to M$ extends continuously to $\overline{H} : [0, 1] \times [a, c] \to M$.*

Proof. There is certainly a value of $\epsilon > 0$ such that H extends continuously to $[0,1] \times (a,c) \cup [0,\epsilon] \times \{c\}$. For this, one appeals to the product structure in a biregular chart about c. If one cannot choose $\epsilon = 1$, then there is a number τ, $0 < \tau \leq 1$, such that H extends to $[0,1] \times (a,c) \cup [0,\tau) \times \{c\}$, but not to $[0,1] \times (a,c) \cup [0,\tau] \times \{c\}$. Using the fact that every leaf of $\mathcal{F}$ through Σ is without holonomy and is dense in U, we deduce a contradiction.

Let $t_n \uparrow \tau$ in $[0,\tau)$. Using the compactness of M and passing to a subsequence, if necessary, we assume that $H(t_n, c) \to z$ in M as $n \to \infty$. Fix a biregular atlas $\mathcal{U}$, let S denote the corresponding 1–dimensional transverse manifold, oriented by the transverse orientation of $\mathcal{F}$, and fix $(W, x.y) \in \mathcal{U}$ such that $z \in W$. On $\overline{W}$, we assume that $0 \leq y \leq 1$. We can suppose that $H(t_n, c) \in W$, $n \geq 1$. Since $\lim_{t \to \tau} H(t,x)$ exists for all $x \in (a,c)$ but does not exist for $x = c$, the product structure of W forces the arcs $H(\{t_n\} \times [b,c])$, for n sufficiently large, to enter the "bottom" of $\overline{W}$ before terminating at $H(t_n, c)$. In fact, there is $c_n \in (b,c)$ such that $c_n \uparrow c$ as $n \to \infty$, the arc $H(\{t_n\} \times [c_n, c])$ lies in $\overline{W}$ and $y(H(t_n, c_n)) = 0$. This is an elementary but crucial point and the reader should ponder it carefully. By throwing away at most finitely many terms of the sequence, we assume that c_n is defined for every $n \geq 1$. Finally, we can suppose that the sequence $y(H(t_n, c)) = y_n$ is either constant or converges strictly monotonically to $y(z)$ in S.

Consider first the case that the sequence is strictly increasing. In fact, we will only use this hypothesis to ensure that $y_1 < y_n$, $\forall n > 1$. Choose n so large that $H(\{t_1\} \times [c_n, c]) \subset W$ and set $y_0 = y(H(t_1, c_n))$. Thus, $H|([t_1, t_n] \times [c_n, c])$ defines a holonomy transformation h of the arc $[0, y_n]$ onto $[y_0, y_1] \subset (0, y_n)$ (moving "backwards" along the fence by letting t decrease from t_n to t_1). By the Brouwer theorem, there is a fixed point of h, hence a maximal such point $u < y_n$. The normal fence then gives a loop on the leaf L_u through $u \in S$ that has nontrivial germinal holonomy on the positive side of L_u. But u is the y–coordinate of a point in the normal fence; hence L_u passes through $\Sigma \subset U$. That is, we have found a leaf of $\mathcal{F}|U$ having nontrivial holonomy.

The cases that $\{y_n\}_{n=1}^{\infty}$ is constant or is strictly decreasing can be handled simultaneously by assuming that the sequence is weakly decreasing. In this case, $0 < y(z) \leq y_1$, implying that the leaf through z lies in U. Since the leaf L_c of $\mathcal{F}$ through c is dense in U, it meets the arc $H(\{t_1\} \times [c_1, c])$ in a point z_0 with y–coordinate $y_0 < y(z) \leq y_n$, $\forall n \geq 1$. A path in L_c from $H(t_1, c)$ to z_0 defines an element g of holonomy on a neighborhood of y_1 in S onto a neighborhood of y_0. Choose n so large that $g(y(H(t_1, c_n))) = y_*$ is defined. Combining g with the holonomy transformation h defined by $H|([t_1, t_n] \times [c_n, c])$ (again moving backwards along the fence by letting t

decrease from t_n to t_1) defines a homeomorphism

$$g \circ h : [0, y_n] \to [y_*, y_0] \subset (0, y_n).$$

The contradiction is now obtained exactly as in the previous paragraph.

We conclude that H extends continuously to $[0,1] \times (a, c]$. In exactly the same way, we obtain a continuous extension $\overline{H}$ to $[0,1] \times [a, c]$. □

We will call $\overline{H}$ a *closed* normal fence and H itself an open one. A subarc $(a, b) \subset \Sigma$ with $a \neq b$ will be called a proper subarc and we consider the set Γ'_Σ consisting of those immersions $g : (a, b) \to \Sigma$, with (a, b) a proper subarc of Σ, having germs $\widehat{g}_x \in \mathcal{G}_\Sigma$, $a < x < b$.

Corollary 9.2.5. *Let $g \in \Gamma'_\Sigma$ with $\operatorname{dom} g = (a, b)$. Then g can be produced by an open normal fence, hence extends to $\overline{g} : [a, b] \to \Sigma$ and thence to $\widetilde{g} \in \Gamma'_\Sigma$ with domain $(\widetilde{a}, \widetilde{b}) \supset [a, b]$.*

Proof. Let Λ_g denote the set of all open normal fences

$$H : [0,1] \times (p, q) \to M$$

such that $(p, q) \subset (a, b)$ and $H(1, x) = g(x), p < x < q$. This set is nonempty. Indeed, if $x \in (a, b)$ the germ $\widehat{g}_x$ at x is an element of $\mathcal{G}_\Sigma$. Hence the restriction of g to some $(x - \epsilon, x + \epsilon) \subset (a, b)$ is a pure word γ in suitable elements $\gamma_{\alpha\beta}$ of the holonomy cocycle. It is easy to see that such γ is represented by an open normal fence H. We put a partial order on Λ_g by setting $F \leq H$ whenever $\operatorname{dom} F \subseteq \operatorname{dom} H$ and $F = H|\operatorname{dom} F$. It is clear that $(\Lambda_g, \leq)$ is an inductive set; so Zorn's lemma provides a maximal $H \in \Lambda_g$. If $[0,1] \times (p, q)$ is the domain of H, we must prove that $p = a$ and $q = b$.

Suppose that $q < b$ and let $\overline{H}$ be the closed normal fence given by the lemma. Then $\sigma(t) = \overline{H}(t, q)$ is a path on the leaf of $\mathcal{F}$ through q with $\sigma(0) = q$ and $\sigma(1) = g(q)$. Let h_σ be an element of holonomy corresponding to σ and defined on a neighborhood of q in (a, b). Also, let $\gamma \in \Gamma_\Sigma$ be a pure word in the elements of the holonomy cocycle, defined on a neighborhood of q in (a, b), with germ $\widehat{\gamma}_q = \widehat{g}_q$. We can assume that γ agrees with g on its entire domain. Then $\gamma^{-1} \circ h_\sigma$ is also defined on a connected neighborhood J of q in (a, b) and fixes q. Since the leaves meeting J are without holonomy, this transformation is the identity on J and we conclude that $h_\sigma|J$ is a pure word in the elements of the holonomy cocycle and agrees with $g|J$. It follows easily that $h_\sigma|J$ is defined by an open normal fence F with $F(t, q) = \sigma(t)$, $0 \leq t \leq 1$. If $J = (q - \epsilon, q + \delta)$, then

$$\widetilde{H}(t, x) = \begin{cases} H(t, x), & p < x < q, \\ F(t, x), & q \leq x < q + \delta \end{cases}$$

defines a normal fence extending H such that $\widetilde{H}(1,x) = g(x)$, $p < x < q+\delta$. This contradiction proves that $q = b$. Similarly, $p = a$. Lemma 9.2.4 gives the extension $\overline{g}$ and essentially the same argument as above gives $\widetilde{g}$. $\square$

9.2.B. The holonomy group G_Σ. Let $p : \mathbb{R} \to \Sigma$ be the universal covering and remark that the lifts of elements $g \in \Gamma'_\Sigma$ to this covering space are one–to–one, hence generate a pseudogroup $\Gamma_\mathbb{R}$ on $\mathbb{R}$. Remark also that Corollary 9.2.5 implies that such a lift has image a bounded interval.

Lemma 9.2.6. *The group $G_\mathbb{R} = \Gamma_\mathbb{R} \cap \operatorname{Diff}^2_+(\mathbb{R})$ generates $\Gamma_\mathbb{R}$ as a pseudogroup. The only element of $G_\mathbb{R}$ that fixes a point is the identity.*

Proof. For the first assertion, we show that every $g \in \Gamma_\mathbb{R}$ with connected domain extends to an element $\widetilde{g} \in G_\mathbb{R}$. Indeed, let $(E_g, \leq)$ denote the partially ordered set of all elements of $\Gamma_\mathbb{R}$ that have connected domain and extend g, where $g_1 \leq g_2$ means that $\operatorname{dom} g_1 \subseteq \operatorname{dom} g_2$ and $g_1 = g_2|\operatorname{dom} g_1$. This is clearly an inductive set and we take $f : (p,q) \to \mathbb{R}$ to be a maximal element. If $q < \infty$, then there is $\epsilon > 0$ such that $f|(q-\epsilon, q)$ is a lift of an element of Γ'_Σ and Corollary 9.2.5 implies that f has an extension $\widetilde{f}$ to a properly larger interval. This contradiction proves that $q = \infty$ and, similarly, $p = -\infty$. If $\widetilde{f}$ is not onto, then we consider $\widetilde{f}^{-1} : (c,d) \to \mathbb{R}$ where either c or d is finite. Say that $d < \infty$. Then there is $\epsilon > 0$ such that $\widetilde{f}^{-1}|(d-\epsilon, d)$ is a lift of an element of Γ'_Σ. But $\lim_{x\to d-} \widetilde{f}^{-1}(x) = \infty$ then contradicts the fact that lifts of elements of Γ'_Σ have bounded images.

For the second assertion, suppose that $g(t) = t$, some $g \in G_\mathbb{R}$ and some $t \in \mathbb{R}$. The restriction of g to a suitable neighborhood of t is a lift of some $\gamma \in \Gamma'_\Sigma$, implying that γ fixes $p(t)$. Thus, γ is the identity in a neighborhood of $p(t)$, implying that g is the identity in a neighborhood of t. Thus, the set of fixed points of g is open. By the continuity of g, this set is also closed, hence must be all of $\mathbb{R}$. $\square$

This lemma allows us to define a total ordering of $G_\mathbb{R}$, setting $g < h$ if and only if $g(t) < h(t)$, for some (hence every) $t \in \mathbb{R}$. Since $G_\mathbb{R}$ preserves orientation, it follows immediately that, if $g < h$, then

$$fg < fh,$$
$$gf < hf,$$

for arbitrary $f \in G_\mathbb{R}$. Furthermore, by considering the values of g, g^{-1}, h and gh on some convenient point, say 0, one checks that

$$g > \mathrm{id} \text{ and } h > \mathrm{id} \Rightarrow gh > \mathrm{id},$$
$$g > \mathrm{id} \Rightarrow g^{-1} < \mathrm{id};$$

so we are dealing with an *ordered group*.

Lemma 9.2.7. *The ordered group $(G_{\mathbb{R}}, <)$ is Archimedean.*

Proof. Let $g > \mathrm{id}$ and $h > \mathrm{id}$. We must show that, for some integer $n \geq 1$, $g^n > h$. If not, then

$$0 < g(0) < g^2(0) < \cdots < g^n(0) < \cdots \leq h(0).$$

Then, there is a least upper bound $0 < t \leq h(0)$ of $\{g^n(0)\}_{n=1}^{\infty}$. This implies that $g(t) = t$, contradicting the fact that $g > \mathrm{id}$. □

We need the following classical result.

Theorem 9.2.8 (Hölder). *Every Archimedean ordered group $(G, <)$ is order isomorphic to an additive subgroup of $\mathbb{R}$. In particular, G is abelian.*

Exercise 9.2.9. Use Dedekind cuts to prove Hölder's theorem.

For another elementary proof, see [**65**, pp. 186–190]. In our case, this implies that $G_{\mathbb{R}}$ projects homomorphically onto an abelian subgroup G_Σ of $\mathrm{Diff}^2_+(\Sigma)$. Indeed, the covering transformations for $p : \mathbb{R} \to \Sigma$ are locally lifts of identity maps in Γ'_Σ, hence form a subgroup $\mathbb{Z} \subset G_{\mathbb{R}}$, and the quotient group $G_{\mathbb{R}}/\mathbb{Z} = G_\Sigma$ has a well–defined induced action on Σ. The following should be clear.

Proposition 9.2.10. *The subgroup $G_\Sigma \subset \Gamma_\Sigma$ of globally defined C^2 diffeomorphisms of Σ is abelian and generates Γ_Σ. This is called the (total) holonomy group of $\mathcal{F}|U$ on Σ.*

9.2.C. A G_Σ–invariant probability measure. We consider probability measures μ on Σ. That is, μ is a Borel measure with $\mu(\Sigma) = 1$. If $g \in G_\Sigma$, then $g_*\mu$, defined by $g_*\mu(B) = \mu(g^{-1}(B))$ on arbitrary Borel sets $B \subseteq \Sigma$, is also a probability measure. We will say that μ is holonomy invariant if $g_*\mu = \mu$, $\forall\, g \in G_\Sigma$. In order to construct such a measure, we need the following observation.

Lemma 9.2.11. *The group G_Σ is at most countably infinite.*

Indeed, the biregular cover $\mathcal{U}$ for $\mathcal{F}$ induces a countable biregular cover $\mathcal{V}$ of $\widehat{U}$ for $\widehat{\mathcal{F}}$ and every element $g \in G_\Sigma$ is locally a word in the countable set $\{\gamma_{ij}\}_{i,j \in \mathfrak{I}}$ of elements of the holonomy cocycle associated to $\mathcal{V}$. Since Σ is compact, it takes only finitely many such words to define g.

We also need the standard fact that the set of probability measures on the compact space Σ is weakly compact. That is, if $\{\mu_q\}_{q=1}^{\infty}$ is a sequence of probability measures, then there is a probability measure μ and a subsequence $\{\mu_{q_k}\}_{k=1}^{\infty}$ such that, for every continuous function $f : \Sigma \to \mathbb{R}$,

$$\lim_{k \to \infty} \int_\Sigma f \, d\mu_{q_k} = \int_\Sigma f \, d\mu.$$

Proposition 9.2.12. *There is a G_Σ–invariant probability measure μ_Σ on the space Σ.*

Proof. Enumerate $G_\Sigma = \{g_i\}_{i=1}^N$. If N is finite, let ν be any probability measure on Σ and set

$$\mu = \frac{g_{1*}\nu + \cdots + g_{N*}\nu}{N},$$

clearly a G_Σ–invariant probability measure. Assume that $N = \infty$, let ν be any probability measure and remark that, for each $k \geq 1$,

$$\nu_k = \frac{\nu + g_{1*}\nu + \cdots + g_{1*}^k\nu}{k}$$

is also a probability measure. Let μ_1 be the weak limit of a suitable subsequence of these measures. Since

$$\nu_k - g_{1*}\nu_k = \frac{\nu - g_{1*}^{k+1}\nu}{k}$$

converges weakly to 0 as $k \to \infty$, it follows that $g_{1*}\mu_1 = \mu_1$. Inductively, suppose that a probability measure μ_n has been found such that $g_{i*}\mu_n = \mu_n$, $1 \leq i \leq n$. Again construct a sequence

$$\frac{\mu_n + g_{n+1*}\mu_n + \cdots + g_{n+1*}^k\mu_n}{k},$$

and pass to a weakly convergent subsequence limiting on a measure μ_{n+1} that is invariant under g_{n+1}. The fact that g_i commutes with g_{n+1}, $1 \leq i \leq n$ implies that each measure in the above sequence is invariant by these same g_i's, hence that μ_{n+1} is g_i–invariant, $1 \leq i \leq n+1$. Finally, passing to a weakly convergent subsequence of $\{\mu_n\}_{n=1}^\infty$ produces the desired G_Σ–invariant probability measure μ_Σ. □

9.2.D. The Sacksteder flow. We take up the proof of Theorem 9.2.1. The first move in the construction of Φ is to transfer the G_Σ–invariant measure μ_Σ to each leaf of $\mathcal{L}|U$. As above, we have a countable biregular cover $\mathcal{V} = \{(V_i, x_i, y_i)\}_{i \in \mathfrak{I}}$ for $(\widehat{U}, \widehat{\mathcal{F}}, \widehat{\mathcal{L}})$ and we let $S_\mathcal{V}$ denote the corresponding transverse space.

Definition 9.2.13. A measure $\mu_\mathcal{V}$ on $S_\mathcal{V}$ is a *holonomy–invariant measure* if it is finite on compact sets and, whenever $\gamma \in \Gamma_\mathcal{V}$ and $B \subseteq \operatorname{dom}\gamma$ is a Borel set, then $\mu_\mathcal{V}(\gamma(B)) = \mu_\mathcal{V}(B)$.

Exercise 9.2.14. Using the fact that the leaves of $\mathcal{F}|U$ are each dense, prove that a holonomy–invariant measure $\mu_\mathcal{V}$ is necessarily *continuous*. That is, it is nonatomic and strictly positive on nonempty, open arcs. (Hint. Define the *support* $\operatorname{supp}\mu$ of the measure μ to be the set of points x such that $\mu(U_x) > 0$, for all neighborhoods $U_x \subset S_{\mathcal{V}}$ of x and investigate its properties.)

Definition 9.2.15. Let μ be a measure that is defined on each leaf of $\mathcal{L}|U$ and is finite on compact subsets of these leaves. We say that μ is a transverse, (holonomy) invariant measure for $\mathcal{F}|U$ if there is a holonomy–invariant measure $\mu_{\mathcal{V}}$ on $S_{\mathcal{V}}$ such that y_i is a measure–preserving homeomorphism of each $\mathcal{L}$–plaque of V_i onto S_i, $\forall\, i \in \mathfrak{I}$.

Lemma 9.2.16. *There is a transverse, invariant measure μ for $\mathcal{F}|U$ that is a continuous measure on each leaf of $\mathcal{L}|U$ and is finite on compact sets.*

The straightforward proof proceeds by transferring the measure μ_Σ locally to a measure $\mu_{\mathcal{V}}$ on $S_{\mathcal{V}}$ via holonomy translations. The G_Σ–invariance of μ_Σ implies that $\mu_{\mathcal{V}}$ is well defined and holonomy invariant. Each $\mathcal{L}$–plaque in V_i identifies canonically with the component S_i of $S_{\mathcal{V}}$; so $\mu_{\mathcal{V}}$ transfers canonically to a measure μ on these plaques. On overlaps, holonomy invariance again guarantees that these local measures on $\mathcal{L}|U$ agree. Details are left to the reader.

Lemma 9.2.17. *Let $x \in \partial\widehat{U}$ and let $[x, y)$ be a subarc of the leaf of $\widehat{\mathcal{L}}$ out of x. Then $\mu(x, y) = \infty$.*

Proof. By Lemma 9.1.8, there is a holonomy contraction

$$h_\sigma : [x, u) \to [x, u),$$

for $u \in (x, y)$ sufficiently near x. We can assume that $[x, u) \subset S_{\mathcal{V}}$, remarking that $h_\sigma|(x, u) \in \Gamma_{\mathcal{U}}$. Let $z \in (x, u)$ and set $z_k = h_\sigma^k(z)$, $k \geq 0$. By holonomy invariance and continuity of the measure, $\mu[z_{k+1}, z_k] > 0$ is independent of $k \geq 0$; so $\mu(x, y) = \infty$. □

Proof of Theorem 9.2.1. On each leaf R of $\mathcal{L}|U$, define a continuous local flow as follows. Using the transverse orientation of $\mathcal{F}$, obtain an orientation on R, hence an ordering that permits us to use interval notation for subarcs of R. If $x \in R$, $\epsilon > 0$ is small enough and $t \in (-\epsilon, \epsilon)$, define $x_t \in R$ as follows.

1. If $t > 0$, the continuity of μ implies that there is a unique point $x_t \in R$, $x_t > x$, such that $\mu[x, x_t) = t$.
2. If $t < 0$, there is a unique $x_t \in R$, $x_t < x$, such that $\mu(x_t, x] = -t$.
3. Set $x_0 = x$.

It is trivial to verify that $x_{t+s} = (x_t)_s = (x_s)_t$, whenever all terms are defined. The continuity of μ implies that this local flow on R is continuous.

Given a pair of compact biregular charts $\overline{V} \subset \operatorname{int}(\overline{W}) \subset \widehat{U}$, let J be a compact $\mathcal{L}$–plaque of $\overline{V}$ contained in the interior of the compact $\mathcal{L}$–plaque K

of $\overline{W}$. The two components of $K \setminus \text{int}(J)$ are nondegenerate; so we choose $\epsilon > 0$ smaller than the μ–measure of each. By the transverse invariant character of μ, this ϵ works simultaneously for all $\mathcal{L}$–plaques J of $\overline{V}$. By the choice of ϵ, we see that $x \in J$ and $t \in (-\epsilon, \epsilon)$ implies that $x_t \in K$. This gives a continuous local flow

$$\Phi : (-\epsilon, \epsilon) \times V \to W \subset M.$$

Because of the holonomy invariance of μ, this flow takes $\mathcal{F}$–plaques into $\mathcal{F}$–plaques and, on overlaps, two such local flows will agree.

Since the measure grows without bound near $\partial\widehat{U}$, we can fix $\epsilon > 0$ uniformly for a (generally infinite) covering of U by charts $\overline{V}$ as above. Thus, we obtain a local flow

$$\Phi : (-\epsilon, \epsilon) \times U \to U.$$

By a standard argument in the theory of flows, this globalizes to a C^0 flow

$$\Phi : \mathbb{R} \times U \to U.$$

By holonomy invariance of the measure, this flow preserves $\mathcal{F}|U$ locally, hence globally. Flow lines that converge to points of $\partial\widehat{U}$ in forward or backward time cannot reach those points in finite time (Lemma 9.2.17). So this flow can be extended to a C^0 flow

$$\Phi : \mathbb{R} \times \widehat{U} \to \widehat{U}$$

by requiring that points of $\partial\widehat{U}$ be fixed. □

Exercise 9.2.18. Remark that the closed transversal Σ is a flow line of period 1. Show that there is an injective homomorphism $\rho : G_\Sigma \to \mathbb{R}/\mathbb{Z}$ such that $g = \Phi_{\rho(g)}|\Sigma$, $\forall\, g \in G_\Sigma$.

Exercise 9.2.19. Let $g \in \text{Homeo}_+(S^1)$. By the proof of Proposition 9.2.12, there is a g–invariant probability measure μ on S^1. Recall that $\operatorname{supp}\mu$ is the set of points x such that $\mu(U_x) > 0$, for all neighborhoods U_x of x in S^1.

1. If $[x, g(x))$ denotes the positively–oriented, half–open arc from x to $g(x)$ in S^1, show that $\mu[x, g(x))$ is locally constant in x, hence constant. (Note that, if $g(x) = x$, $[x, g(x)) = \emptyset$.) Let $\rho(g) \in \mathbb{R}/\mathbb{Z}$ be the residue class of $\mu[x, g(x))$ (mod $\mathbb{Z}$).
2. Show that g has a fixed point $\Leftrightarrow \rho(g) = 0$.
3. Prove that $\rho(g^n) = n\rho(g)$, $\forall\, n \in \mathbb{Z}$.
4. Show that there is a periodic point of minimal period $q > 1 \Leftrightarrow \rho(g)$ has representative $p/q \in (0, 1)$, p and q relatively prime.
5. If $\rho(g)$ is irrational, prove that μ is nonatomic and that $\operatorname{supp}\mu$ is the unique minimal, closed, nonempty, g–invariant subset. (Hint. A

minimal set is either exceptional or all of S^1. In the exceptional case, consider orbits of gaps.)

6. If $\rho(g)$ is irrational and $g \in \mathrm{Diff}^2_+(S^1)$, show that g is topologically conjugate to a rotation through $2\pi\rho(g)$ radians (Denjoy's theorem).

In fact it can be shown that $\rho(g)$ is independent of the choice of g–invariant measure μ. It is called the *Poincaré rotation number* of g.

9.3. A Closed, Nonsingular 1–Form

We turn to the case that $U = M$. That is, the compact, C^2–foliated manifold $(M, \mathcal{F})$ itself is without holonomy. To avoid the completely uninteresting case of a product foliation, we assume that $\partial_\tau M = \emptyset$. Let $\mathcal{U} = \{(U_\alpha, x_\alpha, y_\alpha)\}_{\alpha=1}^q$ and $\mathcal{W} = \{(W_\lambda, u_\lambda, v_\lambda)\}_{\lambda=1}^m$ be biregular atlases such that, whenever $U_\alpha \cap U_\beta \neq \emptyset$, then $U_\alpha \cup U_\beta \subset W_\lambda$, for some λ. Fix a choice of $\mathcal{F}$–plaque $P_\alpha \subset U_\alpha$, $1 \le \alpha \le q$. Choose a new last coordinate y^*_α in U_α by letting the value of y^*_α on the $\mathcal{F}$–plaque $P \subset U_\alpha$ be $\pm$ the μ–measure of the arc in S_α joining P and P_α, the sign being chosen to respect the transverse orientation. Hereafter, we will refer to this as the *signed distance* from P_α to P. This notion of signed distance is equally valid for plaques in W_λ.

Lemma 9.3.1. *The atlas $\{U_\alpha, x_\alpha, y^*_\alpha\}_{\alpha=1}^q$ determines a possibly new C^∞ structure on M and is biregular for $\mathcal{F}$ and $\mathcal{L}$. Indeed, if $U_\alpha \cap U_\beta \neq \emptyset$, there is a constant $c_{\alpha\beta}$ such that $y^*_\alpha = y^*_\beta + c_{\alpha\beta}$ on $U_\alpha \cap U_\beta$.*

Proof. All assertions follow from the last one. To prove this, let $U_\alpha \cap U_\beta \neq \emptyset$ and let $U_\alpha \cup U_\beta \subset W_\lambda$. Let Q_α and Q_β be the $\mathcal{F}$–plaques in W_λ containing P_α and P_β respectively. Let $c_{\alpha\beta}$ be the signed distance from Q_α to Q_β. If $z \in U_\alpha \cap U_\beta$, let y^*_α and y^*_β be the nth coordinates of z in the two charts. Let $z' \in P_\beta \subset Q_\beta$ such that $\Phi_{y^*_\beta}(z') = z$. Similarly, let $z'' \in P_\alpha \subset Q_\alpha$ satisfy $\Phi_{y^*_\alpha}(z'') = z$. Then $\Phi_{c_{\alpha\beta}}(z'') = z'$ and so

$$\Phi_{y^*_\alpha}(z'') = z = \Phi_{y^*_\beta}(z') = \Phi_{y^*_\beta}(\Phi_{c_{\alpha\beta}}(z'')) = \Phi_{y^*_\beta + c_{\alpha\beta}}(z'').$$

This implies that $y^*_\alpha = y^*_\beta + c_{\alpha\beta}$. □

Corollary 9.3.2. *In the new differentiable structure, there is a closed, nonsingular 1–form ω that defines the foliation $\mathcal{F}$, has nonsingular restriction $\omega|\partial_\pitchfork M$, and has nontrivial cohomology class $[\omega] \in H^1(M;\mathbb{R})$.*

Proof. The exact forms $dy^*_\alpha \in A^1(U_\alpha)$ define $\mathcal{F}|U_\alpha$. This is because the $\mathcal{F}$–plaques are the level sets of y^*_α. On $U_\alpha \cap U_\beta$, $dy^*_\alpha = dy^*_\beta$ and these forms assemble to a well–defined, global, nonsingular form ω that defines $\mathcal{F}$. Since ω is locally exact, it is closed. Finally, it is not possible that $\omega = df$. Indeed, $f : M \to \mathbb{R}$ would assume an extreme value at some point $x \in M$. Such a point is either a critical point, contradicting the fact that $\omega_x \neq 0$, or $x \in \partial M$. But we have assumed that $\partial M = \partial_{\pitchfork} M$; so $\omega|\partial M$ is also nonsingular and the same contradiction arises. □

Exercise 9.3.3. This exercise requires some knowledge of Čech cohomology. Show that the assignment $U_\alpha \cap U_\beta \mapsto c_{\alpha\beta}$ in Lemma 9.3.1 is a Čech 1–cocycle c defining the same cohomology class on M as the closed form ω of Corollary 9.3.2. More precisely, assume that the regular cover has been sufficiently refined that its Čech cohomology is the Čech cohomology of M and use the de Rham theorem to relate $[c]$ and $[\omega]$.

Remarks. V. I. Arnol′d [**4**, Section 1] has constructed smooth, orientation–preserving diffeomorphisms of S^1 that have irrational rotation number (and so are C^0 conjugate to a rotation by Exercise 9.2.19) but are not C^1 conjugate to a rotation. Suspending such a diffeomorphism over T^n produces a smooth, dense leaved foliation of T^{n+1} without holonomy that cannot be defined by a closed, nonsingular 1–form. Thus, the change of differentiable structure in Corollary 9.3.2 cannot generally be avoided. The transverse, invariant measure μ in these examples, while continuous, is not *absolutely* continuous. This means that μ and Lebesgue measure do not have exactly the same sets of measure zero.

Recall that, when $\mathcal{F}$ is defined by a closed, nowhere vanishing 1–form, it is obvious that $\mathrm{gv}(\mathcal{F}) = 0$. If $\mathcal{F}$ has trivial holonomy, we have shown that it is *homeomorphic* to a foliation defined by a closed, nowhere vanishing 1–form, but homeomorphisms do not respect smooth forms and we cannot conclude that $\mathrm{gv}(\mathcal{F}) = 0$. This is true, however, being a corollary to Duminy's theorem [**38**]. Indeed, such foliations cannot have resilient leaves. The first advance on this question was made by M. Herman [**67**] and G. Wallet [**145**] for foliated bundles over T^2 without holonomy. Subsequently, the theorem for all foliations without holonomy was demonstrated by S. Morita and T. Tsuboi [**93**].

Since the holonomy–invariant measure μ defines a closed 1–form $\omega = d\mu$ after the change of C^∞ structure, we can define *line integrals* $\int_s d\mu$ for all continuous paths $s : [a, b] \to M$. Indeed, having changed the differentiable structure of M and replaced s by a smooth path s', homotopic to s by a homotopy keeping the endpoints fixed, we define

$$\int_s d\mu = \int_{s'} \omega.$$

Any other smooth path s'' that is homotopic to s mod the endpoints is smoothly homotopic to s' mod the endpoints; so

$$\int_{s'} \omega = \int_{s''} \omega$$

and the line integral is well defined. The following is evident.

Proposition 9.3.4. *The line integral $\int_s d\mu$ depends only on the homotopy class of s mod the endpoints. In particular, μ defines a group homomorphism*

$$[\mu] : \pi_1(M) \to \mathbb{R},$$
$$[\mu]([\sigma]) = \int_\sigma d\mu,$$

and, as an element of $H^1(M;\mathbb{R})$, $[\mu] = [\omega]$ (where $[\omega]$ is as in Corollary 9.3.2). *Finally, if $s : [a,b] \to M$ has image a subarc C of a leaf of $\mathcal{L}$, then, relative to the linear order in C determined by the transverse orientation of $\mathcal{F}$,*

$$\int_s d\mu = \mu(C), \qquad s(a) < s(b),$$
$$\int_s d\mu = -\mu(C), \qquad s(a) > s(b).$$

Definition 9.3.5. The group $P(\mu)$ of *periods* of the measure μ is the image of the homomorphism $[\mu] : \pi_1(M) \to \mathbb{R}$. The rank $p(\mu)$ of μ is the rank of the group $P(\mu)$.

Since M is compact, $\pi_1(M)$ is finitely generated and $P(\mu)$ must be free abelian on finitely many generators. That is, $p(\mu)$ is a nonnegative integer.

Let Φ_t be the foliation–preserving flow associated to μ (Theorem 9.2.1).

Lemma 9.3.6. *The period group $P(\mu)$ is exactly the set of $\tau \in \mathbb{R}$ such that Φ_τ carries each leaf of $\mathcal{F}$ onto itself.*

Exercise 9.3.7. Prove Lemma 9.3.6. The harder direction is to prove that, if $\tau \in P(\mu)$, then Φ_τ carries each leaf of $\mathcal{F}$ to itself. Proceed as follows.

1. Let σ be a loop such that $\int_\sigma d\mu = \tau$ and show that this integral depends only on the *free* homotopy class of σ.
2. Let L be a leaf of $\mathcal{F}$ and show that σ is freely homotopic to a loop
$$s = s_0 + s_1 + \cdots + s_{2p+1},$$
such that $s(0) = s(1) \in L$ and s_{2i} lies in a leaf of $\mathcal{F}$ and s_{2i+1} is a subarc of a leaf of $\mathcal{L}$, $0 \le i \le p$. Let $t_i = \int_{s_{2i+1}} d\mu$, a positive number if s_{2i+1} has orientation agreeing with that of $\mathcal{L}$, negative otherwise.
3. Thus, $\tau = t_0 + t_1 + \cdots + t_p$ and you must prove that $\Phi_\tau(L) = L$.

Corollary 9.3.8. *The foliation $\mathcal{F}$ is a fibration if and only if $p(\mu) = 1$.*

This is almost immediate. An additive subgroup of $\mathbb{R}$ has rank greater than one if and only if it contains nontrivial elements arbitrarily close to 0. By Lemma 9.3.6, this happens if and only if every leaf of $\mathcal{F}$ is nonproper.

Corollary 9.3.9. *A continuous path $s : [a,b] \to M$ has endpoints in the same leaf of $\mathcal{F}$ if and only if $\int_s d\mu \in P(\mu)$.*

Proof. If $s(a)$ and $s(b)$ lie on the same leaf L, let s' be a path on L from $s(b)$ to $s(a)$. Then $\int_{s'} d\mu = 0$ and $\sigma = s + s'$ is a loop; so

$$\int_s d\mu = \int_s d\mu + \int_{s'} d\mu = \int_\sigma d\mu \in P(\mu).$$

Conversely, suppose that $\int_s d\mu = t_1 \in P(\mu)$ and let L be the leaf through $s(a)$. By Theorem 9.2.1, there is a value $t_0 \in \mathbb{R}$ such that $\Phi_{t_0}(s(b)) \in L$. Let $s'(t) = \Phi_t(s(b))$, $0 \le t \le t_0$, and remark that $s + s'$ is a path with endpoints on the same leaf L. Thus, by the argument given above,

$$t_1 + t_0 = \int_{s+s'} d\mu \in P(\mu).$$

Since $P(\mu)$ is a group, it follows that $t_0 \in P(\mu)$, hence, by Lemma 9.3.6, that $s(b) = s'(0) \in L$. □

9.4. Tischler's Theorem

Assume that $\omega \in A^1(M)$ is a closed, nonsingular 1–form and that $\omega \pitchfork \partial M$. That is, $\partial M = \partial_\pitchfork M$.

Let $\{[\eta_1], [\eta_2], \dots, [\eta_p]\}$ be a free abelian basis of $H^1(M;\mathbb{Z}) \subset H^1(M;\mathbb{R})$. By the canonical one–to–one correspondence between $H^1(M;\mathbb{Z})$ and the set of homotopy classes of maps $f : M \to S^1$, we write

$$[\eta_i] = f_i^*[\theta], \quad 1 \le i \le p,$$

for a suitable choice of the smooth maps f_i, θ being the standard volume form on S^1. At the form level, take $\eta_i = f_i^*(\theta)$, $1 \le i \le p$.

Lemma 9.4.1. *If $[\omega] = \sum_{i=1}^p c_i[\eta_i]$, the numbers $\{c_1, c_2, \dots, c_p\}$ generate the group $P(\omega)$ of periods of ω.*

Proof. Let $\sigma_1, \sigma_2, \dots, \sigma_p$ be smooth loops in M representing the basis of $H_1(M;\mathbb{Z})/\text{torsion}$ dual to $\{[\eta_1], [\eta_2], \dots, [\eta_p]\}$. Since the homomorphism

$$\omega : \pi_1(M) \to \mathbb{R},$$

defined by integration around loops, has abelian image, it is trivial on commutators, hence factors through $H_1(M;\mathbb{Z})/\text{torsion}$ (the Hurewicz theorem). That is, the numbers $c_i = \int_{\sigma_i} \omega$ generate the image $P(\omega)$ of this homomorphism. □

Remark that any finitely generated, additive subgroup P of the rational number field $\mathbb{Q}$ has rank 1. Indeed, there is an integer $k \ne 0$ such that $kP \subseteq \mathbb{Z}$. Similarly, if $P \subset \mathbb{R}$ is an additive subgroup of rank 1, there is a real number ρ such that $\rho P = \mathbb{Z}$. By Corollary 9.3.8, the foliation $\mathcal{F}_\omega$ is a fibration of M over S^1 if and only if $p(\omega) = 1$.

In the general case that $p(\omega) \geq 1$, we can write

$$\omega = \sum_{i=1}^{p} c_i \eta_i \ + dg,$$

where $g : M \to \mathbb{R}$ is smooth and the coefficients c_i are real numbers. If the number c_i' is sufficiently near c_i, $1 \leq i \leq p$, it follows that the closed form

$$\omega' = \sum_{i=1}^{p} c_i' \eta_i \ + dg$$

is as close to ω in the C^∞ topology as desired. In particular, this form can be taken so close to ω that it is also nonsingular. Finally, since $\mathbb{Q} \subset \mathbb{R}$ is dense, we can choose all c_i' to be rational. Equivalently, $P(\omega') \subset \mathbb{Q}$ and $p(\omega') = 1$. We have proven the following theorem [**136**].

Theorem 9.4.2 (Tischler). *If $\omega \in A^1(M)$ is closed and nonsingular, and if $\omega|\partial M$ is also nonsingular, then ω can be arbitrarily well approximated (in the C^∞ topology) by closed, nonsingular 1–forms of rank 1.*

Geometrically, this means that foliations defined by closed, nowhere vanishing 1–forms can be smoothly and uniformly "well approximated" by fibrations over S^1. As an intuitive example, consider a linear foliation $\mathcal{F}$ of T^2 of irrational slope r and a sequence $\{\mathcal{F}_k\}_{k=1}^\infty$ of linear foliations with rational slopes $r_k \to r$. For large values of k, $\mathcal{F}_k$ would be indistinguishable from $\mathcal{F}$ to the naked eye. For much larger values of k, even an electron microscope would fail to see the difference.

By Corollary 9.3.2, foliations without holonomy can be "approximated" in some C^0 sense by fibrations. In particular, we note the following.

Corollary 9.4.3. *A compact manifold M admits a foliation without holonomy and transverse to ∂M if and only if it admits a fibration over S^1 with fibers transverse to ∂M.*

Remark. For compact 3–manifolds, a remarkable theorem of J. Stallings [**125**], characterizes the ones that fiber over S^1 with fibers not homeomorphic to the projective plane. The necessary and sufficient condition is that M be irreducible (meaning that every imbedded 2–sphere bounds an imbedded 3–ball) and that $\pi_1(M)$ have a finitely generated normal subgroup $G \neq \mathbb{Z}_2$ such that $\pi_1(M)/G \cong \mathbb{Z}$.

9.5. Isotopy

We will prove the following analogue of Corollary 3.6.7.

Proposition 9.5.1. *Let $\mathcal{F}_0$ and $\mathcal{F}_1$ be transversely–oriented, C^2 foliations of M without holonomy, both transverse to the same 1–dimensional foliation $\mathcal{L}$. Suppose that these foliations have respective continuous, transverse, holonomy–invariant measures μ_0 and μ_1 such that $[\mu_0] = [\mu_1]$. Then there is a C^0 isotopy of $\mathcal{F}_0$ to $\mathcal{F}_1$.*

Fix the hypotheses of Proposition 9.5.1. For $0 \le \tau \le 1$, define

$$\mu_\tau = \tau\mu_1 + (1-\tau)\mu_0,$$

a continuous, strictly positive measure on the leaves of $\mathcal{L}$. This measure has line integrals, well defined by

$$\int_s d\mu_\tau = \tau \int_s d\mu_1 + (1-\tau)\int_s d\mu_0.$$

Since μ_0 and μ_1 have equal line integrals around loops σ, $\int_\sigma d\mu_\tau$ takes this same value. Similarly, for any continuous path s in M, the value of $\int_s d\mu_\tau$ depends only on the homotopy class of s mod the endpoints. In this way, μ_τ defines a cohomology class and $[\mu_\tau] = [\mu_0] = [\mu_1]$, $0 \le \tau \le 1$. One suspects that μ_τ is a holonomy–invariant measure for a C^0 foliation $\mathcal{F}_\tau$ without holonomy and that this defines the desired isotopy. We will construct $\mathcal{F}_\tau$.

Fix a leaf L of $\mathcal{F}_0$ and a basepoint $x \in L$. For fixed but arbitrary $\tau \in [0,1]$, we are going to define a continuous map

$$\varphi_\tau : L \to M,$$

the image of which will be one leaf of the foliation $\mathcal{F}_\tau$ that we are trying to construct. For each $z \in L$, let σ_z be a path in L from x to z and let

$$a_\tau(z) = \int_{\sigma_z} d\mu_\tau.$$

Remark that $a_0(z) = 0$, $\forall\, z \in L$.

Lemma 9.5.2. *The real–valued function $a_\tau(z)$ is well defined on $[0,1] \times L$, depending only on the basepoint x, not on the choices of the paths σ_z in L. This function is continuous on $[0,1] \times L$, both with respect to the manifold topology and the relative topology of $[0,1] \times L$ as a subset of $[0,1] \times M$.*

Proof. If σ'_z is another path in L from x to z, then $\sigma_z - \sigma'_z$ is a loop in L; so

$$\int_{\sigma'_z} d\mu_\tau - \int_{\sigma_z} d\mu_\tau = \int_{\sigma_z - \sigma'_z} d\mu_\tau = \int_{\sigma_z - \sigma'_z} d\mu_0 = 0,$$

proving that $a_\tau(z)$ is well defined. The continuity assertions will be part of Exercise 9.5.5. □

For $\tau \in [0, 1]$, let

$$\Phi^\tau : \mathbb{R} \times M \to M$$

be the flow defined along the leaves of $\mathcal{L}$ by the measure μ_τ. The following is also left as part of Exercise 9.5.5.

Lemma 9.5.3. *The map*

$$[0, 1] \times \mathbb{R} \times M \to M,$$
$$(\tau, t, z) \mapsto \Phi^\tau_t(z)$$

is continuous.

Finally, define

$$\varphi_\tau : L \to M,$$
$$\varphi_\tau(z) = \Phi^\tau_{-a_\tau(z)}(z),$$

remarking that, for each leaf ℓ of $\mathcal{L}$, φ_τ carries $L \cap \ell$ into ℓ and that $\varphi_0 = \mathrm{id}_L$. Again, the continuity is left for Exercise 9.5.5.

Lemma 9.5.4. *The map*

$$[0, 1] \times L \to M,$$
$$(\tau, z) \mapsto \varphi_\tau(z)$$

is continuous, both for the manifold topology and the relative topology of $[0, 1] \times L$.

Exercise 9.5.5. Check the continuity assertions in Lemmas 9.5.2, 9.5.3 and 9.5.4.

A leaf ℓ of $\mathcal{L}$ is a connected 1–manifold without boundary, hence is diffeomorphic either to $\mathbb{R}$ or S^1. In the first case, the orientation of $\mathcal{L}$ induces a linear order on ℓ. In the second case, the induced order is a *cyclic* one. More precisely, if z, y, and w are distinct points of $\ell \cong S^1$, then y lies either in the positively–oriented arc of ℓ from z to w or in the positively–oriented arc from w to z, but not both. In the first case, we write $z < y < w$ and, in the second, $w < y < z$. Of course, $z < y < w$, $y < w < z$ and $w < z < y$ are all equivalent, hence the cyclic character of the order. If $E \subseteq \ell \cong S^1$, a map $\psi : E \to \ell$ preserves the cyclic order if $z < y < w$ in $E \Rightarrow \psi(z) < \psi(y) < \psi(w)$.

Lemma 9.5.6. *If ℓ is a leaf of $\mathcal{L}$, then $\varphi_\tau : L \to M$ restricts to an order–preserving map of the dense subset $\ell \cap L$ of ℓ onto a dense subset of ℓ. Consequently, $\varphi_\tau|\ell \cap L$ extends uniquely to a homeomorphism $\varphi_\tau : \ell \to \ell$ and $\varphi_\tau(z)$ is continuous on $[0, 1] \times \ell$.*

Proof. Suppose that $z, w \in \ell \cap L$ and that s_{zw} is an imbedded, positively–oriented arc in ℓ from z to w. (If $\ell \cong \mathbb{R}$, this means that $z < w$.) Let σ_z and σ_w be paths in L from x to z and from x to w, respectively, and consider the loop $\alpha = \sigma_z + s_{zw} - \sigma_w$ in M. Since $[\mu_\tau] = [\mu_0]$,

$$\int_\alpha d\mu_0 = \int_\alpha d\mu_\tau,$$

implying that

$$a_\tau(z) - a_\tau(w) + \mu_\tau(s_{zw}) = \mu_0(s_{zw}).$$

Thus,

$$\begin{aligned}
\Phi^\tau_{\mu_0(s_{zw})}(\varphi_\tau(z)) &= \Phi^\tau_{a_\tau(z)-a_\tau(w)+\mu_\tau(s_{zw})}(\Phi^\tau_{-a_\tau(z)}(z)) \\
&= \Phi^\tau_{-a_\tau(w)}(\Phi^\tau_{\mu_\tau(s_{zw})}(z)) \\
&= \Phi^\tau_{-a_\tau(w)}(w) \\
&= \varphi_\tau(w).
\end{aligned}$$

Since $E = \ell \cap L$ is dense in ℓ, z and w can be chosen so that $\mu_0(s_{zw})$ is as small as desired. Thus, one consequence of the above is that $\varphi_\tau(E)$ is also dense in ℓ. Furthermore, if $\ell \cong \mathbb{R}$, the fact that $\mu_0(s_{zw}) > 0$ implies that the unique imbedded arc in ℓ from $\varphi_\tau(z)$ to $\varphi_\tau(w)$ is positively oriented. That is, φ_τ preserves the linear order. If $\ell \cong S^1$, suppose that $z < y < w$ in ℓ. In the above notation,

$$s_{zw} = s_{zy} + s_{yw},$$

these being positively–oriented, imbedded arcs in ℓ, and

$$\begin{aligned}
\Phi^\tau_{\mu_0(s_{zy})}(\varphi_\tau(z)) &= \varphi_\tau(y), \\
\Phi^\tau_{\mu_0(s_{yw})}(\varphi_\tau(y)) &= \varphi_\tau(w), \\
\Phi^\tau_{\mu_0(s_{zw})}(\varphi_\tau(z)) &= \varphi_\tau(w).
\end{aligned}$$

Since $\mu_0(s_{zw}) = \mu_0(s_{zy}) + \mu_0(s_{yw}) < \mu_0(\ell) = \mu_\tau(\ell)$, it follows that

$$\varphi_\tau(z) < \varphi_\tau(y) < \varphi_\tau(w)$$

in the cyclic order of ℓ. It is standard and elementary that an order–preserving map of a dense subset of ℓ onto a dense subset of ℓ extends uniquely to a homeomorphism of ℓ. The continuity of the extension of $\varphi_\tau(z)$ to $[0, 1] \times \ell$ is a consequence of this and Lemma 9.5.4. □

Since ℓ ranges over all the leaves of $\mathfrak{L}$, we have extended $\varphi_\tau : L \to M$ to a bijective map $\varphi_\tau : M \to M$. The continuity of $\varphi_\tau(z)$ on $[0, 1] \times M$ follows from Lemma 9.5.3, Lemma 9.5.4 and Lemma 9.5.6. Thus, the compactness of M implies that φ_τ is a homeomorphism, $\forall\, \tau \in [0, 1]$, and that this defines a C^0 isotopy of $\varphi_0 = \mathrm{id}$ to φ_1, hence of the foliation $\mathcal{F}_0 = \varphi_0(\mathcal{F}_0)$ to $\varphi_1(\mathcal{F}_0)$.

Proposition 9.5.7. *The foliation $\varphi_1(\mathcal{F}_0)$ is identical with $\mathcal{F}_1$; hence $\mathcal{F}_0$ and $\mathcal{F}_1$ are topologically isotopic.*

Proof. For $z \in L$, let s_z be the directed subarc of the leaf ℓ of $\mathfrak{L}$ through z with initial point z and terminal point $\varphi_1(z)$. Let σ_z be a path in L from x to z and set $\sigma = \sigma_z + s_z$. Then

$$\int_\sigma d\mu_1 = a_1(z) - a_1(z) = 0$$

and, by Corollary 9.3.9, the endpoints x and $\varphi_1(z)$ of σ lie in the same leaf of $\mathfrak{F}_1$. That is, φ_1 carries L into the leaf L' of $\mathfrak{F}_1$ through x. Also note that $\varphi_1(x) = x$. By a connectivity argument, $\varphi_1 : L \to L'$ is the unique continuous map that projects along the leaves of $\mathfrak{L}$ and carries x to x. By interchanging the roles of $\mathfrak{F}_0$ and $\mathfrak{F}_1$ throughout the above discussion, we likewise obtain a unique continuous projection $L' \to L$ along the leaves of $\mathfrak{L}$ that carries x to x. These invert one another; so $\varphi_1 : L \to L'$ is a homeomorphism. Since these leaves are dense in M, it follows that $\varphi_1(\mathfrak{F}_0) = \mathfrak{F}_1$. □

The proof of the following is postponed to Section 10.3, where it will be a consequence of Lemma 10.2.8.

Proposition 9.5.8. *Let $(M, \mathfrak{F})$ be a compact, C^2–foliated manifold without holonomy, let $\mathfrak{L}$ be a smooth, 1–dimensional foliation transverse to $\mathfrak{F}$, and let μ be a continuous, transverse, holonomy–invariant measure for $\mathfrak{F}$. Then there is a smooth foliation $\mathfrak{F}'$ of M, transverse to $\mathfrak{L}$ and defined by a closed, nonsingular 1–form ω such that $[\omega] = [\mu]$.*

Corollary 9.5.9. *If $(M, \mathfrak{F})$ is a compact, C^2–foliated manifold without holonomy, then $\mathfrak{F}$ is topologically isotopic to a smooth foliation $\mathfrak{F}'$ of M that is defined by a closed, nonsingular 1–form.*

By Corollary 9.3.2, $(M, \mathfrak{F})$ is topologically conjugate to a smoothly foliated manifold $(M', \mathfrak{F}')$ in which $\mathfrak{F}'$ is defined by a closed, nonsingular 1–form and the differentiable manifold M' is homeomorphic but possibly not diffeomorphic to M. In Corollary 9.5.9, however, there is no change of differentiable structure and the topological conjugacy is a C^0 isotopy.

Remark. For compact 3–manifolds, foliations defined by closed, cohomologous 1–forms are smoothly isotopic, whether or not they are transverse to a common 1–dimensional foliation. This is a deep theorem of Laudenbach and Blank [**86**] (see also [**108**] and [**17**]). By Corollary 9.5.9, it follows that C^2 foliations of M without holonomy that have cohomologous measures are C^0 isotopic.

Part 3

Arbitrary Codimension

Foreword to Part 3

As in the case of dynamical systems, the methods for investigating foliations of arbitrary codimension differ greatly from those employed in codimension one. "Hands on" techniques such as the (generalized) Poincaré–Bendixson theory fail, giving way, for example, to the "statistical" techniques of (generalized) ergodic theory.

The main new ideas in this part of the book are foliation cycles and entropy of foliations. In fact, foliation cycles were encountered in Part 2 under the guise of transverse, holonomy–invariant measures, but here we follow D. Sullivan [**127**], developing the theory in the context of de Rham currents and treating some of the striking applications to geometry and topology due to Sullivan, J. F. Plante [**107**], *et al.* Finally, foliation entropy, a concept due to E. Ghys, R. Langevin and P. Walczak [**50**], will be introduced and a number of results of these authors will be treated in detail.

For the most part, results proven by these methods hold also for *laminations*, these being partial foliations of a manifold. Indeed, beginning with Chapter 11, most of the discussion will be carried out for "abstract laminations", the presence of an ambient manifold being immaterial. Examples show that this level of generality is not frivolous.

Chapter 10

Foliation Cycles

De Rham p–currents are *distributional* p–forms. That is, when the real vector space $A^p(M)$ has been suitably topologized, p–currents are elements of the topological dual space. The theory of p–currents is exactly analogous to L. Schwartz's theory of distributions [**118**], Schwartzian distributions being 0–currents. In this chapter, we will use certain standard results from this theory without proof.

Following S. Schwartzmann [**119**] and D. Sullivan [**127**], we use homology built on de Rham currents to study the cone $\mathcal{C}_{\mathcal{F}}$ of p–currents that are "tangent" to an oriented, p–dimensional foliation or lamination and are oriented coherently with the orientation of the leaves. These are called the *foliation currents.* The relative position of $\mathcal{C}_{\mathcal{F}}$ and the spaces of closed currents and exact currents determines whether there are closed p–forms and/or exact p–forms transverse to $\mathcal{F}$. This is proven by an application of the Hahn–Banach theorem to the topological vector space $\mathcal{D}'_p$ of p–currents on M. We will see that such information carries interesting consequences. For example, when $p = n - 1$, we will prove that the foliation $\mathcal{F}$ is taut (Definition 6.3.3) if and only if there is a closed p–form transverse to $\mathcal{F}$. We will see that this property is equivalent to the existence of a Riemannian metric on M relative to which every leaf is a minimal submanifold.

10.1. Basic Notions

Let M be a smooth, compact n–manifold. We will topologize the space $A^p(M)$ of p–forms on M, denoting this topological vector space by $\mathcal{D}_p(M)$ or $\mathcal{D}_p$. This is de Rham's notation for the space of compactly supported forms on arbitrary manifolds.

Fix a finite coordinate atlas $\mathcal{V} = \{(V_\ell, z_\ell)\}_{\ell=1}^N$ on M. For each integer $k \geq 0$, let $\|\cdot\|_k$ denote the C^k norm on $\mathfrak{D}_p$ relative to the atlas $\mathcal{V}$. To be more precise, if $\psi \in \mathfrak{D}_p$, then $\|\psi\|_k$ denotes the supremum of the sum of the absolute values of the mixed partials of orders $\leq k$ (including the 0th–order partials) of the coefficients of the z_ℓ–coordinate representations of $\psi|V_\ell$, $1 \leq \ell \leq N$. This defines a locally convex, Hausdorff topology $\mathfrak{T}_k$ on $\mathfrak{D}_p$ relative to which the linear operations are continuous. One sees that $\mathfrak{T}_k \subseteq \mathfrak{T}_{k+1}$, $\forall\, k \geq 0$, and it follows easily that the topology $\mathfrak{T}$, generated by $\bigcup_{k=0}^\infty \mathfrak{T}_k$, makes $\mathfrak{D}_p$ into a locally convex, topological vector space. We will denote the topological vector spaces $(\mathfrak{D}_p, \mathfrak{T}_k)$ by $\mathfrak{D}_p^k$ and, by an abuse of notation, $(\mathfrak{D}_p, \mathfrak{T})$ by $\mathfrak{D}_p$.

Lemma 10.1.1. *A sequence $\{\psi_m\}_{m=1}^\infty$ converges to ψ in $\mathfrak{D}_p$ if and only if it converges to ψ in $\mathfrak{D}_p^k$, $\forall\, k \geq 0$.*

Proof. Since $\mathfrak{T}_k \subset \mathfrak{T}$, $\forall\, k \geq 0$, convergence in $\mathfrak{D}_p$ implies convergence in every $\mathfrak{D}_p^k$. For the converse, every open neighborhood W of ψ in $\mathfrak{D}_p$ contains a neighborhood of ψ of the form $W_0 \cap \cdots \cap W_q$, for some integer $q \geq 0$, where W_k is an open neighborhood of ψ in $\mathfrak{D}_p^k$, $0 \leq k \leq q$. Convergence in every $\mathfrak{D}_p^k$ implies that all but finitely many terms ψ_m lie in $W_0 \cap \cdots \cap W_q$, hence in W. Since $W \in \mathfrak{T}$ is an arbitrary neighborhood of ψ, $\psi_m \to \psi$ in $\mathfrak{D}_p$. □

Recall that a function $\varphi : X \to Y$ between metric spaces is continuous if and only if it is sequentially continuous (*i.e.*, φ maps sequences converging to a point x onto sequences converging to $\varphi(x)$). This criterion for continuity holds, therefore, for the normed linear spaces $\mathfrak{D}_p^k$. Although $\mathfrak{D}_p$ is not a normed space, it inherits the same property.

Lemma 10.1.2. *A function $\varphi : \mathfrak{D}_p \to \mathfrak{D}_r$ or $\varphi : \mathfrak{D}_p \to \mathbb{R}$ is continuous if and only if it is sequentially continuous.*

Proof. Clearly, continuity implies sequential continuity. Suppose, then, that $\varphi : \mathfrak{D}_p \to \mathfrak{D}_r$ is sequentially continuous. If φ is not continuous at $\psi \in \mathfrak{D}_p$, there is a neighborhood V of $\varphi(\psi)$ in $\mathfrak{D}_r$ such that $\varphi(W) \not\subseteq V$ for all neighborhoods W of ψ in $\mathfrak{D}_p$. In particular, let $W_k \in \mathfrak{T}_k \subset \mathfrak{T}$ be the ball centered at ψ of radius $1/k$ in the C^k norm and choose $\psi_k \in W_k$ such that $\varphi(\psi_k) \notin V$, $\forall\, k \geq 0$. Since

$$\|\psi_k - \psi\|_0 \leq \|\psi_k - \psi\|_1 \leq \cdots \leq \|\psi_k - \psi\|_k \leq \frac{1}{k}, \quad \forall\, k \geq 1,$$

Lemma 10.1.1 implies that $\lim_{m\to\infty} \psi_m = \psi$ in $\mathfrak{D}_p$, contradicting the assumption that φ is sequentially continuous. The proof for $\varphi : \mathfrak{D}_p \to \mathbb{R}$ is entirely the same. □

Definition 10.1.3. A subset $S \subset \mathcal{D}_p$ is *bounded* if it is bounded relative to the C^k norm, $\forall\, k \geq 0$. Likewise, a linear map $\varphi : \mathcal{D}_p \to \mathcal{D}_r$ or $\varphi : \mathcal{D}_p \to \mathbb{R}$ is bounded if it is bounded relative to the C^k norm, $\forall\, k \geq 0$.

Lemma 10.1.4. *A bounded linear map $\varphi : \mathcal{D}_p \to \mathcal{D}_r$ or $\mathbb{R}$ is continuous. A continuous linear map $\varphi : \mathcal{D}_p \to \mathbb{R}$ is bounded.*

Proof. If φ is a bounded linear map, it is bounded relative to all C^k norms, hence is continuous in these norms, hence is sequentially continuous in them. By Lemma 10.1.1, it follows that φ is sequentially continuous in the topology $\mathfrak{T}$; hence Lemma 10.1.2 asserts that φ is continuous in that topology. Conversely, if φ is not bounded, there is a bounded sequence $\{\psi_k\}_{k=1}^{\infty}$ in $\mathcal{D}_p$ such that $\varphi(\psi_k) \geq k$, $1 \leq k < \infty$. Thus, $\lim_{k\to\infty} \psi_k/k = 0$ in the topology $\mathfrak{T}$ while $\varphi(\psi_k/k) \geq 1$, for all $k \geq 1$, showing that φ is not continuous. □

Corollary 10.1.5. *Exterior differentiation*

$$d : \mathcal{D}_p \to \mathcal{D}_{p+1}$$

is a continuous linear map.

Definition 10.1.6. The strong dual $\mathcal{D}'_p$ of $\mathcal{D}_p$ is the space of continuous linear functionals on $\mathcal{D}_p$. The elements of $\mathcal{D}'_p$ are the *p*–currents on M.

Example 10.1.7. Let $x \in M$ and $v_x \in \Lambda^p(T_x(M))$. Define $\varphi_{v_x} : \mathcal{D}_p \to \mathbb{R}$ by

$$\varphi_{v_x}(\omega) = \omega_x(v_x).$$

It is clear that φ_{v_x} is a bounded linear functional, hence $\varphi_{v_x} \in \mathcal{D}'_p$. It is called a *Dirac current* and is often denoted by v_x.

Example 10.1.8. Let

$$c = \sum_{i=1}^{m} c_i \sigma_i,$$

where $c_i \in \mathbb{R}$ and $\sigma_i : \Delta^p \to M$ is a smooth singular p–simplex, $1 \leq i \leq m$. Define $\varphi_c : \mathcal{D}_p \to \mathbb{R}$ by

$$\varphi_c(\omega) = \int_c \omega = \sum_{i=1}^{m} c_i \int_{\Delta^p} \sigma_i^*(\omega).$$

Again, this defines a bounded linear functional, called a *singular current.*

Example 10.1.9. Assume that M is oriented. Let $\eta \in \mathcal{D}_{n-p}$ and define $\varphi_\eta : \mathcal{D}_p \to \mathbb{R}$ by

$$\varphi_\eta(\omega) = \int_M \omega \wedge \eta.$$

This bounded linear functional is called a *diffuse current.*

Example 10.1.10. Let $N \subset M$ be a smooth, oriented, compact submanifold of dimension p, $\partial N = \emptyset$. More generally, let $f : N \to M$ be a smooth map of such a manifold into M (called a singular submanifold). Define $\varphi_f : \mathcal{D}_p \to \mathbb{R}$ by

$$\varphi_f(\omega) = \int_N f^*(\omega).$$

If f is an imbedding, one also writes φ_N or $\int_N$ for this linear functional. Again, it is bounded, hence a p–current.

Example 10.1.11. Let ν be a Borel measure on M such that $\nu(M) < \infty$ and let v be a continuous field of p–vectors on M. Define $\varphi_{v,\nu} : \mathcal{D}_p \to \mathbb{R}$ by

$$\varphi_{v,\nu}(\omega) = \int_M \omega(v)\, d\nu.$$

This current is also denoted by

$$\varphi_{v,\nu} = \int v\, d\nu$$

and is called a *current of integration*. The Dirac current v_x (Example 10.1.7) is an example of a current of integration where v is any continuous extension of v_x and ν is the probability measure concentrated on the singleton $\{x\}$. Remark that a current of integration can be evaluated on p–forms ω that are only *continuous*. The formula $\int_M \omega(v)\, d\nu$ makes perfectly good sense for such forms.

Example 10.1.12. Analogous to a current of integration is the p–current φ_μ defined by a holonomy–invariant measure μ for a compact foliated manifold $(M, \mathcal{F})$, $\dim \mathcal{F} = p$, with $T(\mathcal{F})$ oriented. More precisely, assume that $\partial M = \partial_\tau M$ and fix a regular foliated atlas $\mathcal{U} = \{U_\alpha\}_{\alpha \in \mathfrak{A}}$ respecting the leafwise orientation. Let $\Gamma = \Gamma_{\mathcal{U}}$ be the associated holonomy pseudogroup, acting on the transverse space S. The measure μ is defined on the Borel subsets of S and satisfies the following (*cf.* Definition 9.2.13).

Definition 10.1.13. A measure μ on S such that $\mu(S) < \infty$ is a (transverse) holonomy–invariant measure if, whenever $\gamma \in \Gamma$ and $B \subseteq \operatorname{dom} \gamma$ is a Borel set, then $\mu(\gamma(B)) = \mu(B)$.

Choose a partition of unity $\lambda = \{\lambda_\alpha\}_{\alpha \in \mathfrak{A}}$, subordinate to $\mathcal{U}$. Let $\omega \in \mathcal{D}_p$. In a foliated chart $U_\alpha \in \mathcal{U}$, one can integrate ω on each $\mathcal{F}$–plaque $P \in S_\alpha$, obtaining a continuous, real–valued function on S_α. This function can then be integrated against the measure $\mu | S_\alpha$. This is an iterated integral

$$\int_{S_\alpha} \left(\int_P \omega \right) d\mu(P).$$

Now define

$$\varphi_\mu(\omega) = \sum_{\alpha\in\mathfrak{A}} \int_{S_\alpha} \left(\int_P \lambda_\alpha \omega \right) d\mu(P).$$

One can use the holonomy invariance of μ to show that this bounded linear functional (p–current)

$$\varphi_\mu : \mathcal{D}_p \to \mathbb{R}$$

is independent of the choice of λ (Exercise 10.1.15). In Exercise 10.1.16, you will show that φ_μ is a p–cycle (*cf.* Definition 10.1.22 *ff.*). This is an example of a "foliation cycle". In fact, it will turn out that all foliation cycles are holonomy–invariant measures (Theorem 10.2.12). Remark that this current is defined for foliations of class at least $C^{1,0+}$ and can be evaluated on merely *continuous* p–forms.

Example 10.1.14. An important generalization of transverse, holonomy–invariant measures for foliations is the analogous concept for *laminations.* Here, a lamination $\mathcal{L}$ in M will be a foliated space $(X, \mathcal{L})$ of dimension p, defined as indicated in the remark on page 32. A careful definition will be given in Chapter 11. Here we only need note that X is topologically imbedded in M so that each leaf is a C^1–immersed submanifold of M without boundary. If M has boundary, we allow no leaf of $\mathcal{L}$ to meet ∂M without being contained in ∂M. The naturally defined bundle $T(\mathcal{L})$ over X of vectors tangent to the leaves is to be a subbundle of $T(M)|X$. We can also consider compact laminations (X, L) in open n–manifolds M since there will always be a compact, n–dimensional submanifold $N \subset M$ containing X in its interior.

The laminated structure is given by a regular "laminated atlas"

$$\mathcal{U} = \{U_\alpha, x_\alpha, y_\alpha\}_{\alpha\in\mathfrak{A}},$$

a finite family of honest coordinate charts in M, together with subcharts $(V_\alpha, x_\alpha, y_\alpha)$ such that $\overline{V}_\alpha \subset U_\alpha$ and $\{V_\alpha\}_{\alpha\in\mathfrak{A}}$ covers X. For each $\alpha \in \mathfrak{A}$, there is a compact subset $Z_\alpha \subset \mathbb{R}^q$ $(q = n - p)$ such that

$$X \cap \overline{V}_\alpha = \bigcup_{z\in Z_\alpha} y_\alpha^{-1}(z) \cap \overline{V}_\alpha.$$

The sets $y_\alpha^{-1}(z)$, $z \in Z_\alpha$, are the plaques of $\mathcal{L}$ in U_α that meet $\overline{V}_\alpha$, each homeomorphic to an open subset of $\mathbb{R}^p$, and the change of x–coordinates on overlapping plaques are smooth of class at least C^1. The change of coordinate formula $y_\beta = y_\beta(x_\alpha, y_\alpha)$ is required to take the form

$$y_\beta = y_\beta(y_\alpha) = \gamma_{\beta\alpha}(y_\alpha)$$

on $X \cap U_\alpha \cap U_\beta$. We also require that the lamination be oriented. That is, the local coordinate changes on overlapping plaques are orientation preserving. The transverse space $Z = \coprod_{\alpha\in\mathfrak{A}} Z_\alpha$ will be a compact subset of $\mathbb{R}^q$, holonomy will be the pseudogroup of local homeomorphisms of this space defined by the cocycle $\{\gamma_{\beta\alpha}\}_{\alpha,\beta\in\mathfrak{A}}$, and μ will be a finite, positive measure on Z that is invariant under this pseudogroup. The plaque structure of the lamination allows us to integrate forms $\omega \in \mathcal{D}_p$ along plaques, producing real–valued functions on subsets of Z, which are then integrated via μ. These local integrals are assembled via a partition of unity, exactly as in Example 10.1.12, and the finiteness and holonomy invariance of μ again show that the resulting linear functional

$$\varphi_\mu : \mathcal{D}_p \to \mathbb{R}$$

is bounded and independent of the choice of partition of unity. As for the case of foliations, this current can be evaluated on merely *continuous* p–forms.

Exercise 10.1.15. In Example 10.1.12, prove that the current φ_μ, determined by a holonomy–invariant measure μ, does not depend on the choice of partition of unity subordinate to the regular, foliated atlas. Note that your proof works equally well for laminations (Example 10.1.14).

Exercise 10.1.16. If φ_μ is the current defined by a holonomy–invariant measure for a p–dimensional foliation $\mathcal{F}$ or lamination $\mathcal{L}$, prove that, for all $(p-1)$–forms ζ, $\varphi_\mu(d\zeta) = 0$.

Remark. In Examples 10.1.11, 10.1.12, and 10.1.14, the currents extend naturally to bounded linear functionals on the space of C^0 forms on M. By the Riesz representation theorem, such currents must be currents of integration. They are said to be of degree 0.

Definition 10.1.17. If $\epsilon > 0$ and $B \subset \mathcal{D}_p$ is bounded, we define

$$V_{B,\epsilon} \subset \mathcal{D}'_p$$

to be the set of p–currents φ such that $\varphi(B) \subseteq [-\epsilon, \epsilon]$. A subset $S_0 \subset \mathcal{D}'_p$ is a neighborhood of 0 if $V_{B,\epsilon} \subseteq S_0$ for some such B and ϵ. If $c \in \mathcal{D}'_p$, the neighborhoods of c are the sets of the form $S_c = c + S_0$, where S_0 is a neighborhood of 0. A subset $W \subseteq \mathcal{D}'_p$ is open if, for each $c \in W$, there is a neighborhood $S_c \subseteq W$.

The following is a straightforward exercise, left to the reader.

Lemma 10.1.18. *The above definition of the term "open set" defines a locally convex, Hausdorff topology $\mathfrak{T}'$ on $\mathcal{D}'_p$. Evaluation of currents on forms defines a continuous bilinear map*

$$e : \mathcal{D}'_p \times \mathcal{D}_p \to \mathbb{R},$$
$$(\varphi, \omega) \mapsto \varphi(\omega).$$

Proposition 10.1.19. *The bilinear evaluation map e identifies $\mathcal{D}_p$ with the space $\mathcal{D}''_p$ of continuous linear functionals on $\mathcal{D}'_p$. Thus, $\mathcal{D}_p$ and $\mathcal{D}'_p$ are strong duals of one another.*

For example, see [**34**, p. 89, Théorème 13].

Definition 10.1.20. A subset $V \subset \mathcal{D}'_p$ is bounded if, for each bounded subset $B \subset \mathcal{D}_p$, the set $\{\varphi(b) \mid \varphi \in V, b \in B\}$ is a bounded subset of $\mathbb{R}$.

Theorem 10.1.21. *The topological vector spaces $\mathcal{D}_p$ and $\mathcal{D}'_p$ are Montel (that is, a subset is bounded if and only if it is precompact).*

See [**118**, p. 70, Théorème VII and p. 74, Théorème XII], where the Montel property is established for the case $p = 0$. The general case is entirely similar.

A bounded linear transformation $L : \mathcal{D}_p \to \mathcal{D}_r$ has adjoint

$$L^* : \mathcal{D}'_r \to \mathcal{D}'_p,$$
$$L^*(\varphi) = \varphi \circ L.$$

It is easily checked that L^* is also a continuous linear transformation. In particular, we apply this observation to the exterior derivative operator.

Definition 10.1.22. The boundary operator

$$\partial : \mathcal{D}'_p \to \mathcal{D}'_{p-1}$$

is the adjoint $\partial = d^*$ of the exterior derivative.

Since $d^2 = 0$, it is evident that $\partial^2 = 0$. We can define the pth de Rham homology of M by setting

$$\mathcal{Z}_p = \ker \partial,$$
$$\mathcal{B}_p = \operatorname{im} \partial,$$
$$H_p^{\mathrm{DR}}(M) = \mathcal{Z}_p/\mathcal{B}_p.$$

We call $\mathcal{Z}_p$ the space of closed p–currents or *p–cycles* and $\mathcal{B}_p$ the space of exact p–currents or *p–boundaries.*

Remark. By Exercise 10.1.16, a holonomy–invariant measure μ for a foliation or lamination defines a p–cycle φ_μ.

The pairing

$$H_p^{\mathrm{DR}}(M) \times H^p(M; \mathbb{R}) \to \mathbb{R},$$
$$([\varphi], [\omega]) \mapsto \varphi(\omega)$$

is clearly well defined and determines a homomorphism

$$j_* : H_p^{\mathrm{DR}}(M) \to H^p(M; \mathbb{R})'$$

into the dual space of the de Rham cohomology of M. By the de Rham theorem, this dual space is canonically identified with the standard singular homology $H_p(M;\mathbb{R})$.

A smooth, singular p–chain c in M is a p–current (Example 10.1.8). We denote the vector space of such p–chains by $C_p(M;\mathbb{R})$. If ∂_s is the singular boundary operator, the combinatorial version of Stokes' theorem states that, for a singular p–chain c and $(p-1)$–form ω,

$$\int_{\partial_s(c)} \omega = \int_c d\omega.$$

That is, under the natural inclusion map

$$i : C_p(M;\mathbb{R}) \hookrightarrow \mathcal{D}'_p,$$

the boundary operator ∂ on currents restricts to the singular boundary operator ∂_s on singular currents. It is well known that the singular homology of a manifold can be defined via smooth singular chains. So the inclusion map

$$i : C_p(M;\mathbb{R}) \hookrightarrow \mathcal{D}'_p$$

passes to a well–defined linear map

$$i_* : H_p(M;\mathbb{R}) \to H_p^{\mathrm{DR}}(M)$$

from singular homology to de Rham homology.

Theorem 10.1.23 (de Rham). *The homomorphisms j_* and i_* are mutually inverse. In particular, singular homology and de Rham homology are canonically isomorphic.*

For details, see [**34**, Chapitre IV].

Lemma 10.1.24. *The subspaces $\mathcal{Z}_p$ and $\mathcal{B}_p$ are closed as subsets of $\mathcal{D}'_p$.*

Proof. Since d is bounded, the operator ∂ is continuous; so $\mathcal{Z}_p = \partial^{-1}(0)$ is closed in $\mathcal{D}'_p$. By Theorem 10.1.23, a class $[\varphi] \in H_p^{\mathrm{DR}}(M)$ is trivial if and only if $\varphi(\omega) = 0$, for every *closed* form $\omega \in \mathcal{D}_p$. That is,

$$\mathcal{B}_p = \bigcap_{\substack{\omega\in\mathcal{D}_p \\ d\omega=0}} \omega^{-1}(0).$$

Since the intersection of closed sets is closed, $\mathcal{B}_p$ is closed in $\mathcal{D}'_p$. □

Exercise 10.1.25. Suppose that M is orientable, connected, noncompact and that $\partial M = \emptyset$. Define the spaces $\mathcal{D}_p$ and $\mathcal{D}'_p$ using compactly supported forms $\omega \in A^p_c(M)$. If $\dim M = n$, show that

$$z(\omega) = \int_M \omega, \quad \forall \omega \in A^n_c(M)$$

defines a cycle $z \in \mathcal{D}'_n$ and that $\mathcal{Z}_n = \{cz \mid c \in \mathbb{R}\}$.

10.2. Foliation Currents

Let $(M, \mathcal{F})$ be a compact, foliated n–manifold of smoothness class at least $C^{1,0+}$. Assume that $T(\mathcal{F})$ is oriented and that $\partial M = \partial_\tau M$.

Remark. Although our primary focus is on foliations, everything in this section (with obvious modifications) will apply equally well to a leafwise oriented lamination $(X, \mathcal{L})$, where $X \subset M$ is compact, the leaves are C^1–immersed without boundary and every leaf meeting ∂M is contained in ∂M.

In both cases, the leaf dimension will be p and the codimension will be $q = n - p$. Choose the regular, finite, foliated atlas

$$\mathcal{U} = \{(U_\alpha, x_\alpha, y_\alpha)\}_{\alpha \in \mathfrak{A}}$$

so that, for each $\alpha \in \mathfrak{A}$, the coordinate p–tuple $(x^1_\alpha, \dots, x^p_\alpha)$ determines the correct orientation of $\mathcal{F}|U_\alpha$.

If $x \in M$, let L be the leaf of $\mathcal{F}$ through x and consider the p–vectors

$$v_x = v_{1\,x} \wedge \cdots \wedge v_{p\,x} \in \Lambda^p(T_x(L)),$$

where $(v_{1\,x}, \dots, v_{p\,x})$ is an ordered basis of $T_x(L)$ that gives the correct orientation. Since $\dim \Lambda^p(T_x(L))$ is 1–dimensional, the p–vectors v_x form a ray in $\Lambda^p(T_x(M))$, $\forall x \in M$. Let φ_{v_x} be the associated Dirac current (Example 10.1.7). By abuse of notation, we will denote this current by v_x or $v_{1\,x} \wedge \cdots \wedge v_{p\,x}$. We call these the Dirac foliation currents (for $\mathcal{F}$).

10.2.A. The cone $\mathcal{C}_\mathcal{F}$.

We show that the Dirac foliation currents generate a convex cone $\mathcal{C}_\mathcal{F} \subset \mathcal{D}'_p$ with compact base and establish basic consequences.

Lemma 10.2.1. *There exists a closed vector subspace $H \subset \mathcal{D}'_p$ of codimension 1 such that all the Dirac foliation currents lie on the same side of H.*

Proof. For each $\alpha \in \mathfrak{A}$, set

$$\widetilde{\omega}_\alpha = dx^1_\alpha \wedge \cdots \wedge dx^p_\alpha.$$

Since $\mathcal{F}$ is only assumed to be of class $C^{1,0+}$, the p–form $\widetilde{\omega}_\alpha$ may only be of class C^0, but it can be approximated uniformly by a smooth p–form ω_α. In particular, we can assume that Dirac foliation currents v_x will be positive on (any extension of) ω_α, $\forall\, x \in U_\alpha$. Let $\{\lambda_\alpha\}_{\alpha\in\mathfrak{A}}$ be a smooth partition of unity subordinate to $\mathcal{U}$ and define

$$\omega = \sum_{\alpha\in\mathfrak{A}} \lambda_\alpha \omega_\alpha.$$

Every Dirac foliation current will be positive on this form and, viewing $\omega \in \mathcal{D}_p$ as a continuous linear functional on $\mathcal{D}'_p$, we see that $H = \omega^{-1}(0)$ has all the desired properties. □

Remark. The p–form ω, constructed in the above proof, is said to be transverse to $\mathcal{F}$. There are infinitely many p–forms transverse to $\mathcal{F}$ and any one of them will provide a subspace $H \subset \mathcal{D}'_p$ with the desired properties.

Definition 10.2.2. The topological closure in $\mathcal{D}'_p$ of the set of all finite linear combinations, with positive coefficients, of Dirac foliation currents, will be denoted by $\mathcal{C}_\mathcal{F}$. The elements of $\mathcal{C}_\mathcal{F}$ are the foliation currents (for $\mathcal{F}$). The set $\mathcal{Z}_\mathcal{F} = \mathcal{Z}_p \cap \mathcal{C}_\mathcal{F}$ is the set of foliation cycles and $\mathcal{B}_\mathcal{F} = \mathcal{B}_p \cap \mathcal{C}_\mathcal{F}$ is the set of foliation boundaries.

It should be clear that $\mathcal{C}_\mathcal{F}$ is a closed, convex cone in $\mathcal{D}'_p$.

Lemma 10.2.3. *Let $\omega \in \mathcal{D}_p$ be transverse to $\mathcal{F}$ and take $H = \omega^{-1}(0)$ as above. Then $\mathcal{C}_\mathcal{F} \cap H = \{0\}$ and $\widehat{\mathcal{C}}_\mathcal{F} = \mathcal{C}_\mathcal{F} \cap \omega^{-1}(1)$ is compact.*

Proof. Fix a Riemannian metric on $\Lambda^p(T(M))$ and define $A \subset \Lambda^p(T(M))$ to be the set of Dirac foliation currents v_x with $\|v_x\| = 1$. Evidently, A is compact as a subset of the manifold $\Lambda^p(T(M))$. If $\eta \in \mathcal{D}_p$, then

$$\eta : \Lambda^p(T(M)) \to \mathbb{R}$$

is a continuous map. So $\eta(A) \subset \mathbb{R}$ is compact and we find $\mu(\eta) \geq 0$ such that $|\eta(a)| \leq \mu(\eta)$, $\forall\, a \in A$. Similarly, there is a real number $\lambda > 0$ such that $\omega(a) > \lambda$, $\forall\, a \in A$.

Suppose that $\theta \in \mathcal{C}_\mathcal{F} \cap H$. Thus, $\omega(\theta) = 0$. Write

$$\theta = \lim_{n\to\infty} \theta_n,$$

$$\theta_n = \sum_{i=1}^{r_n} c_{i,n} a_{i,n},$$

where each $c_{i,n} > 0$ and each $a_{i,n} \in A$. Then

$$0 = \lim_{n\to\infty} \omega(\theta_n) \geq \lambda \lim_{n\to\infty} \sum_{i=1}^{r_n} c_{i,n} \geq 0.$$

That is,

$$0 = \lim_{n\to\infty} \sum_{i=1}^{r_n} c_{i,n}.$$

If $\eta \in \mathcal{D}_p$,

$$|\eta(\theta)| \leq \mu(\eta) \lim_{n\to\infty} \sum_{i=1}^{r_n} c_{i,n} = 0$$

and, since η is arbitrary, $\theta = 0$.

We prove that $\widehat{\mathcal{C}}_{\mathcal{F}} = \mathcal{C}_{\mathcal{F}} \cap \omega^{-1}(1)$ is compact. Since $\mathcal{C}_{\mathcal{F}}$ and $\omega^{-1}(1)$ are closed in $\mathcal{D}'_p$, so is $\widehat{\mathcal{C}}_{\mathcal{F}}$. Since $\mathcal{D}'_p$ is Montel, we only need to show that $\widehat{\mathcal{C}}_{\mathcal{F}}$ is bounded. That is, for each bounded subset $S \subset \mathcal{D}_p$, $S(\widehat{\mathcal{C}}_{\mathcal{F}}) \subset \mathbb{R}$ is bounded. This will be true if, for each $\eta \in \mathcal{D}_p$, $\eta(\widehat{\mathcal{C}}_{\mathcal{F}}) \subset \mathbb{R}$ is bounded [**118**, p. 72]. Let K be the dense subset of $\kappa \in \widehat{\mathcal{C}}_{\mathcal{F}}$ defined by the conditions

$$1 = \omega(\kappa),$$

$$\kappa = \sum_{i=1}^{r} c_i a_i,$$

where each $c_i > 0$ and each $a_i \in A$. For all $\kappa \in K$,

$$1 = \omega(\kappa) = \sum_{i=1}^{r} c_i \omega(a_i) > \sum_{i=1}^{r} c_i \lambda.$$

Thus, $\sum_{i=1}^{r} c_i < 1/\lambda$ and

$$|\eta(\kappa)| = \left|\sum_{i=1}^{r} c_i \eta(a_i)\right| \leq \sum_{i=1}^{r} c_i |\eta(a_i)| \leq \sum_{i=1}^{r} c_i \mu(\eta) < \frac{\mu(\eta)}{\lambda}.$$

Since K is dense in $\widehat{\mathcal{C}}_{\mathcal{F}}$, we conclude that $\eta(\widehat{\mathcal{C}}_{\mathcal{F}}) \subseteq [-\mu(\eta)/\lambda, \mu(\eta)/\lambda]$. □

Definition 10.2.4. If $\omega \in \mathcal{D}_p$ is transverse to $\mathcal{F}$, then $\widehat{\mathcal{C}}_{\mathcal{F}} = \mathcal{C}_{\mathcal{F}} \cap \omega^{-1}(1)$ will be called a *base* of the cone $\mathcal{C}_{\mathcal{F}}$.

We are going to need the following classical theorem from analysis, which is a consequence of the Hahn–Banach theorem (see [**83**, pp. 243–244]).

Theorem 10.2.5. *Let W be a locally convex, topological vector space over $\mathbb{R}$, $X \subseteq W$ a nonempty, compact, convex subset, and $V \subseteq W$ a nonempty, closed, convex subset. If $X \cap V = \emptyset$, there is a continuous linear functional $\theta : W \to \mathbb{R}$ and a real number d such that $\theta(V) < d < \theta(X)$.*

Remark. If, in this theorem, V is a nontrivial vector subspace, we must have $\theta(V) = \{0\}$. The only alternative is $\theta(V) = \mathbb{R}$ and the number d could not exist. In this case, $d > 0$, and so the compact set $\theta(X) \subset \mathbb{R}$ is positive and bounded away from 0.

Sullivan uses Lemma 10.2.3 and Theorem 10.2.5 to prove some fundamental lemmas concerning the existence of closed and/or exact p-forms transverse to the foliation.

Lemma 10.2.6. *There is an exact p-form transverse to $\mathcal{F}$ if and only if* 0 *is the only foliation cycle for $\mathcal{F}$.*

Proof. Suppose that $\omega = d\eta$ is transverse to $\mathcal{F}$, $\eta \in A^{p-1}(M) = \mathcal{D}_{p-1}$. Thus, if $c \in \mathcal{C}_\mathcal{F} \smallsetminus \{0\}$, $\omega(c) > 0$. If $z \in \mathcal{C}_\mathcal{F} \cap \mathcal{Z}_p$,

$$\omega(z) = d\eta(z) = \eta(\partial z) = \eta(0) = 0;$$

so $z = 0$. For the converse, assume that $\mathcal{C}_\mathcal{F} \cap \mathcal{Z}_p = \{0\}$. Since $\widehat{\mathcal{C}}_\mathcal{F}$ is a nonempty, compact, convex subset of the locally convex, topological vector space $\mathcal{D}'_p$ and $\widehat{\mathcal{C}}_\mathcal{F}$ is disjoint from $\mathcal{Z}_p$, Theorem 10.2.5 implies that there is a continuous linear functional ω on $\mathcal{D}'_p$ such that

$$\omega(\mathcal{Z}_p) = \{0\},$$
$$\omega(\widehat{\mathcal{C}}_\mathcal{F}) > 0.$$

But $\mathcal{D}_p$ is the topological dual of $\mathcal{D}'_p$; so $\omega \in A^p(M)$. The fact that $\omega > 0$ on $\widehat{\mathcal{C}}_\mathcal{F}$ implies that ω is positive on all of $\mathcal{C}_\mathcal{F} \smallsetminus \{0\}$, hence is positive on all Dirac foliation currents. That is, ω is transverse to $\mathcal{F}$. Since

$$(d\omega)(c) = \omega(\partial c) = 0,$$

for every current $c \in \mathcal{D}'_{p+1}$, ω is a closed form. In fact, ω is exact, since $[\omega] \in H^p(M;\mathbb{R})$ vanishes on $[z] \in H_p(M;\mathbb{R})$, $\forall\, z \in \mathcal{Z}_p$. That is, $[\omega]$ is the zero functional on $H_p(M;\mathbb{R})$, hence $[\omega] = 0$. □

We illustrate Lemma 10.2.6 in Figure 10.2.1. There is an exact transverse form precisely when the cone of foliation currents touches $\mathcal{Z}_p$ only at the point 0.

Lemma 10.2.7. *There is a closed p-form transverse to $\mathcal{F}$ if and only if* 0 *is the only foliation boundary for $\mathcal{F}$.*

Proof. Let ω be a closed p-form transverse to $\mathcal{F}$, $z = \partial c \in \mathcal{C}_\mathcal{F} \cap \mathcal{B}_p$. Then

$$\omega(z) = \omega(\partial c) = (d\omega)(c) = 0.$$

Since ω is transverse to $\mathcal{F}$, it follows that $z = 0$. Conversely, suppose that $\mathcal{C}_\mathcal{F} \cap \mathcal{B}_p = \{0\}$. Then $\widehat{\mathcal{C}}_\mathcal{F}$ and the subspace $\mathcal{B}_p$ are disjoint and Theorem 10.2.5 implies that there is a p-form ω with $\omega(\widehat{\mathcal{C}}_\mathcal{F}) > 0$ and $\omega(\mathcal{B}_p) = 0$. The inequality implies that ω is transverse to $\mathcal{F}$. The equality implies that ω is closed. Indeed, if $c \in \mathcal{D}'_{p+1}$,

$$(d\omega)(c) = \omega(\partial c) = 0.$$

Since c is arbitrary, $d\omega = 0$. □

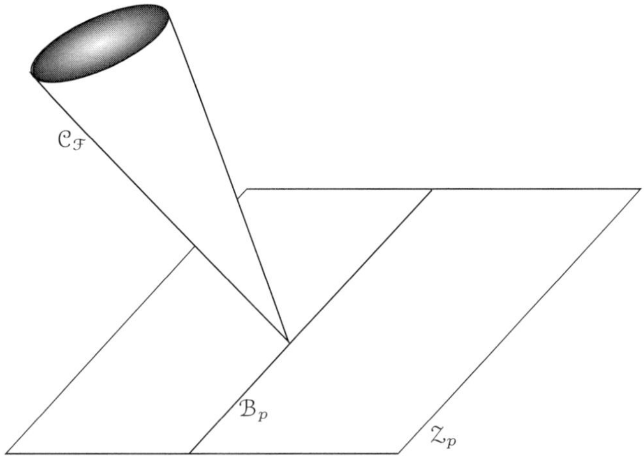

Figure 10.2.1. An exact p–form is transverse to $\mathcal{F}$.

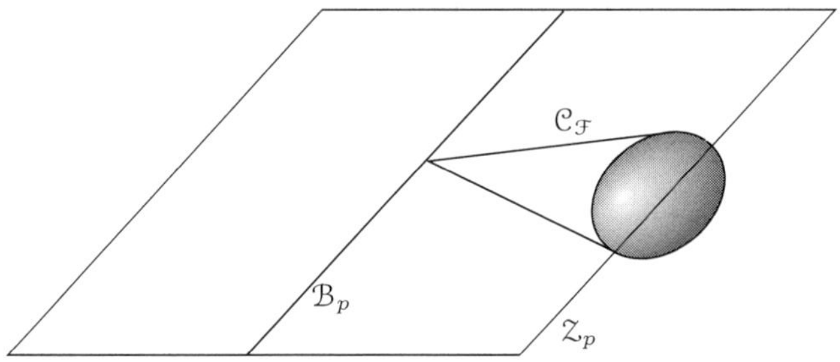

Figure 10.2.2. A closed (but not exact) p–form is transverse to $\mathcal{F}$.

Figures 10.2.2 and 10.2.1, together, illustrate Lemma 10.2.7. There is a closed, transverse form exactly when the cone of foliation currents touches $\mathcal{B}_p$ only at the point 0. The situation in which there are no closed p–forms transverse to $\mathcal{F}$ is illustrated in Figure 10.2.3. Figure 10.2.2 illustrates exactly the situation in which $\mathcal{F}$ admits both nontrivial foliation cycles and closed, transverse p–forms.

Lemma 10.2.8. *If there are closed p–forms transverse to $\mathcal{F}$ and nontrivial foliation cycles as well, the natural map $\mathcal{Z}_{\mathcal{F}} \to H_p(M;\mathbb{R})$ carries every nontrivial foliation cycle to a nontrivial homology class. The image of this map*

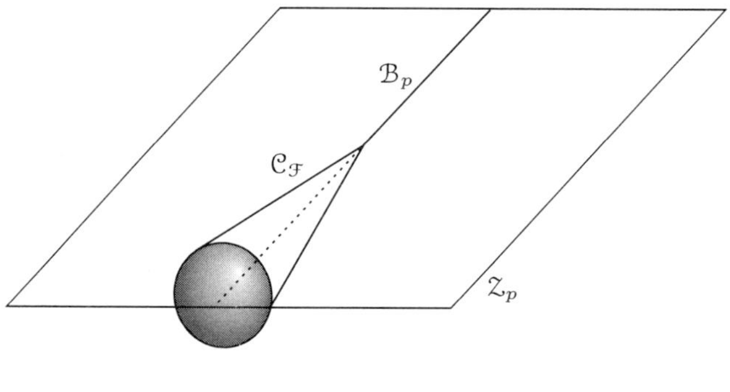

Figure 10.2.3. No closed p–form is transverse to $\mathcal{F}$.

is a convex cone $\mathbf{C}_\mathcal{F} \subset H_p(M;\mathbb{R})$ having a compact base. The dual cone $\mathbf{C}^*_\mathcal{F} \subset H^p(M;\mathbb{R})$, defined by

$$\mathbf{C}^*_\mathcal{F} = \{\gamma \mid z(\gamma) \geq 0, \forall\, z \in \mathbf{C}_\mathcal{F}\},$$

is also convex and has nonempty interior consisting of exactly those cohomology classes that can be represented by a closed p–form transverse to $\mathcal{F}$.

Proof. This map is the restriction to $\mathcal{Z}_\mathcal{F}$ of the bounded linear projection

$$\mathcal{Z}_p \to \mathcal{Z}_p/\mathcal{B}_p = H_p(M;\mathbb{R}),$$

hence carries $\mathcal{Z}_\mathcal{F}$ onto a convex cone $\mathbf{C}_\mathcal{F}$. The base $\widehat{\mathcal{C}}_\mathcal{F}$ of the cone of foliation currents can be defined by a closed p–form ω transverse to $\mathcal{F}$. Hence the compact set $\widehat{\mathcal{C}}_\mathcal{F} \cap \mathcal{Z}_p$ is carried onto the base $\widehat{\mathbf{C}}_\mathcal{F} = [\omega]^{-1}(1) \cap \mathbf{C}_\mathcal{F}$. As the continuous image of a compact set, this base is compact. Also, our hypothesis forces $\mathcal{C}_\mathcal{F} \cap \mathcal{B}_p = \{0\}$; so $[z] \neq 0$ for every nontrivial foliation cycle z. Choose closed p–forms $\eta_1, \dots, \eta_r$ whose cohomology classes are a basis of $H^p(M;\mathbb{R})$. If ω is a closed form, transverse to $\mathcal{F}$, then so is

$$\omega + \sum_{i=1}^{r} a_i \eta_i,$$

for all sufficiently small values of a_i, $1 \leq i \leq r$. Therefore, the classes of these forms range over an open neighborhood of $[\omega]$ in $H^p(M;\mathbb{R})$ lying in $\mathbf{C}^*_\mathcal{F}$. This proves that the classes represented by closed forms transverse to $\mathcal{F}$ lie in the interior of $\mathbf{C}^*_\mathcal{F}$ (and, in particular, this interior is nonempty). We prove that these are the only classes in the interior of the dual cone. If $[\gamma] \in \mathbf{C}^*_\mathcal{F}$ and γ vanishes on some nontrivial foliation cycle z, then some $\eta_i(z) \neq 0$; so there are arbitrarily small values of $a \in \mathbb{R}$ such that $(\gamma + a\eta_i)(z) < 0$. That is, a class $[\gamma]$ in the dual cone is in the interior of that cone only if γ

takes strictly positive values on all nontrivial foliation cycles. Equivalently, the subspace $V = \gamma^{-1}(0) \cap \mathcal{Z}_p \subset \mathcal{Z}_p$ meets $\mathcal{Z}_{\mathcal{F}}$ only at the vertex 0. This subspace corresponds exactly to the ray through $[\gamma]$ in the dual cone. By Theorem 10.2.5, applied to the closed subspace $V \subset \mathcal{D}'_p$ and the compact base of the cone $\mathcal{C}_{\mathcal{F}}$, there is a p–form ω, strictly positive on $\mathcal{C}_{\mathcal{F}} \setminus \{0\}$, such that $\omega^{-1}(0) \cap \mathcal{Z}_p = \gamma^{-1}(0) \cap \mathcal{Z}_p$. This subspace contains $\mathcal{B}_p$; so ω is a closed form, transverse to $\mathcal{F}$, and $[\omega]$ lies in the ray through $[\gamma]$. It follows that γ is cohomologous to the form $b\omega$, for some $b > 0$. □

10.2.B. Foliation cycles are holonomy–invariant measures.

We turn to a useful characterization of foliation cycles. Let μ be a transverse, holonomy–invariant measure for $\mathcal{F}$ that is finite on compact sets. In Exercise 10.1.16 on page 236, you showed that μ defines a cycle $\varphi_\mu \in \mathcal{Z}_p$.

Proposition 10.2.9. *The cycle φ_μ defined by a holonomy–invariant measure is a foliation cycle.*

Exercise 10.2.10. Prove Proposition 10.2.9. (Hint. In each $(U_\alpha, x_\alpha, y_\alpha) \in \mathcal{U}$, approximate φ_μ by a suitable family of Riemann sums and interpret these as positive linear combinations of Dirac foliation currents.)

Exercise 10.2.11. Let L be a compact leaf of $\mathcal{F}$ and define the current φ_L, as in Example 10.1.10. Construct explicitly a holonomy–invariant measure μ, finite on compact sets, such that $\varphi_L = \varphi_\mu$.

The above exercise is a special case of the following converse of Proposition 10.2.9.

Theorem 10.2.12 (Sullivan). *Every foliation cycle is of the form φ_μ, for some holonomy–invariant measure μ.*

The proof requires some preliminary discussion. We are going to show first that foliation currents are currents of integration. Let ω be a smooth p–form transverse to $\mathcal{F}$ and choose a continuous field v of nonzero p–vectors tangent to $\mathcal{F}$ such that $\omega(v) \equiv 1$. Given $c \in \mathcal{C}_{\mathcal{F}}$, define a continuous linear functional

$$J_c : C^\infty(M) \to \mathbb{R}$$

by

$$J_c(f) = c(f\omega).$$

Lemma 10.2.13. *There is a unique, bounded, nonnegative measure ν_c on M such that*

$$J_c(f) = \int_M f\, d\nu_c, \quad \forall\, f \in C^\infty(M).$$

Proof. If $c = w_x = av_x$ is a Dirac foliation current, then

$$J_c(f) = f(x)\omega(w_x) = af(x).$$

Thus, the atomic measure ν, concentrated at x with $\nu(\{x\}) = a$, is the unique measure ν_c. More generally, if c is a finite linear combination of Dirac foliation currents with positive coefficients, ν_c will be the sum of the corresponding atomic measures. In the completely general case, $c = \lim_{n\to\infty} c_n$, where each c_n is a finite linear combination, with positive coefficients, of Dirac foliation currents. The corresponding bounded sequence $\{\nu_{c_n}\}_{n=1}^{\infty}$ of measures has a subsequence converging to a measure ν_c with the required properties. By the Riesz representation theorem [**112**, pp. 42–43] and the fact that $C^\infty(M)$ is dense in $C^0(M)$, the linear functional J_c determines the measure ν_c uniquely. □

Write $c = \lim_{n\to\infty} c_n$ as in the proof of Lemma 10.2.13 and let ν_n be the (discrete) measure ν_{c_n}, $\forall\, n \geq 1$. Then,

$$\lim_{n\to\infty} \int_M f\, d\nu_n = \int_M f\, d\nu_c, \quad \forall\, f \in C^0(M).$$

Since v is the continuous field of p–vectors tangent to $\mathcal{F}$ such that $\omega(v) \equiv 1$, $\eta \in \mathcal{D}_p$ can be decomposed uniquely as

$$\eta = \eta_0 + \eta(v)\omega,$$

where $\eta_0 | \Lambda^p(T(\mathcal{F})) \equiv 0$. We emphasize that η_0 and $\eta(v)$ may only be continuous since $\mathcal{F}$ may be of class $C^{1,0+}$. It is evident that

$$c_n(\eta) = \int_M \eta(v)\, d\nu_n, \quad \forall\, n \geq 1;$$

so

$$c(\eta) = \lim_{n\to\infty} c_n(\eta) = \lim_{n\to\infty} \int_M \eta(v)\, d\nu_n = \int_M \eta(v)\, d\nu_c.$$

We denote this current by $\int v\, d\nu_c$ and remark that the following is now proven.

Proposition 10.2.14. *Foliation currents are currents of integration. More precisely, given a continuous field v of nowhere zero p–vectors tangent to $\mathcal{F}$, the foliation currents c correspond one–to–one to the bounded, nonnegative measures ν_c on M by $c = \int v\, d\nu_c$.*

Let $0 \neq z \in \mathcal{Z}_{\mathcal{F}}$ and fix a partition of unity $\{\lambda_\alpha\}_{\alpha\in\mathfrak{A}}$ subordinate to the regular foliated atlas $\mathcal{U} = \{(U_\alpha, x_\alpha, y_\alpha)\}_{\alpha\in\mathfrak{A}}$. By Proposition 10.2.14, we find a nontrivial measure ν on M and a continuous, nowhere zero field v of p–vectors tangent to $\mathcal{F}$, such that

$$z = \int v\, d\nu,$$

a current of integration. Fix some open, foliated chart (V, x, y), not necessarily in $\mathcal{U}$, and let ν denote $\nu|V$. Let $S \subset \mathbb{R}^q$ be the transverse space $y(V)$, points of S being identified with plaques $P \subset V$. Use the transverse coordinate map

$$y : V \to S$$

to push ν forward to a finite–valued, nonnegative measure μ' on S. That is, if $X \subset S$,

$$\mu'(X) = \nu(y^{-1}(X)).$$

As in Exercise 10.2.15 below, we "disintegrate" the measure ν, finding a nonnegative probability measure ν_P on μ'–almost every $\mathcal{F}$–plaque $P \in S$ such that

$$\int_V h\, d\nu = \int_S \left(\int_P h\, d\nu_P \right) d\mu'(P),$$

whenever $h \in C^\infty(V)$ is ν–integrable. If $\{\nu'_P\}_{P \in S}$ is another family of measures satisfying the above, then $\nu_P = \nu'_P$ for μ'–almost every $P \in S$.

Exercise 10.2.15. Verify the claim above that ν can be disintegrated uniquely into a family of plaque measures. Proceed as follows.

1. Let μ' be the y–projected measure on S, μ'' the x–projected measure on an open set $D \subset \mathbb{R}^p$. Prove that ν is absolutely continuous with respect to the product measure $\mu'' \times \mu'$. That is, if $Z \subset V$ and $(\mu'' \times \mu')(Z) = 0$, then $\nu(Z) = 0$.
2. Use the Radon–Nikodym theorem to find a density function $f(x, y)$, defined and unique ν–almost everywhere on V, such that

   $$d\nu = f\, d(\mu'' \times \mu').$$

   That is, for ν–integrable functions h on V,

   $$\int_V h(x, y)\, d\nu(x, y) = \int_V h(x, y) f(x, y)\, d(\mu'' \times \mu')(x, y).$$
3. Use Fubini's theorem to conclude the proof.

We remark that this is a special case of a more general disintegration theorem for measures [**14**, Ch. 6, pp. 58–59].

Let $\mathcal{D}_p^c(V) \subset \mathcal{D}_p$ be the subspace of forms compactly supported in V. Let $\mathcal{D}_p^c(V)'$ be the dual space of continuous linear functionals on $\mathcal{D}_p^c(V)$. For each $P \in S$, define θ_P and $\sigma_P \in \mathcal{D}_p^c(V)'$ by

$$\theta_P(\eta) = \int_P \eta, \quad \forall\, \eta \in \mathcal{D}_p^c(V),$$

$$\sigma_P(\eta) = \int_P \eta(v)\, d\nu_P, \quad \forall\, \eta \in \mathcal{D}_p^c(V).$$

Lemma 10.2.16. *For $\eta \in \mathcal{D}_p^c(V)$,*

$$\int_S \sigma_P(\eta)\, d\mu'(P) = z(\eta).$$

Proof. Indeed,

$$\begin{aligned} z(\eta) &= \int_M \eta(v)\, d\nu \\ &= \int_V \eta(v)\, d\nu \\ &= \int_S \left(\int_P \eta(v)\, d\nu_P \right) d\mu'(P) \\ &= \int_S \sigma_P(\eta)\, d\mu'(P). \end{aligned}$$

□

Lemma 10.2.17. *The current θ_P is closed, for all $P \in S$, and σ_P is closed for μ'–almost every P.*

Proof. The standard application of Stokes' theorem shows that θ_P is closed. Let $\tau \in \mathcal{D}_{p-1}^c(V)$ and define

$$\begin{gathered} g_\tau : S \to \mathbb{R}, \\ g_\tau(P) = \partial\sigma_P(\tau) = \sigma_P(d\tau), \quad \forall P \in S. \end{gathered}$$

We will show that the measure $g_\tau \mu'$ vanishes, hence that

$$\partial\sigma_P(\tau) = g_\tau(P) = 0,$$

for μ'–almost every P. Indeed, if $h \in C^\infty(S)$, interpret h as a smooth function on V, constant along each plaque P (not, of course, compactly supported). Then $dh \wedge \tau$ is compactly supported, $(dh \wedge \tau)|P \equiv 0$ and

$$\sigma_P(dh \wedge \tau) = \int_P (dh \wedge \tau)(v)\, d\nu_P = 0,$$

since v is tangent to P. Consequently,

$$\sigma_P(h(P)\,\tau) = \sigma_P(h\,\tau) = \sigma_P(dh \wedge \tau + h\,\tau) = \sigma_P(d(h\tau)).$$

Therefore,

$$\begin{aligned}\int_S h(P)\,d[g_\tau\mu'](P) &= \int_S h(P)\sigma_P(d\tau)\,d\mu'(P)\\ &= \int_S \sigma_P(h(P)d\tau)\,d\mu'(P)\\ &= \int_S \sigma_P(d(h\tau))\,d\mu'(P)\\ &= z(d(h\tau)) \quad \text{(By Lemma 10.2.16)}\\ &= \partial z(h\tau)\\ &= 0.\end{aligned}$$

Since the C^∞ functions on S are dense in the space of continuous functions (relative to the uniform norm), the above continues to hold when h is merely continuous. Thus, the measure $g_\tau\mu'$ vanishes as claimed and $\partial\sigma_P(\tau)$ vanishes for μ'–almost every P. The topological space $\mathcal{D}^c_{p-1}(V)$ is separable; so we choose a countable dense subset $\{\tau_i\}_{i=1}^\infty$, remarking that all $\partial\sigma_P(\tau_i)$ simultaneously vanish for μ'–almost every P. By continuity of the currents $\partial\sigma_P$ on $\mathcal{D}^c_{p-1}(V)$, the assertion follows. □

Remark. In the first line of the above proof, we tacitly assume that the leaves of $\mathcal{F}$ have empty boundary. For this, the hypothesis (made at the beginning of this section) that $\partial M = \partial_\tau M$ is essential. In the lamination case, we assumed the leaves to be without boundary for the same reason.

By Lemma 10.2.17, we can assume that σ_P is a closed current, $\forall\, P \in S$. By Exercise 10.1.16 on page 236, applied to the noncompact p–manifold P without boundary ($\partial M = \partial_\tau M$), we conclude that there is a number $\kappa(P)$, $\forall\, P \in S$, such that

$$\sigma_P = \kappa(P)\theta_P.$$

The function $\kappa : S \to \mathbb{R}$ is μ'–measurable and we define $\mu = \kappa\mu'$ on S. If $\eta \in \mathcal{D}^c_p(V)$ and $\eta(v) \geq 0$, then $\theta_P(\eta) \geq 0$ and $\sigma_P(\eta) \geq 0$. Thus, the function κ takes nonnegative values and μ is a nonnegative measure.

Lemma 10.2.18. *The measure μ is the unique nonnegative measure on S, finite on compact sets, such that*

$$z(\eta) = \int_S\left(\int_P \eta\right) d\mu(P), \quad \forall\, \eta \in \mathcal{D}^c_p(V).$$

Proof. Given $\eta \in \mathcal{D}_p^c(V)$, we compute

$$\begin{aligned} z(\eta) &= \int_S \sigma_P(\eta)\, d\mu'(P) \\ &= \int_S \kappa(P)\theta_P(\eta)\, d\mu'(P) \\ &= \int_S \theta_P(\omega)\, d[\kappa\mu'](P) \\ &= \int_S \left(\int_P \eta\right) d\mu(P). \end{aligned}$$

It is easy enough to find a form $\zeta \in A^p(V)$, compactly supported in the x–coordinates but not in the y–coordinates, such that $\int_P \zeta = 1$, for every plaque $P \subset V$. If $h : S \to \mathbb{R}$ is smooth and compactly supported, view h as a function on V, constant along plaques, and obtain $h\zeta \in \mathcal{D}_p^c(V)$. Then

$$\int_S h\, d\mu = \int_S h\left(\int_P \zeta\right) d\mu = z(h\zeta)$$

is finite. It follows that μ is finite on compact sets.

Suppose that $\widetilde{\mu}$ is another such measure. Since $\mu - \widetilde{\mu}$ is finite on compact sets, it makes sense as a signed measure and

$$\int_S \left(\int_P \eta\right) d[\mu - \widetilde{\mu}](P) = 0, \quad \forall \eta \in \mathcal{D}_p^c(V).$$

Thus, as above,

$$0 = \int_S h\, d[\mu - \widetilde{\mu}],$$

for every compactly supported, C^∞ function h on S. That is, $\mu - \widetilde{\mu}$ vanishes on compact sets, and so $\mu = \widetilde{\mu}$. □

Proof of Theorem 10.2.12. For each $(U_\alpha, x_\alpha, y_\alpha) \in \mathfrak{U}$, the above construction determines a nonnegative measure μ_α on S_α. By the definition of regular foliated atlases, there is a foliated chart (W, x, y) such that $\overline{U}_\alpha \subset W$, where distinct plaques in $\overline{U}_\alpha$ lie in distinct plaques in W. Thus, if $\widetilde{S}$ is the transverse space for W, there is a natural inclusion

$$\overline{S}_\alpha \hookrightarrow \widetilde{S}.$$

The measure $\widetilde{\mu}$ on $\widetilde{S}$, obtained as above, restricts to a measure $\widetilde{\mu}_\alpha$ on S_α with all the right properties. So Lemma 10.2.18 guarantees that $\mu_\alpha = \widetilde{\mu}_\alpha$. In particular, since $\widetilde{\mu}(\overline{S}_\alpha) < \infty$, $\mu_\alpha(S_\alpha) < \infty$ and we can define a finite measure

$$\mu = \coprod_{\alpha \in \mathfrak{A}} \mu_\alpha$$

on

$$S = \coprod_{\alpha \in \mathfrak{A}} S_\alpha.$$

We can assume that the atlas is such that, whenever $U_\alpha \cap U_\beta \neq \emptyset$, there is a biregular foliated chart (W, x, y) such that $\overline{U}_\alpha \cup \overline{U}_\beta \subset W$ (*cf.* the proof of Lemma 1.2.17). So the uniqueness of μ_α and μ_β implies that they both agree with the restrictions of $\widetilde{\mu}$, hence agree under the cocycle transformation $\gamma_{\alpha\beta}$. Since these transition functions generate the holonomy pseudogroup, the measure μ is $\Gamma_{\mathfrak{U}}$–invariant. Finally, if $\{\lambda_\alpha\}_{\alpha \in \mathfrak{A}}$ is a partition of unity subordinate to $\mathfrak{U}$, each $\eta \in \mathcal{D}_p$ decomposes into

$$\eta = \sum_{\alpha \in \mathfrak{A}} \lambda_\alpha \eta,$$

where each $\lambda_\alpha \eta \in \mathcal{D}_p^c(U_\alpha)$. Thus,

$$\begin{aligned} z(\eta) &= \sum_{\alpha \in \mathfrak{A}} z(\lambda_\alpha \eta) \\ &= \sum_{\alpha \in \mathfrak{A}} \int_{S_\alpha} \left(\int_P \lambda_\alpha \eta \right) d\mu(P) \\ &= \varphi_\mu(\eta). \end{aligned}$$

□

Exercise 10.2.19. Identifying the set $\mathcal{F}_p(M)$ of smooth p–dimensional foliations with integrable p–plane fields on M and identifying p–plane fields as sections of $\Lambda^p(T(M))$, use a Riemannian metric on this bundle to put a topological metric ρ on $\mathcal{F}_p(M)$. Define a proposition $\mathbf{P}(\mathcal{F})$ about the foliation $\mathcal{F} \in \mathcal{F}_p(M)$ to be a stable property if, whenever $\mathbf{P}(\mathcal{F}_0)$ is true, there is an $\epsilon > 0$ such that

$$\rho(\mathcal{F}_0, \mathcal{F}) < \epsilon \Rightarrow \mathbf{P}(\mathcal{F}).$$

Use foliation cycles to prove the following.

1. The proposition $\mathbf{P}(\mathcal{F})$ = "$\mathcal{F}$ admits no nontrivial, holonomy–invariant measure" is a stable property.
2. The proposition $\mathbf{P}(\mathcal{F})$ = "$[\mu] \neq 0$, $\forall$ nontrivial, holonomy–invariant measure μ for $\mathcal{F}$" is a stable property.

10.2.C. Averaging sequences. Let L be a leaf of $\mathcal{F}$ and let $A \subseteq L$ be a compact, p–dimensional submanifold. For simplicity, assume that ∂A is piecewise C^1. All we will really need is that the boundary be sufficiently regular that Stokes' theorem holds. At any rate, we define a p–current by

$$\omega \mapsto \int_A \omega, \quad \forall \omega \in \mathcal{D}_p,$$

denoting this current by $\int_A$. Since A is contained in a leaf, one expects that $\int_A \in \mathcal{C}_{\mathcal{F}}$.

Proposition 10.2.20. *The current $\int_A$ is a foliation current for $\mathcal{F}$.*

Proof. Fix a Riemannian metric on M and let v be the corresponding field of p–vectors tangent to $\mathcal{F}$, defining the correct orientations of the leaves, and having Riemannian norm 1. Let Ω be a p–form such that $\Omega(v) \equiv 1$. Of course, if $\mathcal{F}$ is only of class $C^{1,0+}$, v and possibly Ω will only be continuous, but $\Omega|A$ will be the volume form. For arbitrary $\omega \in \mathcal{D}_p$, it should be clear that $\omega|A = \omega(v)\Omega|A$. Thus, if ν is the Borel measure on M which, on open sets W, takes the value $\nu(W) = \int_{W\cap A} \Omega$, we see that

$$\int_A \omega = \int_M \omega(v)\, d\nu.$$

By Proposition 10.2.14, it follows that $\int_A \in \mathcal{C}_{\mathcal{F}}$. □

A useful way of generating foliation cycles is by *averaging sequences.* This idea, having its roots in the "Fölner condition for amenability", is due to S. Goodman and J. F. Plante [**55**]. In what follows, fix a Riemannian metric on M and let vol_m denote Riemannian volume of m–dimensional submanifolds of M.

Definition 10.2.21. Let $\mathcal{F}$ be a p–dimensional foliation. An averaging sequence for $\mathcal{F}$ is a sequence $\{A_k\}_{k=1}^{\infty}$ of compact, p–dimensional submanifolds (with piecewise C^1 boundary) of leaves such that

$$\lim_{k\to\infty} \frac{\mathrm{vol}_{p-1}(\partial A_k)}{\mathrm{vol}_p(A_k)} = 0.$$

Theorem 10.2.22 (Goodman and Plante). *Let $\{A_k\}_{k=1}^{\infty}$ be an averaging sequence for $\mathcal{F}$. Then there is a subsequence $\{A_{k_r}\}_{r=1}^{\infty}$ such that*

$$\lim_{r\to\infty} \frac{1}{\mathrm{vol}_p(A_{k_r})} \int_{A_{k_r}} = \varphi$$

exists in $\mathcal{C}_{\mathcal{F}}$ and is a foliation cycle.

Proof. By Proposition 10.2.20, this is a sequence in $\mathcal{C}_{\mathcal{F}}$ and is clearly bounded. By the Montel property of $\mathcal{D}'_p$, there is a convergent subsequence. By passing to this subsequence, we can assume that

$$\lim_{k\to\infty} \frac{1}{\mathrm{vol}_p(A_k)} \int_{A_k} = \varphi \in \mathcal{C}_{\mathcal{F}}.$$

By Stokes' theorem and the compactness of M, there is a constant $B \geq 0$, depending only on the $(p-1)$–form η, such that

$$|\varphi(d\eta)| = \left| \lim_{k\to\infty} \frac{1}{\mathrm{vol}_p(A_k)} \int_{A_k} d\eta \right| \leq \lim_{k\to\infty} \frac{B\,\mathrm{vol}_{p-1}(\partial A_k)}{\mathrm{vol}_p(A_k)} = 0.$$

Since η is arbitrary, $\partial\varphi = 0$. □

10.3. Foliations of Dimension One

Foliation cycles for oriented 1–dimensional foliations $\mathcal{L}$ were first studied by S. Schwartzmann [**119**], who called them *asymptotic cycles.* A fundamental fact is the following.

Lemma 10.3.1. *If $\mathcal{L}$ is a $C^{1,0+}$ foliation by curves, there is a nontrivial foliation cycle for $\mathcal{L}$.*

Proof. By Lemma 10.2.6, the assertion will follow if we show that no exact p–form is transverse to $\mathcal{L}$. Suppose that $\omega \in \mathcal{D}_1$ is exact. That is, $\omega = df$ for a suitable smooth function $f : M \to \mathbb{R}$. Since M is compact, f has an extremal $x \in M$. Since $df_x = 0$, ω cannot be transverse to $\mathcal{L}$ at x. □

Fix a Riemannian metric on M and let ω be a 1–form that defines the volume (length) element along the leaves. Let v be the vector field tangent to $\mathcal{L}$ such that $\omega(v) \equiv 1$. This field is only C^0 on M but it is the velocity field of the flow $\Phi : \mathbb{R} \times M \to M$ with trajectories the leaves of $\mathcal{L}$ and with arclength as flow parameter. Recall (Proposition 10.2.14) that foliation currents c correspond one–to–one to bounded, nonnegative Borel measures ν_c on M by $c = \int v\, d\nu_c$.

Proposition 10.3.2. *The foliation current c is a cycle if and only if ν_c is Φ–invariant.*

Exercise 10.3.3. Here is a sketch of the proof of Proposition 10.3.2. The exercise is to fill in the details. In proving Theorem 10.2.12, we "disintegrated" ν_c in a flow box into a measure ν_P along each plaque P and a transverse measure which we can call μ_c. By our choice of the vector field v, the measure ν_P is just the Riemannian length measure on P, hence is Φ–invariant. It follows that ν_c is Φ–invariant if and only if μ_c is holonomy invariant. Theorem 10.2.12 then gives the proposition.

Proposition 10.3.2 reduces the study of foliation cycles in dimension one to the classical ergodic theory of flow–invariant measures. Essentially, a foliation cycle is that aspect of a flow–invariant measure that neglects the parametrization of the flow. We will need some results from ergodic theory.

Definition 10.3.4. A Φ–invariant measure ν is ergodic if, whenever $A \subseteq M$ is a Φ–invariant set, either $\nu(A) = 0$ or $\nu(A) = \nu(M)$.

The Φ–invariant probability measures form a compact, convex set $\mathcal{M}_\Phi$ and the ergodic elements $\mathcal{E}_\Phi$ of this set are the extreme points. The ergodic decomposition theorem allows an element $\mu \in \mathcal{M}_\Phi$ to be decomposed into a "continuous sum" $\mu = \int_{\mathcal{E}_\Phi} m\, d\tau(m)$ of the ergodic measures $m \in \mathcal{E}_\Phi$ relative to a suitable probability measure τ on $\mathcal{E}_\Phi$. For this theory, which uses Choquet's theorem, see [**100**]. Since bounded measures can be normalized to probability measures, the following is much weaker than the ergodic decomposition theorem, but adequate for our purposes.

Proposition 10.3.5. *If ν is a Φ–invariant measure, it can be arbitrarily well approximated (in the space of 0–currents) by a finite linear combination of ergodic measures, the coefficients all being positive.*

The well–known ergodic theorem for flows [**32**, pp. 11–12] has the following consequence when the invariant measure is ergodic.

Theorem 10.3.6. *If ν is an ergodic, Φ–invariant measure, then for ν–almost every $x \in M$ and all $f \in C^0(M)$,*

$$\lim_{\tau\to\infty} \frac{1}{\tau} \int_0^\tau f(\Phi_t(x))\, dt = \int_M f\, d\nu.$$

Example 10.3.7. Let $x \in M$ be a periodic point of Φ. The orbit of x traces out a compact leaf σ of $\mathcal{L}$. This leaf is a foliation cycle, the corresponding transverse, holonomy–invariant measure μ_σ being the one that deposits zero mass on all $\mathcal{L}$–plaques not in σ and unit mass on the plaques in σ. The corresponding Φ–invariant measure ν_σ is the one that assigns to every Borel set B the Riemannian length measure of $B \cap \sigma$. The ergodicity of ν_σ and the assertion of Theorem 10.3.6 are rather obvious in this case.

Let $x \in M$ be a nonperiodic point of the flow and, for each $\tau > 0$, let $\Gamma_\tau = \{\Phi_t(x) \mid 0 \le t \le \tau\}$. If $\tau_k \uparrow \infty$, we set $\Gamma_k = \Gamma_{\tau_k}$, obtaining an averaging sequence since $\partial\Gamma_k$ consists of just two points. After passing to a subsequence, we obtain a foliation cycle

$$z = \lim_{k\to\infty} \frac{1}{\tau_k} \int_{\Gamma_k} \cdot$$

By compactness of M, the point x in the orbit $\Phi_{\mathbb{R}}(x)$ can be chosen so that $\Phi_t(x)$ repeatedly enters a foliation chart U containing x as $t \uparrow \infty$. If we choose the sequence $\{\tau_k\}_{k=1}^\infty$ so that $x_k = \Phi_{\tau_k}(x) \in U$, $\forall\, k \ge 1$, we can close up Γ_k to a loop Γ_k^* (a "long, almost closed orbit") by adjoining an arc in U from x_k to x. These arcs can be kept uniformly bounded in length. Hence the sequence of *singular cycles* $(1/\tau_k)\Gamma_k^*$ (generally not foliation cycles) also converges to z in $\mathcal{D}_1'$.

Definition 10.3.8. If a foliation cycle z is either a closed orbit or

$$z = \lim_{k\to\infty} \frac{1}{\tau_k} \Gamma_k^*$$

is obtained from long, almost closed orbits Γ_k^* as above, then z is called a *homology direction.*

Since $\omega(v) \equiv 1$,

$$\frac{1}{\tau_k} \int_{\Gamma_k} \omega = 1, \quad \forall\, k.$$

This gives the following.

Lemma 10.3.9. *If z is a homology direction, then $z(\omega) = 1$.*

Proposition 10.3.10. *If a foliation cycle z for $\mathcal{L}$ corresponds to an ergodic Φ–invariant measure ν_z, then z is a homology direction.*

Proof. If ν_z corresponds to a closed orbit, we are done. By Theorem 10.3.6, choose $x \in M$ such that

$$\lim_{\tau \to \infty} \frac{1}{\tau} \int_0^\tau f(\Phi_t(x))\, dt = \int_M f\, d\nu_z.$$

By a suitable choice of $\tau_k \uparrow \infty$, we obtain that, for every 1–form η,

$$z(\eta) = \int_M \eta(v)\, d\nu_z = \lim_{\tau_k \to \infty} \frac{1}{\tau_k} \int_0^{\tau_k} \eta_{\Phi_t(x)}(v_{\Phi_t(x)})\, dt = \lim_{k \to \infty} \frac{\Gamma_k^*(\eta)}{\tau_k},$$

as claimed. □

The following consequence of Propositions 10.3.5 and 10.3.10 is rather obvious.

Proposition 10.3.11. *Every foliation cycle z for $\mathcal{L}$ can be arbitrarily well approximated in $\mathcal{D}_1'$ by a finite linear combination of homology directions, all the coefficients being positive. Similarly, z can be arbitrarily well approximated by such linear combinations of long, almost closed orbits. In both cases, if $z \neq 0$, the sum of these coefficients can be kept bounded from 0.*

The last assertion is a consequence of Lemma 10.3.9. Indeed, suppose that $\sum_{i=1}^m a_i z_i$ is a linear combination of homology directions approximating z. Then

$$\sum_{i=1}^m c_i = \sum_{i=1}^m c_i z_i(\omega)$$

and this approximates $z(\omega) > 0$.

Suppose that $\mathcal{L}$ is transverse to a codimension–one, C^2 foliation $\mathcal{F}$ without holonomy and let $\mu_{\mathcal{F}}$ be a continuous, holonomy–invariant measure for $\mathcal{F}$ as in Lemma 9.2.16. This can be viewed as a continuous measure along the leaves of $\mathcal{L}$. Recall also that this measure defines a class $[\mu_{\mathcal{F}}] \in H^1(M; \mathbb{R})$ by integration on singular 1–cycles. We will denote this class by $[\mathcal{F}]$. Actually, the class is only defined by $\mathcal{F}$ up to a positive constant multiple. If $\mu_{\mathcal{F}}$ is required to be a *probability* measure, $[\mathcal{F}]$ is unique.

Lemma 10.3.12. *Under the above hypotheses, there is a constant $b > 0$ such that $[z]([\mathcal{F}]) \geq b$, for all homology directions z of $\mathcal{L}$.*

Proof. Fix a biregular atlas for $\mathcal{F}$ and $\mathcal{L}$. Any two $\mathcal{L}$–plaques have connected intersection. The pieces of orbit Γ_k in the definition of homology directions can always be chosen so that there is a subdivision

$$0 = t_0 < t_1 < \cdots < t_{n+1} = \tau_k$$

such that $\Phi_{[t_i,t_{i+2}]}(x)$ is an $\mathcal{L}$–plaque, $1 \le i \le n-1$. By the finiteness of the biregular atlas, there is a universal upper bound on the Riemannian length measures of the segments $\Phi_{[t_i,t_{i+1}]}(x)$ of $\mathcal{L}$–plaques and a positive lower bound on their $\mu_{\mathcal{F}}$ measures. It follows that there is a universal constant $b > 0$ such that

$$\int_{\Gamma_k} d\mu_{\mathcal{F}} \ge b\tau_k.$$

Thus, if z is a homology direction,

$$[z]([\mathcal{F}]) = \lim_{k\to\infty} \frac{1}{\tau_k} \int_{\Gamma_k^*} d\mu_{\mathcal{F}} = \lim_{k\to\infty} \frac{1}{\tau_k} \int_{\Gamma_k} d\mu_{\mathcal{F}} \ge b$$

as claimed. □

Corollary 10.3.13. *If $\mathcal{L}$ is transverse to a codimension–one, C^2 foliation $\mathcal{F}$ without holonomy, then $[z]([\mathcal{F}]) > 0$, for all nontrivial foliation cycles z of $\mathcal{L}$.*

Proof. If z is a nontrivial foliation cycle, Lemma 10.3.12 and Proposition 10.3.11 imply that $[z]([\mathcal{F}]) > 0$. □

We give various properties of a 1–dimensional foliation that are equivalent to the existence of a transverse foliation without holonomy.

Proposition 10.3.14. *If $\mathcal{L}$ is a $C^{1,0+}$ foliation by curves, the following are equivalent.*

(1) *$\mathcal{L}$ admits no nontrivial foliation boundary.*
(2) *There is a closed 1–form transverse to $\mathcal{L}$.*
(3) *$\mathcal{L}$ is transverse to the fibers of a smooth fibration $\pi : M \to S^1$.*
(4) *$\mathcal{L}$ admits a global section.*
(5) *$\mathcal{L}$ is transverse to a C^2 foliation of codimension one without holonomy.*

Proof. The implications (1) ⇔ (2) are by Lemma 10.2.7. If θ is a closed 1–form transverse to $\mathcal{L}$, then θ is nonsingular. By Theorem 9.4.2, θ can be arbitrarily well approximated by a closed, nonsingular 1–form of rank 1. By Corollary 9.3.8, it follows that $\mathcal{L}$ is transverse to a fibration $\pi : M \to S^1$. Conversely, if $\mathcal{L}$ is transverse to such a fibration and if ds denotes the element of length on S^1, then $\theta = \pi^*(ds)$ is a closed 1–form transverse to $\mathcal{L}$; so (2) ⇔ (3). The implications (3) ⇔ (4) are the content of Example 2.1.1.

Since a fibration is a foliation without holonomy, it is trivial that (3) $\Rightarrow$ (5). Corollary 10.3.13 shows that (5) $\Rightarrow$ (1). □

Finally, the postponed proof of a proposition that was stated and used in Chapter 9 can now be given.

Proof of Proposition 9.5.8. We are given that $\mathfrak{L}$ is transverse to a C^2 foliation of codimension one without holonomy. In our current language, we are to prove that $[\mathcal{F}] = [\theta]$, where θ is a closed 1–form transverse to $\mathfrak{L}$. By Proposition 10.3.14, $\mathfrak{L}$ satisfies the hypotheses of Lemma 10.2.8. By Corollary 10.3.13, $[\mathcal{F}]$ is in the interior of the dual cone $\mathbf{C}^*_{\mathfrak{L}} \subset H^1(M;\mathbb{R})$; hence $[\mathcal{F}] = [\theta]$ as desired. □

10.4. Taut Foliations of Codimension One

Let $(M, \mathcal{F})$ be a compact, connected, oriented and transversely–oriented $C^{1,0+}$–foliated manifold of codimension one. Let $n = \dim M$ and assume that $\partial_{\pitchfork} M = \emptyset$; hence $\partial_\tau M = \partial M$. Recall that the foliation is taut (Definition 6.3.3) if each leaf either meets a closed transversal or meets a transverse arc with endpoints in ∂M.

Proposition 10.4.1. *With the above hypotheses, the following properties are equivalent.*

1. *The foliation $\mathcal{F}$ is taut.*
2. *No nontrivial foliation cycle for $\mathcal{F}$ bounds.*
3. *There is a closed $(n-1)$–form transverse to $\mathcal{F}$.*

Proof. Assume that $\mathcal{F}$ is taut and that μ is a nontrivial, holonomy–invariant measure. If L is a leaf in the support (*cf.* Exercise 9.2.14) of μ, let σ be a closed transversal through L or a transverse arc through L from one component of ∂M to another. Let $N(\sigma) \cong \sigma \times B$ be a normal neighborhood of σ with each fiber $\{x\} \times B$ an open ball in the leaf of $\mathcal{F}$ through x. The Poincaré dual of $[\sigma] \in H_1(M, \partial M; \mathbb{R})$ is represented by a closed $(n-1)$–form ω, supported in $N(\sigma)$ and such that

$$\int_{\{x\}\times B} \omega = 1, \quad \forall x \in \sigma.$$

Evidently, $[\mu] \in H_{n-1}(M;\mathbb{R})$ has positive value on $[\omega]$, proving that, as a foliation cycle, μ is not a boundary. This proves that (1) $\Rightarrow$ (2). The implication (2) $\Rightarrow$ (3) is given by Lemma 10.2.7. We prove that (3) $\Rightarrow$ (1).

By Corollary 6.3.4, $\mathcal{F}$ fails to be taut exactly if there is a finite collection of compact leaves $L_1, \dots, L_q$ such that

$$\zeta = \sum_{i=1}^{q} [L_i] = 0 \in H_{n-1}(M; \mathbb{R}).$$

But then the existence of a closed $(n-1)$–form transverse to $\mathcal{F}$ would give the contradiction that $0 < \int_\zeta \omega = \int_0 \omega = 0$. □

Exercise 10.4.2. Give a direct proof that (1) ⇒ (3). (Hint. Cover M with finitely many tubular neighborhoods of transversals σ as in the above proof.)

We investigate the phenomenon of "exploding plateaus", a generalization of the "exploding disks" that play a key role in Novikov's proof that foliations of S^3 have Reeb components.

Example 10.4.3. Let $(R, \mathcal{G})$ be a Reeb component in $(M, \mathcal{F})$. That is, R is a compact, $\mathcal{F}$–saturated subset of M, homeomorphic to $S^1 \times D^{n-1}$ (where $n = \dim M$), and $\mathcal{G} = \mathcal{F}|R$ is the Reeb foliation. Because of this Reeb component, the foliation is not taut. Fix a transverse, 1–dimensional foliation $\mathcal{L}$. Since $\mathcal{G}$ fibers the interior R_0 of R over S^1, one defines a map

$$\Phi : R \times [0,1) \to R, \quad \Phi(x,t) = \Phi_t(x),$$

by reparametrizing the leaf–preserving flow along $\mathcal{L}$ so that, for $x \in R_0$ sufficiently near ∂R, $\Phi_t(x) \to \partial R$ as $t \to 1$. This is stationary along ∂R and carries leaves to leaves. If $S \subset \partial R$ is an imbedded copy of S^{n-2} bounding a factor $\{z\} \times D^{n-1}$, $z \in S^1$, projection along the leaves of $\mathcal{L}$ imbeds S in one of the leaves L of $\mathcal{G}|R_0$. This imbedded sphere $S' \subset L$ bounds an $(n-1)$–dimensional ball $D' \subset L$ and Φ restricts to a map

$$\varphi : D' \times [0,1) \to R$$

such that $\varphi_t(D')$ is contained in a leaf L_t of $\mathcal{G}|R_0$, $0 \le t < 1$, and $\varphi_t(S') \to S$ uniformly as $t \to 1$. Notice that some of the leaves of $\mathcal{L}$ meeting the interior of D' never meet ∂R. So there are points $x \in \operatorname{int} D'$ such that $\varphi_t(x)$ remains bounded away from ∂R, $0 \le t < 1$. Since $\varphi_t(\partial D') \to S \subset \partial R$, this forces $\varphi_t(D')$ to "spread out" in $\varphi_t(L)$, becoming arbitrarily large as $t \to 1$, while $\varphi_t(\partial D')$ remains bounded. This "exploding ball" exemplifies the general exploding plateau phenomenon in the following discussion.

Let P be a smooth, compact, connected, orientable $(n-1)$–manifold with nonempty boundary. We are interested in smooth isotopies, called "moving plateaus",

$$\varphi : P \times [0,1) \to M$$

such that

$$\varphi_t : P \to M$$

is a smooth imbedding into a leaf L_t of $\mathcal{F}$, $0 \le t < 1$, and such that φ restricts to a regular map of $\{x\} \times [0,1)$ into a leaf of $\mathcal{L}$, $\forall x \in P$, this being an imbedding if $x \in \partial P$. We set $P_t = \varphi_t(P)$, $0 \le t < 1$, and let $C_1, C_2, \dots, C_r$ denote the components of ∂P.

Definition 10.4.4. The moving plateau $\varphi : P \times [0,1) \to M$ is an *exploding plateau* in the foliated manifold $(M, \mathcal{F})$ if the following conditions are fulfilled:

(1) $\varphi|\partial P \times [0,1)$ extends to a smooth imbedding $\widetilde{\varphi} : \partial P \times [0,1] \to M$ so as to carry $C_i \times \{1\}$ into a leaf F_i of $\mathcal{F}$, $1 \le i \le r$;

(2) φ does not extend to a continuous map $\overline{\varphi} : P \times [0,1] \to M$, agreeing with $\widetilde{\varphi}$ and carrying $P \times \{1\}$ into a leaf of $\mathcal{F}$.

Remark. If $F_i \neq F_j$ for suitable $i, j \le r$, then property (2) in Definition 10.4.4 is redundant.

We will see that an exploding plateau really does "explode" in the sense that $\limsup_{t \to 1} \operatorname{vol}(P_t) = \infty$, where $\operatorname{vol}(P_t)$ denotes the $(n-1)$–dimensional volume relative to some fixed Riemannian metric. It is clear, however, that the $(n-2)$–dimensional volume of ∂P_t remains bounded.

Theorem 10.4.5. *If $\mathcal{F}$ is taut, there can be no exploding plateau in $(M, \mathcal{F})$.*

The basic idea of the proof is to produce an averaging sequence from the exploding plateau, showing that the resulting foliation cycle is nontrivial and bounds. We will proceed via some lemmas.

Lemma 10.4.6. *The set $U = \{x \in P \mid \lim_{t \to 1} \varphi_t(x)$ exists$\}$ is an open neighborhood of ∂P but is not all of P.*

Proof. Clearly, $\partial P \subseteq U$. If $x \in U$ and $y = \lim_{t \to 1} \varphi_t(x)$, let $W \subset M$ be a *biregular neighborhood* of y. This means that $W \cong Q \times J$ is an open neighborhood such that $\{z\} \times J$ is an $\mathcal{L}$–plaque, $\forall z \in Q$, and $Q \times \{s\}$ is an $\mathcal{F}$–plaque, $\forall s \in J$. It is clear, then, that some neighborhood $U_x \subset P$ of x is carried to a subset $\varphi_t(U_x)$ of the $\mathcal{F}$–plaque $Q_t \subset W$ through $\varphi_t(x)$, for $t < 1$ and sufficiently close to 1. Since Q_t limits uniformly on the plaque $Q_1 \subset W$ through y, as $t \to 1$, and since $\varphi(\{z\} \times [0,1))$ lies in a leaf of $\mathcal{L}$, $\forall z \in U_x$, it follows easily that $U_x \subseteq U$. This argument shows also that, on each component U' of U, $\varphi|U' \times [0,1)$ extends continuously to $U' \times [0,1]$ so as to carry $U' \times \{1\}$ into a leaf of $\mathcal{F}$. It follows immediately that $U \neq P$. □

Let $x \in P \setminus U$ and let $\{t_k\}_{k=1}^{\infty} \subset [0,1)$ be a sequence converging monotonically to 1. Since M is compact, we can pass to a convergent subsequence of $\{\varphi_{t_k}(x)\}_{k=1}^{\infty}$; so we assume that this sequence converges to $y_* \in M$. Let W be a biregular neighborhood of y_* and let $Q \subset W$ be the $\mathcal{F}$–plaque containing y_*. Since infinitely many of the points $\varphi_{t_k}(x)$ must lie in W, although

$\lim_{t\to 1}(\varphi_t(x))$ does not exist, the curve $\lambda(t) = \varphi_t(x)$, $0 \le t < 1$ must pass through W infinitely often as $t \to 1$. But this curve lies in a leaf of $\mathcal{L}$; hence it must cross the plaque Q infinitely often as $t \to 1$. Thus, we can assume that the sequence $\{t_k\}_{k=1}^{\infty}$ was so chosen that $y_k = \varphi_{t_k}(x) \in Q$, $\forall k \ge 1$.

In particular, L_{t_k} is independent of k; so we denote this leaf by L_*. Let $P_k = \varphi_{t_k}(P)$, an imbedded submanifold of L_*.

Lemma 10.4.7. *After passing to a subsequence, we can assume that, for all $k \ge 1$,*

1. $P_k \subset \operatorname{int}(P_{k+1})$;
2. $\bigcup_{k=1}^{\infty} P_k = L_*$.

Proof. For k sufficiently large, $y_* \in P_k$. Indeed, $y_k \in P_k$ and these points converge to y_* in the bounded $\mathcal{F}$–plaque $Q \subset L_*$. If $y_* \notin P_k$, then ∂P_k separates y_* from y_k. If this occurred for infinitely many values of k, we could pass to a subsequence, finding $b_{k_j} \in \partial P_{k_j}$ such that $\lim_{j\to\infty}(b_{k_j}) = y_*$ and the convergence takes place within Q. But $\{\partial P_{k_j}\}_{j=1}^{\infty}$ converges uniformly in M to $\zeta = \widetilde{\varphi}(\partial P \times \{1\})$ and the convergence is along $\mathcal{F}$–transverse directions. It follows that all but finitely many b_{k_j} lie outside of Q, a contradiction.

Fix $k \ge 1$. Since $\{\partial P_n\}_{n=1}^{\infty}$ clusters uniformly onto ζ along the $\mathcal{F}$–transverse directions, we conclude that $\partial P_n \cap P_k = \emptyset$, for n sufficiently large. Since $y_* \in P_k \cap P_n$, we conclude that $P_k \subset \operatorname{int}(P_n)$. Passing once more to a subsequence, we obtain $P_k \subset \operatorname{int}(P_{k+1})$, $\forall\, k \ge 1$.

Finally, given any compact subset $K \subset L_*$, it is clear that, for k sufficiently large, $\partial P_k \cap K = \emptyset$. If we choose K to be connected and such that $y_* \in K$, it follows that $K \subset \operatorname{int}(P_k)$. But it is clearly possible to choose an increasing nest $K_1 \subset K_2 \subset \cdots \subset K_n \subset \cdots$ of such compact sets such that $L_* = \bigcup_{n=1}^{\infty} K_n$; so $L_* = \bigcup_{k=1}^{\infty} P_k$. □

In particular, L_* cannot be compact and $\lim_{k\to\infty} \operatorname{vol}(P_k) = \infty$. Since ∂P_k remains bounded, we can pass again to a subsequence, obtaining an *averaging sequence* (Definition 10.2.21) converging to a nontrivial foliation cycle μ (Theorem 10.2.22), viewed as a finite, strictly positive, transverse, holonomy–invariant measure μ.

Lemma 10.4.8. *The foliation cycle μ bounds.*

Proof. We must show that, for arbitrary closed 2–forms ω on M, $\mu(\omega) = 0$. Set $B_k = P \times [t_1, t_k]$. The isotopy φ restricts to an immersion $f : B_k \to M$. Since ω is closed, Stokes' theorem implies that

$$0 = \int_{\partial B_k} f^*(\omega) = \int_{P_k} \omega - \int_{P_1} \omega + \int_{\varphi(\partial P \times [t_1, t_k])} \omega.$$

Here,

$$\varphi(\partial P \times [t_1, t_k]) \subset \widetilde{\varphi}(\partial P \times [0,1])$$

and the integral over this set is uniformly bounded as $k \to \infty$. Consequently,

$$0 = \lim_{k\to\infty} \frac{\int_{\partial B_k} f^*(\omega)}{\operatorname{vol}(P_k)} = \lim_{k\to\infty} \frac{\int_{P_k} \omega}{\operatorname{vol}(P_k)} = \mu(\omega).$$

□

Together with Proposition 10.4.1, these lemmas complete the proof of Theorem 10.4.5.

10.5. Taut Foliations of Arbitrary Codimension

For simplicity, throughout this section everything in sight will be smooth of class C^∞. It is not hard to reduce this requirement to smoothness of class C^2. We treat only foliations (not laminations) and require that $\partial M = \partial_\tau M$.

In codimension one, topological tautness (Definition 6.3.3) was shown to be equivalent to the existence of a closed $(n-1)$–form transverse to the foliation (Proposition 10.4.1). One consequence of this result, to be proven in this section, is that topological tautness is equivalent, in codimension one, to *geometric tautness*, defined (in arbitrary codimension) as follows.

Definition 10.5.1. The foliated manifold $(M, \mathcal{F})$ is geometrically taut if there exists a Riemannian metric g on M relative to which all leaves are minimal submanifolds (mean curvature zero). In this case, $\mathcal{F}$ is also said to be geometrically taut or *harmonic* and the triple $(M, \mathcal{F}, g)$ is called a *harmonic foliated manifold.*

Our goal is to show that a rather subtle homological condition involving foliation currents is equivalent to geometric tautness. This theorem of Sullivan requires some results of H. Rummler [**114**] that express geometric tautness in terms of differential forms. In the case of codimension one, Rummler's criterion reduces to the existence of a closed $(n-1)$–form transverse to $\mathcal{F}$.

10.5.A. Rummler's criterion. We give a characterization of harmonic foliated manifolds $(M, \mathcal{F}, g)$ of arbitrary codimension in terms of differential forms. First, we review the notion of a minimal submanifold.

In what follows, we write $\langle \xi, \zeta \rangle$ for $g(\xi, \zeta)$. The Levi–Civita connection for g will be denoted by ∇. If $V \subset T_z(M)$ is a vector subspace,

$$\pi_V : T_z(M) \to V$$

will denote orthogonal projection relative to g. Let $L \hookrightarrow M$ be a smoothly immersed submanifold of dimension $p < n = \dim M$ and let $N(L)$ denote the normal bundle (relative to g) of the immersion. This is a q–plane bundle over L, $q = n-p$, and the fiber $N_z(L)$ is the orthogonal complement of $T_z(L)$ in $T_z(M)$, for each $z \in L$. We recall that the Weingarten map

$$W_z^\nu : T_z(L) \to T_z(L)$$

is defined, for each $z \in L$ and each $\nu \in N_z(L)$, by

$$W_z^\nu(\xi) = \pi_{T_z(L)}(\nabla_\xi \nu),$$

where ν is extended locally to a normal field along L and the Weingarten map is independent of the choice of this extension. (Sign conventions vary.) The Weingarten map is diagonalized via an orthonormal basis of eigenvectors.

Definition 10.5.2. The *mean curvature* at z in the normal direction ν of the submanifold L is the trace $\operatorname{tr} W_z^\nu$ of the Weingarten map. We say that the submanifold is *minimal* if the mean curvature vanishes everywhere and in all normal directions.

Example 10.5.3. If $\dim L = 1$, we parametrize L by arclength, obtaining a curve σ with $\langle \dot\sigma, \dot\sigma \rangle \equiv 1$. Given $z \in L$, we can choose this parametrization such that $z = \sigma(0)$. If ν is a normal field along σ near z, then

$$0 = \dot\sigma \langle \nu, \dot\sigma \rangle = \langle \nabla_{\dot\sigma} \nu, \dot\sigma \rangle + \langle \nu, \nabla_{\dot\sigma} \dot\sigma \rangle .$$

It follows that

$$\left\langle W_z^{\nu(0)}(\dot\sigma(0)), \dot\sigma(0) \right\rangle = - \left\langle \nu(0), \nabla_{\dot\sigma(0)} \dot\sigma \right\rangle .$$

Since $z \in L$ and $\nu(0) \in N_z(L)$ are arbitrary, it follows that L is minimal if and only if the field $\nabla_{\dot\sigma}\dot\sigma$ is tangent to L. But

$$0 \equiv \dot\sigma \langle \dot\sigma, \dot\sigma \rangle = 2 \langle \nabla_{\dot\sigma} \dot\sigma, \dot\sigma \rangle ;$$

so $\nabla_{\dot\sigma}\dot\sigma$ is tangent to L if and only if $\nabla_{\dot\sigma}\dot\sigma \equiv 0$. In other words, 1–dimensional submanifolds are minimal if and only if, when parametrized by arclength, they are geodesics.

Exercise 10.5.4. If the p–dimensional, immersed submanifold $L \hookrightarrow M$ is totally geodesic, prove that it is minimal.

Example 10.5.5. In $\mathbb{R}^n$, the totally geodesic, connected submanifolds are the p–planes, $0 \le p \le n$. Similarly, in S^n, the totally geodesic, connected submanifolds are the great p–spheres. Generally, minimal submanifolds are much more plentiful than totally geodesic ones.

Let $(M, \mathcal{F})$ be a foliated manifold of arbitrary codimension $q = n - p$ and let g be a Riemannian metric on M. Let $\omega \in A^p(M)$ be chosen so that

(1) for each leaf L of $\mathcal{F}$, the volume form of $g|L$ is $\omega|L$;

(2) if $\eta_1, \dots, \eta_{p-1} \in \mathfrak{X}(M)$ are tangent to $\mathcal{F}$ and $\nu \in \mathfrak{X}(M)$ is a vector field everywhere perpendicular to $\mathcal{F}$, then $\omega(\nu, \eta_1, \dots, \eta_{p-1}) \equiv 0$.

If $\operatorname{codim}\mathcal{F} = 1$, these properties uniquely determine $\omega \in A^{n-1}(M)$. For larger codimension, ω is not unique, but always exists. It is easy to choose such forms locally in foliated charts and piece these local choices together by a smooth partition of unity.

Remark that property (2) of ω can be rephrased in terms of the interior product $i_\nu\omega \in A^{p-1}(M)$. If ν is everywhere perpendicular to $\mathcal{F}$, the restriction of $i_\nu\omega$ to an antisymmetric, $(p-1)$–linear form on $T(\mathcal{F})$ is trivial.

Lemma 10.5.6. *Let $\omega \in A^p(M)$ have the above properties and let U be an open neighborhood in M that trivializes $T(\mathcal{F})$. Let $\xi_1, \dots, \xi_p \in \mathfrak{X}(U)$ form an orthonormal basis of $T(\mathcal{F})|U$, compatible with the orientation of the leaves. Let $\nu \in \mathfrak{X}(U)$ be everywhere perpendicular to $\mathcal{F}|U$ and let W^ν be the corresponding Weingarten map for the leaves of $\mathcal{F}|U$. Then,*

$$d\omega(\nu, \xi_1, \dots, \xi_p) = \operatorname{tr} W^\nu.$$

Proof. We use the well–known formula

$$\begin{aligned} d\omega(\eta_1, \dots, \eta_{p+1}) = &\sum_{i=1}^{p+1} (-1)^{i+1} \eta_i(\omega(\eta_1, \dots, \widehat{\eta}_i, \dots, \eta_{p+1})) \\ &+ \sum_{i<j} (-1)^{i+j} \omega([\eta_i, \eta_j], \eta_1, \dots, \widehat{\eta}_i, \dots, \widehat{\eta}_j, \dots, \eta_{p+1}), \end{aligned}$$

where $\widehat{\eta}_i$ and $\widehat{\eta}_j$ indicate omissions. Take $\nu = \eta_1$ and $\xi_i = \eta_{i+1}$, $1 \le i \le p$, in this formula. Straightforward computations, using properties (1) and (2) of ω and the integrability of $T(\mathcal{F})$, yield

$$d\omega(\nu, \xi_1, \dots, \xi_p) = -\sum_{k=1}^{p} \omega(\xi_1, \dots, \xi_{k-1}, [\nu, \xi_k], \xi_{k+1}, \dots, \xi_p).$$

Since ∇ is torsion free and $\langle \nabla_\nu \xi_k, \xi_k \rangle = 0$,

$$\begin{aligned} \langle [\nu, \xi_k], \xi_k \rangle &= \langle \nabla_\nu \xi_k, \xi_k \rangle - \langle \nabla_{\xi_k} \nu, \xi_k \rangle \\ &= - \langle W^\nu(\xi_k), \xi_k \rangle \end{aligned}$$

and it follows that

$$d\omega(\nu, \xi_1, \dots, \xi_p) = \sum_{k=1}^{p} \langle W^\nu(\xi_k), \xi_k \rangle = \operatorname{tr} W^\nu.$$

The second equality is true for any orthonormal basis of $T(\mathcal{F}|U)$ since there exists an orthonormal basis of eigenvectors of W^ν. □

In terms of the interior product, the conclusion of the above lemma can be written

$$i_\nu d\omega(\xi_1, \dots, \xi_p) = \operatorname{tr} W^\nu.$$

Example 10.5.7. Given $z_0 \in M$, choose a foliated chart (U, x, y) about z_0 and let $\nu_0 \in \mathfrak{X}(U)$ be a vector field everywhere perpendicular to the $\mathcal{F}$–plaques which is carried by the differential of $y : U \to \mathbb{R}^q$ to a well–defined vector field ζ on $y(U)$. Such a field is easily constructed by arbitrarily choosing the field ζ on $y(U)$ and, for each $z \in U$, taking ν_{0z} to be the unique vector at $z \in U$ that is perpendicular to the plaque through z and is such that $y_{*z}(\nu_z) = \zeta_{y(z)}$. Such vector fields ν_0, called $\mathcal{F}$*-parallel* fields, are exactly the normal fields to $\mathcal{F}|U$ that generate local flows Φ_t that preserve $\mathcal{F}|U$. Let $\omega \in A^p(U)$ be as above. If $P = P_0 \subset U$ is the $\mathcal{F}$–plaque through z_0, set $P_t = \Phi_t(P)$ for small values of $|t|$ and let $V(t)$ denote the Riemannian volume of P_t. Then,

$$\begin{aligned}\frac{dV}{dt}(0) &= \frac{d}{dt}\bigg|_{t=0} \int_{P_t} \omega \\ &= \frac{d}{dt}\bigg|_{t=0} \int_{P_0} \Phi_t^*(\omega) \\ &= \int_{P_0} \frac{d}{dt}\bigg|_{t=0} \Phi_t^*(\omega) \\ &= \int_{P_0} L_{\nu_0}\omega,\end{aligned}$$

where L_{ν_0} denotes the Lie derivative by the vector field ν_0. The standard identity

$$L_{\nu_0}\omega = i_{\nu_0} d\omega + d i_{\nu_0}\omega$$

and the fact that $i_{\nu_0}\omega|P_0 \equiv 0$ imply

$$\frac{dV}{dt}(0) = \int_{P_0} i_{\nu_0} d\omega.$$

This vanishes for all choices of $z_0 \in M$, of foliated chart about z_0, and $\mathcal{F}$–parallel field ν_0 if and only if $i_\nu d\omega$ vanishes along the leaves for all choices of normal field ν ($\mathcal{F}$–parallel or not). By Lemma 10.5.6, it follows that $(M, \mathcal{F}, g)$ is harmonic if and only if the derivatives dV/dt vanish.

Definition 10.5.8. A form $\theta \in A^p(M)$ is $\mathcal{F}$*–closed* if $d\theta(\eta_1, \dots, \eta_{p+1}) = 0$ whenever at least p of the fields η_i are tangent to $\mathcal{F}$.

Remark. Equivalently, the p–form θ is $\mathcal{F}$–closed if and only if

$$i_\nu d\theta(\xi_1, \dots, \xi_p) = 0,$$

whenever ν is orthogonal to $\mathcal{F}$ and $(\xi_1, \dots, \xi_p)$ is a (locally–defined) orthonormal frame field tangent to $\mathcal{F}$. If $\operatorname{codim}\mathcal{F} = 1$, an $(n-1)$–form is $\mathcal{F}$–closed if and only if it is closed.

Recall that a p–form θ is said to be transverse to $\mathcal{F}$ if $\theta|L$ is nonsingular, for each leaf L of $\mathcal{F}$. For example, property (1) of the form ω implies that this form is transverse to $\mathcal{F}$. Here is Rummler's criterion.

Theorem 10.5.9 (Rummler [**114**]). *The dimension–p foliation $\mathcal{F}$ is geometrically taut if and only if there exists an $\mathcal{F}$–closed p–form on M that is transverse to $\mathcal{F}$.*

Proof. Let $(M, \mathcal{F}, g)$ be harmonic. The form ω, as in the above discussion, is transverse to $\mathcal{F}$ and Lemma 10.5.6 shows that it is $\mathcal{F}$–closed. Conversely, let $\theta \in A^p(M)$ be $\mathcal{F}$–closed and transverse to $\mathcal{F}$. Let g_0 be any smooth, positive definite bilinear form on $T(\mathcal{F})$ and let ω_0 be the smooth section of $\Lambda^p(T^*(\mathcal{F}))$ that is the volume form of g_0 along the leaves. There is a smooth, strictly positive function f on M such that $f^p\omega_0 = \theta|\Lambda^p(T(\mathcal{F}))$. Let $g = f^2 g_0$. Then g is a smooth, positive definite bilinear form on $T(\mathcal{F})$ having $\theta|L$ as volume form along L, for each leaf L of $\mathcal{F}$.

We need to extend g suitably to a Riemannian metric on M. It is important to determine the correct orthogonal complement $N_z \subset T_z(M)$ to $T_z(\mathcal{F})$, $\forall\, z \in M$. Fix an orthonormal basis $\xi_1, \dots, \xi_p$ of $T_z(\mathcal{F})$ and define 1–forms (at z)

$$\theta_k = i_{\xi_1\wedge\cdots\wedge\widehat{\xi}_k\wedge\cdots\wedge\xi_p}\theta, \quad 1 \le k \le p.$$

Then

$$\theta_k(\xi_j) = \pm\delta_{kj}$$

and this implies that the space

$$N_z = \{\eta \in T_z(M) | \theta_k(\eta) = 0, 1 \le k \le p\}$$

is complementary to $T_z(\mathcal{F})$ and of dimension $q = n - p$. This defines the normal bundle $N(\mathcal{F})$ of $\mathcal{F}$ and we extend g to be positive definite on this bundle and such that $N(\mathcal{F}) \perp T(\mathcal{F})$. Thus, θ satisfies the properties, relative to g, that ω satisfied relative to the Riemannian metric in the above discussion. Lemma 10.5.6 and the fact that θ is $\mathcal{F}$–closed imply that $(M, \mathcal{F}, g)$ is harmonic. □

Corollary 10.5.10. *If the foliation $\mathcal{F}$ is of codimension one, the foliated manifold $(M, \mathcal{F})$ is geometrically taut if and only if it is topologically taut.*

Indeed, by Theorem 10.5.9, geometric tautness is equivalent to the existence of a closed $(n-1)$–form transverse to $\mathcal{F}$. By Proposition 10.4.1, this is equivalent to topological tautness.

10.5.B. Sullivan's criterion. In order to state Sullivan's homological criterion for geometric tautness, we need a couple of definitions.

Definition 10.5.11. A smooth, singular $(r+1)$–chain

$$c = \sum_{i=1}^{m} a_i \sigma_i,$$
$$\sigma_i : \Delta_{r+1} \to M, \quad 1 \le i \le m,$$

is tangent to the foliation $\mathcal{F}$ if

$$T_{\sigma_i(x)}(\mathcal{F}) \subseteq \sigma_{i*x}(T_x(\Delta_{r+1})), \quad 1 \le i \le m, \ \forall x \in \Delta_{r+1}.$$

In this case, ∂c is called a *tangent r–homology.*

In particular, it is required in this definition that $r+1 \ge p = \dim \mathcal{F}$. Note that, even at points $x \in \partial\Delta_{r+1}$, the space $T_x(\Delta_{r+1})$ is $(r+1)$–dimensional.

Definition 10.5.12. A p–dimensional foliation $\mathcal{F}$ is *homologically taut* if no nontrivial foliation cycle φ_μ can be arbitrarily well approximated in $\mathcal{D}'_p$ by tangent p–homologies.

Theorem 10.5.13 (Sullivan [**128**]). *The foliation $\mathcal{F}$ is geometrically taut if and only if it is homologically taut.*

Since homological tautness is a rather nonintuitive property, we will give some corollaries and examples before proving Theorem 10.5.13.

Example 10.5.14. An early question in this direction was raised by H. Gluck [**51**]. His question, "Can space be filled by geodesics?", amounted to asking which 1–dimensional foliations are geometrically taut. As he pointed out, a 1–dimensional foliation of T^2 is not geometrically taut if it has a Reeb component. It is easy to see that 2–dimensional Reeb components violate homological tautness. Indeed, a triangulation of the annulus underlying the Reeb component produces a 2–chain c that is tangent to $\mathcal{F}$. But $\partial c = \sigma_1 + \sigma_2$, the sum of the oriented boundary circles (see Figure 10.5.1). The holonomy–invariant measure μ with $\operatorname{supp} \mu = \sigma_1 \cup \sigma_2$, giving weight 1 to each of these leaves, is nontrivial and $\varphi_\mu = \sigma_1 + \sigma_2 = \partial c$ in $\mathcal{D}'_1$. So this *is* a tangent 1–homology, hence is arbitrarily well approximated by such.

More generally, in manifolds of arbitrary dimension n, a Reeb component R gives a tangent $(n-1)$–homology. Indeed, when $\dim M \ge 3$, the boundary of R is a compact leaf, hence a nontrivial foliation cycle φ_μ. But again, a triangulation of R provides an n–chain c that is tangent to the foliation $\mathcal{F}$ and $\partial c = \varphi_\mu$. That is, φ_μ is a tangent $(n-1)$–homology.

Example 10.5.15. If the cone $\mathcal{C}_{\mathcal{F}}$ of foliation currents is positioned as in Figure 10.2.1, there is no nontrivial foliation cycle; so the foliation is homologically taut by default. By Lemma 10.2.6, this is exactly the situation in which there exists an exact p–form transverse to $\mathcal{F}$. In Figure 10.2.2,

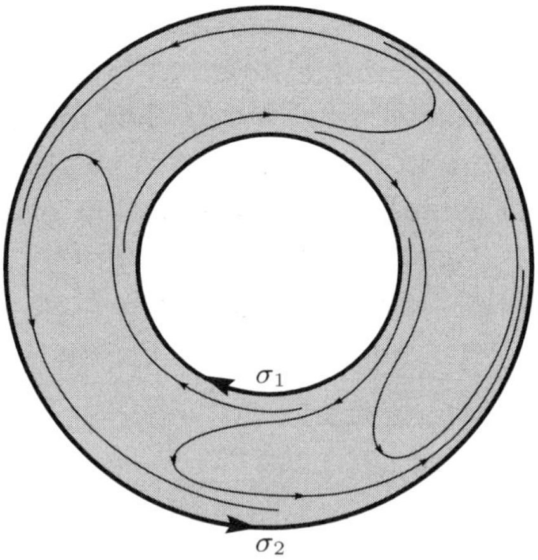

Figure 10.5.1. The tangent homology $\sigma_1 + \sigma_2$ of a Reeb annulus

there are nontrivial foliation cycles, but none of them are exact. Since the subspace $\mathcal{B}_p \subset \mathcal{D}'_p$ is closed, a foliation cycle that is the limit of tangent homologies would, in fact, have to be an exact current. Thus, if Figure 10.2.2 holds, the foliation is homologically taut. By Lemma 10.2.7, this is exactly the situation in which there exists a closed p–form transverse to $\mathcal{F}$. Since exact forms are closed, we can sum up these observations as follows.

Corollary 10.5.16. *If there is a closed p–form transverse to $\mathcal{F}$, then $\mathcal{F}$ is geometrically taut. Equivalently, the condition is that $[\varphi_\mu] \neq 0$ in $H_p(M;\mathbb{R})$, for every nontrivial, holonomy–invariant measure μ. This condition is fulfilled if there are no nontrivial, holonomy–invariant measures.*

Example 10.5.17. If $\operatorname{codim} \mathcal{F} = 1$, we have seen that topological tautness allows us to construct a closed $(n-1)$–form transverse to $\mathcal{F}$, implying homological tautness. Conversely, if $\mathcal{F}$ is not topologically taut, there is a compact n–manifold $Q \subset M$, with $\partial Q \neq \emptyset$ a union of compact leaves $L_1, \dots, L_r$, all oriented inwardly to Q by the transverse orientation of $\mathcal{F}$ (Corollary 6.3.4). A triangulation of Q provides an n–chain c tangent to $\mathcal{F}$ and

$$\partial(c) = \sum_{i=1}^{r} \varphi_{L_i}.$$

This is a nontrivial foliation cycle and a tangent $(n-1)$–homology; so $\mathcal{F}$ is not homologically taut. These observations give a direct proof that, in codimension one, topological tautness is equivalent to homological tautness.

The submanifold Q of Corollary 6.3.4 is often referred to as a "deadend component". In arbitrary codimension p, a tangent $(p+1)$–chain c with

$\partial c = \varphi_\mu$, a nontrivial foliation cycle, should be thought of as a generalized deadend component.

Example 10.5.18. The situation pictured in Figure 10.2.3 is that in which some nontrivial foliation cycle is a boundary. For foliations of codimension one, this prevents the foliation from being geometrically taut, but not in higher codimension. For example, every foliation of S^3 by curves has a nontrivial foliation cycle (Corollary 10.3.14), but $H_1(S^3;\mathbb{R}) = 0$; so every foliation cycle is a boundary. Nevertheless, the Hopf fibration of S^3 is a foliation by geodesics. This is constructed by viewing

$$S^3 = \{(u,v) \in \mathbb{C}^2 \mid |u|^2 + |v|^2 = 1\}$$

and defining the flow Φ_t on S^3 by

$$\Phi_t(u,v) = (e^{it}u, e^{it}v).$$

The flow lines are the intersections of S^3 with the complex, 1–dimensional subspaces of $\mathbb{C}^2$, hence are great circles. These are geodesics in the standard metric. By Theorem 10.5.13, none of these circles can be arbitrarily well approximated by tangent 1–homologies.

Example 10.5.19. We construct a 1–dimensional foliation of S^3 that is not geodesible. In the usual way, decompose

$$S^3 = S^1 \times D^2 \cup D^2 \times S^1.$$

The core circles

$$L_1 = S^1 \times \{0\},$$
$$L_2 = \{0\} \times S^1$$

are to be leaves. There is an "open book" decomposition of $D^2 \times S^1$ by annuli, all meeting L_2 along one common boundary circle (the "spine" of the open book). This is a singular foliation. It can be constructed by "foliating" the solid cylinder $D^2 \times [0,1]$ by rectangles as in Figure 10.5.2 and then gluing $D^2 \times \{0\}$ to $D^2 \times \{1\}$ by $(z,0) \equiv (z,1)$. We can foliate the annulus A by curves as in Figure 10.5.3, one boundary circle being $\partial_\tau A$ and the other $\partial_\pitchfork A$.

Suitably reproducing this foliation in each of the annular leaves, in such a way that $\partial_\tau A$ is always the spine L_2, gives a 1–dimensional foliation of $D^2 \times S^1$, transverse to the boundary and having L_2 as the sole compact leaf. Notice that the outer boundaries of the annular leaves are longitudes on $D^2 \times S^1$ and match up with meridians on $S^1 \times D^2$. In order to continue our 1–dimensional foliation into this solid torus, we first produce a singular foliation of $S^1 \times D^2$ by half–open annuli, having the core L_1 as the singular set and meeting the boundary in meridians. These half–open annuli will look like infinite trumpets, the mouthpiece of each becoming arbitrarily small as it

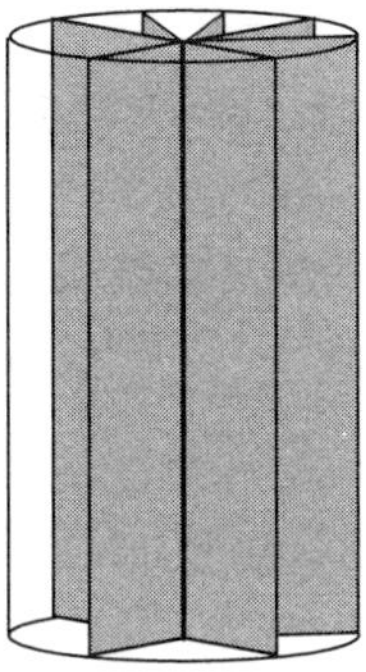

Figure 10.5.2. An open book decomposition

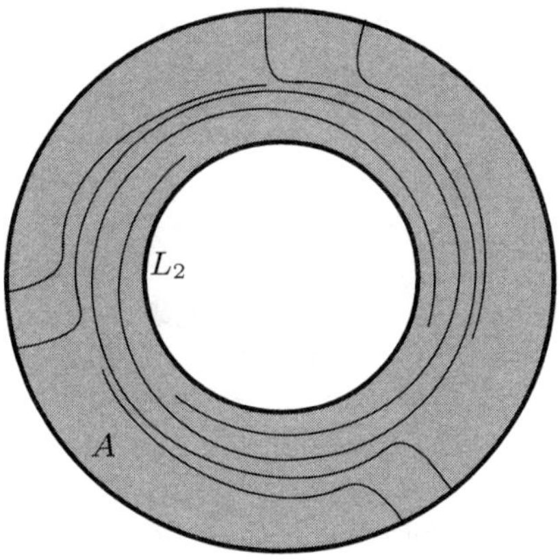

Figure 10.5.3. Foliation of the annulus A

stretches to infinity and asymptotically approaches L_1. In Figure 10.5.4, we give the preimage of this singular foliation in the universal covering space $\mathbb{R} \times D^2$. Each trumpet A', viewed as a half–open annulus, is foliated by radial spokes as in Figure 10.5.5. These spokes become infinitely long in $S^1 \times D^2$, approaching L_1 asymptotically; so we have foliated $S^1 \times D^2$ by curves with exactly the one compact leaf L_1. This foliation is transverse to the boundary. Gluing the two solid tori together, so that the longitudes of $\partial D^2 \times S^1$ match the meridians of $\partial D^2 \times S^1$, produces a foliation of S^3 by curves, exactly two of which, L_1 and L_2, are closed. This can be done so that the resulting one–dimensional foliation $\mathcal{F}$ is of class C^∞, as the reader can verify. The annuli A in $D^2 \times S^1$ and the half–open annuli A' in

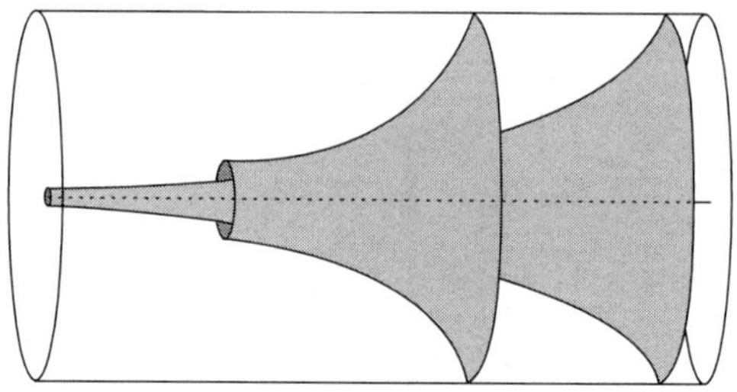

Figure 10.5.4. Singular foliation by trumpets

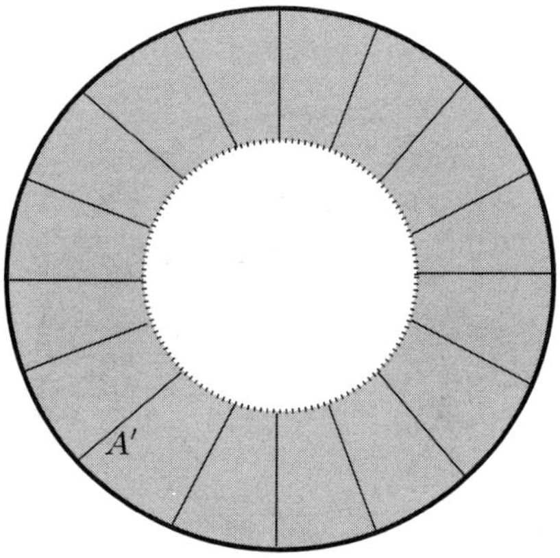

Figure 10.5.5. Foliation of the half–open annulus A'

$S^1 \times D^2$ fit together to give half–open annuli, foliated as in Figure 10.5.6. In that figure, we have shaded a compact subannulus A_t, cobounded by L_2 and a concentric circle C_t, parametrized so that C_t expands to the missing boundary circle (represented by the dotted circle) as $t \uparrow \infty$. The circle C_t is actually shrinking in length to 0 in S^3 as $t \uparrow \infty$. If we triangulate the annulus A_t, we obtain a 2–chain a_t tangent to $\mathcal{F}$, and $\partial a_t = L_2 - C_t$. If $\omega \in A^1(S^3)$,

$$\lim_{t\to\infty} \int_{C_t} \omega = 0$$

since the length of C_t goes to 0. That is, as a current,

$$\lim_{t\to\infty} C_t = 0.$$

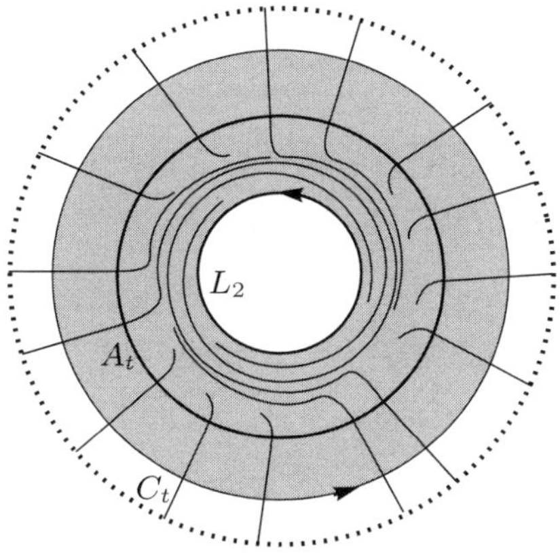

Figure 10.5.6. A tangent homology that approximates L_2

Consequently,

$$\lim_{t\to\infty} \partial a_t = L_2,$$

and this is a nontrivial foliation cycle. Since the currents ∂a_t are tangent 1–homologies, this foliation of S^3 by curves is not homologically taut. By Theorem 10.5.13, $\mathcal{F}$ is not geodesible.

We turn to the proof of Sullivan's criterion.

Lemma 10.5.20. *The p–form ω is $\mathcal{F}$–closed if and only if*

$$\int_{\partial c} \omega = 0,$$

for each tangent p–homology ∂c.

Proof. Suppose that $\omega \in \mathcal{D}_p$ is $\mathcal{F}$–closed and let

$$c = \sum_{i=1}^{r} a_i \sigma_i,$$

where the $\sigma_i : \Delta_{p+1} \to M$ are tangent to $\mathcal{F}$. Clearly, $\sigma_i^*(d\omega) = 0$; so

$$\int_{\partial c} \omega = \int_c d\omega = 0.$$

Conversely, suppose that

$$\int_{\partial c} \omega = 0,$$

for all tangent p–homologies ∂c. For arbitrary $z \in M$, arbitrary p–frames $(\xi_1, \ldots, \xi_p)$ in $T_z(\mathcal{F})$, and arbitrary $\eta \in T_z(M)$, we need to show that

$$d\omega_z(\eta, \xi_1, \ldots, \xi_p) = 0.$$

If not, work in a foliated chart W about z and produce a smooth imbedding $\iota : \mathbb{R}^{p+1} \hookrightarrow W$ such that

- $\iota(0) = z$;
- $\iota_{*w}(T_w(\mathbb{R}^{p+1})) \supset T_{\iota(w)}(\mathcal{F})$, $\forall\, w \in \mathbb{R}^{p+1}$;
- $\eta, \xi_1, \dots, \xi_p \in \iota_{0*}(T_0(\mathbb{R}^{p+1}))$;
- $\iota^*(d\omega)$ is nowhere zero on $\mathbb{R}^{p+1}$.

Then, $\sigma = \iota|\Delta_{p+1}$ is a tangent $(p+1)$–chain and

$$\int_{\partial\sigma} \omega = \int_\sigma d\omega = \int_{\Delta_{p+1}} \iota^*(d\omega) \neq 0.$$

This contradiction proves that ω is $\mathcal{F}$–closed. □

Proof of Theorem 10.5.13. Let $\mathcal{B}_p^\tau \subset \mathcal{B}_p$ be the closed vector subspace generated by all tangent p–homologies. The closed set $\mathcal{B}_p^\tau \cap \mathcal{C}_\mathcal{F}$ is exactly the set of foliation cycles that can be arbitrarily well approximated by tangent p–homologies. Thus, $\mathcal{F}$ is homologically taut if and only if this set reduces to $\{0\}$. A sufficient condition for this is that there be $\omega \in \mathcal{D}_p$ such that

$$\omega|\mathcal{B}_p^\tau \equiv 0,$$
$$\omega|(\mathcal{C}_\mathcal{F} \smallsetminus \{0\}) > 0.$$

By an application of Theorem 10.2.5, these also constitute a necessary condition. By the above lemma, the first of these conditions is equivalent to ω being $\mathcal{F}$–closed, while it is clear that the second is equivalent to ω being transverse to $\mathcal{F}$. □

We close with two exercises on volume–preserving flows that give another application of de Rham currents to Riemannian geometry in the spirit of Theorem 10.5.13.

Exercise 10.5.21. A nonsingular flow Φ_t on the n–manifold M is said to be *volume preserving* if there is a nonsingular n–form Ω on M such that the Lie derivative $L_\nu\Omega = 0$, where $\nu \in \mathfrak{X}(M)$ is the infinitesimal generator of the flow. Show that this is equivalent to the existence of a Riemannian metric on M such that Φ_t preserves the Riemannian volumes of subsets of M.

Exercise 10.5.22. Let $F \subset T(M)$ be an oriented $(n-1)$–plane distribution. Just as for a foliation, one can define the closed, convex cone $\mathcal{C}_F \subset \mathcal{D}'_{n-1}$ of currents generated by the Dirac currents $v_{1\,x} \wedge \cdots \wedge v_{n-1\,x}$ formed from orientation–respecting $(n-1)$–frames in the fibers F_x of F. Show that there is a volume–preserving flow, transverse to F, if and only if $\mathcal{C}_F \cap \mathcal{B}_{n-1} = \{0\}$.

Chapter 11

Foliated Spaces

As previously noted, the theory of foliation cycles works quite well for compact laminations, these being partial foliations of a manifold. Transversely, laminations are modeled on locally compact, metrizable spaces, rather than on $\mathbb{R}^q$, and the ambient manifold is only used as a convenient setting for cohomology, de Rham currents and Poincaré duality. In fact, as will be indicated in Section 11.5, much of the theory can be developed without reference to this manifold. Furthermore, with a few isolated exceptions, the ambience of the lamination becomes irrelevant for the remainder of this volume. So we introduce the notion of an "abstract lamination" or "foliated space". We prefer the latter terminology, reserving the word "lamination" for a foliated space embedded in a manifold. We will see, however, that compact foliated spaces can be realized as laminations in separable Hilbert space.

As we will see, many classical constructions in mathematics can be viewed as foliated spaces; so the notion is not artificial. Nevertheless, the reader who prefers to ignore this level of generality can easily do so, skipping most of this chapter and treating the term "foliated space" as a synonym for "foliated manifold" throughout.

11.1. An Example

A theorem of M. Bebutov, to be discussed later, implies that nonsingular flows on compact metric spaces are foliated spaces. To illustrate the fact that one should be ready to accept the appearance of non–manifolds we provide an example.

Since Riemann, it has been known that the series

$$\sum_{n=1}^{\infty} \frac{1}{n^s}$$

defines a holomorphic function on $\Re(s) > 1$ that admits a meromorphic extension $\zeta(s)$ to the plane of the complex variable $s = \sigma + it$ having a single pole at $s = 1$. Since Euler, it has been known that, for $\Re(s) = \sigma > 1$, this function can be represented as a product

$$\zeta(s) = \prod_{p \in \mathcal{P}} (1 - p^{-s})^{-1},$$

where $\mathcal{P} = \{2, 3, 5, \cdots\}$ is the collection of integer primes. This representation shows that $\zeta(s)$ has no zeroes on $\sigma > 1$.

Fix the restriction that $\sigma > 1$. Using the branch of the logarithm that is real on the positive real axis, we write

$$\log \zeta(s) = -\sum_{p \in \mathcal{P}} \log(1 - p^{-s}).$$

In terms of the splitting $s = \sigma + it$, this becomes

$$\log \zeta(\sigma + it) = -\sum_{p \in \mathcal{P}} \log(1 - p^{-\sigma} e^{-it \log p}).$$

This expression suggests that we consider the function

$$f_\sigma(\{z_p\}) = -\sum_{p \in \mathcal{P}} \log(1 - p^{-\sigma} z_p)$$

on the Cartesian product $\mathbb{T}^{\mathcal{P}} = \prod_{p \in \mathcal{P}} S^1$.

The space $\mathbb{T}^{\mathcal{P}}$ is a compact topological group and, as such, it admits many flows. A particularly interesting one is given by the action

$$\varphi : \mathbb{R} \times \mathbb{T}^{\mathcal{P}} \to \mathbb{T}^{\mathcal{P}}$$

defined by

$$\varphi_t(\{z_p\}) = \{e^{-it \log p} z_p\}.$$

This flow is minimal and uniquely ergodic. That is, there is a probability measure μ on $\mathbb{T}^{\mathcal{P}}$ (in fact, Haar measure) that is invariant under φ_t, and such that, for any continuous function h on $\mathbb{T}^{\mathcal{P}}$ and any $z \in \mathbb{T}^{\mathcal{P}}$, the limit

$$\lim_{T \to \infty} \frac{1}{T} \int_0^T h(\varphi_t(z))\, dt = \int_{\mathbb{T}^{\mathcal{P}}} h(w)\, d\mu(w).$$

Since $\log \zeta(\sigma + it) = f_\sigma(\varphi_t(\{1\}))$, one may expect to apply the ergodic theorem to φ_t and f_σ and obtain properties of the averages of the function ζ along vertical lines $\sigma + it$, $t \in \mathbb{R}$. Although not exactly phrased in terms of the ergodic theorem, the study of these averages and those of other quantities associated to ζ (such as the number of zeroes of $\zeta(s) - a$ for ζ and other Dirichlet series) received much attention in the work of H. Bohr and B. Jessen. Not only can one obtain information for $\sigma > 1$, but also for σ in the critical strip $1/2 < \sigma \leq 1$. References for this work can be found in [**10**].

11.2. Foliated Atlases

We generalize the notion of a foliated atlas, defining the term "foliated space" via coherence classes of such atlases. For the most part, proofs are essentially unchanged from the corresponding ones in Section 1.2; so we refer the reader to that section for the details.

The transverse model of a foliated space, on which the holonomy pseudogroup is defined, will be a "reasonable" topological space, usually denoted by letters such as Z or Y, rather than the "S" that we have been using, in order to emphasize the lack of much transverse regularity. On the other hand, the topological space underlying the foliation will usually be denoted by "manifold" letters, such as M or N, to emphasize the analogy with foliated manifolds. However, we do not commit to any rigidly standardized set of symbols.

Example 11.2.1. Let $\{0, 1\}$ have the discrete topology and give the infinite Cartesian product $K = \{0, 1\}^{\mathbb{Z}}$ the standard product topology. It is well known and easily checked that K is compact, totally disconnected, 2nd countable and perfect; hence K is homeomorphic to the Cantor set. Let $f : K \to K$ be the "shift" homeomorphism, defined by

$$f(\dots, i_k, i_{k+1}, \dots) = (\dots, i'_k, i'_{k+1}, \dots),$$

where $i'_k = i_{k+1}$, $\forall\, k \in \mathbb{Z}$. In $[0, 1] \times K$, make the identifications

$$(1, \iota) \sim (0, f(\iota)), \quad \forall\, \iota \in K,$$

producing a bundle $\pi : M \to S^1$ with fiber K. The intervals $[0, 1] \times \{\iota\}$ are hooked together to define 1–dimensional leaves of a "foliation" $\mathcal{F}$ of M, in perfect analogy with the suspension construction in Example 2.1.1. Note that $f : K \to K$ has infinitely many periodic points of arbitrarily large periods, resulting in infinitely many closed leaves of arbitrarily large lengths. There are also f–orbits limiting on periodic orbits in forward and/or backward time, as well as a bewildering collection of minimal sets that are not closed orbits and, finally, there are orbits that are dense in K in forward and/or backward time. All of these asymptotic properties are mirrored by the corresponding leaves of $\mathcal{F}$. Finally, if we cover S^1 with a finite family of open subarcs $A_1, A_2, \dots, A_N$, any two of which intersect, if at all, in an open arc, then the sets $U_i = \pi^{-1}(A_i) \cong A_i \times K$ form an open cover of M. In U_i, the induced foliation $\mathcal{F}|U_i$ has as leaves the open arcs $A_i \times \{\iota\}$, $\iota \in K$. These open sets $U_i \cong A_i \times K$ are to be thought of as "foliated charts", with coordinates $x_i : U_i \to A_i$ and $y_i : U_i \to K$ the projections onto the respective factors. The leaves of $\mathcal{F}|U_i$ are the plaques and it is clear that a plaque of U_i meets at most one plaque of U_j, $1 \leq i, j \leq N$. We will say that $\mathfrak{U} = \{U_i, x_i, y_i\}_{i=1}^N$ is a regular foliated atlas. We choose a base point $z_i \in A_i$ and identify the set of plaques of U_i with $Z_i = \pi^{-1}(z_i) \cong K$. Notice that, whenever $U_i \cap U_j \neq \emptyset$, $\gamma_{ij} = y_i \circ y_j^{-1} : Z_j \to Z_i$ is a well–defined homeomorphism and that $\{\gamma_{ij}\}_{i,j=1}^N$ satisfies the standard cocycle property. This holonomy cocycle generates the total holonomy group $G(\mathcal{F}) \subset \mathrm{Homeo}(K)$, as well as the holonomy pseudogroup $\Gamma_{\mathfrak{U}}$ on the transverse space $Z = \bigcup_{i=1}^N Z_i$.

Exercise 11.2.2. Let $S^1 \subset \mathbb{C}$ be the unit circle and let $f_n : S^1 \to S^1$ be the map $z \to z^n$, $n \geq 2$. The inverse limit Σ_n of this map is, by definition, the subset of the infinite Cartesian product $S^1 \times S^1 \times \cdots \times S^1 \times \cdots$ consisting of the points $(z_1, z_2, \ldots, z_k, \ldots)$ such that $z_k = f_n(z_{k+1})$, for all $k \geq 1$. If d is the standard distance function on S^1, we define a distance d_n on Σ_n by

$$d_n((z_1, z_2, \ldots, z_k, \ldots), (w_1, w_2, \ldots, w_k, \ldots)) = \sum_{k=1}^{\infty} \frac{d(z_k, w_k)}{2^k}.$$

Prove that, in this metric, Σ_n is compact, that projection $\pi : \Sigma_n \to S^1$ onto the first coordinate is a continuous surjection, and that $\pi^{-1}(z)$ is a Cantor set, for all $z \in S^1$. Show informally that there is a 1–dimensional foliation $\mathcal{F}$ of Σ_n such that $(\Sigma_n, \mathcal{F}, \pi)$ is a foliated bundle.

Definition 11.2.3. Let M and Z be locally compact, second countable, metrizable spaces. A *foliated chart* in M of leaf dimension p, modeled transversely on Z, is a pair (U, φ), where $U \subseteq M$ is open and $\varphi : U \to B_\tau \times B_\pitchfork$ is a homeomorphism, $B_\pitchfork$ being an open, precompact neighborhood in Z and B_τ an open ball in $\mathbb{R}^p$ of finite radius. The set $P_y = \varphi^{-1}(B_\tau \times \{y\})$, $y \in B_\pitchfork$, is called a *plaque* of this foliated chart. For each $x \in B_\tau$, the set $S_x = \varphi^{-1}(\{x\} \times B_\pitchfork)$ is called a *transversal* of the foliated chart.

Definition 11.2.4. Let M and Z be as above and let $\mathfrak{U} = \{U_\alpha, \varphi_\alpha\}_{\alpha \in \mathfrak{A}}$ be a family of foliated charts in M of dimension p, modeled transversely on Z and covering M. Assume further that the foliated charts are *coherently foliated* in the sense that, whenever P and Q are plaques in distinct charts of $\mathfrak{U}$, then $P \cap Q$ is open both in P and Q. Then $\mathfrak{U}$ is called a foliated atlas on M of leaf dimension p, modeled transversely on Z.

Generally, the transverse model space Z and leaf dimension p will be understood from the context; so we will just say that $\mathfrak{U}$ is a foliated atlas. We will write $\varphi_\alpha = (x_\alpha, z_\alpha)$, where

$$x_\alpha : U_\alpha \to \mathbb{R}^p,$$
$$z_\alpha : U_\alpha \to Z.$$

On $\varphi_\beta(U_\alpha \cap U_\beta)$, the formula for the change of coordinates is

$$g_{\alpha\beta}(x_\beta, z_\beta) = \varphi_\alpha \circ \varphi_\beta^{-1}(x_\beta, z_\beta) = (x_\alpha(x_\beta, z_\beta), z_\alpha(x_\beta, z_\beta)).$$

Exactly as remarked on page 24, the condition that the charts be coherently foliated means that each point $z \in U_\alpha \cap U_\beta$ has a neighborhood in which the change of transverse coordinate takes the form

$$z_\alpha = z_\alpha(z_\beta).$$

Usually we will be interested in considering objects along the leaves, *e.g.* differential forms, and their derivatives with respect to the leaf parameters. For this, we need to define what is meant by smoothness.

Definition 11.2.5. Let Z be topological space, and let $U \subset \mathbb{R}^p \times Z$ be an open set. A function $f : U \to \mathbb{R}$ is *smooth of class* C^r, $0 \leq r \leq \infty$, on U if all the partial derivatives of order $\leq r$ of f in the $\mathbb{R}^p$ variables are continuous functions on U. That is, if $(x_1, \cdots, x_p)$ are coordinates on $\mathbb{R}^p$, and if $(i_1, \cdots, i_p)$ is a multi–index with $i_1 + \cdots + i_p \leq r$, then the map

$$(x, z) \in U \mapsto \frac{\partial^{i_1+\cdots+i_p}}{\partial^{i_1}x_1 \cdots \partial^{i_p}x_p} f(x, z) \in \mathbb{R}$$

is continuous.

A map $f : U \to \mathbb{R}^q$ is smooth of class C^r if the coordinate maps

$$y_j \circ f : U \to \mathbb{R}, \quad j = 1, \ldots, q,$$

are of class C^r.

A function $f : U \to \mathbb{R}^q \times W$ (W a topological space) is smooth of class C^r if it is locally of the form $f(x, z) = (f_1(x, z), f_2(z))$ (*i.e.*, preserves leaves), where the map f_2 is continuous and the map $f_1 : U \to \mathbb{R}^q$ is of class C^r. Finally, a map $f : U \to U' \subset \mathbb{R}^p \times Z$ is a C^r–diffeomorphism if it is smooth of class C^r and it has an inverse that is also smooth of class C^r.

Remark. In order to emphasize the partial nature of this sort of smoothness, we will sometimes use the term "leafwise smooth", meaning exactly what has been defined above.

Exercise 11.2.6. Let $M = D \times Z$ be a trivial foliated space, with Z metrizable. If $f_1 : D \to \mathbb{R}$ is smooth and $f_2 : Z \to \mathbb{R}$ is continuous, show that $f(x, z) = f_1(x) f_2(z)$ is a smooth function, as in Definition 11.2.5, and that finite linear combinations of functions of this kind are dense in the space $C(M)$ of continuous functions on M (with the compact–open topology). Similarly, show that compactly supported functions of this type form a dense subspace of $C(M)$. Finally, show that, if $B \subset D$ is a closed ball in D and $Y \subset Z$ is a compact subset, then there is a smooth function f on M that is 1 on $B \times Y$ and 0 outside a neighborhood of $B \times Y$ in M.

Definition 11.2.7. The foliated atlas $\mathfrak{U}$ is of class C^r, $0 \le r \le \infty$, if on every nonempty intersection $U_\alpha \cap U_\beta$ of charts in $\mathfrak{U}$, the coordinate change

$$g_{\alpha\beta} = \varphi_\alpha \circ \varphi_\beta^{-1} : \varphi_\beta(U_\alpha \cap U_\beta) \subset \mathbb{R}^p \times Z \to \varphi_\alpha(U_\alpha \cap U_\beta) \subset \mathbb{R}^p \times Z$$

is of class C^r. We will also say that $\mathfrak{U}$ is a C^r–foliated atlas.

When M is a manifold, this definition is equivalent to our definition of class $C^{r,0}$ (Definition 1.2.22). Since there is no question of transverse smoothness in general foliated atlases, we omit the double superscript.

One defines *coherence* of foliated atlases $\mathfrak{U}$ and $\mathfrak{V}$ on M exactly as in Definition 1.2.8 and proves that coherence is an equivalence relation exactly as in Lemma 1.2.9.

As remarked on page 26, whenever (U, φ) and (W, ψ) are foliated charts such that $\overline{U} \subset W$ and $\varphi = \psi | U$, we can speak of the (closed) plaques and transversals of $\overline{U}$.

Definition 11.2.8. A foliated atlas $\mathfrak{U} = \{U_\alpha, \varphi_\alpha\}_{\alpha \in \mathfrak{A}}$ on M is said to be *regular* if

(1) for each $\alpha \in \mathfrak{A}$, $\overline{U}_\alpha$ is a compact subset of a foliated chart (W_α, ψ_α) and $\varphi_\alpha = \psi_\alpha | U_\alpha$;
(2) the cover $\{U_\alpha\}_{\alpha \in \mathfrak{A}}$ is locally finite;
(3) if $(U_\alpha, \varphi_\alpha)$ and (U_β, φ_β) are elements of $\mathfrak{U}$, then the interior of each closed plaque $P \subset \overline{U}_\alpha$ meets at most one plaque in $\overline{U}_\beta$.

Lemma 11.2.9. *Every foliated atlas has a coherent refinement that is regular.*

The proof mimics that of Lemma 1.2.17. This is the first time that we make full use of the hypotheses that M is locally compact, second countable and metrizable.

Given a regular foliated atlas $\mathfrak{U}$, we define an equivalence relation on M by setting $x \sim y$ if and only if either there is a $\mathfrak{U}$–plaque P_0 such that $x, y \in P_0$ or there is a sequence $\{P_0, P_1, \dots, P_p\}$ of $\mathfrak{U}$–plaques such that

$$\begin{aligned} &x \in P_0, \\ &y \in P_p, \\ &P_i \cap P_{i-1} \neq \emptyset, \qquad 1 \le i \le p. \end{aligned}$$

This will be called a *plaque chain of length* p connecting x and y. In the case that $x, y \in P_0$, we say that $\{P_0\}$ is a plaque chain of length 0 connecting x and y. The fact that $\sim$ is an equivalence relation is clear. It is also clear that each equivalence class L is a union of plaques. Exactly as in the proof of Lemma 1.2.16, we obtain the following.

Lemma 11.2.10. *Each equivalence class L associated to a regular foliated atlas has a natural p–dimensional topological manifold structure. If the foliated atlas is of class C^r, this manifold structure is smooth of class C^r.*

We will say that the equivalence class L is a *leaf* associated to the regular foliated atlas $\mathfrak{U}$. As in the proof of Lemma 1.2.10, one checks the following.

Lemma 11.2.11. *Regular foliated atlases $\mathfrak{U}$ and $\mathcal{V}$ on M are coherent if and only if they give rise to exactly the same set of leaves.*

Definition 11.2.12. A p–dimensional foliated space $(M, \mathfrak{F})$ of class C^r is a locally compact, separable, metrizable space M, together with a coherence class $\mathfrak{F}$ (called a *foliation* of dimension p) of foliated atlases of class C^r. The leaves associated to any regular, C^r–foliated atlas representing $\mathfrak{F}$ are said to be the leaves of the foliation $\mathfrak{F}$.

As in the case of foliated manifolds, the term "foliation" can be defined, equivalently, as a maximal foliated atlas and as an equivalence relation on M with equivalence classes the leaves.

Definition 11.2.13. Let $(M, \mathfrak{F})$ be a C^r–foliated space. Then $C^r(M)$ will denote the set of real–valued functions on M that, on each foliated chart, are of class C^r as in Definition 11.2.5. Usually, $C^0(M)$ will be denoted by $C(M)$.

If $r \geq 1$, the tangent spaces $T_w(L)$ to the leaves L at points $w \in L$ fit together to form a topological vector bundle $T(\mathfrak{F})$ over M that is of class C^{r-1} along the leaves. Since there is little chance of confusion, one can also denote this bundle by $T(M)$ and call it the tangent bundle of the foliated space. The sections of $T(M)$ of class C^{r-1} (leafwise) are vector fields on M that are tangent to $\mathfrak{F}$. The space of such may be denoted by $\mathfrak{X}(M)$ or $\mathfrak{X}(\mathfrak{F})$. Similarly, the space of differential k–forms, denoted by $A^k(M)$ or $A^k(\mathfrak{F})$, makes sense.

Definition 11.2.14. A C^r–foliated space $(M, \mathfrak{F})$, $r \geq 1$, is said to be orientable if the tangent bundle $T(M)$ is an orientable vector bundle over M. An orientation of $(M, \mathfrak{F})$ is a choice of orientation for $T(M)$.

Equivalently, $(M, \mathfrak{F})$ is orientable if it admits a foliated atlas for which the leafwise Jacobian determinants of the coordinate changes are all positive. If M is oriented, such an atlas is said to belong to the orientation class. As in the case of smoothness for foliated spaces, one sometimes says that $(M, \mathfrak{F})$ (or simply M) is *leafwise* oriented, meaning exactly what is defined above, but emphasizing the partial nature of the orientation.

Several definitions and results appearing later will require that the foliated space involved be oriented. Usually this is not a very restrictive requirement, for we always have available the "double cover" trick.

Exercise 11.2.15. Let $(M,\mathcal{F})$ be a C^r–foliated space, $r \geq 1$. Prove that there is an orientable foliated space $(M',\mathcal{F}')$ and a smooth map of foliated spaces $f : M' \to M$ that is a two–to–one local homeomorphism.

We conclude this section with a description of the *holonomy pseudogroup* of a foliated space $(M,\mathcal{F})$ or, more properly, of a regular foliated atlas $\mathfrak{U}$ representing $\mathcal{F}$. By the third property in Definition 11.2.8, the change of transverse coordinate on $U_\alpha \cap U_\beta$ has the form $z_\alpha = z_\alpha(z_\beta)$, which can be viewed as a homeomorphism

$$\gamma_{\alpha\beta} : z_\beta(U_\alpha \cap U_\beta) \to z_\alpha(U_\alpha \cap U_\beta).$$

These local homeomorphisms satisfy the cocycle conditions. Because of regularity of the atlas $\mathfrak{U}$, each $\gamma_{\alpha\beta}$ extends to a homeomorphism on an open subset of Z containing the compact closure of $z_\beta(U_\alpha \cap U_\beta)$. In particular, $\gamma_{\alpha\beta}$ extends to a homeomorphism $\overline{\gamma}_{\alpha\beta}$ on the compact closure of its domain. Definition 11.2.8 also allows us to produce another regular, foliated atlas $\mathfrak{U}' = \{U'_\alpha, x'_\alpha, z'_\alpha\}_{\alpha\in\mathfrak{A}}$ such that $\overline{U}_\alpha \subset U'_\alpha$, $x_\alpha = x'_\alpha|U_\alpha$ and $z_\alpha = z'_\alpha|U_\alpha$ $\forall\, \alpha \in \mathfrak{A}$. We call $\mathfrak{U}'$ an *expansion* of $\mathfrak{U}$.

Definition 11.2.16. The set $\gamma = \{\gamma_{\alpha\beta}\}_{\alpha,\beta\in\mathfrak{A}}$ is the holonomy cocycle associated to the regular foliated atlas $\mathfrak{U}$. The pseudogroup $\Gamma = \Gamma_{\mathfrak{U}}$ of local homeomorphisms on Z generated by γ is called the holonomy pseudogroup of the atlas $\mathfrak{U}$.

It is convenient in some situations to view the subsets $z_\alpha(U_\alpha) \subset Z$ as disjoint. A device that can be used for this is to choose an expansion $\mathfrak{U}'$ of $\mathfrak{U}$ and set

$$Z_\alpha = z_\alpha(U_\alpha) \times \{\alpha\} \subset y'_\alpha(U'_\alpha) \times \{\alpha\} = Z'_\alpha.$$

Then sets with indices $\alpha \neq \beta$ are disjoint, the union Z' of the Z'_α's contains the union Z of the Z_α's and the closure $\overline{Z}$ of Z in Z' is compact. The cocycle γ can be viewed in this setting and generates the pseudogroup $\Gamma_{\mathfrak{U}}$ on the new transverse space Z, $\overline{Z}$ or Z'. One also views the space Z_α (respectively, $\overline{Z}_\alpha$) as the set of plaques in U_α (respectively, $\overline{U}_\alpha$) and, when convenient, as a transversal embedded in U_α (respectively, in $\overline{U}_\alpha$). In this last version, the embedded transversals can be chosen to be pairwise disjoint, allowing us to view Z (respectively, $\overline{Z}$) as embedded in M. Notice that pure compositions f of elements of the holonomy cocycle extend canonically to homeomorphisms $\overline{f}$ on the compact closures of their domains.

Exactly as in Section 2.3, the holonomy pseudogroup assigns to each leaf L of $\mathcal{F}$ (and to each base point $y \in L \cap Z$) a *holonomy homomorphism*

$$\widehat{h} : \pi_1(L, y) \to \mathcal{H}_y(L) \subset \mathcal{G}^y_y,$$

where $\mathcal{G}_y^y$ is the group of germs at y of local homeomorphisms of Z that fix y. The image $\mathcal{H}_y(L)$ is called the (germinal) holonomy group of the leaf L at the base point y.

11.3. Some Constructions of Foliated Spaces

A justification for the study of foliations of manifolds is that they are a generalization of flows and may be viewed as the geometric study of differential equations on a manifold. There are however many other constructions in mathematics which, when geometrically interpreted, produce foliated spaces. In this section we will describe two general constructions and examples. Since our purpose is motivational, many details will be exercises.

11.3.A. Inverse limits. A popular family of examples generalizes Exercise 11.2.2. Recall that an inverse system of spaces consists of a collection of topological spaces X_i, indexed by $i \in I$, where $(I, \leq)$ is a partially ordered set, together with continuous maps

$$f_{ij} : X_j \to X_i, \quad \text{whenever } i, j \in I \text{ and } i \leq j.$$

The maps f_{ij} are called the bonding maps of the system, and are required to satisfy the composition rule $f_{ij} \circ f_{jk} = f_{ik}$, whenever $i \leq j \leq k$. To an inverse system $\{X_i, f_{ij}\}_{i,j \in I}$ there is associated a topological space, the *inverse limit*

$$X = \varprojlim_{i \in I} X_i.$$

This is defined to be the subset of the topological product space $\mathfrak{X} = \prod_{i \in I} X_i$ consisting of those elements $x = (x_i)_{i \in I}$ such that $f_{ij}(x_j) = x_i$, whenever $i \leq j$. The space X is a closed subset of $\mathfrak{X}$ with the relative topology. In case the spaces X_i are compact manifolds of a fixed dimension n and the bonding maps are submersions, one expects the inverse limit to carry a natural foliation of leaf dimension n. We will sketch this in the case that the inverse system is sequential.

The inverse limit space of an inverse system may well be empty. It is very easy to produce such examples. A condition guaranteeing a nonempty limit is given in the following exercise.

Exercise 11.3.1. Let $\{X_i, f_{ij}\}_{i \in I}$ be an inverse system and suppose that all the spaces X_i are compact, Hausdorff and nonempty. Then the inverse limit space X is compact, Hausdorff and nonempty.

If $I = \mathbb{Z}$ with the natural order, an inverse system $\{X_i, f_{ij}\}_{i,j \in \mathbb{Z}}$ is completely determined by the data $\{X_i, f_i\}_{i \in \mathbb{Z}}$, where $f_i = f_{i,i+1}$. The limit is just the set $\{(x_i) \mid f_i(x_{i+1}) = x_i\}$. One can also restrict this system to the "half infinite" inverse system $\{X_i, f_i\}_{i \in \mathbb{Z}^+}$, indexed by the nonnegative integers.

Exercise 11.3.2. If $\{i_k\}_{k=0}^{\infty}$ is a strictly increasing sequence of integers, prove that

$$\varprojlim_{i\in\mathbb{Z}} X_i = \varprojlim_{k\in\mathbb{Z}^+} X_{i_k}.$$

In particular,

$$\varprojlim_{i\in\mathbb{Z}} X_i = \varprojlim_{i\in\mathbb{Z}^+} X_i.$$

Consider an inverse system $\{X_i, f_i\}_{i\in\mathbb{Z}^+}$ in which the spaces X_i are compact and the bonding maps f_i are covering maps. We also suppose that the spaces are connected, locally path connected and locally simply connected. This includes compact, connected manifolds and simplicial complexes and guarantees that the spaces have universal covering spaces. As usual, we subscribe to the convention that universal coverings are principal bundles, the covering group acting on the total space from the right.

Exercise 11.3.3. Under the above hypotheses, prove the following.

1. The universal covering spaces of the X_i's can all be identified with a single space $\widetilde{X}$, together with covering maps $p_i : \widetilde{X} \to X_i$ such that $p_i = f_i \circ p_{i+1}$.
2. If G_i is the group of covering transformations of $p_i : \widetilde{X} \to X_i$, then $G_{i+1} \subset G_i$, $i \geq 0$.
3. For $y \in \widetilde{X}$ and $i \geq 0$, the set G_i/G_{i+1} corresponds one–to–one with the finite set $f_i^{-1}(p_i(y))$ via the map
$$gG_{i+1} \mapsto p_{i+1}(ygG_{i+1}).$$

We suppose that infinitely many of the covering maps $f_i : X_{i+1} \to X_i$ are not bijective. By Exercise 11.3.2, we lose no generality in assuming this for all $i \geq 0$. Give each of the groups G_i the discrete topology and let

$$G = G_0 \times G_1 \times \cdots \times G_i \times \cdots,$$

a topological group with the product topology, and set $\widehat{X} = \widetilde{X} \times G$. Define a right action of G on $\widehat{X}$ by

$$(y, g_0, g_1, \dots, g_i, \dots) \cdot (h_0, h_1, \dots, h_i, \dots) \\ = (yh_0, h_0^{-1}g_0h_1, \dots, h_i^{-1}g_ih_{i+1}, \dots).$$

Finally, define $\varphi : \widehat{X} \to \mathfrak{X}$ by

$$\varphi(y, g_0, g_1, \dots, g_i, \dots) = (p_0(yg_0), p_1(yg_0g_1), \dots, p_i(yg_0g_1\cdots g_i), \dots).$$

This is clearly continuous.

Exercise 11.3.4. Prove that the image of φ is exactly the inverse limit $X \subset \mathfrak{X}$ and that $\varphi(y, g) = \varphi(y', g')$ if and only if (y, g) and (y', g') lie in the same G–orbit.

Proposition 11.3.5. *The space*

$$X = \varprojlim{}_{i\in\mathbb{Z}^+} X_i$$

fibers canonically over X_0 with fiber a Cantor set.

Proof. The projection $\pi : X \to X_0$ onto the first factor is a continuous surjection. Let $x \in X_0$ and fix a simply connected neighborhood $V \subset X_0$ of x and a compact neighborhood $C_x \subset V$ of x. Then, for all $k \geq 0$, C_x is evenly covered by

$$f_0 \circ f_1 \circ \cdots \circ f_k : X_{k+1} \to X_0.$$

This compact neighborhood is also evenly covered by $p_0 : \widetilde{X} \to X_0$. So we choose $y \in p_0^{-1}(x)$ and the unique compact neighborhood $C_y \subset \widetilde{X}$ of y that is carried homeomorphically by p_0 onto C_x. For each $k \geq 0$, let

$$K_k = \{g_k^1, g_k^2, \ldots, g_k^{n_k}\} \subset G_k$$

be formed by choosing one representative from each element of G_k/G_{k+1}. Exercise 11.3.3 guarantees that K_k is finite, hence compact in the discrete topology. Then the product topology makes

$$K = K_0 \times K_1 \times \cdots \times K_k \times \cdots$$

compact, totally disconnected, 2nd countable and perfect. That is, K is a Cantor set and $C_y \times K$ is a compact subset of $\widehat{X}$. By Exercise 11.3.4 and the fact that C_x is evenly covered by all $f_0 \circ f_1 \circ \cdots \circ f_k$, we see that φ carries $C_y \times K$ one–to–one onto the subset $\pi^{-1}(C_x)$. In particular, this latter set is compact and, all the spaces being Hausdorff, φ is a homeomorphism between these sets. The diagram

$$\begin{array}{ccc} C_y \times K & \xrightarrow{\ \varphi\ } & \pi^{-1}(C_x) \\ {\scriptstyle p}\big\downarrow & & \big\downarrow{\scriptstyle \pi} \\ C_y & \xrightarrow[p_0]{} & C_x \end{array}$$

clearly commutes, where p is just projection onto the first factor. Since $x \in X_0$ is arbitrary, the assertion follows. □

It is now trivial to explain the relation of inverse limits with foliated spaces.

Proposition 11.3.6. *Let $\{X_i, f_i\}_{i\in\mathbb{Z}^+}$ be an inverse system, the spaces X_i being compact manifolds of the same dimension n and each of the bonding maps being a submersion but not a diffeomorphism. Then the inverse limit space has the structure of a foliated space of leaf dimension n. In fact, this is a foliated bundle over X_0 with fiber a Cantor set.*

Indeed, the bundle structure $\pi : X \to X_0$ given by Proposition 11.3.5 provides a covering of X by foliated charts $\pi^{-1}(U)$, where U ranges over a suitable finite cover of X_0 by open Euclidean balls. The fact that any two of these charts are coherently foliated follows from the fact that the fiber is totally disconnected. Leafwise smoothness is guaranteed by the fact that the base is a smooth manifold (*cf.* the remark on page 52).

Remark. Foliated spaces constructed as inverse limits have important applications in dynamics. One of the first such uses is the theory developed by R. Williams [**149**] to model expanding attractors of diffeomorphisms of manifolds. His theory involves the construction of foliated spaces via inverse limits of branched manifolds and expanding maps. It starts with a branched Riemannian manifold M and a map $f : M \to M$ that is expanding. A set of axioms is imposed on f implying in particular that the inverse limit space of the system

$$\Sigma(M, f) = \varprojlim\{M \xleftarrow{f} M \xleftarrow{f} \cdots\}$$

has the structure of a foliated space.

The foliated space $\Sigma = \Sigma(M, f)$ is very interesting. The expanding map f of the branched manifold M induces a homeomorphism of Σ that preserves the foliated structure. This homeomorphism has a dense set of periodic points. It sends each leaf diffeomorphically to some other leaf, in an expanding manner. It follows from this that each leaf containing a periodic point is homeomorphic to Euclidean space and has polynomial growth (hence that the foliated space admits an invariant measure). For some of these facts, see also [**129**].

11.3.B. Group actions. Another natural way in which a foliated space arises is given by a continuous action

$$\varphi : M \times G \to M$$

of a connected Lie group G on a locally compact, separable metric space M. That is to say, φ is a continuous map that satisfies the following conditions:

- $\varphi(p, gh) = \varphi(\varphi(p, g), h)$; and
- $\varphi(p, e) = p$, for all $p \in M$, e the identity of G and $g, h \in G$.

Usually, the notation $\varphi(p, g)$ will be abbreviated to pg.

The orbit of a point $p \in M$ is the subset $pG = \{pg \mid g \in G\}$ of M. Each orbit pG of the action can be given the structure of a manifold in a natural way, namely as the quotient space $G_p\backslash G$, where G_p is the *isotropy* group of p (*i.e.*, the subgroup of G that stabilizes p), which is a closed subgroup of G. (But note that the topology on pG induced from M generally has nothing to do with that of the homogeneous space $G_p\backslash G$.)

The orbits decompose M into a family of manifolds. But, since the orbits may not all have the same dimension, this decomposition is in general incompatible with that of a foliated space structure. However, for a natural class of actions, those that are *locally free*, the orbits are locally homeomorphic to G and it is reasonable to expect them to be the leaves of a foliation.

Definition 11.3.7. The action of G on M is locally free if, for each $p \in M$, there is a neighborhood V of e in G such that the map $g \in V \mapsto pg \in M$ is injective; equivalently, the isotropy group G_p of every point $p \in M$ is discrete.

Exercise 11.3.8. Let $e \in G$ denote the identity of the group. If the action $M \times G \to M$ is locally free, prove that, for each $x \in M$, there is a neighborhood $U \subset M \times G$ of (x, e) such that

$$(y, g) \in U \text{ and } yg = y \Rightarrow g = e.$$

Theorem 11.3.9. *If $M \times G \to M$ is a locally free action of a connected Lie group G on a locally compact, separable metric space M, then the G–orbits are the leaves of a foliation $\mathcal{F}$ of M.*

In particular, nonsingular flows on compact metric spaces give one–dimensional foliations (Bebutov's theorem [**97**]). An interesting example of such a flow was given at the beginning of this chapter. The following exercises outline other examples.

Exercise 11.3.10. Consider the space $C_b(\mathbb{R})$ of continuous, bounded, real–valued functions on $\mathbb{R}$ with the uniform topology. A function $f \in C_b(\mathbb{R})$ is said to be *almost periodic* if the family of translated functions

$$f_t(x) = f(x+t), \quad t \in \mathbb{R},$$

has compact closure M in $C_b(\mathbb{R})$. Then M is a compact, metrizable space. If $g \in M$ and $t \in \mathbb{R}$, we can write $g = \lim_{n\to\infty} f_{t_n}$, for some sequence $\{t_n\}$ of real numbers, and define $\Phi_t(g) = \lim_{n\to\infty} f_{t+t_n}$. Prove that, if f is nonconstant, this gives a well–defined, nonsingular flow on M, hence that M carries a one–dimensional foliation.

A trivial example of an almost periodic function is a periodic one, *i.e.*, $f : \mathbb{R} \to \mathbb{R}$ such that $f(x + t_0) = f(x)$ for all $x \in \mathbb{R}$ and some $t_0 \neq 0$. In this case the foliated space M of Exercise 11.3.10 is simply a circle. However, there are more interesting examples, the following exercise being one such (and note the example in the first section of this chapter).

The following exercise requires more than a superficial knowledge of almost periodic functions. A good reference is [**9**].

Exercise 11.3.11. Let $f : \mathbb{R} \to \mathbb{R}$ be the function defined by

$$f(x) = \sum_{n=0}^{\infty} \frac{1}{2^n} \cos \frac{x}{2^n}.$$

Show that f is almost periodic and that the corresponding foliated space is canonically the same as the one obtained in Exercise 11.2.2 as the inverse limit of the squaring map $f(z) = z^2$ of the circle S^1.

Example 11.3.12. Let $C_b^*(\mathbb{R})$ be the space of real–valued, bounded, continuous functions on $\mathbb{R}$, endowed with the compact–open topology. The action of $\mathbb{R}$ on $C_b^*(\mathbb{R})$, defined as in Exercise 11.3.10, is also continuous. This is called Bebutov's dynamical system. A theorem of his, with an improvement due to S. Kakutani, states that this $\mathbb{R}$-action is universal in the following sense: a flow on a compact metric space can be equivariantly embedded in Bebutov's system if and only if its set of stationary points is homeomorphic to a subset of $\mathbb{R}$.

Definition 11.3.13. Let M and G be as above. A compact subset T of M is a *local cross section* through the point $p \in M$ if $p \in T$ and there is a neighborhood V of e in G such that the orbit map $(x, g) \in T \times V \to xg \in M$ is a homeomorphism onto a neighborhood of p.

Theorem 11.3.14. *Every locally free action $M \times G \to M$ of a Lie group on a locally compact metric space admits a local cross section through each point $p \in M$.*

This implies Theorem 11.3.9. Indeed, it follows immediately that there is a foliated atlas with local G–orbits as plaques; hence the corresponding foliation has the global G–orbits as leaves. Theorem 11.3.14 is proven in [**71**] in the more general context of local actions (with trivial isotropy) of local Lie groups. A different proof is given in [**109**]. For the group $G = \mathbb{R}$, it is proven by M. Bebutov [**97**, p. 333] and the proof is easily adapted to actions of $\mathbb{R}^n$. Of course, if M is a manifold and the action of G is smooth, then the existence of local cross sections follows from the Frobenius theorem (Theorem 1.3.8), since the action naturally defines an involutive subbundle of $T(M)$.

Exercise 11.3.15. Show that a regular foliated atlas, produced by local cross sections of the G–action, is C^∞ in the sense of Definition 11.2.7. (Hint. Local leafwise coordinates x_α can be identified as local coordinates on G. The coordinate change $x_\alpha = x_\alpha(x_\beta, z_\beta)$ is identified as a local left translation in G by an element $\gamma(z_\beta)$ that depends continuously on z_β.)

If $p \in M$ and $U \subset M$ is a neighborhood of p, a map $f : U \to G$ is said to be equivariant near p if U can be chosen small enough that there is a neighborhood V of the identity $e \in G$ such that $f(xg) = f(x)g$, for all $(x,g) \in U \times V$ with $xg \in U$. The next proposition shows that the existence of such a map guarantees the existence of a local cross section through the point p.

Proposition 11.3.16. *Let U be a neighborhood of p in M, $f : U \to G$ a continuous map such that $f(p) = e$ and f is equivariant near p. Then there exists a local cross section through the point p.*

Proof. Let V and U be as above. Let $U' \subset \operatorname{int} U$ be a compact neighborhood of p such that $f(U') \subset V$. Set $T = U' \cap f^{-1}(e)$. Since the action is locally free and f is equivariant near p, $f(U')$ is a neighborhood of e. Let $W \subset f(U')$ be a compact, symmetric neighborhood of e, so small that, for each $g \in W$, $U'g \subset U$. Consider the continuous map

$$\varphi : (x,g) \in T \times W \mapsto xg \in M.$$

We will prove that φ maps $T \times W$ homeomorphically onto a neighborhood of p. Indeed, there is an open neighborhood $U'' \subset U'$ of p such that, for each $y \in U''$, $g = f(y) \in W = W^{-1}$, $yg^{-1} \in U'$ and $f(yg^{-1}) = f(y)g^{-1} = e$. Here we are allowed to use the equivariance because $yg^{-1} \in U$. In fact, $yg^{-1} \in U' \cap f^{-1}(e) = T$. Thus, $(yg^{-1}, g) \in T \times W$ and $\varphi(yg^{-1}, g) = y$, proving that U'' is in the image of φ. Thus, $\varphi(T \times W)$ is a neighborhood of p. If (x,g) and $(x',g') \in T \times W$ and $xg = x'g'$, then this point is in U by the choice of W. Thus, the equivariance of f applies, giving

$$g = f(x)g = f(xg) = f(x'g') = f(x')g' = g',$$

and so $x = x'$ also. Thus, φ is a continuous bijection of the compact set $T \times W$ onto the neighborhood $\varphi(T \times W)$ of p, hence a homeomorphism. □

Every Lie group G admits a nontrivial, left–invariant Borel measure that is finite on compact sets and strictly positive on open sets (Haar measure). In terms of the corresponding integral $\int_G \varphi \, dg$, left invariance means that, for each $k \in G$,

$$\int_G \varphi(kg)\, dg = \int_G \varphi(g)\, dg.$$

This integral is defined and left invariant not only for real–valued, integrable functions φ, but for $\mathfrak{gl}(n,\mathbb{R})$–valued ones as well, where we integrate the matrix–valued function φ in each entry φ_{ij}. Also note that, if $k \in \mathfrak{gl}(n,\mathbb{R})$, then

$$\int_G \varphi(g)k\, dg = \left(\int_G \varphi(g)\, dg\right)k.$$

Here, the indicated multiplication is the usual multiplication of matrices.

In order to communicate the basic ideas, we will construct the equivariant map f in Proposition 11.3.16 for the case $G = \mathrm{Gl}_+(n,\mathbb{R})$ (the identity component of $\mathrm{Gl}(n,\mathbb{R})$), thereby avoiding technical complications. In particular, this includes Bebutov's theorem ($\mathbb{R} \cong \mathrm{Gl}_+(1)$). This proof is based on the one in [**71**]. It will be left as an exercise to adapt it to general Lie groups (Exercise 11.3.17).

The action of $\mathrm{Gl}_+(n,\mathbb{R})$ on M allows us to define *convolution* of functions

$$F : M \to \mathfrak{gl}(n,\mathbb{R}) \text{ and } \psi : \mathrm{Gl}_+(n,\mathbb{R}) \to \mathfrak{gl}(n,\mathbb{R})$$

by

$$F * \psi(x) = \int_{\mathrm{Gl}_+(n,\mathbb{R})} F(xg)\psi(g^{-1})\,dg, \quad \forall x \in M.$$

As an important example, let $V \subset \mathrm{Gl}_+(n,\mathbb{R})$ be a relatively compact, symmetric neighborhood of I and define $\chi_V : \mathrm{Gl}_+(n,\mathbb{R}) \to \mathfrak{gl}(n,\mathbb{R})$ by

$$\chi_V(g) = \begin{cases} g, & g \in V, \\ 0, & \text{otherwise.} \end{cases}$$

This makes sense and is integrable since $\mathrm{Gl}_+(n,\mathbb{R}) \subset \mathfrak{gl}(n,\mathbb{R})$ is an open subset and V is relatively compact. Let $F : M \to \mathfrak{gl}(n,\mathbb{R})$ be continuous and fix $k \in \mathrm{Gl}_+(n,\mathbb{R})$. Then the change of variables $h = kg$, the symmetry $V = V^{-1}$ and the left invariance of the measure imply

$$(*) \qquad \begin{aligned} F * \chi_V(xk) &= \int_{\mathrm{Gl}_+(n,\mathbb{R})} F(xkg)\chi_V(g^{-1})\,dg \\ &= \int_V F(xkg)g^{-1}\,dg \\ &= \left(\int_{kV} F(xh)h^{-1}dh\right)k. \end{aligned}$$

Since the action is locally free, we can choose V as above and a compact neighborhood V' of $\overline{V}$ in $\mathrm{Gl}_+(n,\mathbb{R})$ such that the orbit map

$$g \in V' \to pg \in M$$

is injective. Let U' be a neighborhood of p in M such that $pV' \cap U' \subset pV$. This is possible since $V \subset V'$ is a neighborhood of I. Let W be a symmetric neighborhood of I such that $W^2 \subset V$ and let U be a neighborhood of p such that $UW^2 \subset U'$. Choosing U even smaller, if necessary, we require that

$$g \in V \smallsetminus W \Rightarrow pg \in U' \smallsetminus \overline{U}.$$

Since pV' is a compact subset of M, the map

$$pg \in pV' \mapsto g \in V' \subset \mathrm{Gl}_+(n,\mathbb{R})$$

is continuous. Viewing this as a map into $\mathfrak{gl}(n,\mathbb{R})$, we apply the Tietze extension theorem to extend it to a continuous map $F' : M \to \mathfrak{gl}(n,\mathbb{R})$. Let $W' \subset W$ be a compact neighborhood of the identity in $\mathrm{Gl}_+(n,\mathbb{R})$ such that

$pW' \subset U$, let $\lambda : M \to [0,1]$ be a continuous map such that $\lambda | pW' \equiv 1$ and $\lambda | (M \smallsetminus U) \equiv 0$, and let $F = \lambda F'$.

Let $\widehat{F} = F * \chi_V : M \to \mathfrak{gl}(n, \mathbb{R})$. Then

$$\widehat{F}(p) = \int_V \lambda(pg) F'(pg) g^{-1}\, dg = \int_V \lambda(pg) I\, dg = cI,$$

where $c = \int_V \lambda(pg) > 0$ by the way λ is defined. Thus, we define

$$f(x) = \widehat{F}(p)^{-1} \widehat{F}(x),$$

a continuous function $f : M \to \mathfrak{gl}(n, \mathbb{R})$ such that $f(p) = I$. Since $\mathrm{Gl}_+(n, \mathbb{R})$ is an open subset of $\mathfrak{gl}(n, \mathbb{R})$, f maps a neighborhood of p into $\mathrm{Gl}_+(n, \mathbb{R})$. We will show that $\widehat{F}$, hence f, is equivariant near p.

For $x \in U$, let $A_x = \{g \in V \mid xg \in U\}$, a neighborhood of I. Then, by the choice of U,

$$Z = \{x \in U \mid A_x \subset W\}$$

is a neighborhood of p and there exists a neighborhood $N \subset Z$ of p and a symmetric neighborhood $K \subset W$ of I such that $NK \subset Z$. We note the following fact.

Claim 1. *If $x \in N$ and $k \in K$, then*

$$A_{xk} = k^{-1} A_x.$$

Proof. If $g \in A_x \subset W$, then $k^{-1} g \in W^2 \subset V$ and $xkk^{-1}g = xg \in U$; hence $k^{-1}g \in A_{xk}$. Similarly, since $xk \in Z$, if $g \in A_{xk}$, then $g \in W$ and $h = kg \in W^2 \subset V$. Since $xh = xkg \in U$, we see that $h \in A_x$. That is, $g = k^{-1}h \in k^{-1}A_x$. □

The following claim completes the proof.

Claim 2. *The map $f : M \to \mathfrak{gl}(n, \mathbb{R})$ is equivariant near p. More precisely, if $x \in N$ and $k \in K$, then*

$$f(xk) = f(x)k.$$

Proof. As already noted, it will be enough to prove this for the map $\widehat{F}$. By definition,

$$\widehat{F}(x) = \int_{\mathrm{Gl}_+(n,\mathbb{R})} F(xg) \chi_V(g^{-1})\, dg.$$

If $x \in N$ and $k \in K$, then

$$g \notin A_{xk} \Rightarrow xkg \notin U \Rightarrow F(xkg) = 0.$$

Thus,

$$\widehat{F}(xk) = \int_{A_{xk}} F(xkg) \chi_V(g^{-1})\, dg = \int_{A_{xk}} F(xkg) g^{-1}\, dg.$$

Then, by the identity $(*)$ and Claim 1,

$$\begin{aligned}\widehat{F}(xk) &= \left(\int_{kA_{xk}} F(xh)h^{-1}\,dh\right)k \\ &= \left(\int_{A_x} F(xh)h^{-1}\,dh\right)k \\ &= \widehat{F}(x)k.\end{aligned}$$

The last equality is again true by the definition of A_x and the fact that $F(xh) = 0$ if $xh \notin U$. □

This proves Theorem 11.3.9 for the Lie group $G = \mathrm{Gl}_+(n, \mathbb{R})$. For the general case, recall Ado's theorem [**76**, p. 202] that every finite–dimensional Lie algebra has a faithful representation in $\mathfrak{gl}(n, \mathbb{R})$, for some $n \geq 1$. It follows that every Lie group G has a neighborhood V of the identity that is isomorphic to a local Lie subgroup of $\mathrm{Gl}_+(n, \mathbb{R})$. One now adapts the above construction as outlined in the following exercise.

Exercise 11.3.17. If G has a locally free action on M, if $V \subset \mathrm{Gl}_+(n, \mathbb{R})$ is a local Lie subgroup isomorphic to a neighborhood of e in G, and if $p \in M$, construct a continuous map $f : M \to \mathfrak{gl}(n, \mathbb{R})$ such that $f(p) = I$ and $f(xg) = f(x)g$, for all x in a suitable neighborhood N of p in M and all g in a suitable neighborhood W of I in V. By the Frobenius theorem, there is a local cross section $T \subset \mathrm{Gl}_+(n, \mathbb{R})$ through I for the action of V by right translations. Show that $f^{-1}(T) \cap N$ is a local section through p for the action of G.

11.3.C. Examples. Locally free actions of Lie groups on metric spaces arise in a number of interesting contexts. The examples of foliated spaces sketched here are mostly of this type. We begin with tilings.

Example 11.3.18. Let us consider the case of tilings of the Euclidean plane $\mathbb{R}^2$ by a finite family $\mathcal{P} = \{P_1, P_2, \ldots, P_n\}$ of compact, polygonal domains. Such a tiling covers $\mathbb{R}^2$ by translates of these polygons (the tiles) in such a way that two tiles intersect, if at all, only in a union of some of their boundary edges. Of course, there may be no tilings associated to a particular choice of $\mathcal{P}$. We do not identify tilings Π and Π' that differ by a translation unless this translation carries each tile of Π onto a tile of Π'.

Let $M_{\mathcal{P}} \neq \emptyset$ denote the set of all tilings associated to the family $\mathcal{P}$. We would like to have a criterion for saying that a tiling $\Pi \in M_{\mathcal{P}}$ is "close" to another tiling Π'. For instance, one can view the two tilings through a "window" $D_0(R)$, this being the disk of radius R, centered at the origin $0 \in \mathbb{R}^2$. If the intersections of Π and Π' with $D_0(R)$ are identical for some large value of R, the tilings are close. In fact, if a small translation moves

Π' to a tiling agreeing with Π in $D_0(R)$, the tilings are thought of as close. More formally, a basic neighborhood $N_\Pi(\epsilon, R)$ of Π in $M_{\mathcal{P}}$, where ϵ and R are positive, is defined to be the set of tilings Π' such that Π agrees in $D_0(R)$ with a translate of Π' by a vector $v \in \mathbb{R}^2$ with $\|v\| < \epsilon$. It is easily seen that the intersection of basic neighborhoods $N_\Pi(\epsilon, R)$ and $N_{\Pi'}(\epsilon', R')$ is a union of basic neighborhoods; hence these neighborhoods form a base of a topology on $M_{\mathcal{P}}$. It can be shown that, with this topology, $M_{\mathcal{P}}$ is compact and metrizable.

The Lie group $\mathbb{R}^2$ acts continuously on $M_{\mathcal{P}}$ by translations. If $a > 0$ is less than the distance between any two vertices of any polygon in $\mathcal{P}$, it should be clear that translation by a vector v with $0 < \|v\| < a$ cannot carry any tiling to itself; so the action is locally free. In the resulting foliation of $M_{\mathcal{P}}$, there are three possible types of leaves, namely planes, cylinders and tori. The type of leaf shows the type of periodicity of any tiling lying in that leaf. The famous Penrose tilings exhibit no periodicity; so a leaf through such a tiling must be planar. A tiling that is periodic under translation by a certain nonzero vector v, but not under translation by any vector linearly independent of v, will lie on a cylindrical leaf. Finally, toroidal leaves represent doubly periodic tilings.

Exercise 11.3.19. Let $\mathcal{P}$ be a finite set of polygonal tiles that tile the plane. Let $M_{\mathcal{P}}$ denote the foliated space associated to $\mathcal{P}$. Show that if $M_{\mathcal{P}}$ has a cylindrical leaf, then it also has a leaf homeomorphic to a torus.

Example 11.3.20. Tilings of the Euclidean plane are not the only interesting ones; the hyperbolic plane also admits interesting tilings. One of the authors has devised a method of representing such tilings by compact, non–Hausdorff surfaces. We illustrate this with an example.

Using the upper half plane model for $\mathbf{H}$, consider the geodesic pentagon P depicted in Figure 11.3.1. One easily checks that the angles at i and $i+1$ are equal to $\alpha = \arctan 4$, that the angle at $i+1/2$ is 2α, and that the angles at $2i$ and $2i+1$ are $\pi - \alpha$. The hyperbolic plane can be tiled by isometric copies of this pentagon, one such tiling being illustrated in Figure 11.3.2. Notice that any row of tiles in Figure 11.3.2 determines all rows below it, but there are two essentially different choices of the row above. It follows that, given one row, there is a Cantor set of tilings, not conjugate under $\mathrm{PSl}(2, \mathbb{R})$, agreeing with each other on and below that row.

Consider the topological space Σ whose underlying set consists of the following pieces: the interior of the geodesic pentagon P, three open geodesic arcs and two points. This is to be assembled as indicated in Figure 11.3.3, the final result being a "hyperbolic surface", *i.e.*, a space that is locally modeled in $\mathbf{H}$, the coordinate changes being isometries of $\mathbf{H}$. It is not an honest surface because it fails to satisfy the Hausdorff separation axiom. (The two

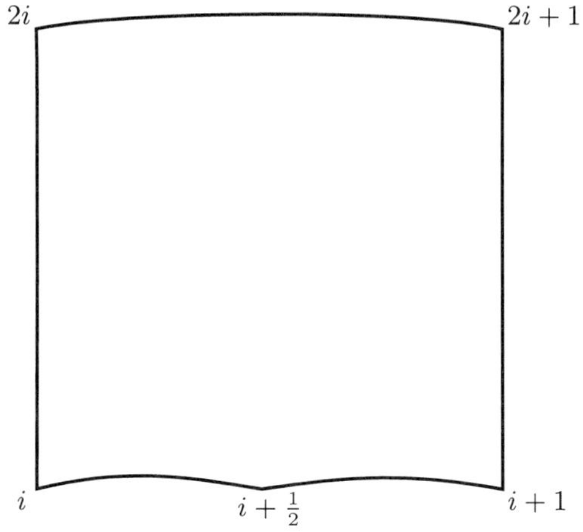

Figure 11.3.1. A hyperbolic pentagon P

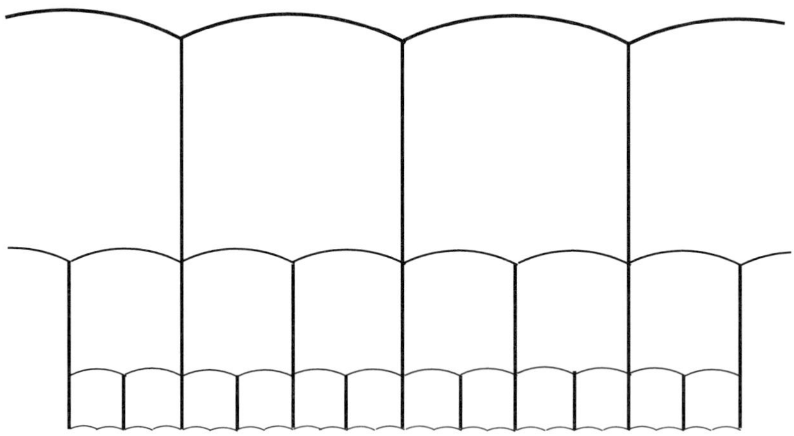

Figure 11.3.2. Tiling of the upper half plane

special points in the construction remain distinct, but all neighborhoods of one overlap all neighborhoods of the other.) As a "surface" it is a very curious one indeed, for it is compact and has both a hyperbolic and an affine structure (and it is well known that no compact surface that is Hausdorff can have both types of structure). With a little thought, it becomes obvious that the tiling in Figure 11.3.2 determines a canonical projection $u : \mathbf{H} \to \Sigma$ that

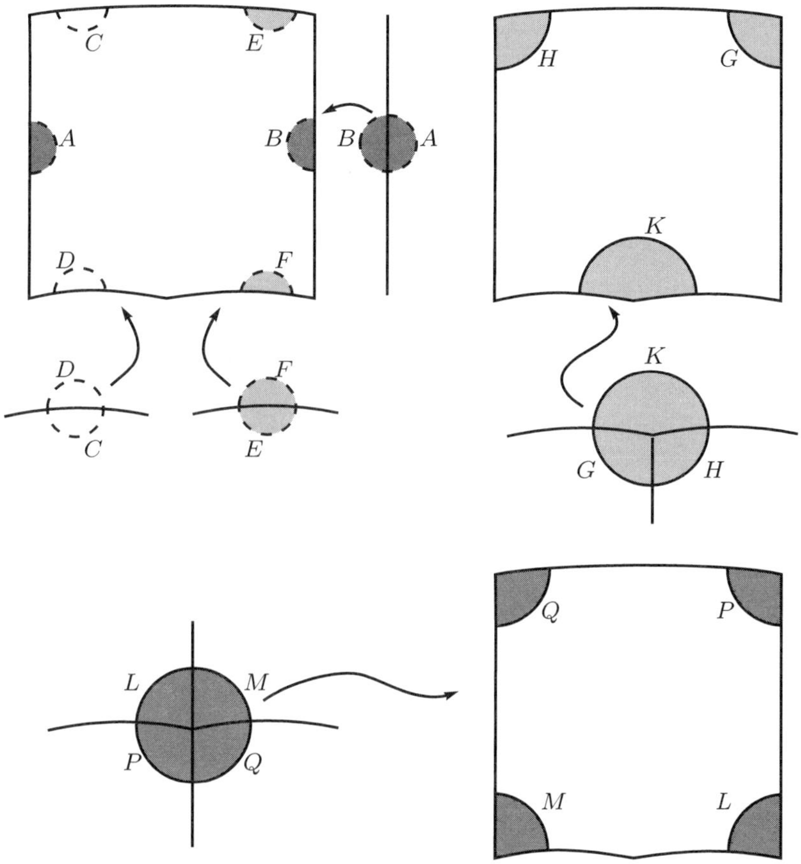

Figure 11.3.3. Coordinate charts of a non–Hausdorff hyperbolic surface

is locally isometric and orientation preserving. It is easily shown that any tiling of the hyperbolic plane by copies of P determines such a projection.

Exercise 11.3.21. Show that the space M of all such maps $\mathbf{H} \to \Sigma$ is a compact Hausdorff space (with the compact–open topology).

The space M has a natural action of $G = \mathrm{PSl}(2, \mathbb{R})$ by composition. If $u \in M$ and $g \in G$, then $u \cdot g = u \circ g$. This action is continuous and locally free, therefore provides M with the structure of a foliated space. It is straightforward to interpret the leaves of this foliated space as equivalence classes of tilings of the plane $\mathbf{H}$, one single leaf representing a tiling up to translation by an element of G. If $K \subset G$ is the compact subgroup consisting of rotations about the point $0 \in \mathbf{H}$ (the unit disk model), then the quotient space M/K is a foliated space whose leaves are surfaces.

Exercise 11.3.22. Show that the leaves of M/K are either planes $\mathbf{H}$ or cylinders $\langle g \rangle \backslash \mathbf{H}$, $g \in G$ a hyperbolic isometry.

Two references for the theory of tilings are [**57**] and [**121**]. These also contain a large bibliography on this popular topic.

Example 11.3.23. A Riemann surface is a connected, one–dimensional complex manifold. Let L be a connected, orientable surface. It is a well–known fact that the choice of an orientation and a metric tensor on L give L the structure of a Riemann surface. (See, for example, [**3**, p. 125] for a proof in the real analytic case of the so–called Gauss lemma, as well as references for more general proofs.) If the orientation is fixed, two metric tensors g_1 and g_2 determine the same complex structure if they are *conformal* (that is, $g_2 = \lambda g_1$ for some positive function λ on L).

A map $f : L \to N$ between Riemann surfaces, with respective metric tensors g_L and g_N, is said to be conformal if it is orientation–preserving and the metrics g_L and f^*g_N are conformal on L.

Another well–known fact from the theory of Riemann surfaces (see [**3**]) is the uniformization theorem of Koebe–Poincaré–Klein.

Theorem 11.3.24. *Let L be a simply connected Riemann surface. Then L is conformally diffeomorphic to one of the following surfaces (with the obvious Riemann surface structure):*

(1) *the complex line $\mathbb{C}$;*
(2) *the Riemann sphere $\mathbb{P} = S^2$;*
(3) *the hyperbolic plane $\mathbf{H}$.*

Therefore, if L is an oriented surface with metric tensor, its universal cover (with the induced orientation and metric tensor) is conformally equivalent to one of the three surfaces listed in the theorem. Those surfaces with universal cover conformal to $\mathbf{H}$ are called *hyperbolic*. Hyperbolic surfaces are quite numerous, the exceptions being the sphere, the plane $\mathbb{C}$, the cylinder $\mathbb{C}^*$, and the torus. It follows from this theorem that, if L is a Riemann surface with metric tensor g, then there is a function $\lambda > 0$ on L such that $(L, \lambda g)$ has constant curvature $1, 0$ or -1. Such a function λ is called a *uniformization function.*

Owing to the Schwarz lemma of complex analysis, the group G of conformal automorphisms of the hyperbolic plane $\mathbf{H}$ is the same as the group of orientation–preserving isometries of the Poincaré metric, which is isomorphic to the Lie group $\mathrm{PSl}(2,\mathbb{R})$ (in the upper half–plane model). The unit tangent bundle to $\mathbf{H}$ can be identified with the group G and, if (L,g) is a hyperbolic surface, then the unit tangent bundle $T^1(L)$ can be identified with the space of conformal covering maps (uniformizations) $u : \mathbf{H} \to L$. This identification takes a uniformization u to the unit vector tangent to L at $u(0)$ in the direction of $du_0(\partial/\partial z)$, where $0 \in \mathbf{H}$ is the origin in the unit disc model of the hyperbolic plane. The uniformization function is uniquely determined by $\lambda(u(0)) = \|du_0\|^2$.

Let M be a compact, oriented foliated space with leaves of dimension 2, and with a metric tensor making all the leaves hyperbolic. Let $T^1(M)$ denote the circle bundle over M consisting of unit tangent vectors to the leaves. Then $T^1(M)$ is a foliated space with leaves of the form $T^1(L)$, L a leaf of M. Let $\mathcal{O}(\mathbf{H}, M)$ denote the space of conformal covering maps $u : \mathbf{H} \to L$, as L runs through the leaves of M. Applying the previous discussion to each leaf of M, we see that there is a one–to–one correspondence between $T^1(M)$ and $\mathcal{O}(\mathbf{H}, M)$.

The natural topology on the space $\mathcal{O}(\mathbf{H}, M)$ is the compact–open topology. There is a natural action of G on $\mathcal{O}(\mathbf{H}, M)$ on the right, defined as follows. If $u \in \mathcal{O}(\mathbf{H}, M)$ and $g \in G$, then $u \cdot g$ is the composition

$$u \circ g : \mathbf{H} \to L.$$

Proposition 11.3.25. *The action*

$$\mathcal{O}(\mathbf{H}, M) \times G \to \mathcal{O}(\mathbf{H}, M)$$

is continuous and locally free.

Proof. With the compact–open topology, $\mathcal{O}(\mathbf{H}, M)$ is a metrizable space. To show continuity, it then suffices to show that, if $u_n \to u$ in $\mathcal{O}(\mathbf{H}, M)$ and $g_n \to 1$ in G, then $u_n \circ g_n \to u$.

Let K be a compact subset of $\mathbf{H}$ and let U be an open set in M such that $u(K) \subset U$. Then there is a larger compact set K' containing K in its interior and such that $u(K') \subset U$. Since $g_n \to 1$, then $g_n(K) \subset K'$ for sufficiently large n. Similarly, $u_n(K') \subset U$ for all sufficiently large n. Therefore $u_n \circ g_n(K) \subset u_n(K') \subset U$, implying $u_n \circ g_n \to u$.

That the action is locally free follows from the fact that each $g \in G$ is an isometry of $\mathbf{H}$ and each $u \in \mathcal{O}(\mathbf{H}, M)$ is a covering map of a leaf, hence locally injective. □

It can be shown that $\mathcal{O}(\mathbf{H}, M)$ is compact when M is compact. Therefore $\mathcal{O}(\mathbf{H}, M)$ is naturally a foliated space. It can also be shown that the correspondence between $T^1(M)$ and $\mathcal{O}(\mathbf{H}, M)$ is a smooth map of foliated spaces. It follows that M can be identified with the quotient foliated space

$\mathcal{O}(\mathbf{H}, M)/\operatorname{PSO}(2)$, for the map $\mathcal{O}(\mathbf{H}, M) \to T^1(M)$ is equivariant with respect to the action of PSO(2) via rotations. The following statement summarizes this discussion (the details can be found in [**16**]).

Theorem 11.3.26. *Let $(M, \mathcal{F})$ be an oriented compact foliated space with 2–dimensional leaves and with a metric tensor making all leaves hyperbolic surfaces. Then $T^1(M)$ is diffeomorphic to the foliated space $\mathcal{O}(\mathbf{H}, M)$, and M to $\mathcal{O}(\mathbf{H}, M)/\operatorname{PSO}(2)$.*

The interest of this fact is that it is very general, 2–dimensional foliated spaces with all leaves hyperbolic being quite numerous, and in fact there is a purely topological condition characterizing them (see Section 12.6).

Foliated spaces whose leaves have a Riemann surface structure have interesting applications in dynamics. The following is an example of D. Sullivan.

Example 11.3.27. Let $f : S^1 \to S^1$ be a smooth, orientation–preserving, expanding map of degree $d > 1$. The expanding property means that there is a constant $k > 1$ such that the norm of the derivative is $\|df\| \geq k$. For example, $f_d : z \mapsto z^d$ defines such a map.

To f one associates the inverse limit space Σ. By Proposition 11.3.6, this space Σ is foliated by lines. One also notes that f induces a map $f : \Sigma \to \Sigma$ which is now a homeomorphism.

While the foliation of Σ can be given a metric under which each leaf is isometric to the line $\mathbb{R}$, the map f is not an isometry on the leaves. However, it can be shown that the leaves of Σ are naturally endowed with an affine structure, and that f preserves this structure.

Let $T(\Sigma)$ denote the tangent space to the leaves of Σ. This space is naturally foliated by planes $T(\mathbb{R})$. Since Σ is oriented, the complement of the zero section of the bundle $T(\Sigma) \to \Sigma$ has two components. Let $T^+(\Sigma)$ denote the positive component. Then f induces a map

$$\begin{aligned} df : T^+(\Sigma) &\to T^+(\Sigma), \\ (x, v) &\mapsto (f(x), df(v)), \end{aligned}$$

hence an action of $\mathbb{Z}$ on $T^+(\Sigma)$. Note that each leaf of $T^+(\Sigma)$ is of the form $\mathbb{R} \times (0, \infty)$ and inherits a complex structure making it conformally equivalent to $\mathbf{H}$. The fact that f is affine on the leaves of Σ implies that df is a conformal map on the leaves of $T^+(\Sigma)$. Therefore, the action of $\mathbb{Z}$ on $T^+(\Sigma)$ is by conformal automorphisms; hence the quotient space $M(f) = \mathbb{Z}\backslash T^+(\Sigma)$ is a space foliated by surfaces.

The outcome of this discussion is the following version, due to D. Sullivan, of a theorem of Shub and Sullivan [**122**].

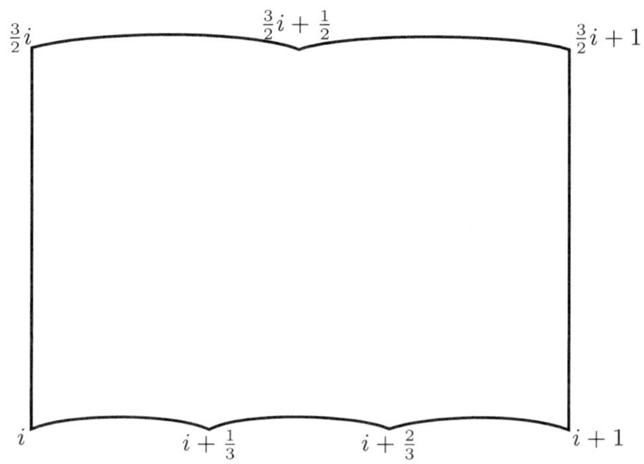

Figure 11.3.4. A tile not associated to an expanding map of S^1

Theorem 11.3.28. *Let f_1, f_2 be two expanding C^r maps ($r \geq 2$) of the circle of degree $d > 1$. Then the foliated spaces $M(f_1)$ and $M(f_2)$ are conformally equivalent if and only if the maps f_1 and f_2 are C^r–conjugate.*

The idea behind this statement is that the conjugacy class of an expanding map such as f is completely determined by the eigenvalues of its derivative at the periodic points. On the other hand, the leaves of $M(f)$ are planes $\mathbf{H}$ or cylinders $\langle z \mapsto \lambda z \rangle \backslash \mathbf{H}$, where λ is such an eigenvalue. A detailed account of the construction of $M(f)$ and of the proof of the theorem can be found in the book of de Melo and van Strien [**33**].

Curiously, the foliated space $M(f)$ associated to the squaring map $f(z) = z^2$ is the same as the foliated space associated to the pentagonal tiling of $\mathbf{H}$ in Example 11.3.20. For $d > 2$ something similar holds for $f(z) = z^d$. The tiles are polygons analogous to P with d edges on the bottom. On the other hand, the foliated space associated to the tile in Figure 11.3.4 does not arise in this fashion.

Exercise 11.3.29. Show that the foliated space $M(f)$, where $f(z) = z^2$ is the squaring map of S^1, admits a locally free action of the affine group $\{x \mapsto ax + b\}$, the orbits of which are the leaves.

Example 11.3.30. Let D be a domain in the complex plane $\mathbb{C}$, and let $G = \mathrm{Aut}(D)$ denote the group of holomorphic homeomorphisms of D. A holomorphic function $f : D \to \mathbb{P}$ is said to be *normal* (see [**87**], for example) if the family of functions $\mathcal{O}(f) = \{f \circ g \mid g \in G\}$ is a normal family in the sense of complex analysis. That is to say, with the compact–open topology in the space of holomorphic functions $\mathcal{O}(D, \mathbb{P})$, $\mathcal{O}(f)$ is relatively compact. Its closure $M(f)$ is a compact Hausdorff space, which supports an action of G given by precomposition. The general domain D has a very small group $G = \mathrm{Aut}(D)$ but, for simply connected domains, it is fairly large and, in this case, it is a Lie group: $G = \mathrm{PSl}(2, \mathbb{R})$ if $D \equiv \mathbf{H}$, and $G = \mathbb{C}^* \rtimes \mathbb{C}$ if $D \equiv \mathbb{C}$ (the $az + b$ group).

Conditions on f guaranteeing that the action of G on $M(f)$ is locally free are easy to describe. For instance:

Exercise 11.3.31. Let $f : D \to \mathbb{P}$ be normal, where D is simply connected. Then the action of G on $M(f)$ is locally free if and only if $M(f)$ contains no constant functions.

In those cases we obtain a foliated space, and the function f naturally defines a function on this space with values in $\mathbb{P}$. Here are some particular examples.

Exercise 11.3.32. It is well known that the series

$$\wp(z) = \frac{1}{z^2} + \sum_{\omega \in \mathbb{Z}^2 \smallsetminus (0,0)} \left(\frac{1}{(z-\omega)^2} - \frac{1}{\omega^2} \right)$$

defines a meromorphic function in the plane $\mathbb{C}$. Discuss its normality and describe the space $M(\wp)$. Consider also the group $\{az+b \mid |a| = 1\}$, instead of the larger $\mathrm{Aut}(D)$, and describe $M(\wp)/\{|a| = 1\}$ in this case.

Exercise 11.3.33. A compact Riemann surface S can be obtained in the form $\Gamma\backslash\mathbf{H}$, where $\Gamma \subset G = \mathrm{PSl}(2, \mathbb{R})$ is a discrete group. Let $f : \mathbf{H} \to \mathbb{P}$ be a meromorphic function that is Γ–invariant. That is to say, $f(\gamma(z)) = f(z)$. Show that f is a normal meromorphic function and describe the space $M(f)$.

11.4. Geometry on Foliated Spaces

In this section, we gather some technical tools which highlight the analogies between foliated spaces and foliated manifolds.

As in the case of C^r manifolds, one has the concept of a C^r partition of unity $\{\lambda_\alpha\}_{\alpha\in\mathfrak{A}}$ subordinate to an open cover $\{V_\alpha\}_{\alpha\in\mathfrak{A}}$. The proof of the following, which is an immediate consequence of the standard proof in manifolds, together with Lemma 11.2.9 and Exercise 11.2.6, is left to the reader (see [**92**]).

Theorem 11.4.1. *Let $(M, \mathcal{F})$ be a C^r–foliated space. Then every open cover of M admits a subordinate C^r partition of unity.*

Definition 11.4.2. If $(M, \mathcal{F})$ is a C^r–foliated space, a (leafwise) Riemannian metric on M is a positive definite metric tensor g on $T(M)$ such that $g(X, Y) \in C^{r-1}(M)$ whenever $X, Y \in \mathfrak{X}(\mathcal{F})$.

The following is a standard consequence of Theorem 11.4.1.

Theorem 11.4.3. *If $(M, \mathcal{F})$ is a C^r–foliated space, $r \geq 1$, then there exists a Riemannian metric on M.*

Such a Riemannian metric defines a notion of distance within each of the leaves. Generally, this will not be the relativization to the leaves of a distance function on M. In fact, such a relativization does not generally give the correct manifold topology to the leaves.

If d_M is a distance function on M, there is a direct way to define a leafwise distance based on paths. First, if $\alpha : [0,1] \to M$ is a path in M contained in a leaf, let

$$\text{length}(\alpha) = \sup_{\mathcal{P}} \sum_{i=1}^{n} d_M(\alpha(t_{i-1}), \alpha(t_i))$$

where $\mathcal{P} = \{0 = t_0 < t_1 < \cdots < t_n = 1\}$ is a partition of $[0,1]$. Then, if x and y belong to the same leaf L, let

$$d_L(x,y) = \inf\{\text{length}(\alpha) \mid \alpha : [0,1] \to L, \alpha(0) = x, \alpha(1) = y\}.$$

If $(M, \mathcal{F})$ is a leafwise–smooth lamination of a manifold W, then a Riemannian metric on W induces a topological metric d_M on M as well as a leafwise Riemannian metric g. In this case, the distance function d_L along each leaf L, constructed from d_M as in the preceding paragraph, coincides with the Riemannian distance along the leaves. We are about to see that, if one is willing to take for ambient manifold W the separable Hilbert space $\mathbb{E}$, these observations generalize to all compact, C^r–foliated spaces, $1 \leq r \leq \infty$. This fact will prove to be useful in other situations also.

Theorem 11.4.4. *If $(M, \mathcal{F})$ is a compact, C^r–foliated space, then there is a leafwise C^r, topological imbedding φ of M into the separable real Hilbert space $\mathbb{E}$.*

Proof. Take a regular foliated atlas $\mathcal{U} = \{(U_i, \varphi_i)\}_{i=1}^N$ of M, where

$$\varphi_i : U_i \to D \times Z_i$$

and $D \subset \mathbb{R}^p$ is a bounded open ball. The metrizable spaces Z_i are second countable; hence they embed in $\mathbb{E}$ [**37**, p. 195]. Thus, we may assume that the maps $\varphi_i : U_i \to D \times Z_i \subset \mathbb{R}^p \times \mathbb{E}$ are C^r embeddings. Let $\{\lambda_i\}_{i=1}^N$ be a partition of unity subordinate to $\mathcal{U}$ and, for each i, define a C^r mapping $h_i : M \to R^p \times \mathbb{E}$ by

$$h_i(w) = \begin{cases} \lambda_i(w)\varphi_i(w) & \text{if } w \in U_i \ , \\ 0 & \text{otherwise.} \end{cases}$$

Let $\varphi : M \to (\mathbb{R}^p \times \mathbb{E})^n \times \mathbb{R}^n \cong \mathbb{E}$ be the map

$$\varphi(w) = (h_1(w), \dots, h_n(w), \lambda_1(w), \dots, \lambda_n(w)).$$

If $\varphi(w) = \varphi(u)$, then $\lambda_i(w) = \lambda_i(u) \neq 0$, for some index i; hence $w, u \in U_i$ and $\varphi_i(w) = \varphi_i(u)$. That is, $w = u$. Also, for each $w \in M$, there is an index i such that $\lambda_i(w) \neq 0$; so h_i restricts to a C^r imbedding of a neighborhood of w in the plaque through w and φ does also. By compactness of M, the continuous injection φ is a homeomorphism onto its image. □

Proposition 11.4.5. *Let $(M, \mathcal{F})$ be a compact, C^r–foliated space, $1 \leq r \leq \infty$. Given a Riemannian metric g on the leaves of M, there is a distance function on M that induces the Riemannian distance on each leaf.*

Proof. By Theorem 11.4.4, we can consider $M \subset \mathbb{E}$ as a compact lamination of $\mathbb{E}$. The tangent spaces to $\mathbb{E}$ are given the natural inner product of the Hilbert space. The bundle $T(M)$ is a finite–dimensional subbundle of $T(\mathbb{E})|M$, and it therefore has an orthogonal complement that we denote by $T^\perp(M)$. If $T(M)$ is given a positive definite inner product, then the direct sum decomposition $T(\mathbb{E})|M = T(M) \oplus T^\perp(M)$ is given the inner product that is the direct sum of g and the Hilbert inner product. This provides a section over M of the bundle of positive definite symmetric bilinear forms on $T(\mathbb{E})|M$. Since the fibers of this bundle are absolute neighborhood retracts and M is compact, this section extends over all of $\mathbb{E}$. This defines a distance function on $\mathbb{E}$ that induces the given distance along the leaves. □

In foliated or laminated manifolds of codimension q, a choice of Riemannian metric determines a normal $\mathbb{R}^q$–bundle over each leaf and over the entire foliated space. For abstract laminations, the choice of an imbedding in $\mathbb{E}$ also gives rise to a normal vector bundle, but the fibers are isomorphic to $\mathbb{E}$. With a little care, one makes use of this normal bundle in close analogy with the finite–dimensional case.

Theorem 11.4.6. *Let M be a compact, foliated space embedded in $\mathbb{E}$ as above. Let $i : L \to \mathbb{E}$ denote the inclusion of a leaf L in $M \subset \mathbb{E}$. Then there is a vector bundle $p : N \to L$ and a neighborhood W of the zero section of N such that the following hold.*

1. *The map $i : L \to \mathbb{E}$ extends to a local diffeomorphism $\varphi : W \to \mathbb{E}$.*
2. *There is a foliated space $F \subset W$, of the same dimension as M, having L as a leaf and transverse to the fibers of N.*
3. *The foliated space $F = \varphi^{-1}(M \cap \varphi(W))$ and the restriction of p to each leaf of F is a local diffeomorphism into L.*

Proof. The statement is pure normal bundle theory, only complicated by the appearance of manifolds of infinite dimension. Fortunately, Lang has written his textbook [**84**] in a way that standard results apply to infinite–dimensional manifolds.

Let N denote the normal bundle to the submanifold $L \subset \mathbb{E}$. Since the tangent bundle $T(L)$ is finite dimensional, there is an orthogonal decomposition $T(\mathbb{E})|L = N \oplus T(L)$. Since $\mathbb{E}$ admits partitions of unity, the construction of W follows from the first part of the theorem in [**84**, p. 96] and from the existence of a finite regular foliated atlas for M. □

Definition 11.4.7. Given $M \subset \mathbb{E}$ a compact lamination, L a leaf of M, the collection (N, p, L, W, F, φ) is called a *quasi–foliated bundle* over L with respect to the pair $(M, \mathbb{E})$.

Remark. This concept is analogous to that of a foliated pseudobundle introduced in [**65**].

One more technical fact on the geometry of foliated spaces is the following local Reeb stability theorem.

Proposition 11.4.8. *Let D be a manifold, and $f : D \to M$ a smooth map whose image has compact closure contained in a leaf L of M. Let $x \in D$ and suppose that the induced homomorphism $\pi_1(D, x) \to \pi_1(L, f(x))$ takes $\pi_1(D, x)$ into the kernel of the holonomy representation of L. Then there is a transversal $X \subset M$ through $f(x)$ and a smooth map F from the product foliation $D \times X$ into M such that $F|D \times x = f$ and $F(D \times y)$ is contained in the leaf though $F(y)$, $\forall\, y \in X$. If $f : D \hookrightarrow L$ is an imbedding, X can be chosen so that F imbeds $D \times X$ in M.*

Exercise 11.4.9. Prove Proposition 11.4.8.

11.5. Holonomy–Invariant Measures

On an oriented manifold of dimension p, one integrates differential forms of degree p. On an oriented, foliated space $(M, \mathcal{F})$ with leaf dimension p, one can also consider differential p–forms that are smooth along leaves and continuous on M. Integrating such a form along each leaf of $\mathcal{F}$ will produce a function of the leaves and, if the leaf space had some reasonable measure, we could in turn integrate against such a measure. But there are foliated spaces whose leaf space is not measurably standard in any reasonable way. So this plan has to be carried out locally, using holonomy–invariant measures. The definition of such measures for foliated manifolds (Definition 10.1.13) adapts unchanged to foliated spaces. That definition seems to depend on the choice of regular foliated atlas, but the following exercise shows that this is not so. Recall that a foliation $\mathcal{F}$ is formally defined to be a coherence class of regular foliated atlases.

Exercise 11.5.1. Let $\mathcal{U} = \{U_\alpha\}_{\alpha \in \mathfrak{A}}$ and $\mathcal{V} = \{V_\beta\}_{\beta \in \mathfrak{B}}$ be regular foliated atlases in the same coherence class $\mathfrak{F}$ and let $Z_\mathcal{U}$ and $Z_\mathcal{V}$ be their respective transverse spaces. Let $\mu_\mathcal{U}$ be a $\Gamma_\mathcal{U}$–invariant measure on $Z_\mathcal{U}$, $\mu_\mathcal{V}$ a $\Gamma_\mathcal{V}$–invariant measure on $Z_\mathcal{V}$. We say that these measures are *coherent* if, whenever $\alpha \in \mathfrak{A}$, $\beta \in \mathfrak{B}$ and $X \subset U_\alpha \cap V_\beta$ is an imbedded (partial) transversal to the two foliations by plaques, then $\mu_\mathcal{U}$ and $\mu_\mathcal{V}$ induce the same measure on X. Prove that coherence is an equivalence relation and that a $\Gamma_\mathcal{U}$–invariant measure for one regular foliated atlas $\mathcal{U}$ induces a unique coherent, $\Gamma_\mathcal{V}$–invariant measure for each regular atlas $\mathcal{V} \in \mathfrak{F}$. Finally, if $\mu_\mathcal{U}$ and $\mu_\mathcal{V}$ are coherent, prove that the set of leaves meeting $\operatorname{supp} \mu_\mathcal{U}$ is exactly the same as the set of leaves meeting $\operatorname{supp} \mu_\mathcal{V}$.

In light of this exercise, we place the following definition.

Definition 11.5.2. If $(M, \mathfrak{F})$ is a foliated space, a (transverse) holonomy–invariant measure for $\mathfrak{F}$ is a coherence class μ of $\Gamma_\mathcal{U}$–invariant measures as $\mathcal{U}$ ranges over the regular foliated atlases $\mathcal{U} \in \mathfrak{F}$. The support $\operatorname{supp} \mu$ is the union of leaves meeting the support of some (hence every) representative measure $\mu_\mathcal{U}$.

Remark. In the literature one also finds the term "transverse invariant measure". It will not be uncommon for us to use this terminology also.

As in Definition 10.1.13, if M is compact, we always assume that a representative measure $\mu_\mathcal{U}$ is finite on the transverse space $Z = Z_\mathcal{U}$. In the noncompact case, one requires that the measure be finite on compact sets.

Exactly as in leafwise–oriented foliated manifolds, such measures can be used to integrate compactly supported differential forms of degree p. Indeed, let $(M, \mathfrak{F})$ be an oriented foliated space and let $\mathcal{U} = \{U_\alpha\}_{\alpha \in \mathfrak{A}}$ be a regular foliated atlas in the orientation class. If $\omega \in A^p_c(M)$ is a differential form and μ a holonomy–invariant measure, then a choice of partition of unity $\{\rho_\alpha\}_{\alpha \in \mathfrak{A}}$ subordinate to the atlas $\mathcal{U}$ permits us to write ω as a finite sum

$$\omega = \sum_{\alpha \in \mathfrak{A}} \rho_\alpha \omega = \sum_{\alpha \in \mathfrak{A}} \omega_\alpha,$$

where each ω_α has compact support in $U_\alpha \cong D_\alpha \times Y_\alpha$ and is identically zero for all but finitely many $\alpha \in \mathfrak{A}$.

With this, we can define the integral of ω with respect to μ by adding up the result of first integrating each ω_α over each plaque $D_\alpha \times \{y\}$ (with respect to the orientation determined by the atlas), producing thereby a continuous function on Y_α, and then integrating this function with respect to the measure μ_α. In symbols, define

$$I_\mu(\omega) = \sum_{\alpha \in \mathfrak{A}} \int_{Y_\alpha} \left(\int_{D_\alpha \times \{y\}} \omega_\alpha \right) d\mu_\alpha(y).$$

The standard trick, using holonomy invariance of the measure, shows at once that the definition of $I_\mu(\omega)$ is independent of the choice of oriented atlas and of subordinate partition of unity, so that this integration process defines a continuous linear functional

$$I_\mu : A^p_c(M) \to \mathbb{R}.$$

Remark. When $(M, \mathcal{F})$ is not orientable, one can do integration of twisted differential forms ("formes différentielles d'espèce impaire" of de Rham). The details are left to the reader.

Alternatively, one can bring into play the orientable double M', introduced in Exercise 11.2.15, and work on M' with objects lifted from M. Observe also that if M' admits a holonomy–invariant measure, so does M.

11.5.A. Leafwise de Rham theory and foliation cycles. The (leafwise) de Rham cohomology of M is defined as follows. On a foliated space M there is a sheaf $\mathbb{R}_\ell$. For each $x \in M$, the stalk of $\mathbb{R}_\ell$ at x is the real vector space of germs at x of continuous, real–valued functions, defined on neighborhoods U of x and constant along the leaves of $\mathcal{F}|U$. Associated to such a sheaf of vector spaces are cohomology groups $H^p(M; \mathbb{R}_\ell)$. We recall briefly this standard construction.

The sheaf $\mathbb{R}_\ell$ admits a resolution

$$\mathbb{R}_\ell \hookrightarrow \mathcal{A}^0 \to \mathcal{A}^1 \to \cdots \to \mathcal{A}^p$$

where $\mathcal{A}^q$ is the sheaf of germs of differential forms of degree q along the leaves. The map $\mathbb{R}_\ell \hookrightarrow \mathcal{A}^0$ is induced by the inclusion of the space of leafwise–constant, continuous functions into the space of all leafwise–smooth, continuous functions, and the maps $\mathcal{A}^{p-1} \to \mathcal{A}^p$ are given by the (leafwise) exterior derivative d. We claim that this sequence is exact on the stalks. Indeed, $d^2 = 0$ and the converse is a consequence of the following exercise.

Exercise 11.5.3. Let $M = \mathbb{R}^n \times Z$ be a trivial foliated space. Show that, if $\omega \in \Gamma\mathcal{A}^p$ is a closed p–form, $p \geq 1$, then $\omega = d\alpha$ for some $\alpha \in \Gamma\mathcal{A}^{p-1}$. (Hint. Adapt the standard proof of the version of the Poincaré lemma that asserts that, for $p \geq 1$, every closed p–form on a contractible manifold is exact.)

The induced sequence

$$\Gamma\mathbb{R}_\ell \hookrightarrow \Gamma\mathcal{A}^0 \xrightarrow{d} \Gamma\mathcal{A}^1 \xrightarrow{d} \cdots \xrightarrow{d} \Gamma\mathcal{A}^p \xrightarrow{d} 0$$

on the spaces of continuous sections is not generally exact, but is a cochain complex of real topological vector spaces. The groups $H^q(M; \mathbb{R}_\ell)$ are the cohomology groups of this cochain complex:

$$H^q(M; \mathbb{R}_\ell) = \mathrm{Ker}(d : \Gamma\mathcal{A}^q \to \Gamma\mathcal{A}^{q+1})/d\Gamma\mathcal{A}^{q-1},$$

where, if $q = 0$, $d\Gamma\mathcal{A}^{q-1}$ is replaced by 0. Remark that $\Gamma\mathcal{A}^q$ is naturally identified with the space $A^q(M)$ of leafwise–smooth, continuous q–forms on M; so this cohomology is the natural generalization of de Rham cohomology to foliated spaces. One defines the compactly supported cohomology $H_c^q(M;\mathbb{R}_\ell)$ similarly, using compactly supported sections of these sheaves.

Exercise 11.5.4. Suppose that M is a compact foliated space that has a dense leaf. Show that $H^0(M;\mathbb{R}_\ell) \cong \mathbb{R}$.

Exercise 11.5.5. The cohomology of the trivial foliated space $M = \mathbb{R}^n \times Z$ is $H^0(M;\mathbb{R}_\ell) = C(Z)$, and $H^p(M;\mathbb{R}_\ell) = 0$ if $p > 0$. What is the cohomology $H_c^*(M;\mathbb{R}_\ell)$?

Exercise 11.5.6. Let M be the 3–sphere, foliated by the Reeb foliation. Show that $H^0(M;\mathbb{R}_\ell) \cong \mathbb{R}$ and that $H^2(M;\mathbb{R}_\ell)$ is nontrivial.

If the differential form ω is exact, that is, $\omega = d\sigma$ for some compactly supported $(p-1)$–form σ, then we see that we can write

$$\omega = d\sigma = d\Big(\sum_{\alpha\in\mathfrak{A}} \rho_\alpha\sigma\Big) = \sum_{\alpha\in\mathfrak{A}} d\sigma_\alpha$$

and, by Stokes' theorem, we obtain $I_\mu(d\sigma) = 0$. This suggests how one might extend (at least partially) the theory of foliation cycles to compact, leafwise–oriented, foliated spaces. We sketch this, and refer the reader to the proof of Theorem 10.2.12, and to the book [**92**], where this general version of foliation cycles is amply treated.

The spaces $A^q(M) = \mathcal{D}_q$ are topologized exactly as the spaces $\mathcal{D}_q$ were topologized in Section 10.1, using the concept of C^∞ on foliated spaces. Just as in that section, one introduces the strong dual $\mathcal{D}'_q$, the space of de Rham currents. The adjoint $\partial = d^*$ of the de Rham operator d makes $(\mathcal{D}'_*, \partial)$ into a chain complex and the homology of this complex can be thought of as the "de Rham homology" of the foliated space. Since $\mathcal{F}$ is oriented, the definitions in Section 10.2 then make sense, providing the cone $\mathcal{C}_\mathcal{F} \subset \mathcal{D}'_p$ of foliation currents. As expected, this cone has compact base.

By the above remarks, holonomy–invariant measures are foliation cycles. The reader is now invited to review the proof of Theorem 10.2.12, adapting it to foliated spaces. We state the result.

Theorem 11.5.7. *On a compact, oriented foliated space $(M, \mathcal{F})$, the correspondence $\mu \mapsto I_\mu$ defines a natural bijection between the set of holonomy–invariant measures for $\mathcal{F}$ and the cone $\mathcal{Z}_\mathcal{F}$ of foliation cycles.*

Remark that p–forms are closed by default, p being the dimension of the leaves. Furthermore, the assumption that $\mathcal{F}$ is oriented implies that there exist p–forms that are strictly positive with respect to the leaf orientations.

In terms of the position of the cone $\mathcal{C}_{\mathcal{F}} \subset \mathcal{D}'_p$ with respect to the subspaces of cycles and boundaries, this observation excludes Figure 10.2.3. Equivalently and more formally, we have the following.

Lemma 11.5.8. *In a compact, oriented foliated space, no foliation cycle bounds.*

Similarly, referring to Figures 10.2.2 and 10.2.1, one notes the following.

Lemma 11.5.9. *In a compact, oriented foliated space, the following are equivalent.*

(1) *There exists a holonomy–invariant probability measure.*

(2) *No exact p–form is strictly positive with respect to the leaf orientations.*

In [**92**], the following version of the above theorem is stated. The proof is a good exercise in the theory of foliation cycles, and it will be omitted. See also [**113**], [**59**].

Theorem 11.5.10. *Let M be a compact, oriented foliated space of dimension p. Then there is an isomorphism of topological vector spaces between the space of (signed) transverse invariant measures and the Hausdorff dual of $H^p(M;\mathbb{R}_\ell)$. The isomorphism is given by the Ruelle–Sullivan map which assigns to a (signed) transverse invariant measure μ the linear functional*

$$I_\mu : H^p(M;\mathbb{R}_\ell) \to \mathbb{R}$$

given by integration of a p–form against μ.

Remark. If X is a topological vector space, the Hausdorff dual is the topological vector space of continuous homomorphisms $X \to \mathbb{R}$ modulo the closure of the zero linear functional $0 : X \to \mathbb{R}$. In general, the subspace $\{0\}$ of X^* may not be a closed subspace.

Exercise 11.5.11. Here are some illustrations of this theorem.

1. If the foliated space M has the structure of a fiber bundle over a compact metrizable space B with a manifold F as compact fiber, prove that the topological dual of $H^p(M;\mathbb{R}_\ell)$ is the space of Borel measures on B.
2. If M has the structure of a foliated bundle over a compact p-manifold B with fiber a compact metrizable space F obtained via suspension of a representation $\varphi : \pi_1(B) \to \mathrm{Homeo}(F)$, prove that the Hausdorff dual of $H^p(M,\mathbb{R}_\ell)$ is the space of $\varphi(\pi_1(B))$–invariant measures on F.

11.5.B. Existence of invariant measures. The criterion (2) given in Lemma 11.5.9 for the existence of a nontrivial holonomy–invariant measure is not easily checked. One criterion that is more easily verified is given by Theorem 10.2.22. We proved this for foliated manifolds but, by the ideas outlined in the previous section, it extends readily to foliated spaces. Remark first that the definition of "averaging sequence" (Definition 10.2.21) carries over unchanged to foliated spaces.

Theorem 11.5.12. *Let $(M, \mathcal{F})$ be a compact foliated space and let $\{A_k\}_{k=1}^{\infty}$ be an averaging sequence for $\mathcal{F}$. Then there is a subsequence $\{A_{k_r}\}_{r=1}^{\infty}$ such that*

$$\lim_{r\to\infty} \frac{1}{\mathrm{vol}_p(A_{k_r})} \int_{A_{k_r}}$$

exists in $\mathcal{C}_{\mathcal{F}}$ and is a nontrivial foliation cycle.

This theorem gives a sufficient and visual condition implying the existence of an invariant measure. A closely related criterion of J. F. Plante, nonexponential growth in the foliation, will be proven in Chapter 12 (Theorem 12.3.1). Both exemplify the "Fölner condition for amenability". We will not discuss this concept here because it requires more depth than this chapter can provide. But in essence all concepts that one has for groups and group actions can be adapted to foliated spaces by translating them to the holonomy groupoid of the foliated space. Thus the interested reader will have no trouble in defining properties like "amenability", "property (T)", etc.

Here we describe another geometric property of the foliation equivalent to the existence of an invariant measure. Somehow the property is reminiscent of the Ghys–Langevin–Walczak theorem (Theorem 13.4.2) that foliations with no invariant measure have positive entropy and that resilient leaves imply positive entropy. Its sources are buried in the initial steps of ergodic theory (*e.g.*, the book of P. Halmos [**60**] and the work of E. Hopf mentioned there) and the theory of paradoxical decompositions of geometric figures, starting with the Hausdorff–Banach–Tarski paradox. In fact, the concept of pseudogroup already appears in A. Tarski's work dealing with these problems [**131**]. A recent description of Tarski's work, and a general reference for this section, is the book of S. Wagon [**142**].

Let Γ be a countably generated pseudogroup with underlying space X separable and metrizable. Two Borel subsets A and B of X are said to be countably equivalent under Γ (equivalent for short) if there are countable Borel partitions $A = \bigsqcup_{i=1}^{\infty} A_i$ and $B = \bigsqcup_{i=1}^{\infty} B_i$ and elements $g_i \in \Gamma$ such that $g_i A_i = B_i$. (Here "$\bigsqcup$" denotes disjoint union.) We denote this equivalence by $A \sim_{\Gamma} B$.

Exercise 11.5.13. If $X \sim_\Gamma A$, then there is a Borel map $f : X \to A$ that is injective and maps Γ-orbits to Γ-orbits.

Definition 11.5.14. The pseudogroup Γ is called *paradoxical* if there are disjoint, Borel subsets A and B of X, both of which are countably equivalent under Γ to X.

Theorem 11.5.15. *Let Γ be a countably generated pseudogroup on a separable, metrizable space X. Then Γ is paradoxical if and only if it admits no invariant (countably additive) probability measure.*

The easy part of this theorem is contained in the next exercise.

Exercise 11.5.16. Show that, if Γ admits an invariant probability measure, then it cannot be paradoxical.

Exercise 11.5.17. The pair (X, Γ) is said to be compressible if there is a Borel subset $A \subset X$ such that $X \sim_\Gamma A$ and the complement $X \smallsetminus A$ meets every Γ–orbit. Otherwise (X, Γ) is said to be incompressible.

1. Show that, if (X, Γ) is compressible, then it cannot admit an invariant probability measure.
2. Show that, if (X, Γ) has a finite orbit, then it cannot be compressible.
3. Show that, if (X, Γ) is paradoxical, then it is compressible.

The converse is more complex and it will not be given here. The proof is contained in the union of [**96**] and [**8**]. It is described in the context of countable equivalence relations, and relies on the following theorem of M. Nadkarmi.

Theorem 11.5.18. *Let Y be a standard Borel space and let Γ be a countably generated pseudogroup of Borel transformations of Y. Then the following are equivalent:*

(1) *(Y, Γ) is incompressible;*

(2) *Y has a Γ–invariant probability measure.*

Chapter 12

Growth, Invariant Measures and Geometry of Leaves

The theme of this chapter is the interrelation between the global properties of the foliated space and the geometric structure of the leaves. For example, certain growth properties of leaves imply the existence of holonomy–invariant measures. In codimension one, the well–understood recurrence structure permits very precise results. Finally, homological properties of the foliated space influence the geometry of its leaves, especially when these are 2–dimensional.

12.1. Quasi-Riemannian Geometry of Leaves

A Riemannian metric (perhaps only C^0) on a compact foliated space makes each leaf into a complete Riemannian manifold, the geometry of such a leaf being unique up to bounded distortions. More precisely, any two such metrics on a leaf induced from metrics on the ambient space are equivalent under *quasi–isometry*.

Definition 12.1.1. Two Riemannian metrics, g and g' on the (possibly noncompact) manifold N are quasi–isometric if there exist positive constants A and B such that

$$A\|v\| \leq \|v\|' \leq B\|v\|, \quad \forall\, v \in T(N),$$

where $\|\cdot\|$ and $\|\cdot\|'$ are the norms determined by g and g', respectively.

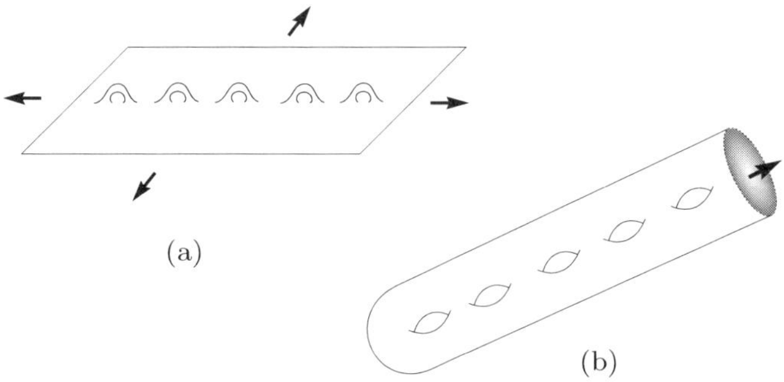

Figure 12.1.1. The Loch Ness monster (a) and Jacob's ladder (b)

Remark. If N is only C^1, there is a C^1–compatible C^∞ structure. The C^0 metric g can be uniformly well approximated by a smooth, quasi–isometric Riemannian metric g'. This remark is useful for applying theorems from Riemannian geometry that require higher–order smoothness.

Quasi–isometry is an equivalence relation on the set of Riemannian metrics on a manifold. We will only be interested in *complete* metrics, although the definition makes sense in general.

Definition 12.1.2. A diffeomorphism $f : N \to N'$ between (complete) Riemannian manifolds (N, g) and (N', g') is a quasi–isometry if $f^*(g')$ is quasi–isometric to g.

A quasi–isometry between Riemannian manifolds has uniformly bounded distortion. Quasi–isometric manifolds are said to "look alike". Quasi–isometry is an equivalence relation on the set of (complete) Riemannian manifolds.

Exercise 12.1.3. Given a leaf L of a compact, C^1–foliated space $(M, \mathcal{F})$, prove that any leafwise Riemannian metric on M restricts to a complete Riemannian metric on L. Prove that any two such metrics on M induce quasi–isometric metrics on L. The moral is that it makes sense to ask what a leaf of a compact foliated space looks like.

Example 12.1.4. Let $\pi : N \to M$ be a covering map, where M is a compact manifold. A Riemannian metric on M lifts to a complete Riemannian metric on N. Again, the fact that M is compact implies that the lifts of any two metrics will be quasi–isometric. One can ask what a covering manifold (especially, the universal covering) of a compact manifold looks like.

Example 12.1.5. The infinite Loch Ness monster and Jacob's ladder (Figure 12.1.1) are diffeomorphic surfaces. With the Riemannian metrics implicit in these pictures, the surfaces are not quasi–isometric. These surfaces do not "look alike". In fact, the infinite Loch Ness monster can appear as a leaf of a C^∞ foliation of S^3, but Jacob's ladder cannot be a leaf of any such foliation of S^3. (Can Jacob's ladder be a leaf of a lamination in S^3?) Two other quasi–isometry types of this surface, the infinite jail cell window and the infinite jungle gym, are depicted in Figure 12.1.2. We will see that these incarnations are also unable to be leaves in foliations of S^3 (but they can be leaves of a lamination in S^3). In fact, the ladder, the jail cell window, and the jungle gym will be examples of leaves that produce a foliation cycle that is not homologous to zero. Hence these quasi–isometry types cannot occur as leaves of 2–dimensional foliations in compact n–manifolds with $H_2(M;\mathbb{R}) = 0$. This is a theorem of Phillips and Sullivan [**101**]. On the other hand, neither the Loch Ness monster nor Jacob's ladder can be covering spaces of a compact surface, while the jail cell window and the jungle gym are easily seen to be covering spaces of the compact surfaces of genus 2 and 3, respectively.

12.2. Growth

An important invariant of the quasi–isometry type of a complete, connected Riemannian manifold (N, g) is the *growth type*. This is related to analogous notions of growth types of groups, orbits of transformation groups and orbits of pseudogroups.

12.2.A. Growth of Riemannian manifolds. Fix $z \in N$ and let

$$D_z(r) = \{w \in N \mid \rho(z, w) \leq r\},$$

where ρ is the topological metric on N induced by g. Let $D_z^\circ(r)$ denote the open metric r–ball and define $G_z : \mathbb{Z}^+ \to \mathbb{R}^+$, the *growth function* of (N, g) at z, by

$$G_z(r) = \operatorname{vol} D_z^\circ(r).$$

This makes sense when N is C^1. If we pass to an underlying C^∞ structure and a smooth, quasi–isometric Riemannian metric g', the growth function changes, but its growth type does not (Exercises 12.2.2 and 12.2.3).

Definition 12.2.1. Let $G, H : \mathbb{Z}^+ \to \mathbb{R}^+$ be nondecreasing functions. Then *G dominates H* (we write $G \succeq H$) if there are positive integers A, B, and C such that

$$AG(Br) \geq H(r), \quad \forall\, r \geq C.$$

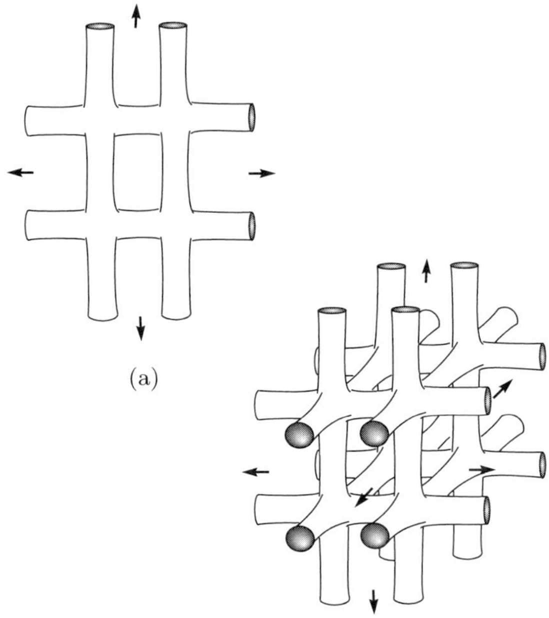

(a)

(b)

Figure 12.1.2. The infinite jail cell window (a) and jungle gym (b)

If $G \succeq H \succeq G$, then G and H are said to *have the same growth type* and we write $\mathrm{gr}(G) = \mathrm{gr}(H)$. If $G \succeq H$, but $H \not\succeq G$ (we write $G \succ H$), we say that G *strictly* dominates H and write $\mathrm{gr}(G) > \mathrm{gr}(H)$.

Mutual dominance is clearly an equivalence relation and "growth type of G" can be taken as the name of the equivalence class of G, denoted by $\mathrm{gr}(G)$.

Exercise 12.2.2. If (N, g) is a complete, connected Riemannian manifold and $z, w \in N$, prove that $\mathrm{gr}(G_z) = \mathrm{gr}(G_w)$.

Exercise 12.2.3. Let (N, g) and (N', g') be complete, connected Riemannian manifolds, $z \in N$ and $w \in N'$. Let the respective growth functions be G_z and G'_w. If these Riemannian manifolds are quasi–isometric, prove that $\mathrm{gr}(G_z) = \mathrm{gr}(G'_w)$.

Definition 12.2.4. The growth type of a connected manifold N, relative to a quasi–isometry class of complete metrics, is $\mathrm{gr}(N) = \mathrm{gr}(G_z)$, where G_z is the growth function of any metric in the given quasi–isometry class.

In particular, if L is a leaf in a compact, foliated space $(M, \mathcal{F})$, $\mathrm{gr}(L)$ is well defined relative to the quasi–isometry class of complete metrics determined by any metric on L relativized from M.

Example 12.2.5. Let $P(r)$ and $Q(r)$ be polynomials with all coefficients nonnegative and respective degrees p and q. If $p > q$, then

$$\lim_{r\to\infty} \frac{P(r)}{Q(r)} = \infty;$$

so there is a constant $C > 0$ such that $P(r) > Q(r)$, $\forall\, r \geq C$. That is, $P \succeq Q$. The reverse dominance fails. Indeed, $AQ(Br)$ will be a polynomial of degree $q < p$, for any positive integers A and B, and

$$\lim_{r\to\infty} \frac{AQ(Br)}{P(r)} = 0.$$

That is, $AQ(Br) < P(r)$ for all large enough values of r; so $P \succ Q$. If $p = q$,

$$\lim_{r\to\infty} \frac{P(r)}{Q(r)} < \infty.$$

In fact, this limit is the ratio a_p/b_p of the coefficients of r^p in the respective polynomials. If the constant integer $B > 0$ is large enough, the polynomial $P(Br)$ has coefficient of r^p larger than that of $Q(r)$ and

$$P(Br) \geq Q(r), \quad \forall\, r \geq C,$$

for a suitable constant C. That is, $P \succeq Q$. Since $q = p$, the same argument shows that $Q \succeq P$; so $\mathrm{gr}(P) = \mathrm{gr}(Q)$. This growth type is said to be *exactly polynomial* of degree p.

Exercise 12.2.6. Let (N, g) be a complete, connected, noncompact Riemannian manifold, ρ the topological metric on N induced by g. Using the Hopf–Rinow theorem, prove that, issuing from an arbitrary point $x \in N$, there is at least one nonclosed geodesic ray $\sigma(t)$ such that $\rho(x, \sigma(t)) = t$, $0 \leq t < \infty$. Using this, prove that, if the injectivity radius of N is bounded below by a positive constant, then $\mathrm{gr}(N)$ dominates the linear growth type (exactly polynomial of degree 1). Is this true without the bound on the injectivity radius?

Exercise 12.2.7. Let (L, g) be a complete Riemannian manifold and let $f : L \to L$ be a uniformly expanding diffeomorphism. That is, there are constants $b > a > 1$ such that all the eigenvalues of $df_x \circ df_x^*$ are in the interval $[a, b]$, for all $x \in L$. Show that L has polynomial growth. (Hint. Show first that f has a unique fixed point.)

Example 12.2.8. A leaf that looks like Jacob's ladder has linear growth. Leaves looking either like the infinite Loch Ness monster or the infinite jail cell window have quadratic growth (exactly polynomial of degree 2). The infinite jungle gym has cubic growth. These assertions are fairly intuitive. In the linear foliation $\mathcal{F}$ of T^3 by dense planes, the leaves appear with the usual quasi–isometry class of $\mathbb{R}^2$, the one determined by the Euclidean metric. This is clear if we give $T^3 = \mathbb{R}^3/\mathbb{Z}^3$ the metric induced by the Euclidean metric in $\mathbb{R}^3$. The projection $p : \mathbb{R}^3 \to T^3$ carries each planar leaf L of $p^{-1}(\mathcal{F})$ isometrically onto $p(L)$. Thus, these leaves grow quadratically. Similar considerations show that each leaf of the linear foliation of T^3 by dense cylinders has linear growth.

Example 12.2.9. At the other extreme, consider the functions

$$\exp_a(r) = a^r,$$
$$\exp_b(r) = b^r,$$

where a and b are constants strictly greater than 1. If we choose a positive integer $B > \log_a b$, then

$$\exp_a(Br) > \exp_b(r), \quad \forall\, r \geq 1.$$

This argument works with the roles of $\exp_a$ and $\exp_b$ reversed; so

$$\operatorname{gr}(\exp_a) = \operatorname{gr}(\exp_b).$$

As usual, exp denotes $\exp_e$ and gr(exp) is called the *exponential* growth type. It is evident that it strictly dominates every polynomial growth type.

Exercise 12.2.10. Prove that the hyperbolic plane grows exponentially.

Exercise 12.2.11. The Euclidean plane $(\mathbb{R}^2, g)$ with the standard metric has quadratic growth. Let $\lambda > 0$ be a positive smooth function on $\mathbb{R}^2$ such that $\lambda(x, y)$ depends only on y and $\lambda(x, y) = |y|^{-2}$ for $|y| \geq 1$. What is the growth type of $(\mathbb{R}^2, \lambda g)$? (Geometrically, this manifold is like two horoballs in the hyperbolic plane zipped up along their boundary.)

Example 12.2.12. In the Roussarie foliation of the unit tangent bundle $T^1(\Sigma_g)$ of the closed surface Σ_g of genus $g > 1$ (Example 1.1.9), there is a metric on $T^1(\Sigma_g)$ in which each leaf has constant curvature -1. Most of the leaves are planes, hence have the quasi–isometry class of the hyperbolic plane. As you proved in Exercise 12.2.10, the hyperbolic plane has exponential growth. We will see that every leaf approaching a resilient leaf has exponential growth (Corollary 12.2.40). Since this foliation has resilient leaves and all leaves are dense, it will follow, in fact, that all leaves of the Roussarie foliation have exponential growth. Of course, there are other ways of proving this.

Proposition 12.2.13. *Let $h : \mathbb{Z}^+ \to \mathbb{R}^+$ be nondecreasing. Then $h \succeq \exp$ if and only if*

$$\liminf_{r\to\infty} \frac{\log h(r)}{r} > 0.$$

Proof. If the lim inf is > 0, there will be positive constants a and C such that

$$\frac{\log h(r)}{r} > a, \quad \forall\, r \geq C.$$

That is, $\log h(r) > ar$ and, if $b = e^a$,

$$h(r) > b^r, \quad \forall\, r \geq C,$$

proving that $h \succeq \exp_b$. Conversely, if there are positive integers $A, B,$ and C such that

$$h(Br) \geq \frac{e^r}{A}, \quad \forall\, r \geq C,$$

consider integers k in the range $Br \leq k \leq B(r+1)$ and conclude that

$$h(k) \geq h(Br) \geq \frac{e^r}{A} \geq \frac{e^{(k-B)/B}}{A}, \quad \forall\, k \geq BC.$$

It follows that

$$\frac{\log h(k)}{k} \geq \frac{1}{B} - \frac{1 + \log A}{k}, \quad \forall\, k \geq BC.$$

Thus,

$$\liminf_{k \to \infty} \frac{\log h(k)}{k} \geq \frac{1}{B} > 0.$$

□

We will see that leaves in a compact, foliated space can never have growth strictly dominating the exponential type (Corollary 12.2.37). This and Proposition 12.2.13 motivate the following definition.

Definition 12.2.14. Let $h : \mathbb{Z}^+ \to \mathbb{R}^+$ be nondecreasing. If

$$\liminf_{r \to \infty} \frac{\log h(r)}{r} = 0,$$

we say that the growth type $\operatorname{gr}(h)$ is *subexponential* or *nonexponential.*

Definition 12.2.15. Let $h : \mathbb{Z}^+ \to \mathbb{R}^+$ be nondecreasing. If

$$\lim_{r \to \infty} \frac{\log h(r)}{r} = 0,$$

we say that the growth type $\operatorname{gr}(h)$ is *quasi–polynomial.*

Typically, a subexponential growth type that is not also quasi–polynomial has arbitrarily long intervals of consecutive integers on which it grows exponentially and arbitrarily long intervals on which it grows quasi–polynomially. In some of the literature, "subexponential" is used for "quasi–polynomial" (*e.g.*, see [**72**]).

Example 12.2.16. The *fractional* types r^s, where $s > 2$ is strictly between consecutive integers, occur for leaves of foliations obtained by suspending a suitable pair of C^∞ diffeomorphisms of S^1 over the 2–holed torus Σ_2 [**19**]. If $k-1 < s < k$, this growth type is strictly dominated by the polynomial type of degree k and strictly dominates the one of degree $k-1$. Such a growth type is said to be *weakly polynomial*. Another weakly polynomial type, represented by $h_s(r) = r^s \log r$, where $s \geq 2$ is a fixed real number, can also be realized by suspending a pair of C^∞ diffeomorphisms of the circle [**28**]. This growth type strictly dominates r^s, but is strictly dominated by r^t, for every real number $t > s$. It is clear that these weakly polynomial types are quasi–polynomial. An example of quasi–polynomial growth that dominates all polynomials is given by $h(r) = r^{\sqrt{r}}$. This, along with uncountably many other quasi–polynomial growth types, uncountably many subexponential types strictly dominating all quasi–polynomial ones, as well as the exponential type, can each be realized by suspensions of suitable $f_1, f_2 \in \mathrm{Diff}^\infty_+(S^1)$ [**20**]. In this and the preceding two references, the leaves with the given growth type are at level 2 and have no holonomy. By allowing the leaves to be at infinite level, it is possible to choose the two diffeomorphisms so that every strictly polynomial type, uncountably many quasi–polynomial growth types, uncountably many subexponential types strictly dominating all quasi–polynomial types, and the exponential type all occur *simultaneously* in the foliated circle bundle [**63**]. In this construction, the polynomial types of degree k occur for leaves at depth k and all the other types occur for leaves at infinite level.

12.2.B. Growth of groups, group actions and coverings. Let Γ be a finitely generated group and let $\Gamma_1 = \{g_i\}_{i=1}^m$ be a finite, symmetric generating set (as for a pseudogroup, this means that Γ_1 is closed under passing to inverses) containing the identity e. Each $g \in \Gamma$ has representations

$$g = g_{i_1} g_{i_2} \cdots g_{i_k}.$$

We can define the *norm* $\|g\|$ of $g \in \Gamma$ to be the smallest integer $k \geq 1$ for which such a representation exists. For each $k \geq 1$, let

$$\Gamma_k = \{g \in \Gamma \mid \|g\| \leq k\}.$$

In particular, $k = 1$ gives back the generating set Γ_1. Relative to this norm, a growth function is defined by

$$G : \mathbb{Z}^+ \to \mathbb{R}^+,$$
$$G(k) = |\Gamma_k|,$$

where the absolute value symbol denotes cardinality.

Lemma 12.2.17. *Let $\Gamma_1' = \{h_j\}_{j=1}^{\ell}$ be another choice of symmetric generating set and let G and G' be the growth functions relative to Γ_1 and Γ_1', respectively. If $\Gamma_1 \subseteq \Gamma_1'$, then $G(k) \leq G'(k)$, $\forall\, k \geq 0$.*

Proof. Let $\|\cdot\|$ and $\|\cdot\|'$ denote the norms defined by Γ_1 and Γ_1', respectively. Clearly, $\|g\|' \leq \|g\|$, $\forall\, g \in \Gamma$, and the assertion follows. □

Remark that, for a fixed integer $N \geq 1$, Γ_N is a symmetric generating set and every word in elements of Γ_N of length $\leq k$ is a word in elements of Γ_1 of length $\leq Nk$. Since $\Gamma_1 \subseteq \Gamma_N$, we conclude that

$$G(k) \leq G_N(k) \leq G(Nk), \quad \forall\, k \geq 0,$$

where G_N is the growth function associated to Γ_N. That is, $\mathrm{gr}(G) = \mathrm{gr}(G_N)$.

Lemma 12.2.18. *The growth type of the growth function of a finitely generated group Γ does not depend on the choice of symmetric generating set. This is called the growth type of the group and is denoted by* $\mathrm{gr}(\Gamma)$.

Proof. For a large enough integer $N \geq 1$, $\Gamma_1' \subseteq \Gamma_N$. By Lemma 12.2.17 and the above remark, $\mathrm{gr}(G') \leq \mathrm{gr}(G)$. By symmetry, $\mathrm{gr}(G') = \mathrm{gr}(G)$. □

Example 12.2.19. A finite group has polynomial growth of degree 0. Indeed, the growth function reaches a maximum value and then becomes constant.

Example 12.2.20. Consider $\Gamma = \mathbb{Z}^n$, the free abelian group on n generators $e_1, \ldots, e_n$. Take

$$\Gamma_1 = \{0, \pm e_i\}_{i=1}^n,$$

a symmetric generating set containing the identity $0 \in \Gamma$ and having cardinality $2n+1$. An easy estimate gives

$$|\Gamma_q| \leq (2q+1)^n \leq |\Gamma_{nq}|,$$

from which we conclude that $\mathbb{Z}^n$ has exactly polynomial growth of degree n.

Lemma 12.2.21. *A finitely generated group Γ has* $\mathrm{gr}(\Gamma) \leq \mathrm{gr}(\exp)$.

Proof. Indeed, let $\Gamma_1 = \{x_1, x_2, \ldots, x_N\}$ be a symmetric generating set. The number of formal words $x_{i_1} x_{i_2} \cdots x_{i_k}$ of norm k in these generators is N^k; so

$$G(k) = |\Gamma_k| \leq 1 + N + N^2 + \cdots + N^k \leq (N+1)^k.$$

Thus, $\mathrm{gr}(G) \leq \mathrm{gr}(\exp_{N+1})$. □

Example 12.2.22. The free group $\Gamma = F_N$ on $N \geq 2$ letters $x_1, x_2, \ldots, x_N$ has exponential growth. Indeed, take $\Gamma_1 = \{e, x_i^{\pm 1}\}_{i=1}^N$ and remark that the number of distinct elements of Γ defined by words of the form $x_{i_1}\cdots x_{i_k}$ is N^k. Therefore,

$$G(k) = |\Gamma_k| \geq 1 + N + N^2 + \cdots + N^k > N^k$$

and, by the above lemma, $\mathrm{gr}(\Gamma)$ is exactly the exponential type.

Lemma 12.2.23. *If Γ is a finitely generated group and $K \subseteq \Gamma$ is a normal subgroup, then* $\mathrm{gr}(\Gamma/K) \leq \mathrm{gr}(\Gamma)$.

Proof. If $\Gamma_1 = \{h_i\}_{i=1}^N$ is a symmetric generating set for Γ, then $\{h_iK\}_{i=1}^N$ is a symmetric generating set Γ_1' for $\Gamma' = H/K$. But the natural projection $\pi : \Gamma \to \Gamma/K$ carries Γ_k onto Γ_k'; so $|\Gamma_k'| \leq |\Gamma_k|$, $\forall\, k \in \mathbb{Z}^+$. $\square$

Example 12.2.24. As is well known, the fundamental group of a compact manifold M is finitely presented; so $\mathrm{gr}(\pi_1(M))$ is defined. Let Σ_g be the compact surface (without boundary) of genus $g \geq 2$. Let M_g be the "solid g–holed torus", remarking that $\partial M_g = \Sigma_g$. Since M_g has the homotopy type of a bouquet of $g \geq 2$ circles, $\pi_1(M_g) = F_g$ is the free group on g letters, hence grows exponentially. The standard set of free generators consists of the images of loops on Σ_g under the inclusion map $\Sigma_g \hookrightarrow M_g$. That is, this inclusion induces a surjection of fundamental groups with kernel $K \subset \pi_1(\Sigma_g)$ a normal subgroup. By Lemma 12.2.21, Example 12.2.22 and Lemma 12.2.23, $\pi_1(\Sigma_g)$ has exponential growth type.

Exercise 12.2.25. Let $\pi : \widetilde{M} \to M$ be a regular covering of a compact, connected manifold M, Γ the group of covering transformations. Then, relative to the Riemannian metric on $\widetilde{M}$ lifted by π from any metric on M, prove that $\mathrm{gr}(\widetilde{M}) = \mathrm{gr}(\Gamma)$.

Example 12.2.26. Let $\pi : \mathbf{H} \to \Sigma_g$ be the universal cover of the g–holed torus, $g \geq 2$. The group of covering transformations is $\Gamma = \pi_1(\Sigma_g)$. Using the metric on Σ_g of constant curvature -1, we identify $\mathbf{H}$ with the hyperbolic plane. By Example 12.2.24 and Exercise 12.2.25, we obtain a combinatorial proof that the hyperbolic plane has exponential growth. Similarly, using the universal cover $\pi : \mathbb{R}^n \to T^n$ and the fact that $\pi_1(T^n) = \mathbb{Z}^n$, together with Exercise 12.2.25 and Example 12.2.20, we get a combinatorial proof that Euclidean n–space has exactly polynomial growth of degree n.

Let Γ be a group with finite, symmetric generating set Γ_1, and suppose that there is a group action

$$\Gamma \times X \to X$$

of Γ on the set X. If $x \in X$, we can define the growth function of the orbit $\Gamma(x)$ by

$$G : \mathbb{Z}^+ \to \mathbb{R}^+,$$
$$G(k) = |\Gamma_k(x)|.$$

As for the growth function of Γ itself, one can check the following.

Lemma 12.2.27. *The growth type of the growth function of an orbit $\Gamma(x)$ of a group action $\Gamma \times X \to X$ does not depend on the choice of finite, symmetric generating set of Γ nor on the choice of $x \in \Gamma(x)$. This is called the growth type of the orbit and is denoted by* $\mathrm{gr}(\Gamma(x))$.

The only new element here is independence of the choice of $x \in \Gamma(x)$. This can safely be left to the reader.

The following is immediate.

Lemma 12.2.28. *If Γ is a finitely generated group, $\Gamma \times X \to X$ a group action, and $x \in X$, then* $\mathrm{gr}(\Gamma(x)) \leq \mathrm{gr}(\Gamma)$.

Example 12.2.29. If $H \subset \Gamma$ is a subgroup, not necessarily normal, the set of cosets Γ/H (called a homogeneous space of Γ) has a well–defined growth type. Indeed, the natural action

$$\Gamma \times \Gamma/H \to \Gamma/H$$

is transitive; so Γ/H is an orbit. When H is a normal subgroup, this coincides with the growth type of the finitely generated group Γ/H.

This example is particularly important for computing the growth of a covering

$$\widehat{\pi} : \widehat{M} \to M$$

of a compact, connected manifold when the covering is not regular. In this case, let

$$H \subset \Gamma = \pi_1(M)$$

be the subgroup that is the image of the injective homomorphism

$$\widehat{\pi}_* : \pi_1(\widehat{M}) \to \pi_1(M).$$

Because the covering is not regular, this subgroup is not normal.

Exercise 12.2.30. Suppose that $\widehat{\pi} : \widehat{M} \to M$ is as above and prove that $\mathrm{gr}(\widehat{M}) = \mathrm{gr}(\Gamma/H)$.

Exercise 12.2.31. Let $\pi : \widehat{N} \to N$ be a finite covering, where N is a complete Riemannian manifold. Give $\widehat{N}$ the lifted metric and prove that $\mathrm{gr}(N) = \mathrm{gr}(\widehat{N})$.

Corollary 12.2.32. *Let $(M, \mathcal{F}, \pi)$ be a C^2–foliated bundle with fiber a compact metric space F, base a compact manifold B, and total holonomy group $\Gamma_{\mathcal{F}}$. Let L be a leaf of $\mathcal{F}$ corresponding to the orbit $\Gamma_{\mathcal{F}}(x)$, $x \in F$. Then* $\operatorname{gr}(L) = \operatorname{gr}(\Gamma_{\mathcal{F}}(x))$.

Proof. Put a leafwise Riemannian metric on M by lifting a Riemannian metric on $T(B)$ to $T(\mathcal{F})$ via the projection π. The total holonomy homomorphism $h : \pi_1(B) \to \operatorname{Homeo}(F)$ has image $\Gamma_{\mathcal{F}}$ and $\pi_1(B)$ is finitely generated, hence $\Gamma_{\mathcal{F}}$ is finitely generated. Let $K \subset \Gamma_{\mathcal{F}}$ be the stabilizer of x. Then h induces a bijection

$$\overline{h} : \pi_1(B)/p_*(\pi_1(L)) \to \Gamma_{\mathcal{F}}/K \cong \Gamma_{\mathcal{F}}(x)$$

and $\Gamma_{\mathcal{F}}(x)$ can also be viewed as an orbit of the action

$$\pi_1(B) \times F \to F$$

induced by h. It is elementary that its growth type as a $\Gamma_{\mathcal{F}}$–orbit is the same as its growth type as a $\pi_1(B)$–orbit; so the assertion follows by Exercise 12.2.30. □

12.2.C. Pseudogroups and growth of leaves. The holonomy pseudogroup of a general foliated space $(M, \mathcal{F})$ was devised to replace the total holonomy group of a foliated bundle. We will generalize Corollary 12.2.32 by showing that, when M is compact, the growth of a leaf L of $\mathcal{F}$ is equal to the growth of the corresponding holonomy orbit.

Let Γ be a finitely generated pseudogroup of homeomorphisms on a topological space X and let $\Gamma_1 = \{h_i\}_{i=1}^m$ be a finite, symmetric generating set (as always, containing id_X). If $x, y \in X$ lie in a common Γ–orbit, we can write

$$y = h_{i_1} \circ h_{i_2} \circ \cdots \circ h_{i_k}(x),$$

where all $h_{i_j} \in \Gamma_1$. We define the *distance* $d(x, y)$ to be the smallest integer k for which such an expression exists. Remark that this is a metric on the orbit and consider the "k–ball" about x

$$\Gamma_k(x) = \{z \in \Gamma(x) \mid d(x, z) \le k\}, \quad \forall\, k \ge 0.$$

Definition 12.2.33. The growth function

$$H_x : \mathbb{Z}^+ \to \mathbb{R}^+$$

of the orbit $\Gamma(x)$ is

$$H_x : \mathbb{Z}^+ \to \mathbb{R}^+,$$
$$H_x(r) = |\Gamma_r(x)|,$$

where the absolute value symbol denotes cardinality. The growth type $\operatorname{gr}(\Gamma(x))$ of the orbit is $\operatorname{gr}(H_x)$.

Evidently, this generalizes the growth function and growth type of an orbit of a group action.

Lemma 12.2.34. *If* Γ_1 *and* Γ_1' *are two finite, symmetric generating sets for the pseudogroup* Γ, *and if* H_x *and* H_y' *are the corresponding growth functions of an orbit* $\Gamma(x) = \Gamma(y)$, *then* $\operatorname{gr}(H_x) = \operatorname{gr}(H_y')$.

The elementary verification is analogous to that in the case of a group action and can safely be left to the reader.

Consider the case in which $\Gamma = \Gamma_{\mathcal{U}}$ is the holonomy pseudogroup of a regular foliated atlas

$$\mathcal{U} = \{(U_\alpha, x_\alpha, y_\alpha)\}_{\alpha \in \mathfrak{A}}$$

for the compact, C^1–foliated space $(M, \mathcal{F})$. Since the atlas is finite, the pseudogroup is finitely generated. The leaves of $\mathcal{F}$ correspond one–to–one to the $\Gamma_{\mathcal{U}}$–orbits in the transverse space Z. We choose the coordinates so that each plaque $P \times \{y_\alpha\} \subset \overline{U}_\alpha$ is coordinatized by $x_\alpha^1, \dots, x_\alpha^p$ as the open unit ball $P \subset \mathbb{R}^p$ about the origin and so that no overlapping plaques have the same origin. Let $P(\epsilon) \subset P$ be the ϵ–ball about the origin, where $\epsilon \in (0,1)$ is chosen so that the ϵ–plaques $P(\epsilon) \times \{y_\alpha\}$ and $P(\epsilon) \times \{y_\beta\}$ never intersect if $\alpha \neq \beta$. We fix a (leafwise) Riemannian metric on M. By the continuity of this metric, the finiteness and regularity of $\mathcal{U}$ and the compactness of M, we have the following.

(1) There is a global lower bound $a > 0$ to $\operatorname{vol}_p(P(\epsilon) \times \{y_\alpha\})$, for all ϵ–plaques in U_α and all $\alpha \in \mathfrak{A}$.
(2) There is a global upper bound $b \in \mathbb{Z}^+$ to $\operatorname{vol}_p(P \times \{y_\alpha\})$, for all plaques in U_α and all $\alpha \in \mathfrak{A}$.
(3) There is a global upper bound $B \in \mathbb{Z}^+$ to $\operatorname{diam}(P \times \{y_\alpha\})$, for all plaques in U_α and all $\alpha \in \mathfrak{A}$.
(4) There is a Lebesgue number $\delta > 0$ (relative to the Riemannian distance function ρ along leaves) for the coverings of the leaves by the plaques of $\mathcal{U}$.

This data will be used to prove the following.

Proposition 12.2.35. *If* L *is the leaf of* $\mathcal{F}$ *containing the plaque* $P_0 \in Z$, *then*

$$\operatorname{gr}(L) = \operatorname{gr}(\Gamma(P_0)),$$

where $\Gamma = \Gamma_{\mathcal{U}}$ *is the holonomy pseudogroup.*

Proof. Fix a plaque $P_0 \in Z$, with corresponding ϵ–plaque $P_0(\epsilon) \subset P_0$ and $x_0 \in P_0(\epsilon)$. We must show that the associated Riemannian growth function G_{x_0} of L and the growth function H_{P_0} of the holonomy orbit $\Gamma(P_0)$ mutually dominate.

Let $P_0, P_1, \ldots, P_k$ be a plaque chain (meaning that successive plaques intersect), $0 \le k \le r$, and let $y \in P_k$. By property (3), $\rho(x_0, y) \le Br$. Choosing $A \in \mathbb{Z}^+$ such that $A \ge 1/a$ and appealing to property (1), we see that

$$AG_{x_0}(Br) \ge H_{P_0}(r).$$

That is, $G_{x_0} \succeq H_{P_0}$.

For the reverse dominance, fix $\eta \in (0, \delta)$, with δ as in property (4), and let L be the leaf through x_0. Let $r \in \mathbb{Z}^+$ and $y \in L$, with $\rho(x_0, y) \le r$. Since L is complete in the Riemannian metric (Exercise 12.1.3), there is a minimal geodesic σ in L from x_0 to y. Choose $x_0, x_1, \ldots, x_k = y$ along σ such that

$$\eta < \rho(x_i, x_{i-1}) < \delta, \quad 1 \le i \le k.$$

Then there is a plaque P_i with $\{x_i, x_{i-1}\} \subset P_i$, $1 \le i \le k$. Thus,

$$P_{i-1} \cap P_i \ne \emptyset$$

and $y \in P_k$. Also,

$$r \ge \rho(x_0, y) = \sum_{i=1}^{k} \rho(x_i, x_{i-1}) > k\eta.$$

Take $D \in \mathbb{Z}^+$ so that $D > 1/\eta$; hence $k < Dr$. By property (2),

$$bH_{P_0}(Dr) = b|\{P_k \mid k \le Dr\}| \ge b|\{P_k \mid k < Dr\}| \ge G_{x_0}(r).$$

That is, $H_{P_0} \succeq G_{x_0}$. □

Corollary 12.2.36. *The growth type of a $\Gamma_{\mathfrak{U}}$-orbit corresponding to a leaf L depends only on L and not on the choice of regular foliated atlas $\mathfrak{U}$. Likewise,* $\mathrm{gr}(L)$ *does not depend on the choice of Riemannian metric on M.*

Of course, the last assertion in the corollary is already known by Exercise 12.2.3.

Corollary 12.2.37. *In a compact foliated space,* $\mathrm{gr}(\exp)$ *dominates the growth type of every leaf.*

Proof. Fix a regular foliated atlas $\mathfrak{U}$ and let N be the number of foliated charts in $\mathfrak{U}$. For any plaque $P_0 \in Z_{\alpha_0}$, we can estimate the growth function by

$$\begin{aligned} H_{P_0}(r) &= |\{P_0, \gamma_{\alpha_m \alpha_{m-1}} \circ \cdots \circ \gamma_{\alpha_1 \alpha_0}(P_0)\}_{1 \le m \le r}| \\ &\le 1 + |\{(\alpha_1, \ldots, \alpha_m)\}_{1 \le m \le r}| \\ &= 1 + N + N^2 + \cdots + N^r \\ &\le (1 + N)^r. \end{aligned}$$

That is, $\exp_{1+N} \succeq H_{P_0}$. □

Corollary 12.2.38. *Let $(M,\mathfrak{F})$ be a compact, transversely–orientable, foliated manifold of codimension one and of class C^2. Let $U \subseteq M$ be an open, $\mathfrak{F}$–saturated set without holonomy. Then all leaves in U have the same growth type.*

Proof. Fix a transverse, one–dimensional foliation $\mathcal{L}$ as usual, requiring that foliated atlases for $\mathfrak{F}$ be biregular for $\mathfrak{F}$ and $\mathcal{L}$. The Sacksteder flow $\Phi_t : \widehat{U} \to \widehat{U}$ has the leaves of $\mathcal{L}|U$ as nondegenerate flow lines, permutes transitively the set of leaves of $\mathfrak{F}|U$ and fixes $\partial\widehat{U}$ pointwise (Theorem 9.2.1). This extends to a C^0 flow on M itself, again denoted by Φ_t, which fixes $M \smallsetminus U$ pointwise. Thus, Φ_t carries a biregular atlas $\mathfrak{U}$ to a biregular atlas $\mathfrak{U}' = \Phi_t(\mathfrak{U})$, carrying $\mathfrak{F}$–plaques to $\mathfrak{F}$–plaques. If L and L' are two leaves of $\mathfrak{F}|U$, choose $t \in \mathbb{R}$ so that $\Phi_t(L) = L'$. Thus, $\operatorname{gr}(L)$, computed via $\mathfrak{U}$, and $\operatorname{gr}(L')$, computed via $\mathfrak{U}'$, are identical. □

Remark. Examples show that this common growth type can be any one of uncountably many possibilities, including the exponential type [**20, 19**].

Recall the definition of a resilient leaf (Definition 8.2.4) for foliations of codimenson one. This can be extended to foliated spaces as follows.

Definition 12.2.39. A loop σ on a leaf F, based at $x_0 \in F$, is contracting if a holonomy transformation $h_\sigma : D \to D$, associated to σ and defined on a suitable compact, transverse metric disk D centered at x_0, is such that $\bigcap_{n=1}^\infty h_\sigma^n(D) = \{x_0\}$. The leaf F of $\mathfrak{F}$ is resilient if it has such a contracting loop and $F \cap (\operatorname{int} D \smallsetminus \{x_0\}) \neq \emptyset$.

Thus, for a suitable power $f_0 = h_\sigma^n$, there is $x_1 \in \operatorname{int} D \smallsetminus f_0(D)$ in the same leaf F as x_0. This provides another holonomy transformation g such that $g(x_0) = x_1$. We make n larger, if necessary, so that

$$f_0(D) = B_0 \subset \operatorname{dom} g,$$
$$B_0 \cap g(B_0) = \emptyset.$$

Finally, set $f_1 = g \circ f_0 : D \to B_1 = g(B_0)$. Thus, f_0 and f_1 are both defined on D with $f_0(D) \cap f_1(D) = \emptyset$. Note also that

$$f_0^2(B_1) \cap f_0(B_1) = \emptyset.$$

Corollary 12.2.40. *Let $(M,\mathfrak{F})$ be a compact, foliated space and let F be a resilient leaf of $\mathfrak{F}$. Then the union of the leaves that are captured by the holonomy contraction f of F is an open, $\mathfrak{F}$–saturated neighborhood of F consisting entirely of leaves having exponential growth.*

Proof. Let $z \in B_1$ and consider equalities of the form

$$f_0^{i_s} \circ f_1 \circ f_0^{i_{s-1}} \circ f_1 \circ \ldots \circ f_0^{i_1} \circ f_1(z)$$
$$= f_0^{j_r} \circ f_1 \circ f_0^{j_{r-1}} \circ f_1 \circ \ldots \circ f_0^{j_1} \circ f_1(z),$$

where all $i_k, j_k \in \{1,2\}$. Since $f_0^2(B_1) \cap f_0(B_1) = \emptyset$, it follows that $i_s = j_r$. Cancelling these terms and the first f_1, we reduce both r and s by 1. By finite repetition, we get to the point that $z \in B_1$ appears alone on one side of the equation, hence on both sides of the equation. That is, $r = s$ and $i_k = j_k$, $1 \le k \le r$. Now consider the subpseudogroup $\Gamma \subset \Gamma_{\mathcal{U}}$ generated by $\{f_0, f_1\}$. The number of distinct points of the form

$$w = f_0^{i_r} \circ f_1 \circ f_0^{i_{r-1}} \circ f_1 \circ \ldots \circ f_0^{i_1} \circ f_1(z),$$

for fixed $z \in B_1$, is 2^r, while the distance (relative to the symmetric generating set $\{f_0^{\pm 1}, f_1^{\pm 1}\}$) of w from z is at most $3r$. Thus, the growth function of $\Gamma(z)$ must dominate $\exp_2$. Extending the set $\{f_0^{\pm 1}, f_1^{\pm 1}\}$ to a finite, symmetric generating set for $\Gamma_{\mathcal{U}}$, we see that $\operatorname{gr}(\Gamma_{\mathcal{U}}(z)) \ge \operatorname{gr}(\exp)$, hence is exactly exponential by Corollary 12.2.37. The $\mathcal{F}$–saturation of B_1 is an open neighborhood of F consisting of leaves with exponential growth. Finally, a leaf is captured by the holonomy contraction if and only if it meets B_1. □

Corollary 12.2.41. *Let L be a proper leaf in a compact, transversely orientable, C^2–foliated manifold $(M, \mathcal{F})$ of codimension one. Then L has growth type*

(a) *exactly polynomial of degree k if and only if L is a leaf at depth k;*

(b) *exponential, otherwise. In this case, $\overline{L}$ contains an exceptional local minimal set.*

Indeed, if $\overline{L}$ contains an exceptional local minimal set, Theorem 8.2.1 implies that L is asymptotic to a resilient leaf. By Corollary 12.2.40, L has exponential growth. In any event, L is at finite level; so $\overline{L}$ is a finite union of local minimal sets (Corollary 8.3.12). None of these can be open; so the only possibilities are proper leaves and exceptional local minimal sets. If none are exceptional, $\overline{L}$ is a finite union of proper leaves, hence is at finite depth. The proof that a leaf at depth k has exactly polynomial growth of degree k is a bit more challenging.

Exercise 12.2.42. You are to prove that, for C^2 foliations, a leaf at depth k has exactly polynomial growth of degree k (this is false for C^1 foliations). Proceed by induction on k, the case $k = 0$ being immediate. Assume the claim for all leaves at depth $\le k$ and prove it for leaves L at depth $k+1$ as follows.

1. Let $L = A \cup B^1 \cup \cdots \cup B^q$ be the "octopus decomposition" of Theorem 8.4.6. If g_i is a growth function for the complete Riemannian manifold B^i, $1 \le i \le q$, set $g = g_1 + \cdots + g_q$ and show that $\mathrm{gr}(L) = \mathrm{gr}(g)$.
2. For $g = g_1 + \cdots + g_q$ as above, if each g_i has exactly polynomial growth, show that $\mathrm{gr}(L)$ is exactly polynomial of degree the maximal degree of $\mathrm{gr}(g_i)$, $1 \le i \le q$.
3. Each B^i spirals on a leaf L^i. If $\mathrm{gr}(L^i)$ is exactly polynomial of degree r, show that $\mathrm{gr}(g_i)$ is exactly polynomial of degree $r+1$. By the inductive hypothesis and properties (3) and (4) in Theorem 8.4.6, the conclusion follows.

Corollary 12.2.43. *Let L be a leaf at infinite level in a compact, transversely–orientable, C^2–foliated manifold $(M, \mathcal{F})$ of codimension one. Then the growth type of L dominates that of every polynomial.*

Exercise 12.2.44. For the proof of Corollary 12.2.43, proceed as follows.

1. Let $U \subset M$ be a foliated product, $L \subset U$ a leaf. If F is a border leaf of U, show that $\mathrm{gr}(L) \ge \mathrm{gr}(F)$.
2. Now use the previous exercise and Lemma 8.3.22 to show that $\mathrm{gr}(L)$ dominates all polynomial types.

Finally, in light of Duminy's vanishing theorem (Example 8.3.18), Corollary 12.2.40 gives the following.

Corollary 12.2.45. *Let the compact, smoothly foliated manifold $(M, \mathcal{F})$ be transversely orientable of codimension one. If no leaf has exponential growth, then* $\mathrm{gv}(\mathcal{F}) = 0$.

12.3. Growth and Holonomy–Invariant Measures

Subexponential growth of leaves or other closely related properties allow averaging procedures that lead to transverse, holonomy–invariant measures (Definition 11.5.2). Our goal is to prove the following existence theorem of J. F. Plante [**107**].

Theorem 12.3.1. *Let $(M, \mathcal{F})$ be a compact foliated space. If L is a leaf of $\mathcal{F}$ having subexponential growth, there is a nontrivial, holonomy–invariant measure μ for $\mathcal{F}$ such that* $\operatorname{supp}\mu \subseteq \overline{L}$.

This will be a consequence of the following.

Theorem 12.3.2. *Let Γ be a pseudogroup on a compact metric space X and suppose there is a finite, symmetric generating set Γ_1 with the property that, whenever $g \in \Gamma_1 \smallsetminus \{\mathrm{id}_X\}$, then $(\operatorname{dom} g) \cap (\operatorname{im} g) = \emptyset$. If $x \in X$ is a* *point such that the orbit $\Gamma(x)$ has nonexponential growth, then there exists a Γ–invariant probability measure μ on X with* $\operatorname{supp}\mu \subseteq \overline{\Gamma(x)}$.

Proof of Theorem 12.3.1 using Theorem 12.3.2. Fix a regular foliated atlas $\mathfrak{U} = \{U_i\}_{i=1}^N$, an expanded atlas $\mathfrak{U}' = \{U_i'\}_{i=1}^N$ (see page 280) and let Z denote the disjoint union of the transversals Z_i, $1 \le i \le N$. This is an open subset of the larger transverse model space Z' (determined by $\mathfrak{U}'$) in which the closure $\overline{Z}$ is compact. The holonomy generators γ_{ij}, $i \ne j$, associated to $\mathfrak{U}$, together with id, extend uniquely to define a finite, symmetric generating system for a pseudogroup $\overline{\Gamma}$ on $\overline{Z}$. This generating set evidently satisfies the hypothesis of Theorem 12.3.2. The existence of a leaf L with nonexponential growth implies that there is an orbit $\overline{\Gamma}(x) = L \cap \overline{Z}$ with nonexponential growth. Indeed, this orbit is contained in the orbit $\Gamma_{\mathfrak{U}'}(x)$ of the pseudogroup determined by $\mathfrak{U}'$ and that orbit has nonexponential growth. By Theorem 12.3.2, we obtain a $\overline{\Gamma}$–invariant probability measure on $\overline{Z}$, supported in the closure $\overline{L} \cap \overline{Z}$ of the orbit $\overline{\Gamma}(x)$. By $\overline{\Gamma}$–invariance, $\operatorname{supp}\mu$ is a union of $\overline{\Gamma}$–orbits, hence meets Z and, in fact, meets Z in a subset of $\overline{L} \cap Z$. Finally, the invariance of μ under $\overline{\Gamma}$ clearly implies invariance of $\mu|Z$ under Γ. □

Thus, we must prove Theorem 12.3.2. Let Γ, Γ_1, X and $x \in X$ satisfy the hypotheses of that theorem. The following is the "Fölner condition" referred to in Section 11.5.B.

Lemma 12.3.3. *There is a sequence $n_k \uparrow \infty$ of positive integers such that*

$$\lim_{k\to\infty} \frac{|\Gamma_{n_k+1}(x) \smallsetminus \Gamma_{n_k-1}(x)|}{|\Gamma_{n_k}(x)|} = 0.$$

Proof. Otherwise, there is an integer $K > 0$ and a real number $a > 0$ such that

$$k \ge 2K \Rightarrow \frac{|\Gamma_{k+1}(x) \smallsetminus \Gamma_{k-1}(x)|}{|\Gamma_k(x)|} > a.$$

Thus,

$$|\Gamma_{k+1}(x)| > a|\Gamma_k(x)| + |\Gamma_{k-1}(x)| \ge (a+1)|\Gamma_{k-1}(x)|,$$

for all $k \ge 2K$. Iterating this estimate, we see that $|\Gamma_{k+1}(x)|$ has the same growth type as $(a+1)^{k/2}$, contradicting the hypothesis of Theorem 12.3.2. □

In the following, we let

$$A \triangle B = A \cup B \smallsetminus A \cap B,$$

the symmetric difference of sets A and B. We also place the following definition.

Definition 12.3.4. Given $g \in \Gamma$ and $f \in C(X)$ with $\operatorname{supp} f \subset \operatorname{im} g$, define $f_g \in C(X)$ by

$$f_g(y) = \begin{cases} f(g(y)) & \text{if } y \in \operatorname{dom} g, \\ 0 & \text{otherwise.} \end{cases}$$

Lemma 12.3.5. *If $g \in \Gamma_1$ and f are as in the definition and $n \geq 1$ is an integer, then*

$$(*) \qquad \sum_{y \in \Gamma_n(x)} (f(y) - f_g(y)) = \sum_{z \in \Gamma_n(x) \triangle g\Gamma_n(x)} \epsilon_z f(z),$$

where

$$\epsilon_z = \begin{cases} 1, & z \in \Gamma_n(x), \\ -1, & z \in g\Gamma_n(x). \end{cases}$$

Proof. If $g = \mathrm{id}_X$, the assertion is trivial (with the convention that an empty sum is 0). By the condition in Theorem 12.3.2 on $g \in \Gamma_1 \smallsetminus \{\mathrm{id}_X\}$,

$$y \in \operatorname{dom} g \Rightarrow y \notin \operatorname{im} g \Rightarrow f(y) = 0.$$

Furthermore,

$$y \notin \operatorname{dom} g \Rightarrow f_g(y) = 0.$$

So the left–hand side of $(*)$ is equal to

$$\sum_{y \in \Gamma_n(x) \cap \operatorname{dom} g} -f(g(y)) \quad + \sum_{y \in \Gamma_n(x) \smallsetminus \operatorname{dom} g} f(y).$$

Since $z \in g\Gamma_n(x)$ if and only if there is a unique $y \in \Gamma_n(x) \cap \operatorname{dom} g$ such that $z = g(y)$, we see that

$$\sum_{y \in \Gamma_n(x) \cap \operatorname{dom} g} -f(g(y)) = \sum_{z \in g\Gamma_n(x)} -f(z).$$

Since $y \in \operatorname{dom} g$ implies that $y \notin \operatorname{im} g$, hence that $f(y) = 0$, we see that

$$\sum_{y \in \Gamma_n(x) \smallsetminus \operatorname{dom} g} f(y) = \sum_{y \in \Gamma_n(x)} f(y) = \sum_{z \in \Gamma_n(x)} f(z).$$

So

$$\sum_{y \in \Gamma_n(x)} (f(y) - f_g(y)) = \sum_{z \in \Gamma_n(x) \bigsqcup g\Gamma_n(x)} \epsilon_z f(z),$$

where "$\bigsqcup$" denotes disjoint union. By the definition of ϵ_z, the two terms of this sum contributed by $z \in \Gamma_n(x) \cap g\Gamma_n(x)$ cancel and $(*)$ follows. $\square$

In order to prove Theorem 12.3.2, we will construct a nonnegative, Γ–invariant, continuous linear functional

$$I : C(X) \to \mathbb{R}$$

such that $I(\mathbf{1}) = 1$ (we say that I is normalized). Here, $C(X)$ is the space of continuous, real–valued functions on X with the uniform norm and $\mathbf{1}$ denotes the function that is constantly equal to 1. By the Riesz representation theorem [**112**, page 42], I will be the integral associated to a Γ–invariant probability measure on X. We should state carefully what it means for I to be Γ–invariant.

Definition 12.3.6. The linear functional $I : C(X) \to \mathbb{R}$ is Γ–invariant if, for each choice of $g \in \Gamma_1$ and $f \in C(X)$ as above, $I(f_g) = I(f)$.

It follows, of course, that $I(f_g) = I(f)$ for arbitrary $g \in \Gamma$ and every $f \in C(X)$ with $\operatorname{supp} f \subset \operatorname{dom} g$. Furthermore, if I is the integral of a measure μ, one sees that μ is Γ–invariant.

Proof of Theorem 12.3.2. In order to construct the functional I, fix the choice of the sequence $\{n_k\}_{k=1}^{\infty}$ as in Lemma 12.3.3 and, for each integer $k \geq 1$, define

$$I_k(f) = \frac{1}{|\Gamma_{n_k}(x)|} \sum_{y \in \Gamma_{n_k}(x)} f(y), \quad \forall\, f \in C(X).$$

These functionals are normalized and, passing to a subsequence if necessary, we assume that they converge in the weak topology to a normalized functional I. (For this, remark that $C(X)$ contains a countable, dense subset and that the sequence of functionals I_k is equicontinuous. A diagonal process gives the assertion.) Since

$$z \in \Gamma_{n-1}(x) \text{ and } f(z) \neq 0 \Rightarrow z \in \Gamma_{n-1}(x) \cap \operatorname{im} g \Rightarrow z = gg^{-1}(z) \in g\Gamma_n(x),$$

we see that

$$\Gamma_n(x) \triangle g\Gamma_n(x) \subseteq \Gamma_{n+1}(x) \smallsetminus \{z \in \Gamma_{n-1}(x) \mid f(z) \neq 0\}.$$

By Lemma 12.3.5, we conclude that

$$\begin{aligned} \left| \sum_{y \in \Gamma_n(x)} (f(y) - f_g(y) \right| &\leq \sum_{z \in \Gamma_{n+1}(x) \smallsetminus \Gamma_{n-1}(x)} |f(z)| \\ &\leq \sup_{z \in X} |f(z)| |\Gamma_{n+1}(x) \smallsetminus \Gamma_{n-1}(x)|. \end{aligned}$$

Finally, by Lemma 12.3.3,

$$|I(f) - I(f_g)| = \lim_{k \to \infty} |I_k(f) - I_k(f_g)| = 0.$$

□

Exercise 12.3.7. Let $(M, \mathfrak{F})$ be a compact foliated space. The *Poincaré recurrence set* $P(\mathfrak{F})$ is defined to be the union of the supports of all holonomy–invariant probability measures. Prove that $P(\mathfrak{F})$ is a compact foliated space.

Exercise 12.3.8. If Γ is a *countably* generated pseudogroup on the compact space X, we say that an orbit $\Gamma(x)$ has nonexponential growth if Γ can be written as an increasing union of finitely generated pseudogroups

$$\Gamma = \bigcup_{k=1}^{\infty} \Gamma^k$$

such that $\Gamma^k(x)$ has nonexponential growth, $\forall\, k \geq 1$. In this case, prove that there is a Γ–invariant probability measure on X.

Exercise 12.3.9. Let $(M, \mathfrak{F})$ be a compact, C^2–foliated manifold of codimension one and without holonomy. Prove that the leaves all have the same exactly polynomial growth type of degree $p \leq b_1(M) - 1$, where $b_1(M)$ is the first Betti number of M. (Hint. This is trivial if $\mathfrak{F}$ fibers M over the circle. In the dense leaved case, use the total holonomy group as defined in Proposition 9.2.10.)

Exercise 12.3.10. Let $(M, \mathfrak{F})$ be a compact, C^2–foliated manifold of codimension one, where $\mathfrak{F}$ is transversely orientable, and let μ be a nontrivial holonomy–invariant measure. Using the theory of levels and the previous exercise, prove that the growth types of the leaves in $\operatorname{supp}\mu$ are exactly polynomial of degrees $p \leq \max\{0, b_1(M) - 1\}$.

Remark. A delicate argument of Plante [**107**] proves, for C^0 foliations, that $\operatorname{supp}\mu$ consists of leaves having weakly polynomial growth types of degrees $p \leq \max\{0, b_1(M) - 1\}$.

Exercise 12.3.11. Let $(M, \mathfrak{F})$ be a compact, orientable and transversely–orientable, C^2–foliated manifold of codimension one with $\partial M = \partial_\tau M \neq \emptyset$. If $\mathfrak{F}|\operatorname{int} M$ has no holonomy, prove that every leaf has exactly polynomial growth.

Exercise 12.3.12. Let $(M, \mathfrak{F})$ be a compact, orientable and transversely–orientable, C^2–foliated manifold of codimension one with $\partial_\pitchfork M = \emptyset$. If a leaf of $\mathfrak{F}$ has weakly polynomial growth that is not exactly polynomial, prove that this growth type strictly dominates the quadratic type. (Hint. Show that such leaves lie in open local minimal sets without holonomy at level ≥ 2.)

Exercise 12.3.13. Let $(M, \mathfrak{F})$ be a compact, oriented and transversely–oriented, foliated manifold of codimension q. Suppose that $N \subset M$ is a compact, connected, oriented q–manifold without boundary, transverse to

$\mathcal{F}$ and meeting every leaf (a global section). Prove that either $[N] \neq 0$ in $H_q(M;\mathbb{R})$ or every leaf has exponential growth. (Hint: There is a closed, $(n-q)$–form ω, compactly supported in a normal neighborhood of N and having $\int_D \omega = 1$ on each normal fiber D [**13**, pp. 61–67]. This is a form whose cohomology class is Poincaré dual to $[N]$.)

12.4. Leaves Closed at Infinity

Sullivan has introduced an interesting quasi–isometry invariant for Riemannian manifolds that is closely related to growth.

Definition 12.4.1. A complete, connected Riemannian p–manifold L is not closed at infinity if there is a bounded $(p-1)$–form η on L and a smooth, positive function $f : L \to \mathbb{R}$, uniformly bounded away from 0, such that $d\eta = f\Omega$, where Ω is the Riemannian volume form. Otherwise, L is closed at infinity.

Exercise 12.4.2. Prove that the hyperbolic plane is not closed at infinity.

Proposition 12.4.3. *If L is not closed at infinity, there is a constant $\gamma > 0$ such that the isoperimetric inequality*

$$\operatorname{vol}_p B \leq \gamma \operatorname{vol}_{p-1} \partial B$$

is valid for all compact, p–dimensional submanifolds $B \subset L$.

Proof. Indeed, let f and η be as in the definition, $f \geq c > 0$ and $\|\eta\| \leq b$. Then

$$c \operatorname{vol}_p B \leq \left| \int_B f\Omega \right| = \left| \int_{\partial B} \eta \right| \leq b \operatorname{vol}_{p-1} \partial B.$$

Take $\gamma = b/c$. □

An interpretation of this result is that a complete Riemannian manifold that is not closed at infinity satisfies a very uniform exponential growth condition. This is a stronger condition than required by exponential growth; so L could grow exponentially and still be closed at infinity. (The example in Exercise 12.2.11 has exponential growth. But it also has an averaging sequence, and hence is closed at infinity.) The following, therefore, is a generalization of Theorem 12.3.1.

Theorem 12.4.4 (Sullivan [**127**]). *Let $(M, \mathcal{F})$ be a smooth, compact foliated space. If some leaf L of $\mathcal{F}$ is closed at infinity, there is a nontrivial holonomy–invariant measure μ with* $\operatorname{supp} \mu \subseteq \overline{L}$.

Proof. Let $Y = \overline{L}$, also a compact foliated space. Thus, Lemma 11.5.9 assures us that the desired measure exists if and only if no exact p–form is transverse to $\mathcal{F}|Y$. Since $L \subseteq Y$, this is guaranteed by the assumption that L is closed at infinity. □

12.5. Dimension Two

We turn to foliated manifolds with 2–dimensional leaves. We will not consider here more general spaces foliated by surfaces. A striking property of such foliations (if they are tangentially orientable) is that a compact leaf of genus ≥ 2 is homologically nontrivial, placing a topological restriction on the n–manifold being foliated. The proof uses the Euler class of $T(\mathcal{F})$ and the fact that the Euler characteristic of such a leaf is nonzero. In this section, we will establish a theorem of Phillips and Sullivan [**101**] which extends this result to certain sorts of noncompact leaves. The key new idea is that of "average Euler characteristic (non)zero". Before defining this notion, we state the central result.

Theorem 12.5.1 (Phillips and Sullivan). *Let $(M, \mathcal{F})$ be a compact, connected, tangentially–oriented, C^∞–foliated manifold with $\partial M = \partial_\tau M$ and 2–dimensional leaves. If a leaf L of $\mathcal{F}$ has nonexponential growth and nonzero average Euler characteristic, then $H_2(M;\mathbb{R}) \neq 0$.*

Remark. Our C^∞ hypothesis is much too strong. The theorem is true for foliations of class $C^{1,0}$, but the proof is sufficiently insightful in the C^∞ case. Two proofs are given in [**101**]. The one that we will treat uses the Euler class of $T(\mathcal{F})$ and the Gauss–Bonnet theorem. The more general proof, valid for $C^{1,0}$ foliations, uses the Poincaré–Hopf theorem, which computes the Euler characteristic of a compact surface F by a formula involving the indices of the zeros of a continuous vector field on F and the rotation number of the vector field (assumed nonsingular at the boundary) on each component of ∂F.

Definition 12.5.2. Let L be a complete, connected, Riemannian 2–manifold. A *topological averaging sequence* in L is an exhaustion

$$B_1 \subset B_2 \subset \cdots \subset B_k \subset \cdots \subset L$$

of L by compact submanifolds with piecewise smooth boundary such that

1. $B_k \subset \operatorname{int} B_{k+1}$, $\forall\, k \geq 1$;
2. there is a basepoint $x \in B_1$, a sequence $\{r_k\}_{k=1}^\infty$ of radii and a constant $Q > 0$ such that $r_k \uparrow \infty$ and

$$D_x(r_k) \subset B_k \subset D_x(Qr_k), \quad \forall\, k \geq 1.$$

Definition 12.5.3. The 2–manifold L, as above, has nonzero average Euler characteristic if, for every topological averaging sequence,

$$\liminf_{k\to\infty} \frac{|\chi(B_k)|}{\operatorname{area} B_k} > 0.$$

Otherwise, L has average Euler characteristic zero.

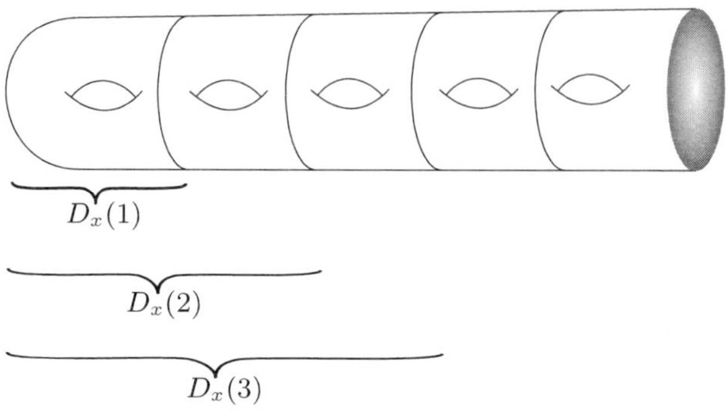

Figure 12.5.1. Jacob's ladder has average Euler characteristic nonzero.

The exact value of the "average Euler characteristic" is not defined, only whether or not it is zero.

Exercise 12.5.4. Prove that the property of having nonzero (respectively, zero) average Euler characteristic is a quasi–isometry invariant, independent of the choice of basepoint x.

Example 12.5.5. Let L be Jacob's ladder and, by uniformly bounded distortions, assume that the metric balls are as indicated in Figure 12.5.1. Clearly, $\chi(D_x(n)) = 1 - 2n$ and, if B is a compact, connected submanifold of L such that $D_x(n) \subset \operatorname{int} B$, then $\chi(B) \leq 1 - 2n$. Thus, if $\{B_k\}_{k=1}^\infty$ is a topological averaging sequence and $\{r_k\}_{k=1}^\infty$ is as in the definition, we obtain

$$|\chi(B_k)| \geq 2[r_k] - 1, \quad \forall\, k \geq 1,$$

where $[r_k]$ is the greatest integer in r_k. If Q is as in the definition, the fact that L has linear growth also gives

$$\operatorname{area} B_k \leq \operatorname{area} D_x(Qr_k) \leq A[r_k], \quad \forall\, k \geq 1,$$

for a suitable constant $A > 0$. Thus,

$$\liminf_{k\to\infty} \frac{|\chi(B_k)|}{\operatorname{area} B_k} \geq \frac{2}{A}.$$

Since the topological averaging sequence is arbitrary, Jacob's ladder has nonzero average Euler characteristic. By Theorem 12.5.1, this quasi–isometry type cannot appear, for instance, as a leaf of a foliation of S^3.

Example 12.5.6. We sketch, in some detail, a cut–and–paste construction of a C^∞ foliation of S^3 having the infinite Loch Ness monster as a leaf. Although diffeomorphic to Jacob's ladder, the "monster" has average Euler characteristic zero; so this construction is in accord with Theorem 12.5.1.

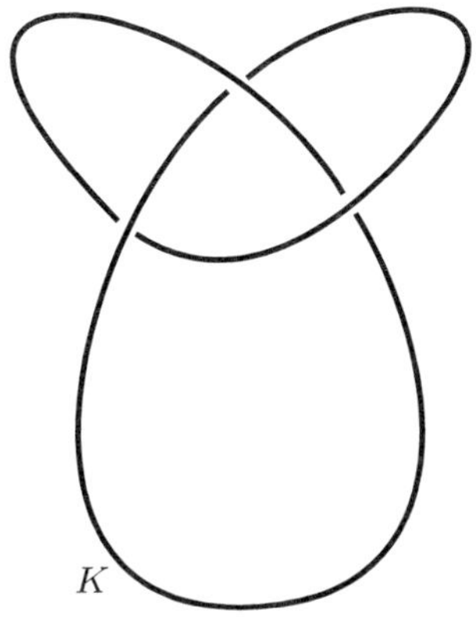

Figure 12.5.2. The trefoil knot

It will be convenient to identify S^3 as the sphere of radius $\sqrt{2}$ in $\mathbb{C}^2$. That is,

$$S^3 = \{(z,w) \in \mathbb{C}^2 \mid |z|^2 + |w|^2 = 2\}.$$

By picking a "point at infinity" in S^3 and stereographically projecting from that point, we make the identification $S^3 = \mathbb{R}^3 \cup \{\infty\}$, allowing us to picture proper subsets of S^3 as subsets of $\mathbb{R}^3$.

If $S^1 \subset \mathbb{C}$ is the standard unit circle, then

$$S^1 \times S^1 \subset S^3$$

is an unknotted imbedding of the torus. Viewed in $\mathbb{R}^3$, this is isotopic to the standard imbedded torus. The simple closed curve $K \subset S^3$, defined by

$$K = \{(e^{6\pi it}, e^{4\pi it}) \mid 0 \le t \le 1\},$$

lies in the imbedded torus, winding twice around one of the circle factors while winding three times around the other as t ranges from 0 to 1. Viewed in $\mathbb{R}^3$, K is the "trefoil knot", pictured (up to isotopy) in Figure 12.5.2. Defining

$$f : S^3 \to \mathbb{C},$$
$$f(z,w) = z^2 - w^3,$$

one easily checks that $K = f^{-1}(0)$.

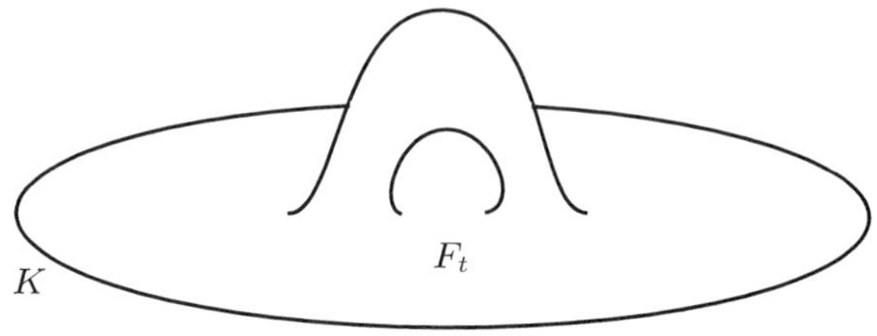

Figure 12.5.3. A typical leaf of the open book

Define the submersion

$$g : S^3 \setminus K \to S^1,$$
$$g(z, w) = \frac{f(z, w)}{|f(z, w)|}.$$

This is actually a fibration, the fibers being homeomorphic to $T^2 \setminus D$, where D is a closed disk in the torus. The closure F_t in S^3 of the fiber $g^{-1}(e^{2\pi i t})$ is obtained by adjoining the knot K. So we obtain an "open book decomposition" of S^3, the spine of the open book being the knot K and the typical leaf being a disk with a handle attached (Figure 12.5.3). Pertinent references include [**7, 89**]. This decomposition is indicated near an arc of K in Figure 12.5.4. Also indicated in that figure is a small circle τ that simply links K and pierces each leaf of the open book once.

Excise the interior of a thin normal neighborhood $N(K)$ of K that misses τ and excise a similar tubular neighborhood $N(\tau)$ of τ. What remains is a compact 3–manifold M with ∂M a union of two tori. The leaves of the open book are modified by these excisions, becoming leaves of an honest foliation $\mathcal{G}$ of M, transverse to both toral boundary components. Excising the neighborhood of K simply trimmed off a small collar neighborhood of the boundary in each F_t, while excising the neighborhood of τ deleted a small open disk from each of the leaves. The typical leaf of $\mathcal{G}$ is the compact surface pictured in Figure 12.5.5. The manifold M can best be pictured by moving the point at infinity in S^3 into the interior of $N(\tau)$. The resulting picture is given in Figure 12.5.6, where we also depict a cross section annulus A that meets each leaf F of $\mathcal{G}$ in an arc from one component of ∂F to the other. These arcs are shown in A, looking like spokes of a wheel. The leaves of $\mathcal{G}$ meet each boundary torus in longitudinal circles that are the leaves of a trivial foliation of each torus. The foliation $\mathcal{G}$ itself cannot be visualized by most of us.

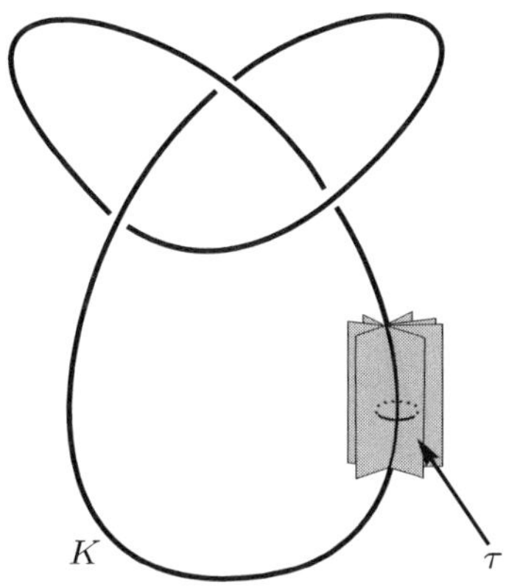

Figure 12.5.4. A local view of the open book

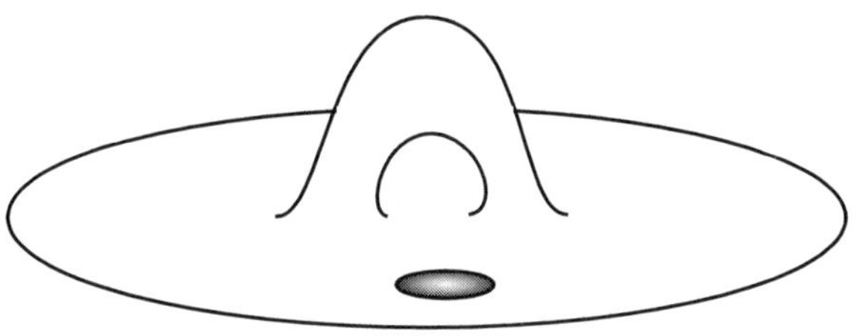

Figure 12.5.5. The typical leaf of $\mathcal{G}$

The next move is to cut M open along the annulus A and reglue by a rotation of the annulus through an irrational multiple of 2π, producing a new foliation $\mathcal{G}'$ of M. Cutting along A cuts each leaf open to "look like" the surface pictured in Figure 12.5.7, the two dotted edges being the parts of the boundary resulting from the cut. This will be a segment of a leaf of $\mathcal{G}'$, these segments being glued together sequentially to form a leaf that one can think of as the infinite Loch Ness monster swimming in an infinite river (Figure 12.5.8). The *irrationality* of the rotation is crucial for this and also guarantees that the "banks" of the river are the leaves of respective irrational slope foliations of the two toral components of ∂M. By

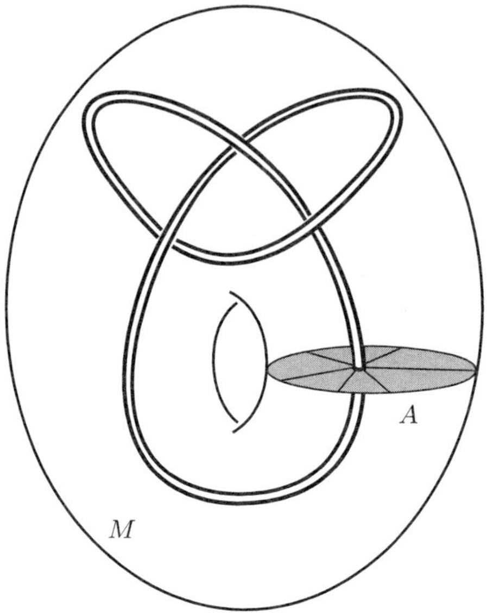

Figure 12.5.6. The manifold M and the cross section A

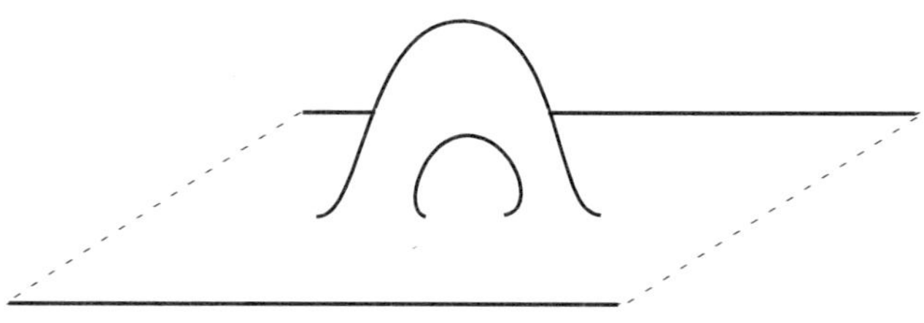

Figure 12.5.7. The typical leaf after cutting along the annulus A

"spinning" this foliation along ∂M, we make each boundary torus a leaf and attach a half plane to each bank of the river. The interior leaves are dense, without holonomy and each has the quasi–isometry type of the infinite Loch Ness monster. This construction is smooth and infinitesimally C^∞–trivial at ∂M because the irrational slope foliations of ∂M are defined by closed, nonsingular 1–forms (Paragraph 3.3.B on page 84 and Subsection 3.4 on page 90). To produce the desired foliation of S^3, glue the solid tori $N(K)$

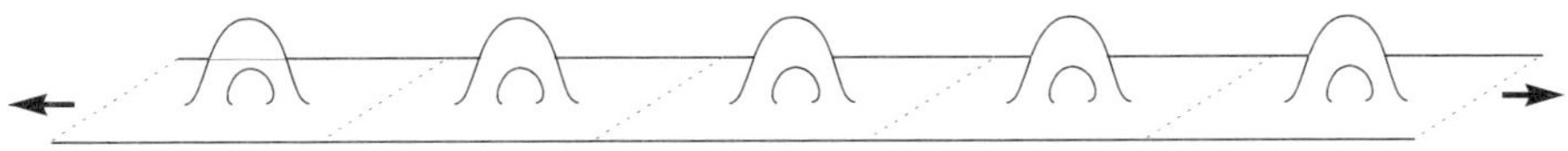

Figure 12.5.8. The Loch Ness monster in a river

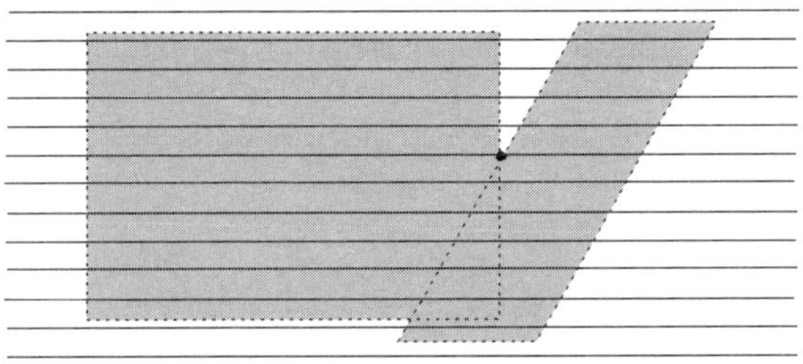

Figure 12.5.9. A tangency of plaque boundaries

and $N(\tau)$ back in, each foliated as a Reeb component that is C^∞–trivial at the boundary.

We turn to the proof of Theorem 12.5.1. Given a regular, foliated atlas $\mathfrak{U}$, as usual, let $\mathfrak{C}$ be the cover of M by the compact sets

$$C_\alpha = \overline{U}_\alpha \in \mathfrak{U},$$

also referred to (somewhat inaccurately) as "flowboxes". These are foliated by compact plaques, assumed diffeomorphic to the standard unit disk D^2. The following is very easy in codimension one, where the flowboxes can be chosen to be *biregular* with respect to a transverse, 1–dimensional foliation. It is harder in arbitrary codimension q, where transverse q–dimensional foliations rarely exist.

Exercise 12.5.7. Prove that the above covering by flowboxes can be chosen so that plaque boundaries intersect transversely, if at all, and so that there are no triple intersection points for plaque boundaries. The proof should be organized inductively. (Hint. The problem is illustrated in Figure 12.5.9, where the tangency of plaque boundaries is only displaced to other plaque boundaries by small perturbations of the foliated charts. The key to the solution (in arbitrary codimension) is suggested in Figure 12.5.10.)

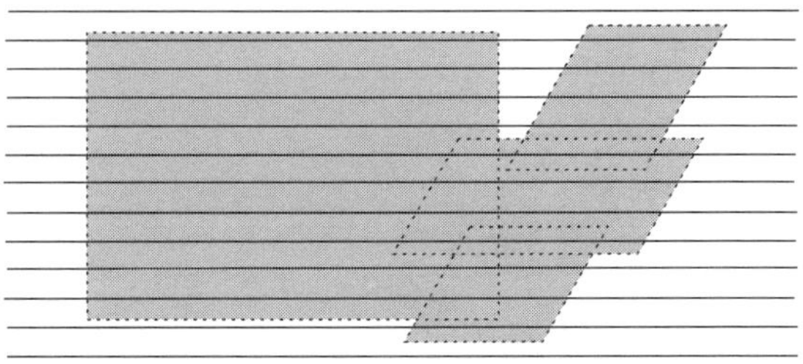

Figure 12.5.10. No tangencies of plaque boundaries

From here on, we assume that the flowboxes are as in Exercise 12.5.7. All plaques are the compact plaques from this system of flowboxes.

Definition 12.5.8. If P is a closed plaque in some flowbox, a sector of P is a nonempty subset of the form

$$P \cap P_1 \cap \cdots \cap P_r \smallsetminus \bigcup_{i=1}^{\ell} Q_i,$$

where the P_j's and Q_i's are other closed plaques in other flowboxes.

Clearly, we have the following:

1. There is a lower bound $\epsilon > 0$ on the length of a segment on the boundary of a plaque joining two points of intersection with boundaries of other plaques.
2. There is a lower bound $\delta > 0$ on the area of sectors of plaques.
3. There is a Lebesgue number $\lambda > 0$ for the covering of leaves by plaques.
4. There is an upper bound c on the circumference of plaques.
5. There is an upper bound d on the diameter of plaques.
6. There is an upper bound $\gamma > 0$ to the absolute value of the geodesic curvature of plaque boundaries.

Let L be a leaf with nonexponential growth and fix $x_0 \in L$, together with a plaque P_0 containing x_0. Set H_m equal to the union of plaques that can be reached from P_0 by plaque chains of length at most m. (Remark that, by Exercise 12.5.7, plaques intersect if and only if their interiors intersect.) The proof of our theorem depends on certain easy claims.

Claim 1. $D_{x_0}(\lambda r) \subseteq H_r \subseteq D_{x_0}(dr)$.

This is almost immediate. The following is a bit trickier.

Claim 2.

$$\liminf_{r\to\infty} \frac{\operatorname{length} \partial H_r}{\operatorname{area} H_r} = 0.$$

Proof. Consider the new plaques added to H_{r-1} to form H_r. We only use those that are not engulfed by $\operatorname{int} H_r$. Suppose there are m of them. Each of these plaques contributes to the area of $H_r \smallsetminus H_{r-1}$ at least one sector that is contributed by no other plaque. Thus,

$$\operatorname{length} \partial H_r \leq cm,$$
$$\operatorname{area}(H_r \smallsetminus H_{r-1}) \geq \delta m,$$

giving

$$\operatorname{length} \partial H_r \leq \frac{c}{\delta} \operatorname{area}(H_r \smallsetminus H_{r-1}).$$

But $\operatorname{gr} L$ is nonexponential and it follows that

$$\liminf_{r\to\infty} \frac{\operatorname{area}(H_r \smallsetminus H_{r-1})}{\operatorname{area} H_r} = 0.$$

Otherwise, there would be an integer $r_0 > 0$ and a constant $A > 0$ such that

$$\frac{\operatorname{area} H_r - \operatorname{area} H_{r-1}}{\operatorname{area} H_r} = \frac{\operatorname{area}(H_r \smallsetminus H_{r-1})}{\operatorname{area} H_r} > A, \quad \forall\, r > r_0.$$

It would follow that, for large r,

$$\operatorname{area} D_{x_0}(dr) \geq \operatorname{area} H_r > (1+A)^{r-r_0} \operatorname{area} H_{r_0},$$

contradicting nonexponential growth. □

By passing to a subsequence $\rho_k \uparrow \infty$ of integers, setting $B_k = H_{\rho_k}$ and $r_k = \lambda\rho_k$, and choosing $Q = d/\lambda$, we obtain a topological averaging sequence (by Claim 1) with the further property that

$$(*) \qquad \lim_{k\to\infty} \frac{\operatorname{length} \partial B_k}{\operatorname{area} B_k} = 0.$$

This, then, is also an averaging sequence (Definition 10.2.21). Thus, passing again to a subsequence, if necessary, we obtain a nontrivial foliation cycle

$$\varphi_\mu = \lim_{k\to\infty} \frac{1}{\operatorname{area} B_k} \int_{B_k}$$

(Theorem 10.2.22).

Claim 3. *The number of corners of ∂B_k is $\leq (\operatorname{length} \partial B_k)/\epsilon$.*

Indeed, each corner is the intersection of two plaque boundaries. Thus, the length of each segment of ∂B_k joining consecutive corners is at least ϵ. The number m of such segments is the number of corners and

$$\operatorname{length} \partial B_k \geq m\epsilon.$$

Proof of Theorem 12.5.1. Assume that $H_2(M;\mathbb{R}) = 0$ and let L, as above, have nonexponential growth. We must show that L has average Euler characteristic zero. Recall that the Euler class of $T(\mathcal{F})$ is represented by the closed 2–form Ω on M such that

$$K = \begin{bmatrix} 0 & \Omega \\ -\Omega & 0 \end{bmatrix}$$

is the curvature matrix for the induced Riemannian connection on $T(\mathcal{F})$. Let $\{B_k\}_{k=1}^{\infty}$ be the topological averaging sequence constructed above and φ_μ the associated foliation cycle. Then, since φ_μ is homologous to zero,

$$(**) \qquad 0 = \varphi_\mu[\Omega] = \lim_{k\to\infty} \frac{1}{\operatorname{area} B_k} \int_{B_k} \Omega.$$

Let $\theta_1, \dots, \theta_m$ be the exterior angles at the corners of ∂B_k and let κ be the geodesic curvature function along the the smooth edges of ∂B_k. By the Gauss–Bonnet formula,

$$2\pi\chi(B_k) = \int_{B_k} \Omega + \int_{\partial B_k} \kappa + \sum_{i=1}^{m} \theta_i.$$

Since $|\theta_i| \le \pi$, $1 \le i \le m$, Claim 3 gives

$$\left|\sum_{i=1}^{m} \theta_i\right| \le \frac{\pi}{\epsilon} \operatorname{length} \partial B_k.$$

Thus,

$$2\pi \frac{|\chi(B_k)|}{\operatorname{area} B_k} \le \left|\frac{1}{\operatorname{area} B_k} \int_{B_k} \Omega\right| + \left(\gamma + \frac{\pi}{\epsilon}\right) \frac{\operatorname{length} \partial B_k}{\operatorname{area} B_k}$$

and $(*)$ and $(**)$ imply that

$$\lim_{k\to\infty} \frac{|\chi(B_k)|}{\operatorname{area} B_k} = 0;$$

so L has average Euler characteristic zero. □

Remarks. The Phillips–Sullivan theory has been generalized by T. Januszkiewicz to other average characteristic numbers [**77**]. Quasi–isometry types of surfaces (complete and with bounded geometry) that cannot be realized in foliations of any closed 3–manifold have been found by P. Schweitzer [**120**]. Theorems of O. Attie and S. Hurder produce uncountably many complete quasi–isometry types of bounded geometry, all diffeomorphic to $S^3 \times S^2 \times \mathbb{R}$, that cannot be realized as leaves in closed, foliated manifolds of codimension one [**5**].

Exercise 12.5.9. Construct a foliated manifold as follows. Let D_k denote the surface obtained by deleting from D^2 the interiors of $k \ge 2$ disjoint compact disks in $\operatorname{int} D^2$. Let $J_1, \dots, J_k$ be disjoint arcs connecting the boundary

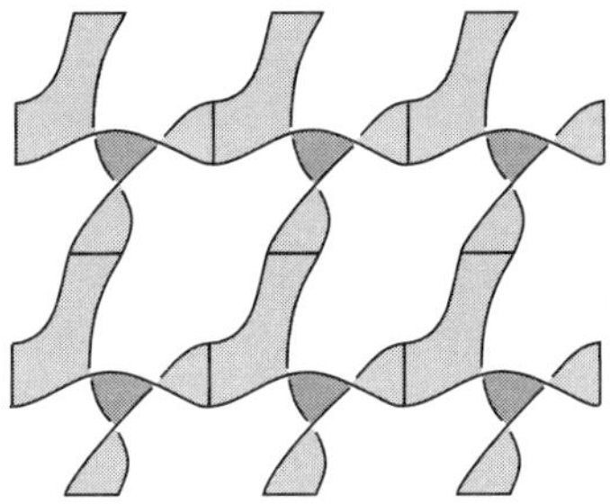

Figure 12.5.11

of each deleted disk to ∂D^2. Let $f_1, \dots, f_k$ be rationally independent rotations of S^1. Foliate the manifold $D_k \times S^1$ by the leaves $D_k \times \{z\}$ and modify this foliation by cutting along the annuli $J_i \times S^1$ and regluing by $\mathrm{id} \times f_i$. (When $k = 2$, the leaves in M look like the surface in Figure 12.5.11. It has quadratic growth and the Euler characteristic also grows quadratically. The average Euler characteristic is nonzero.) This foliation induces an irrational slope foliation of each component of ∂M. Spin along these components and show that the resulting leaves in $\operatorname{int} M$ have exactly polynomial growth of degree $k + 1$. The growth of the Euler characteristic appears to be polynomial of degree k. Incorporate this foliation into a C^∞ foliation of S^3, concluding from Theorem 12.5.1 that the leaves really do have average Euler characteristic zero.

12.6. Uniformization

Here we discuss another result for foliations with two–dimensional leaves. Again the Euler class plays a key role, but this time everything can be carried out for abstract foliated spaces.

By Theorem 11.3.24, the universal cover of a surface L with a Riemannian metric g is conformally diffeomorphic to one of three models: the sphere, the Euclidean plane or the hyperbolic plane. Since the canonical metric of these models has constant curvature and is homogeneous, (L, g) admits a conformal metric of constant curvature. Furthermore, for every compact (oriented) surface with a Riemannian metric of constant curvature, the Gauss–Bonnet theorem implies that the sign of the curvature is the same as the sign of the Euler characteristic.

In order to extend this uniformization theorem to spaces foliated by surfaces, we need the concept of "Euler characteristic" for such a foliated space. Let $(M, \mathcal{F})$ be an oriented, C^∞–foliated space with 2–dimensional leaves and

Riemannian metric g. Let $\Omega \in A^2(M)$ denote the leafwise curvature form of g. If μ is a holonomy–invariant measure, then one defines the integral

$$\int_M \Omega \, d\mu$$

as in Section 11.5.

Definition 12.6.1. Let $(M, \mathfrak{F})$ be a compact, C^∞–foliated space with a leafwise Riemannian metric and associated curvature form Ω. Let μ be a holonomy–invariant measure for $\mathfrak{F}$. Then the Euler characteristic of $(M, \mathfrak{F})$ with respect to the measure μ is

$$\chi_\mu(M, \mathfrak{F}) = \int_M \Omega \, d\mu.$$

Exercise 12.6.2. Prove that $\chi_\mu(M, \mathfrak{F})$ is independent of the conformal class of the Riemannian metric.

The question then arises as to whether, given a 2–dimensional foliated space M with leafwise metric g, it is possible to uniformize its leaves simultaneously. That is, one asks whether there exists a continuous, leafwise–smooth function $\lambda > 0$ on M such that the new metric λg has the same constant curvature $\kappa = 0, +1$ or -1 on all leaves. In general the answer is "no". For example, both a torus and a sphere could occur simultaneously as leaves in a connected foliated space $(M, \mathfrak{F})$. The following theorem was proven in [**16**].

Theorem 12.6.3 (Candel). *Let $(M, \mathfrak{F})$ be a compact, oriented, C^∞–foliated space of leaf–dimension two with a leafwise Riemannian metric. Then there is a conformal, leafwise Riemannian metric such that every leaf has constant curvature -1 if and only if $\chi_\mu(M, \mathfrak{F}) < 0$ for all nontrivial, holonomy–invariant measures μ.*

Note in particular that the condition on the Euler characteristic may be satisfied vacuously (*i.e.*, if there are no invariant measures). The proof of this result falls into two parts, one of which will be discussed here in the following form.

Proposition 12.6.4. *Let $(M, \mathfrak{F})$ be a compact, oriented, C^∞–foliated space of leaf–dimension two with a leafwise Riemannian metric. If some leaf of M fails to be conformally covered by hyperbolic space, then there is a nontrivial holonomy–invariant measure μ such that $\chi_\mu(M, \mathfrak{F}) \geq 0$, where equality holds if L is not a sphere.*

We must construct an invariant measure when a leaf L fails to be conformally covered by the hyperbolic plane. Topologically, such a leaf must be a sphere, a torus, a plane or a cylinder. In the compact cases, we can take μ to be the Dirac measure supported on L, and so we only need to consider the case of a plane or a cylinder. We treat the case in which L is a plane and leave the required modifications for the case that L is a cylinder for Exercise 12.6.6.

Lemma 12.6.5 (Ahlfors). *Let $f : (\mathbb{R}^2, g) \to (L, g_L)$ be a conformal diffeomorphism, where g is the Euclidean metric. Then (L, g_L) has an averaging sequence.*

Proof. Let $B(r)$ and $\partial B(r)$ denote the Euclidean disk and the circle of radius r, respectively, about the origin. That f is conformal means that the pull–back metric is of the form $f^* g_L = \lambda g$, where $\lambda > 0$ is a smooth function. Let $d\eta$ denote the volume form of $\mathbb{R}^2$ and let $d\nu_r$ denote the volume (length) form of the circle $\partial B(r)$ (induced by g). By the change of variables and co–area formulae, the area $A(r)$ of $f(B(r))$ is

$$A(r) = \int_{B(r)} \lambda \, d\eta = \int_0^r \int_{\partial B(s)} \lambda \, d\nu_s \, ds,$$

and differentiating this formula gives

$$A'(r) = \int_{\partial B(r)} \lambda \, d\nu_r.$$

By Schwarz's inequality, the length $L(r)$ of the image $f(\partial B(r))$ satisfies

$$L(r)^2 = \left(\int_{\partial B(r)} \lambda^{1/2} \, d\nu_r \right)^2 \leq \left(\int_{\partial B(r)} d\nu_r \right) \left(\int_{\partial B(r)} \lambda \, d\nu_r \right) = 2\pi r A'(r).$$

Therefore, the integral

$$\int_1^\infty \frac{1}{r} \left(\frac{L(r)}{A(r)} \right)^2 dr \leq 2\pi \int_1^\infty \frac{A'(r)}{A(r)^2} \, dr \leq \frac{2\pi}{A(1)}$$

is finite. The change of variables $r = e^s$ gives

$$\int_1^\infty \frac{1}{r} \left(\frac{L(r)}{A(r)} \right)^2 dr = \int_0^\infty \left(\frac{L(e^s)}{A(e^s)} \right)^2 ds,$$

and so

$$\liminf_{s \to \infty} \frac{L(e^s)}{A(e^s)} = 0,$$

in view of the previously displayed inequality. □

Exercise 12.6.6. This exercise contains the rest of the proof of Proposition 12.6.4.

1. Show that, if the universal cover of one leaf L of $\mathcal{F}$ fails to be conformally diffeomorphic to the hyperbolic plane, then there is a sequence of regions $\{B_k\}$, exhausting L and forming an averaging sequence as

in Definition 10.2.21. Moreover, this sequence has nonnegative average Euler characteristic:

$$\lim_{k\to\infty} \frac{\chi(B_k)}{\operatorname{area} B_k} \geq 0.$$

(Hint. If L is a cylinder, then it is conformally diffeomorphic to $\mathbb{R}^1 \times S^1$ with the standard metric. Modify the argument given for the plane.)

Consider a cover $\mathcal{U}$ by foliated charts as in Exercise 12.5.7, and let H_k be the region obtained as the union of all plaques of $\mathcal{U}$ that cover B_k.

2. Show that this collection of regions $\{H_k\}$ also forms an averaging sequence.

Such an averaging sequence $\{H_k\}$ defines a holonomy–invariant measure μ, the Euler characteristic $\chi_\mu(M, \mathcal{F})$ being easily verified to be

$$\chi_\mu(M, \mathcal{F}) = \lim_{k\to\infty} \frac{1}{\operatorname{area} H_k} \int_{H_k} \Omega.$$

3. Show that $\chi_\mu(M, \mathcal{F}) \geq 0$, with equality if L is not the 2–sphere. (Hint. Assume that L is not compact. By the Gauss–Bonnet theorem, as in the proof of Theorem 12.5.1,

$$\left|\int_{H_k} \Omega\right| \leq 2\pi \left|\chi(H_k)\right| + \left|\int_{\partial H_k} \kappa\right| + \left|\sum_i \theta_i\right|.$$

The last two terms on the right are treated as in Theorem 12.5.1. The first satisfies

$$|\chi(H_k)| \leq 2 + \frac{1}{\epsilon} \operatorname{length} \partial H_k.)$$

The rest of the proof of Theorem 12.6.3 is carried out as follows. When all leaves of M are hyperbolic, the uniformization function of each leaf provided by Theorem 11.3.24 is unique; so there is a well–defined uniformization function on M. One is required to show that this function, which is constructed leaf by leaf, is a smooth function on the foliated space M. The study of this function was started by E. Ghys [**48**] and A. Verjovsky [**141**], and was completed in [**16**].

Keeping in mind Reeb's stability theorem, a similar statement can be made and proven regarding the existence of a conformal metric of constant curvature +1. The proof requires the full force of Theorem 12.6.3.

Proposition 12.6.7. *Let $(M, \mathcal{F})$ be a compact, connected, orientable foliated space with 2–dimensional leaves. Suppose that M admits nontrivial, holonomy–invariant measures. Then every leaf is a 2–sphere if and only if $\chi_\mu(M, \mathcal{F}) > 0$ for every nontrivial, holonomy–invariant measure.*

Proof. Assume that all $\chi_\mu(M, \mathcal{F}) > 0$. First, by the local Reeb stability theorem (Proposition 11.4.8), the union R of the leaves that are spheres is an open saturated subset of M. Furthermore, the union of the supports of the transverse, holonomy–invariant measures of M is a compact saturated set P (Exercise 12.3.7) and $R \subseteq P$. Then $N = P \smallsetminus R$ is a compact foliated space that, if nonempty, admits nontrivial holonomy–invariant measures, but contains no spherical leaf. On the one hand, our hypothesis requires that none of the Euler characteristics for measures supported in N vanish. Hence Proposition 12.6.4 implies that all leaves of $(N, \mathcal{F}|N)$ are hyperbolic. On the other hand, Theorem 12.6.3 then implies that, for each holonomy–invariant measure μ for M with support in $P \smallsetminus R$, we must have $\chi_\mu(M, \mathcal{F}) < 0$, contradicting our hypothesis. Thus $R = P$ is open, closed and nonempty; hence $R = M$ by connectedness. This proves the "if" assertion. The "only if" assertion is elementary, for it follows from Proposition 11.4.8 that, if all leaves are spheres, then M is the total space of a fibration over the space of leaves, and so invariant measures for $(M, \mathcal{F})$ correspond to measures on the base space of this fibration. □

There is a related theorem of A. Connes [**30**] to the effect that, if a measure μ has $\chi_\mu(M, \mathcal{F}) > 0$, then almost all the leaves in its support are spheres.

Under the conditions of the proposition, there is a conformal metric giving each leaf constant curvature $+1$. However, no such uniformization is available for the zero curvature case.

Exercise 12.6.8. Show that every noncompact leaf of the Reeb foliation is conformally diffeomorphic to the Euclidean plane, but that there is no Riemannian metric for which all leaves have curvature zero.

Remark. More interesting counterexamples are due to E. Ghys. In [**49**], he constructed compact manifolds foliated by planes, each one conformal to the Euclidean plane and each one dense in the manifold, for which no simultaneous, continuous (nor even measurable) uniformization exists.

Chapter 13

Entropy of Foliations

In this chapter we will study the generalization of the notion of entropy of a map or flow to more complicated dynamical systems such as foliations. In general terms, entropy measures the rate of creation of information. Roughly, if the states of a system are described by iteration of a map, states that may be indistinguishable at some initial time may diverge into clearly different states as time passes. Entropy measures the rate of creation of states. In the language of dynamics, it measures the rate of divergence of orbits of a map.

There are two concepts of entropy for maps, topological entropy and measure–theoretic entropy. Topological entropy is the one we will extend to foliated spaces. This extension is due to Ghys, Langevin and Walczak [**50**]. The idea is to define the entropy of a holonomy pseudogroup in close analogy to that of a map. Since there is no canonical choice of holonomy pseudogroup, another notion, geometric entropy of a foliation, will be introduced. One of the main theorems relates these two definitions.

We will prove two other important theorems. If one looks at the formal definitions, the vanishing of entropy seems to suggest that the leaves have subexponential growth. This is not so, but we will see that zero entropy implies the existence of a transverse, invariant measure.

The other main theorem concerns foliated bundles of fiber dimension one, asserting that such foliations have positive entropy if and only if there is a resilient leaf. In very imprecise language, the intuition behind this is that a resilient leaf produces a free semigroup (with two generators) in the holonomy group, with some contracting property, hence strong exponential growth (*cf.* Corollary 12.2.40), hence positive entropy. By Duminy's theorem, we then know that foliations with nonvanishing Godbillon–Vey class have positive entropy. The converse is certainly not true. Nonvanishing of the Godbillon–Vey class is related to the existence of free groups in the holonomy pseudogroup. Although this may seem quite similar to the existence of free semigroups, it is not sufficiently so. For example, these properties mark the difference between the group of affine transformations of the line and the group of homographic transformations.

The measure–theoretic entropy of a map is formally similar to the topological version. In defining the measure–theoretic entropy, one distinguishes orbits with respect to a measure invariant by the transformation. For maps of metric spaces, the two concepts are interrelated by a variational formula, the topological entropy being the supremum of the measure entropies with respect to all invariant probability measures. The holonomy pseudogroup of a foliated space does not generally admit an invariant probability measure; so the measure entropy is not readily extendable to foliations. There have been some attempts to do this, but we will not consider these here.

More information on this topic and related matters will be found in the projected book of P. Walczak [**144**]. For another approach to foliation entropy, based on the leafwise geodesic flow, see S. Hurder [**73**]. A general reference for the entropy of continuous maps is [**146**].

13.1. Entropy of Maps

Suppose that f is a map of a compact manifold into itself. To measure the number of orbits one takes an empirical approach, not distinguishing ϵ–close points for a given $\epsilon > 0$. If x and y are two indistinguishable points, then their orbits $\{f^k(x)\}_{k=1}^{\infty}$ and $\{f^k(y)\}_{k=1}^{\infty}$ will be distinguishable provided that, for some k, the points $f^k(x)$ and $f^k(y)$ are at a distance greater than ϵ. Then one counts the number of distinguishable orbit segments of length n for fixed magnitude ϵ, and looks at the growth rate of this function of n. Finally, one improves the resolution arbitrarily well by letting $\epsilon \to 0$. The value obtained is called the entropy of f, and it measures the asymptotic growth rate of the number of orbits of finite length as the length goes to infinity.

The precise definitions are as follows. Let (X, d) be a compact metric space and let $f : X \to X$ be a continuous map. Say that a set $E \subset X$ is (n, ϵ)–separated by f if, whenever x, y are two points of E, then $d(f^k(x), f^k(y)) \geq \epsilon$ for some $0 \leq k \leq n$. The maximum number of pairwise ϵ–distinguishable orbits of length n is

$$S(f, n, \epsilon) = \sup\{\text{card}(E) \mid E \text{ is } (n, \epsilon)\text{–separated by } f\}.$$

This number is finite because X is compact. Its growth rate is

$$h(f,\epsilon) = \limsup_{n\to\infty} \frac{1}{n} \log S(f,n,\epsilon),$$

a number (possibly infinite) that is nondecreasing as $\epsilon \downarrow 0$.

Definition 13.1.1. The entropy of f is the number (nonnegative or ∞)

$$h(f) = \lim_{\epsilon\downarrow 0} h(f,\epsilon).$$

Remark. A nice way to think of (n,ϵ)–separation is to define a new metric d_n on X, compatible with the topology, by

$$d_n(x,y) = \max_{0\le k\le n} d(f^k(x), f^k(y)).$$

Then points x and y are (n,ϵ)–separated by f if and only if $d_n(x,y) \geq \epsilon$.

Here are some examples of entropy.

Exercise 13.1.2. If f is an isometry, show that $h(f) = 0$. In particular, a rotation of the circle has zero entropy.

Exercise 13.1.3. Let C be the Cantor set, identified with the countably infinite product $\prod_{n=1}^{\infty}\{0,1\}$. This becomes a metric space if, for arbitrary points $x = (x_1, x_2, \dots)$ and $y = (y_1, y_2, \dots)$, we define

$$d(x,y) = \sum_{i=1}^{\infty} \frac{|x_i - y_i|}{2^i}.$$

Let f be the shift map of C, defined by $f(x_1, x_2, \dots) = (x_2, x_3, \dots)$ and prove that $h(f) = \log 2$.

Exercise 13.1.4. If A is an $n\times n$ matrix with integer entries, then A defines an endomorphism f of the n-dimensional torus $T^n = \mathbb{R}^n/\mathbb{Z}^n$. If $\lambda_1, \cdots, \lambda_n$ are the eigenvalues of A, show that

$$h(f) = \sum_{|\lambda_i|>1} \log|\lambda_i|.$$

In particular, if $S^1 \subset \mathbb{C}$ is the unit circle and $f_n : S^1 \to S^1$ is defined by $z \to z^n$, $n > 0$, then $h(f_n) = \log n$.

Exercise 13.1.5. Define a subset $E \subset X$ to be (f,n,ϵ)–spanning if, for each $x \in X$, there is $y \in E$ with $d(f^k(x), f^k(y)) < \epsilon$, for some value of $k = 0, 1, 2, \dots, n$. Define $R(f,n,\epsilon)$ to be the infimum of the cardinality of the (f,n,ϵ)–spanning sets and prove that $R(f,n,\epsilon) \leq S(f,n,\epsilon)$, for all n and ϵ. Finally, show that

$$h(f) = \lim_{\epsilon\downarrow 0} \limsup_{n\to\infty} \frac{1}{n} \log R(f,n,\epsilon).$$

It will next be shown that a homeomorphism of the circle has zero entropy. This theorem appeared in [**1**], the first paper dealing with topological entropy. There, the notion of topological entropy was defined via open covers, in a manner analogous to the definition of metric entropy. The proof here is a prelude to the later analysis of the entropy of a group of homeomorphisms of the circle.

Theorem 13.1.6. *A homeomorphism* $f : S^1 \to S^1$ *has entropy* $h(f) = 0$.

Proof. For definiteness, assume that the circle has unit length. Let $\epsilon > 0$ be as small as desired. Let E be an (n, ϵ)–separated set of cardinality $S(f, n, \epsilon)$ and choose a subset $F \subset S^1$ of $1 + [3/\epsilon]$ equally spaced points (where $[\cdot]$ denotes the greatest integer function). Then consecutive points $u, v \in F$ are at distance $d(u, v) < \epsilon/3$ and the set

$$F_n = \bigcup_{k=0}^{n} f^{-k}(F)$$

has cardinality $\leq (n+1)(1 + [3/\epsilon])$.

Let I be an interval of $S^1 \smallsetminus E$ having endpoints $x, y \in E$. Then $d(f^k(x), f^k(y)) \geq \epsilon$ for some $0 \leq k \leq n$. The distance $d(f^k(x), f^k(y)$ is the smallest of the lengths of the interval $f^k(I)$ and its complementary interval $S^1 \smallsetminus f^k(I)$; so $f^k(I)$ has length $\geq \epsilon$ and must contain a point of F. Therefore, I contains a point of F_n.

This shows that the cardinality of E, which is the same as the number of components of $S^1 \smallsetminus E$, is less than or equal to that of F_n. Therefore,

$$S(f, n, \epsilon) \leq (n+1)(1 + [3/\epsilon])$$

and

$$\limsup_{n\to\infty} \frac{1}{n} \log S(f, n, \epsilon) \leq \limsup_{n\to\infty} \frac{1}{n} \log(n+1)(1 + [3/\epsilon]) = 0.$$

□

Our definition of entropy seems to depend on the choice of metric d; hence we should, perhaps, write $S(n, \epsilon, d)$, $h(f, \epsilon, d)$ and $h(f, d)$. The justification for ignoring the choice of metric is the following.

Proposition 13.1.7. *Let* $f : X \to X$ *be a continuous map of a compact metrizable space* X. *Then the entropy* $h(f)$ *is independent of the choice of metric compatible with the topology of* X. *That is, entropy depends only on the uniform structure of* X.

Proof. If d_1 and d_2 are two distances on X inducing the same topology, then they are uniformly continuous. That is, for each $\epsilon > 0$ there exists $\epsilon' > 0$ such that $d_2(x, y) \leq \epsilon' \Rightarrow d_1(x, y) \leq \epsilon$ and there is $\epsilon'' > 0$ such that $d_1(x,y) \leq \epsilon'' \Rightarrow d_2(x,y) \leq \epsilon'$. Thus, an (n,ϵ,d_1)–separated set is (n,ϵ',d_2)–separated, and an (n,ϵ',d_2)–separated set is (n,ϵ'',d_1)–separated. Therefore,

$$S(f,n,\epsilon,d_1) \leq S(f,n,\epsilon',d_2) \leq S(f,n,\epsilon'',d_1), \quad \forall n \geq 0.$$

As ϵ approaches 0, so does ϵ', which in turn forces $\epsilon'' \to 0$. Hence, taking limits as $\epsilon \to 0$ we obtain the desired equality $h(f,d_1) = h(f,d_2)$. □

13.2. Entropy of Groups, Pseudogroups and Foliations

The dynamical properties of a foliated bundle over a compact manifold are described by the (finitely generated) total holonomy group. The dynamics of a compact, foliated space are described by the (finitely generated) holonomy pseudogroup. Therefore we will extend the definition of entropy of a map in the previous section to finitely generated groups G of homeomorphisms of a space X and finitely generated pseudogroups Γ of local homeomorphisms of X.

13.2.A. Entropy of transformation groups. Let G be a group of homeomorphisms of the metric space (X,d), with finite generating set G_1. We suppose that this generating set is *symmetric*, meaning that G_1 contains the identity transformation and is invariant under the operation of passing to the inverse. We set G_n equal to the set of elements of G that can be written as words of length $\leq n$ in elements of G_1. We say that points $x, y \in X$ are (n,ϵ)–separated by G_1 if there is $g \in G_n$ such that $d(g(x),g(y)) \geq \epsilon$. A subset E of X is (n,ϵ)–separated by G_1 if each pair of distinct points in E is so separated. The supremum of the cardinalities of such sets is denoted by $S(G,G_1,n,\epsilon)$ and we define the entropy

$$h(G,G_1) = \lim_{\epsilon\downarrow 0} \limsup_{n\to\infty} \frac{1}{n} \log S(G,G_1,n,\epsilon).$$

Remark. To avoid an orgy of symbols, notation such as (G,G_1,n,ϵ) will often be replaced by the slimmer (n,ϵ) if it is clear from the context what the other two symbols G and G_1 should be. On the other hand, if several (n,ϵ) appear within the same discussion, we may add some extra symbols to tell them apart.

Exercise 13.2.1. Let $f : X \to X$ be a homeomorphism, G the cyclic group generated by f, and set $G_1 = \{f, f^{-1}, \mathrm{id}\}$. Prove that $2h(f) = h(G,G_1)$. (Hint. Show that $E \subset X$ is (n,ϵ)–separated by G_1 if and only if $f^{-n}(E)$ is $(2n,\epsilon)$–separated by f.)

Exercise 13.2.2. If $\varphi = \{\varphi_t\}_{t\in\mathbb{R}}$ is a flow on a compact metric space (X, d), one can define a continuous version of entropy. We say that points $x, y \in X$ are (r, ϵ)–separated by φ if, for some $t \in [-r, r]$, $d(\varphi_t(x), \varphi_t(y)) \geq \epsilon$. A set is (r, ϵ)–separated by φ if any two distinct points of the set are so separated. Let $S(r, \epsilon)$ be the supremum of the cardinalities of such sets and define

$$h(\varphi) = \lim_{\epsilon\downarrow 0} \limsup_{r\to\infty} \frac{1}{r} \log S(r, \epsilon).$$

On the other hand, φ_1 is a map of X. Prove that $h(\varphi) = 2h(\varphi_1)$.

If $(M, \mathcal{F})$ is a foliated bundle over a compact manifold B with fiber Y, the total holonomy group $G(\mathcal{F})$ is the image of the total holonomy homomorphism

$$f : \pi_1(B) \to \mathrm{Homeo}(Y).$$

This group is finitely generated because $\pi_1(B)$ is finitely generated. Relative to a choice of finite, symmetric generating system $G(\mathcal{F})_1$, we interpret the entropy $h(G(\mathcal{F}), G(\mathcal{F})_1)$ as entropy of the foliated bundle, relative to the choice of $G(\mathcal{F})_1$. This motivates a definition of entropy of general foliated spaces that replaces the role of the holonomy group with that of the holonomy pseudogroup.

13.2.B. Entropy of pseudogroups. We fix a metric space (X, d), together with a pseudogroup Γ of local homeomorphisms of X that admits a finite symmetric generating subset Γ_1. It is required that Γ_1 contain id_X. We will not require X to be compact. For each positive integer n, let Γ_n denote the collection of elements of Γ that can be obtained by composition of at most n elements of Γ_1. For notational convenience, we let Γ_0 consist of the identity of X.

Let $\epsilon > 0$. Points $x, y \in X$ are said to be (n, ϵ)–separated by Γ_1 (or, more concisely, (n, ϵ)–separated) if there exists $f \in \Gamma_n$, whose domain contains both x and y, such that $d(f(x), f(y)) \geq \epsilon$. (Remark that, for every real number $\epsilon > 0$, if $d(x, y) > \epsilon$, x and y are automatically (n, ϵ)–separated for every integer $n \geq 0$. This is because $\mathrm{id}_X \in \Gamma_n$.) A subset E of X is defined to be (n, ϵ)–separated if every pair x, y of distinct elements of E are (n, ϵ)–separated. Finally, let $S(n, \epsilon)$ denote the supremum of the cardinalities of (n, ϵ)–separated subsets of X, this being a positive integer or ∞. Then set

$$h(\Gamma, \Gamma_1, \epsilon) = \limsup_{n\to\infty} \frac{1}{n} \log S(n, \epsilon),$$

$$h(\Gamma, \Gamma_1) = \lim_{\epsilon\downarrow 0} h(\Gamma, \Gamma_1, \epsilon).$$

Definition 13.2.3. The number $h(\Gamma, \Gamma_1)$ is called the entropy of Γ with respect to Γ_1. It is a nonnegative real number or ∞.

Definition 13.2.4. The pair (Γ, Γ_1) is *regular* if, for each $g \in \Gamma_1 \setminus \mathrm{id}_X$, $\operatorname{dom} g$ has compact closure contained in an open set U to which g extends as a homeomorphism $\widetilde{g}$ onto an open set $\widetilde{g}(U)$.

Remark that, if (Γ, Γ_1) is regular, every pure composition

$$f = g_1 \circ g_2 \circ \cdots \circ g_k$$

of elements of Γ_1 also has relatively compact domain and a continuous extension $\overline{f}$ to the compact closure of that domain. For a regular pair, we can ignore all but a compact subspace of X. That is, we lose no generality in assuming that X itself is compact.

Exercise 13.2.5. If (Γ, Γ_1) is regular, prove that $S(n, \epsilon)$ is always finite.

Exercise 13.2.6. Let X be a compact metric space, $f : X \to X$ a continuous surjection such that, for each $x \in X$, there is an open neighborhood U_x of x such that f carries U_x homeomorphically onto an open neighborhood of $f(x)$. Then there is a finite open cover $\{U_i\}_{i=1}^N$ of X such that the restrictions $g_i = f|U_i$ generate a pseudogroup Γ on X. Let Γ_1 consist of $g_i^{\pm 1}$, $1 \le i \le N$, together with id_X. If $h(f) > 0$, prove that $h(\Gamma, \Gamma_1) > 0$.

We have the following analogue of Proposition 13.1.7.

Proposition 13.2.7. *The entropy $h(\Gamma, \Gamma_1)$ of a regular pair is independent of the choice of metric d compatible with the topology of X.*

This justifies omitting reference to d in the notation $h(\Gamma, \Gamma_1)$ for the entropy of the regular pair.

Exercise 13.2.8. Suppose that G is a group of homeomorphisms of the compact metric space (X, d) admitting a finite, symmetric generating set G_1. Let $\{U_i\}_{i=1}^N$ be a finite open cover of X and, for each $g \in G_1$, define $g_{ij} : U_j \cap g^{-1}(U_i) \to U_i$ to be the restriction of g.

1. Let Γ_1 be the set of all g_{ij}, together with id_X, clearly a symmetric generating set for a pseudogroup Γ, and (Γ, Γ_1) is a regular pair. Prove that $h(\Gamma, \Gamma_1) = h(G, G_1)$. (Hint. Prove two inequalities. One is almost immediate and, for the other, require the ϵ's in the definition of entropy to be smaller than a Lebesgue number for the open cover $\{\operatorname{dom} g_{ij}\}_{1 \le i,j \le N}$.)
2. Let $Z_i = \overline{U}_i$ and set

$$Z = \bigcup_{i=1}^{N} Z_i \times \{i\},$$

clearly a disjoint union. Any metric d on X induces a metric d' on Z by a standard trick. On each $Z_i \times \{i\}$, $d'((x,i),(y,i)) = d(x,y)$. If K is a number greater than the diameters of all the sets $Z_i \times \{i\}$, define $d'((x,i),(y,j)) = K$ whenever $i \neq j$. Define $g^*_{ij}(x,j) = (g_{ij}(x),i)$, whenever $x \in Z_j \cap \operatorname{dom} g_{ij}$, let Γ^*_1 be the set of g^*_{ij}, together with id_Z, and let Γ^* be the pseudogroup on the compact space Z generated by Γ^*_1. Prove that $h(\Gamma^*, \Gamma^*_1) = h(\Gamma, \Gamma_1)$.

Lemma 13.2.9. *If $S(0,\epsilon) < \infty$, for $\epsilon > 0$ sufficiently small, and if Γ_1 consists of local isometries of (X,d), then $h(\Gamma,\Gamma_1) = 0$.*

The easy proof is left to the reader. One expects that pseudogroups that are *almost isometric* will also have vanishing entropy. Here are a couple of possible meanings of "almost isometric".

Definition 13.2.10. The pseudogroup Γ is said to be *equicontinuous* if, for every $\epsilon > 0$, there exists $\delta > 0$ such that $d(f(x), f(y)) < \epsilon$ for every two points x and y with $d(x,y) < \delta$ and all $f \in \Gamma$ with $x, y \in \operatorname{dom} f$.

Definition 13.2.11. The pseudogroup Γ is said to be *uniformly Lipschitz* if there is a constant $A \geq 1$ such that

$$A^{-1} d(x,y) \leq d(f(x), f(y)) \leq A d(x,y),$$

for all $f \in \Gamma$ and all $x, y \in \operatorname{dom} f$.

Evidently, the uniform Lipschitz property implies equicontinuity. Results of M. Kellum [**79**] on the structure of uniformly Lipschitz foliations strongly suggest that uniformly Lipschitz pseudogroups behave very much like pseudogroups of isometries.

Exercise 13.2.12. Let Γ be an equicontinuous pseudogroup on a compact metric space, Γ_1 a finite, symmetric generating set. Prove that $h(\Gamma, \Gamma_1) = 0$.

Exercise 13.2.13. The number $h(\Gamma,\Gamma_1)$ certainly depends on the generating system. Indeed, if (Γ, Γ_1) is a regular pair, let Γ^2 denote the collection of homeomorphisms obtained from Γ_1 by composition or amalgamation of pairs of elements of Γ_1. Then Γ^2 is a finite, symmetric generating system for Γ. What is the relation between the numbers $h(\Gamma,\Gamma_1)$ and $h(\Gamma, \Gamma^2)$?

A condition guaranteeing the finiteness of entropy is the following:

Proposition 13.2.14. *Let (Y,d) be a complete metric space that satisfies the following homogeneity property: there is a constant $C > 0$ such that, for every real number r, any ball of radius r has a covering by no more than e^C balls of radius r/e. Let X be an open subset of Y with compact closure. Let Γ_1 be a finite, symmetric collection of Lipschitz homeomorphisms between* *open subsets of X generating a pseudogroup Γ. If $K > e$ is an upper bound for the Lipschitz constants of the elements of Γ_1, then*

$$h(\Gamma, \Gamma_1) \leq C \log K.$$

Proof. If x and y are points of X that are (n, ϵ)–separated, then there is $f \in \Gamma_n$ such that

$$\epsilon \leq d(f(x), f(y)) \leq K^n d(x, y).$$

Since X is relatively compact in Y, it has finite diameter and therefore, for any integer $m \geq 0$, it has a covering by no more than e^{mC} balls of radius $\operatorname{diam} X/e^m$. Let $\epsilon > 0$ be so small that

$$\log \operatorname{diam} X - \log \frac{\epsilon}{2} > 0.$$

Then, for any positive integer n, we can find a nonnegative integer m such that

$$(*) \qquad \frac{\epsilon}{2} K^{-(n+1)} \leq \frac{\operatorname{diam} X}{e^m} < \frac{\epsilon}{2} K^{-n}.$$

If x and y are points of X lying in a metric ball of radius $\operatorname{diam} X/e^m$ and $f \in \Gamma_n$, then

$$d(f(x), f(y)) \leq K^n d(x, y) \leq 2K^n \frac{\operatorname{diam} X}{e^m} \leq \epsilon.$$

It follows that the maximal cardinality of an (n, ϵ)–separated set is at most e^{mC}; hence

$$\frac{1}{n} \log S(\Gamma, \Gamma_1, n, \epsilon) \leq \frac{1}{n} \log e^{mC} = \frac{mC}{n}.$$

By the first inequality in $(*)$,

$$\limsup_{n \to \infty} \frac{1}{n} \log S(n, \epsilon) \leq C \log K.$$

This is independent of ϵ and the asserted inequality follows. □

This homogeneity condition can be codified neatly in terms of the metric dimension of Y [**81**]. Obviously, the condition holds if Y is Euclidean space $\mathbb{R}^q$, but there are many interesting examples where Y is a zero–dimensional metric space. One of the most popular is the following.

Exercise 13.2.15. Let p be a positive integer and let Z be the set of infinite sequences $a = (a_0, a_1, \cdots)$, where a_i are integers $1 \leq a_i \leq p$. If $a, b \in Z$, set $d(a, b) = p^{-n}$, where n is the smallest integer such that $a_n \neq b_n$. If there is no such integer, $a = b$ and we set $d(a, b) = 0$. Show that (Z, d) is a compact metric space satisfying the homogeneity property of Proposition 13.2.14.

Exercise 13.2.16. Construct a complete metric space that does not satisfy the homogeneity condition. What about a compact one?

13.2.C. Application to foliations. Let $(M, \mathcal{F}, d)$ be a compact, foliated metric space with distance function d, modeled transversely on a locally compact, separable, metric space (X, ρ) with distance function ρ. If

$$\mathcal{U} = \{U_\alpha, x_\alpha, y_\alpha\}_{\alpha \in \mathfrak{A}}$$

is a regular foliated atlas for $(M, \mathcal{F})$, then the holonomy pseudogroup $\Gamma_{\mathcal{U}}$ on X defines a regular pair $(\Gamma_{\mathcal{U}}, \Gamma_1)$, where

$$\Gamma_1 = \{\gamma_{\alpha\beta}, \mathrm{id}_S\}_{\alpha,\beta \in \mathfrak{A}}.$$

This is the example of main interest to us. As usual, we will write $S(n, \epsilon)$ for the maximal cardinality of sets (n, ϵ)–separated by Γ_1.

Let us consider the geometric meaning of (n, ϵ)–separation for the case of compact foliated spaces. Since M is compact, $\mathcal{U} = \{U_i, x_i, y_i\}_{1 \le i \le N}$ is a finite foliated atlas of M, each U_i is precompact, and the coordinates extend to the compact closure, allowing us to identify

$$\overline{U}_i = D_i \times Z_i.$$

The transverse space $Z \subset X$ is the disjoint union of the finitely many compact spaces Z_i and we obtain a compact metric space (Z, ρ) by restricting the metric ρ on Y to each Z_i and decreeing that the distance between points of distinct compacta Z_i and Z_j be a fixed value

$$K > \max_{1 \le k \le N} \operatorname{diam} Z_k.$$

It is a slight abuse of notation to use the same symbol ρ for this new metric. The holonomy pseudogroup $\Gamma_{\mathcal{U}}$ is defined on Z. As usual, whenever convenient, we view Z as the set of compact plaques $P \subset \overline{U}_i$, $1 \le i \le N$.

Recall that the term "chain" $(P_0, P_1, \dots, P_n)$ of compact plaques includes the condition that the interior of P_{i-1} meets the interior of P_i, $1 \le i \le n$.

Definition 13.2.17. Points $x, y \in M$ are (n, ϵ)–separated by $\mathcal{U}$ if they are distinct and either x and y do not lie in a common chart of $\mathcal{U}$ or there are chains $(P_0, P_1, \dots, P_n)$ and $(Q_0, Q_1, \dots, Q_n)$ of closed $\mathcal{U}$–plaques such that

1. $x \in \operatorname{int} P_0$, $y \in \operatorname{int} Q_0$ and $P_0 \neq Q_0$;
2. P_i and Q_i are plaques of a chart $U_{k_i} \in \mathcal{U}$, $0 \le i \le n$;
3. $\rho(P_i, Q_i) \ge \epsilon$, for some $i = 0, 1, \dots, n$.

A subset $E \subset M$ is (n, ϵ)–separated by $\mathcal{U}$ if every pair of distinct points of E are so separated.

In this case, we will write $S_{\mathcal{U}}(n, \epsilon)$ for the supremum of the cardinalities of (n, ϵ)–separated sets and define the entropy of $\mathcal{F}$ relative to the regular foliated atlas $\mathcal{U}$ to be

$$h_{\mathcal{U}}(\mathcal{F}) = \lim_{\epsilon \downarrow 0} \limsup_{n \to \infty} \frac{1}{n} \log S_{\mathcal{U}}(n, \epsilon).$$

While the reader can check that each (n, ϵ)–separated set is finite, there is no guarantee that $h_{\mathcal{U}}(\mathcal{F})$ will be finite.

On the other hand, we can consider the entropy $h(\Gamma_{\mathcal{U}}, \Gamma_1)$, computed using the metric ρ to define the uniform structure on Z.

Theorem 13.2.18. *If $(M, \mathcal{F})$ is a compact, foliated metric space and $\mathcal{U}$ is a regular foliated atlas, then*

$$h_{\mathcal{U}}(\mathcal{F}) = h(\Gamma_{\mathcal{U}}, \Gamma_1).$$

Proof. The space Z may be viewed as a subset of M that is the disjoint union of the transversals Z_i. Let $E \subset Z$ be (n, ϵ)–separated by Γ_1. Let $E_i = E \cap Z_i$, $1 \leq i \leq N$. The fact that $E_i \subset Z_i$ and is (n, ϵ)–separated by Γ_1 immediately implies that it is (n, ϵ)–separated by $\mathcal{U}$, $1 \leq i \leq N$; so

$$S(n, \epsilon) \leq N S_{\mathcal{U}}(n, \epsilon).$$

This implies that $h(\Gamma_{\mathcal{U}}, \Gamma_1) \leq h_{\mathcal{U}}(\mathcal{F})$.

For the reverse inequality, let $E \subset M$ be (n, ϵ)–separated by $\mathcal{U}$. We may assume that $0 < \epsilon < K$. To each $x \in E$ assign $z(x) \in Z$, one of the finitely many plaques containing x. By the definition of (n, ϵ)–separation of E by $\mathcal{U}$, together with the regularity of $\mathcal{U}$, this map $z : E \to Z$ must be injective. We prove that $z(E)$ is $(n+1, \epsilon)$–separated by Γ_1. Indeed, if $x, y \in E$ and do not lie in a common chart of $\mathcal{U}$, then $\rho(z(x), z(y)) = K > \epsilon$; so these points are $(n+1, \epsilon)$–separated by Γ_1. Even if x and y lie in a common $\mathcal{U}$–chart, it may be that $z(x)$ and $z(y)$ lie in distinct transversals, again implying that $\rho(z(x), z(y)) = K$. By the definition of (n, ϵ)–separation by $\mathcal{U}$, the only remaining possibility is that there is $\gamma \in \Gamma_1$ such that $\gamma(z(x))$ and $\gamma(z(y))$ are (n, ϵ)–separated by Γ_1; hence $z(x)$ and $z(y)$ are $(n+1, \epsilon)$–separated by Γ_1. Thus, $S_{\mathcal{U}}(n, \epsilon) \leq S(n+1, \epsilon)$. Writing

$$\frac{1}{n} \log S(n+1, \epsilon) = \frac{n+1}{n} \frac{1}{n+1} \log S(n+1, \epsilon)$$

and passing to the usual limits, we conclude that $h_{\mathcal{U}}(\mathcal{F}) \leq h(\Gamma_{\mathcal{U}}, \Gamma_1)$. □

Corollary 13.2.19. *If $(M, \mathcal{F})$ is a compact, C^1–foliated manifold, then $h_{\mathcal{U}}(\mathcal{F})$ is finite.*

Proof. The transverse space Z is the compact closure of an open subset of $\mathbb{R}^q$ and $\Gamma_{\mathcal{U}}$ consists of Lipschitz homeomorphisms; so Proposition 13.2.14 implies that $h(\Gamma_{\mathcal{U}}, \Gamma_1) < \infty$. □

Exercise 13.2.20. Let $f : X \to X$ be a homeomorphism of a compact metrizable space and form the suspension of f, a foliated bundle over S^1 with fiber X and foliation $\mathfrak{F}$. Let $A_1, \dots, A_N$ be a covering of S^1 by open arcs such that $A_i \cap A_{i+1}$ is an open arc, $1 \le i \le N$ (mod N), with no other nonempty intersections. Then the union of fibers over A_i is an open set U_i and $\mathfrak{U} = \{U_i\}_{i=1}^N$ is a regular foliated atlas. Show that $Nh_{\mathfrak{U}}(\mathfrak{F}) = 2h(f)$. If we lift the standard Riemannian metric on S^1 to the leaves, we can define $\operatorname{diam} \mathfrak{U}$ to be the maximal length of the plaques of elements of $\mathfrak{U}$ and conclude from the above that

$$\frac{1}{\operatorname{diam} \mathfrak{U}} h_{\mathfrak{U}}(\mathfrak{F}) \le 2h(f).$$

Show that, as one varies $\mathfrak{U}$ over all regular foliated atlases of the above type, the right–hand side of this inequality is the supremum of the left–hand side.

Exercise 13.2.21. While $h_{\mathfrak{U}}(\mathfrak{F})$ depends, in general, on the choice of $\mathfrak{U}$, its vanishing or nonvanishing is independent of that choice. For this, let $\mathfrak{U}'$ be a refinement of $\mathfrak{U}$ and show that there are constants $N \ge 1$ and $K \ge 1$ such that, for all sufficiently small $\epsilon > 0$ and all $n \ge 0$,

$$S_{\mathfrak{U}'}(n, \epsilon) \le S_{\mathfrak{U}}(n, \epsilon) \le N S_{\mathfrak{U}'}(Kn, \epsilon).$$

Show that the assertion follows.

13.3. Geometric Entropy

Following Ghys, Langevin and Walczak [**50**], we define a *geometric* entropy $h_g(\mathfrak{F})$, determined by a choice of leafwise Riemannian metric but independent of the choice of a foliated atlas. This entropy measures the way leaves "peel apart" in various directions. The metric will play a role analogous to that of the parameter t of a flow φ_t in defining $h(\varphi)$ (see Exercise 13.2.2). In fact, if φ_t is nonsingular, the leafwise metric tensor g such that φ_t has unit speed will determine $h_g(\mathfrak{F}) = h(\varphi)$, where $\mathfrak{F}$ is the one–dimensional foliation underlying the flow (Exercise 13.3.17). We will relate geometric entropy to the pseudogroup entropies, exploiting the fact that $h_{\mathfrak{U}}(\mathfrak{F})$ does not depend on the choice of g, while $h_g(\mathfrak{F})$ does not depend on the choice of $\mathfrak{U}$. While this program works well for foliated manifolds, there are technical problems in general foliated spaces. We will comment on this in Subsection 13.3.B.

13.3.A. Geometric entropy for foliated manifolds. In this subsection, we restrict our attention to smooth foliations $\mathfrak{F}$ of compact, connected Riemannian manifolds (M, g). We allow the presence of tangential and/or transverse boundary, possibly separated by corners, as modeled locally in Definition 1.1.17. If there is transverse boundary, all Riemannian metrics should be chosen so that $\partial_{\pitchfork} M$ is totally geodesic in M.

We fix a choice of smooth Riemannian metric g on M. There are various (ultimately equivalent) ways to use g to measure the separation properties of leaves via leafwise paths parametrized by arclength. The definition in [**50**] is rather complicated, but makes it immediately clear that geometric entropy depends only on the metric $g_{\mathcal{F}}$ induced by g along the leaves. A conceptually simpler definition [**85, 144**] uses orthogonal projection of leaf curves to nearby leaves but creates the illusion that the resulting entropy depends on the transverse metric also. We will use this approach, noting that the key result (Theorem 13.3.2) relating geometric and pseudogroup entropies implies independence of transverse metrics.

Remark. In the following discussion, we frequently assert that certain constants, depending on the metric tensor, can be chosen uniformly on M. This is always justified by the compactness of M, together with the fact that all derivatives of components of g in local coordinates are continuous. Details, which are elementary but not always trivial, will be left to the reader.

Let d denote the distance function on M induced by g. Let $\epsilon_1 > 0$ be the injectivity radius of g. Thus, for each point $x \in M$, the exponential map $\exp : T_x(M) \to M$ carries the open ball $B(0_x, \epsilon_1)$ diffeomorphically onto the open ball $B(x, \epsilon_1) \subset M$, centered at x and of d–radius ϵ_1.

One can also define the exponential map $\exp_{\mathcal{F}} : T_x(\mathcal{F}) \to L_x$ associated to the leafwise metric $g_{\mathcal{F}}$ induced by g on the leaves. By the compactness of M and a well–known theorem of J. H. C. Whitehead [**148**], there is a largest $\epsilon_2 > 0$ (possibly infinite) such that $\exp_{\mathcal{F}} : B_{\mathcal{F}}(0_x, \epsilon_2) \to L_x$ imbeds this ball as an open, geodesically convex ball in L_x (relative to $g_{\mathcal{F}}$), $\forall\, x \in M$. Here, of course, $B_{\mathcal{F}}(0_x, \epsilon_2)$ is the open ϵ_2–ball in $T_x(\mathcal{F})$ and we will denote its image in L_x by $B_{\mathcal{F}}(x, \epsilon_2)$.

Let $T_x^{\perp}(\mathcal{F})$ denote the orthogonal complement of $T_x(\mathcal{F})$ in $T_x(M)$, and let $B_{\mathcal{F}}^{\perp}(0_x, \epsilon)$ be the corresponding open ball in this normal space. If $\epsilon \le \epsilon_1$, then

$$B_{\mathcal{F}}^{\perp}(x, \epsilon) = \exp B_{\mathcal{F}}^{\perp}(0_x, \epsilon)$$

is an imbedded q–ball. For a possibly smaller value of $\epsilon > 0$, this q–ball will be transverse to $\mathcal{F}$, $\forall\, x \in M$. It should be clear that the supremum ϵ_3 of all such ϵ is actually the maximal such.

Set $\epsilon_0 = \min\{\epsilon_1, \epsilon_2, \epsilon_3\}$. Clearly, this is an invariant of $(M, \mathcal{F}, g)$. Finally, we can choose a number $\delta > 0$ so small that, if $d(x, y) < \delta$, then

$$B_{\mathcal{F}}(y, \epsilon_0) \cap B_{\mathcal{F}}^{\perp}(x, \epsilon_0) = \{p_y(x)\} \quad \text{(a singleton)}.$$

Again, the supremum δ_0 of all such choices is, in fact, the maximum, giving another well–defined invariant of $(M, \mathcal{F}, g)$.

If $d(x, y) < \delta_0$ and $\gamma : [0, b] \to L_x$ is a continuous leaf curve starting at x, $p_y(\gamma(t)) \in B_{\mathcal{F}}(y, \epsilon_0)$ will be defined and continuous in L_y for some range

$0 \leq t < \eta \leq b$. By a fairly obvious continuation argument, we define a continuous curve $p_y(\gamma) : [0, a) \to L_y$ on a maximal domain with $a \leq b$. It is fairly easy to see that $\lim_{t \uparrow a} p_y(\gamma(t)) = \zeta$ exists in L_y and is the endpoint of a shortest geodesic, emerging from $\gamma(a)$, perpendicular to L_x at that point, and of length at most ϵ_0. If $a < b$, maximality of a implies that this length is exactly ϵ_0. In any event, set $p_y(\gamma(a)) = \zeta$ and obtain a projected curve $p_y(\gamma|[0, a])$. If $a = b$, we say that γ projects to $p_y(\gamma)$. Note that, if γ is piecewise smooth, so are these projected curves.

If γ is a piecewise smooth leaf path, $\ell(\gamma)$ will denote the length of γ, as computed via the metric g.

Definition 13.3.1. Let $R > 0$ and $\epsilon > 0$ be given with $\epsilon < \min\{\epsilon_0, \delta_0\}$. A point $x \in M$ will be said to be (R, ϵ)–separated by g from a point $y \in M$ if either $d(x, y) \geq \epsilon$ or there exists a piecewise smooth path $\gamma : [0, r] \to L_x$, $0 < r = \ell(\gamma) \leq R$, parametrized by arclength and originating at $x = \gamma(0)$, which projects to $p_y(\gamma)$ with

$$d(\gamma(t), p_y(\gamma(t))) \geq \epsilon,$$

for some $t \in [0, r]$. If either x is (R, ϵ)–separated by g from y or y is (R, ϵ)–separated by g from x, we say that x and y are (R, ϵ)–separated by g. A subset $E \subset M$ is (R, ϵ)–separated by g if every two distinct points in E are so separated.

As usual, one sets $S_g(R, \epsilon)$ equal to the supremum of the cardinalities of (R, ϵ)–separated sets, defining

$$h_g(\mathcal{F}) = \lim_{\epsilon \downarrow 0} \limsup_{R \to \infty} \frac{1}{R} \log S_g(R, \epsilon).$$

This is called the *geometric entropy* of $\mathcal{F}$ relative to g. As already noted, the metric g along the leaves plays much the same role as the time parameter t of a flow or as the integer parameter of the iterates of a map; so it is not an artificial device.

Remark. Let positive numbers R and ϵ be given. Then, if $z \in M$, one can find $\eta > 0$ so small that, if $d(z, x) < \eta$, $d(z, y) < \eta$ and $\gamma : [0, r] \to L_x$ is a leaf curve out of x of length at most R, then $d(x, y) < \epsilon$, $p_y(\gamma)$ is defined and $d(\gamma(t), p_y(\gamma(t))) < \epsilon$, $0 \leq t \leq r$. By the compactness of M, one finds a finite cover of M by such η–balls, concluding that $S_g(R, \epsilon)$ is finite, being bounded by the cardinality of this cover. Nevertheless, nothing we have said so far prevents $h_g(\mathcal{F})$ from being infinite (but see Corollary 13.3.16).

Our goal is to relate the entropies $h_g(\mathcal{F})$ and $h_{\mathcal{U}}(\mathcal{F})$. The main theorem of this section is the following, due to Ghys, Langevin and Walczak [**50**]. (Compare this with Exercise 13.2.20.)

Theorem 13.3.2. *Let $(M, \mathcal{F})$ be a compact, C^∞–foliated manifold and let g be a metric along the leaves. Then*

$$h_g(\mathcal{F}) = \sup_{\mathfrak{U}} \frac{1}{\operatorname{diam}_g \mathfrak{U}} h_{\mathfrak{U}}(\mathcal{F}),$$

where the supremum is taken over all regular foliated atlases $\mathfrak{U}$ for $\mathcal{F}$ and $\operatorname{diam}_g \mathfrak{U}$ *denotes the least upper bound of the diameters of the plaques of the elements of $\mathfrak{U}$, measured using the metric g.*

Before proving this, we consider some corollaries and examples.

Corollary 13.3.3. *For a compact, C^∞–foliated manifold $(M, \mathcal{F})$, the following are equivalent:*

(1) $h_g(\mathcal{F}) = 0$, *for some choice of metric tensor g;*
(2) $h_{\mathfrak{U}}(\mathcal{F}) = 0$, *for every choice of regular foliated atlas $\mathfrak{U}$;*
(3) $h_{\mathfrak{U}}(\mathcal{F}) = 0$, *for some choice of regular foliated atlas $\mathfrak{U}$;*
(4) $h_g(\mathcal{F}) = 0$, *for every choice of metric tensor g.*

In this case, we will say that $\mathcal{F}$ has entropy zero.

Indeed, Theorem 13.3.2 gives (1) $\Rightarrow$ (2) $\Rightarrow$ (4), while (4) $\Rightarrow$ (1) is trivial and Exercise 13.2.21 asserts the equivalence of (2) and (3).

Corollary 13.3.4. *If g and g' are two Riemannian metrics on M, then there is a constant $A \geq 1$ such that*

$$\frac{1}{A} h_g(\mathcal{F}) \leq h_{g'}(\mathcal{F}) \leq A h_g(\mathcal{F}).$$

Proof. The metrics are quasi–isometric by compactness of M. Denoting by $\|\cdot\|$ and $\|\cdot\|'$ the respective norms induced on $T(\mathcal{F})$ by g and g', we choose $A \geq 1$ so that

$$\frac{1}{A}\|v\| \leq \|v\|' \leq A\|v\|, \quad \forall\, v \in T(\mathcal{F}).$$

Then

$$\frac{1}{A} \operatorname{diam}_g \mathfrak{U} \leq \operatorname{diam}_{g'} \mathfrak{U} \leq A \operatorname{diam}_g \mathfrak{U};$$

hence

$$\frac{1}{A} \frac{1}{\operatorname{diam}_g \mathfrak{U}} h_{\mathfrak{U}}(\mathcal{F}) \leq \frac{1}{\operatorname{diam}_{g'} \mathfrak{U}} h_{\mathfrak{U}}(\mathcal{F}) \leq A \frac{1}{\operatorname{diam}_g \mathfrak{U}} h_{\mathfrak{U}}(\mathcal{F}),$$

and the assertion follows. □

Example 13.3.5. Let $M = F \times T$ be the Cartesian product of compact manifolds, with $\mathcal{F}$ the product foliation by leaves $F \times \{z\}$. Then $\mathcal{F}$ has entropy zero. There is no constraint on the choice of metric tensor; so we take g to be the product of a Riemannian metric on F with one on T. Thus, $d(\gamma(t), p_y(\gamma(t)))$ will be constant for all leaf curves γ; so $S_g(n, \epsilon)$ is constant in n.

Example 13.3.6. If $\varphi : (M, \mathcal{F}) \to (M', \mathcal{F}')$ is a diffeomorphism of foliated manifolds, then one of these foliations has entropy zero if and only if both do. Indeed, given a Riemannian metric g' on M', its pullback $g = \varphi^*(g')$ is a Riemannian metric on M and $\varphi : (M, g) \to (M', g')$ is an isometry.

Exercise 13.3.7. Prove that geometric entropy is only sensitive to the leafwise metric $g_{\mathcal{F}}$. More precisely, if $\varphi : (M, \mathcal{F}, g) \to (M', \mathcal{F}', g')$ is a homeomorphism between two foliated manifolds that is a smooth isometry along the leaves (relative to $g_{\mathcal{F}}$ and $g'_{\mathcal{F}'}$), prove that $h_g(\mathcal{F}) = h_{g'}(\mathcal{F}')$. If $\varphi : (M, \mathcal{F}) \to (M', \mathcal{F}')$ is a covering map of foliated manifolds and $g = \varphi^* g'$, what is the relation between $h_g(\mathcal{F})$ and $h_{g'}(\mathcal{F}')$?

Exercise 13.3.8. Define a vector field ξ on an open subset $U \subseteq M$ to be *basic* (with respect to $\mathcal{F}$) if, for every vector field ζ tangent to $\mathcal{F}$, the Lie bracket $[\xi, \zeta]$ is tangent to $\mathcal{F}|U$. Say that the Riemannian metric g is *bundle like* (with respect to $\mathcal{F}$) if, whenever ξ and ζ are basic fields on U that are orthogonal to the leaves, the function $g(\xi, \zeta)$ on U is constant along the leaves. Establish the following.

1. If (U, x, y) is a foliated chart, then ξ defined on U is a basic field if and only if the derivative map y_* carries ξ to a well–defined vector field on $y(U)$.
2. If the metric tensor is bundle like and $\mathcal{U}$ is a regular foliated atlas with corresponding transverse space T, then there is a Riemannian metric on T relative to which the holonomy pseudogroup $\Gamma_{\mathcal{U}}$ consists of local isometries.
3. Conversely, if there is a Riemannian metric on T making $\Gamma_{\mathcal{U}}$ a pseudogroup of local isometries, prove that there is a bundle–like metric on M.
4. If M admits a bundle–like metric, prove that $\mathcal{F}$ has entropy zero.

The proof of Theorem 13.3.2 will be divided into a sequence of lemmas and propositions.

The first order of business is to establish a convenient choice of transverse distance function relative to a regular foliated atlas $\mathcal{U} = \{U_i, x_i, y_i\}_{i=1}^N$. Here, the transverse model space is $Y = \mathbb{R}^q$. By the definition of regularity, we fix a second such atlas $\mathcal{U}' = \{U'_i, x'_i, y'_i\}_{i=1}^N$ such that $\overline{U}_i \subset U'_i$, $x_i = x'_i|U_i$ and $y_i = y'_i|U_i$, $1 \le i \le N$. Let Z'_i denote the set of plaques of $\overline{U}'_i$, viewed via y'_i as a compact ball in $\mathbb{R}^q$, $Z_i \subset Z'_i$ the corresponding transversal for $\overline{U}_i$. As usual, it is easily arranged that $Z'_i \cap Z'_j = \emptyset$, $i \ne j$, and we let Z and Z' denote the respective unions of these transversals. The Riemannian distance function on M induces the Hausdorff metric [**37**, p. 205] ρ on the set Z'_i of plaques of $\overline{U}'_i$. Let K be greater than $\operatorname{diam}_\rho Z'_i$, $1 \le i \le N$. If $P \in Z'_i$, $Q \in Z'_j$

and $i \neq j$, define $\rho(P,Q) = K$. One easily checks that this defines a metric ρ on the compact space Z', hence restricts to a metric (denoted also by ρ) on Z. We use this metric to define $h_{\mathfrak{U}}(\mathcal{F})$, recalling that all metrics compatible with the topology of Z give the same entropy (Proposition 13.2.7).

Lemma 13.3.9. *Let P be a plaque of $\overline{U}'_i$ and suppose there is a geodesic arc $\sigma : [0,1] \to M$, orthogonal to P at $\sigma(0) \in \operatorname{int} P$ and realizing the Riemannian distance of $\sigma(1) \in U'_i$ from P. Then*

$$d(\sigma(0), \sigma(1)) \leq \rho(y'_i(\sigma(0)), y'_i(\sigma(1))).$$

Indeed, by its definition, the Hausdorff distance between P and the plaque through $\sigma(1)$ is at least as big as the distance of $\sigma(1)$ from P. This observation will be needed in the proof of Lemma 13.3.14.

Exercise 13.3.10. If g and g^* are two Riemannian metrics on M, prove that the corresponding Hausdorff metrics ρ and ρ^* on Z'_i are quasi–isometric. That is, there is a constant $A > 0$ such that $A^{-1}\rho \leq \rho^* \leq A\rho$.

Lemma 13.3.11. *There is a constant $C > 0$ such that, whenever $u, w \in U'_i$, then $\rho(y'_i(u), y'_i(v)) \leq C d(u,v)$.*

Proof. By regularity, the coordinates (x'_i, y'_i) extend smoothly to coordinates (x''_i, y''_i) defining a slightly larger foliated chart $U''_i \supset \overline{U}'_i$. With this set up, a Riemannian metric g^* on M can be found that, on a neighborhood of $\overline{U}'_i$, agrees with the pullback via (x''_i, y''_i) of the standard Euclidean metric on $\mathbb{R}^p \times \mathbb{R}^q$. It is easily arranged that U'_i is geodesically convex in M (for instance, by making U'_i correspond to the Cartesian product of the ball of radius 1 in $\mathbb{R}^p$ with the ball of radius 1 in $\mathbb{R}^q$), suitably blending the Euclidean metric across U''_i to agree outside a neighborhood of $\overline{U}'_i$ with a given metric g on M. The associated Hausdorff metric $\rho^*(y'_i(u), y'_i(v))$ on Z'_i is simply the Euclidean distance between these points in $\mathbb{R}^q$. Thus,

$$\rho^*(y'_i(u), y'_i(v)) \leq d^*(u,v),$$

where d^* denotes the distance function on M associated to g^*. Since g and g^* are quasi–isometric, there is a constant $B > 0$ such that $d^* \leq Bd$. By Exercise 13.3.10, there is also a constant $A > 0$ such that $\rho \leq A\rho^*$ and the constant $C = AB$ is as advertised, for the element U'_i of the covering $\mathfrak{U}$. Since this consists of finitely many flowboxes, the proof is complete. □

Proposition 13.3.12. *If $\mathfrak{U}$ is a regular foliated atlas for $\mathcal{F}$, then*

$$\frac{1}{\operatorname{diam}_g \mathfrak{U}} h_{\mathfrak{U}}(\mathcal{F}) \leq h_g(\mathcal{F}).$$

Proof. We fix the associated regular foliated atlas $\mathfrak{U}'$ as in the above discussion and carry over notation established there. In particular, $C > 0$ will be as in Lemma 13.3.11. We will show that, for $\epsilon > 0$ sufficiently small,

$$(*) \qquad S_{\mathfrak{U}}(n, C\epsilon) \le S_g(n \operatorname{diam}_g \mathfrak{U}, \epsilon), \quad n \ge 0.$$

The desired inequality follows from $(*)$ by taking two limits.

Choose $\lambda > 0$ so small that the following all hold:

(1) λ is a Lebesgue number for $\mathfrak{U}$;

(2) $d(x, y) < \lambda$ and $x \in \overline{U}_i$ implies that $y \in U'_i$;

(3) $\lambda < \max\{\epsilon_0, \delta_0\}$.

The requirement that ϵ be sufficiently small will mean that $\epsilon < \lambda$.

We assume that x and y are not $(n \operatorname{diam}_g \mathfrak{U}, \epsilon)$–separated by g. In particular,

$$d(x, y) < \epsilon < \lambda.$$

We must prove that x and y are not $(n, C\epsilon)$–separated by $\mathfrak{U}$. The inequality $(*)$ then follows.

Since λ is a Lebesgue number for $\mathfrak{U}$ and $d(x, y) < \lambda$, we can find a chart $U_{k_0} \in \mathfrak{U}$ containing both x and y. Let P_0 and Q_0 be plaques in $\overline{U}_{k_0}$ with $x \in \operatorname{int} P_0$, $y \in \operatorname{int} Q_0$. If $P_0 = Q_0$, we are done; so assume that these plaques are distinct and let $g \in \Gamma_n$ be defined at both P_0 and Q_0. Since $\operatorname{id} \in \Gamma_1$, we can write g as a word of length exactly n in these generators. This corresponds to a sequence of charts $U_{k_0}, U_{k_1}, \dots, U_{k_n} \in \mathfrak{U}$ and plaque chains $(P_0, P_1, \dots, P_n)$ and $(Q_0, Q_1, \dots, Q_n)$ such that P_i and Q_i are plaques of $\overline{U}_{k_i}$, $0 \le i \le n$

Let $\gamma(t)$, $0 \le t \le \ell(\gamma) = R$, be a piecewise smooth path in L_x, parametrized by arclength, such that $\gamma(0) = x$ and, for a suitable partition $0 = t_0 < t_1 < \cdots < t_n = R$, the segment $\gamma_i = \gamma|[t_{i-1}, t_i]$ is a path in $\operatorname{int} P_i$, $1 \le i \le n$. We can also assume that each segment γ_i has length less than $\operatorname{diam}_g \mathfrak{U}$; so $R < n \operatorname{diam}_g \mathfrak{U}$.

Since $d(x, y) < \lambda < \delta_0$, there is $r \in (0, R]$ such that the projected curve $p_y(\gamma(t))$ is defined for $0 \le t \le r$. Since we assume that x and y are not $(n \operatorname{diam}_g \mathfrak{U}, \epsilon)$–separated by g, $d(\gamma(t), p_y(\gamma(t))) < \epsilon < \delta_0$, $0 \le t \le r$. If $r < R$, this projection can be extended to a properly larger interval $[0, r']$. In this way, we see that the entire curve γ projects to $p_y(\gamma)$ and that $d(\gamma(t), p_y(\gamma(t))) < \epsilon$, $0 \le t \le R$. Again appealing to the fact that $\epsilon < \lambda$, we see that, for $1 \le i \le n$ and $t_{i-1} \le t \le t_i$, $p_y(\gamma(t)) \in Q'_i$, a plaque of $\overline{U}'_i$. Clearly, $Q_0 \subset Q'_0$ and one proceeds inductively up the chain to conclude that each $Q_i \subset Q'_i$. Finally, Lemma 13.3.11 implies that

$$\rho(P_i, Q_i) \le C d(\gamma(t_i), p_y(\gamma(t_i))) < C\epsilon.$$

Since $g \in \Gamma_n$ was arbitrary, subject to the condition that its domain contain both P_0 and Q_0, this proves that x and y are not $(n, C\epsilon)$–separated by $\mathcal{U}$. □

Exercise 13.3.13. Let $\mathcal{F}$ be orientable with leaf dimension 1, $\varphi = \{\varphi_t\}_{t\in\mathbb{R}}$ the underlying unit speed flow and recall the definition of $h(\varphi)$ in Exercise 13.2.2. Modify the proof of Proposition 13.3.12 to prove

$$\frac{1}{\operatorname{diam}_g \mathcal{U}} h_{\mathcal{U}}(\mathcal{F}) \leq h(\varphi).$$

By Proposition 13.3.12,

$$(**) \qquad \sup_{\mathcal{U}} \frac{1}{\operatorname{diam}_g \mathcal{U}} h_{\mathcal{U}}(\mathcal{F}) \leq h_g(\mathcal{F}).$$

To prove the reverse inequality we will consider regular foliated atlases $\mathcal{U}_\Delta$ having foliated charts that are transversely very thin relative to plaque diameters. The plaques are to be geodesically convex balls in their respective leaves, all having the same diameter Δ. In particular, $\Delta = \operatorname{diam}_g \mathcal{U}_\Delta$. In codimension 1, these charts look like very thin coins.

Let $\eta \in (0,1)$, let $\mathbb{Z}^+$ denote the set of nonnegative integers and define $\nu : [0,\infty) \to \mathbb{Z}^+$ by

$$\nu(R) = [R/((1-\eta)\Delta)] + 1,$$

where $[\cdot]$ denotes the greatest integer function.

Lemma 13.3.14. *For each $\eta \in (0,1)$, there is a choice of atlas $\mathcal{U}_\Delta$ as above such that, for all sufficiently small $\epsilon > 0$ and all $R \in [0,\infty)$, points $x, y \in M$ that are (R,ϵ)–separated by g are either $(\nu(R),\epsilon)$–separated by $\mathcal{U}_\Delta$ or are at distance $d(x,y) \geq \epsilon$.*

Before giving the proof of this lemma, which is somewhat technical, we show how it completes the proof of the Ghys–Langevin–Walczak theorem.

Corollary 13.3.15. *For each $\eta \in (0,1)$, there is a choice of atlas $\mathcal{U}_\Delta$ as above such that*

$$(1-\eta) h_g(\mathcal{F}) \leq \frac{1}{\Delta} h_{\mathcal{U}_\Delta}(\mathcal{F}).$$

Proof. Let E be a subset of M that is (R,ϵ)–separated by g, where $\epsilon > 0$ is small enough for Lemma 13.3.14 to hold. Let $\{B_k\}_{k=1}^{N(\epsilon)}$ be a finite cover of M by open balls of radius $\epsilon/2$ and set $E_k = E \cap B_k$, $1 \leq k \leq N(\epsilon)$. The lemma then implies that E_k is $(\nu(R),\epsilon)$–separated by $\mathcal{U}_\Delta$; so

$$S_g(R,\epsilon) \leq N(\epsilon) S_{\mathcal{U}_\Delta}(\nu(R),\epsilon).$$

Thus,

$$\frac{1}{R}\log S_g(R,\epsilon) \leq \frac{\log N(\epsilon)}{R} + \frac{\nu(R)}{R}\frac{1}{\nu(R)}\log S_{\mathcal{U}_\Delta}(\nu(R),\epsilon).$$

By the definition of $\nu(R)$,

$$\lim_{R\to\infty} \frac{\nu(R)}{R} = \frac{1}{(1-\eta)\Delta},$$

and the asserted inequality follows. $\square$

Proof of Theorem 13.3.2. Since $\Delta = \operatorname{diam}_g \mathcal{U}_\Delta$ and η can be taken as close to 0 as desired, Corollary 13.3.15 implies that

$$h_g(\mathcal{F}) \le \sup_{\mathcal{U}} \frac{1}{\operatorname{diam}_g \mathcal{U}} h_{\mathcal{U}}(\mathcal{F}),$$

where the supremum is taken over all regular foliated atlases $\mathcal{U}$ for $\mathcal{F}$. By the reverse inequality $(**)$, equality holds. $\square$

We also note another consequence of Corollary 13.3.15 which is proven by taking into account Corollary 13.2.19.

Corollary 13.3.16. *Geometric entropy of compact, foliated manifolds is finite.*

We turn to the proof of Lemma 13.3.14.

Fix a choice of $0 < 2\Delta < \epsilon_0$. Given any $x \in M$,

$$B_{\mathcal{F}}(x, \Delta/2) \subset B_{\mathcal{F}}(x, \Delta) \subset B_{\mathcal{F}}(x, 2\Delta)$$

are geodesically convex balls of respective diameters Δ, 2Δ and 4Δ. Let

$$\begin{aligned} U_x &= \bigcup\{B_{\mathcal{F}}(y, \Delta/2) \mid y \in B_{\mathcal{F}}^{\perp}(x, \delta)\}, \\ U'_x &= \bigcup\{B_{\mathcal{F}}(y, \Delta) \mid y \in B_{\mathcal{F}}^{\perp}(x, \delta')\}, \\ U''_x &= \bigcup\{B_{\mathcal{F}}(y, 2\Delta) \mid y \in B_{\mathcal{F}}^{\perp}(x, \delta'')\}, \end{aligned}$$

where we can choose $\delta'' > \delta' > \delta > 0$, independent of x and so small that these are all foliated charts with compact closures

$$\overline{U}_x \subset U'_x \text{ and } \overline{U}'_x \subset U''_x.$$

By choosing δ and δ' suitably small, we guarantee:

(i) for all $z \in \overline{U}_x$, $B_{\mathcal{F}}^{\perp}(z, \epsilon_0)$ will meet each plaque of U'_x in exactly one point, realizing the distance of that point from the U'_x–plaque through z.

Furthermore, the choice of δ can be made smaller, if necessary, so as to guarantee that, whenever $U_x \cap U_z \neq \emptyset$, we have:

(ii) $\overline{U}_z \subset U'_x$;

(iii) distinct plaques of $\overline{U}_z$ lie in distinct plaques of U'_x.

Indeed, it is clear that, if a plaque of U_x meets a plaque of U_z, the union of these plaques must lie in a plaque of U'_x. It follows from (ii) and (iii) that any finite collection $\{U_{z_i}\}_{i=1}^N$ that covers M will be a regular foliated atlas $\mathfrak{U}$. The only role of the charts U''_x will be to guarantee that the charts U'_x have nice compact closures.

Thus far, our choices have not been concerned with the value of the given $\eta \in (0,1)$. Fix $0 < \eta' < \eta$. We now choose $\mu > 0$, independent of x and small enough that, if $d(x,z) < \mu$, then the following properties hold:

(1) $z \in U_x$;

(2) $B_{\mathcal{F}}(z,(1-\eta')\Delta/2) \subset P_z$, where P_z is the U_x–plaque through z;

(3) if P' is a plaque of U'_x through a point $z' \in B_{\mathcal{F}}^{\perp}(z,\epsilon_0)$ with $d(z,z') < \mu$, then the orthogonal projection w' of any $w \in B_{\mathcal{F}}(z,(1-\eta)\Delta/2)$ into P' has leaf distance $d_{\mathcal{F}}(w',z') < (1-\eta')\Delta/2$.

To create the desired atlas $\mathfrak{U}_\Delta$, we choose a finite set of points $\Lambda \subset M$ such that every point of M is at distance strictly less than $\mu/2$ from some point of Λ. We set

$$\mathfrak{U}_\Delta = \{U_u\}_{u\in\Lambda},$$
$$\mathfrak{U}'_\Delta = \{U'_u\}_{u\in\Lambda}.$$

Proof of Lemma 13.3.14. The condition that $\epsilon > 0$ be sufficiently small will mean that $\epsilon < \mu/2$. We are given $x, y \in M$ that are (R,ϵ)–separated by g. If $d(x,y) \geq \epsilon$, we are done; so we assume that $d(x,y) < \epsilon$ and prove that x and y are $(\nu(R),\epsilon)$–separated by $\mathfrak{U}_\Delta$. For definiteness, we assume that x is (R,ϵ)–separated from y by g.

Let $\gamma(t)$, $0 \leq t \leq r \leq R$, be a piecewise smooth leaf curve, parametrized by arclength, with $\gamma(0) = x$ and such that $d(p_y(\gamma(t)),\gamma(t)) \geq \epsilon$, for some $t \in [0,r]$. Choose a partition $0 = t_0 < t_1 < \cdots < t_k = r$ so that the segments $\gamma_i = \gamma|[t_i,t_{i-1}]$ have lengths at most $(1-\eta)\Delta$, $1 \leq i \leq k$. This can be achieved for some choice of $k \leq \nu(R)$ because $\ell(\gamma) \leq R$. For $0 \leq i \leq k$, choose $x_i \in \Lambda$ such that $d(x_i,\gamma(t_i)) < \mu/2$ and set $U_i = U_{x_i}$. Similarly, set $U'_i = U'_{x_i}$. Condition (1) on μ implies that $\gamma(t_i) \in U_i$, $0 \leq i \leq k$. Let P_i denote the plaque of U_i through $\gamma(t_i)$, $P'_i \supset P_i$ the corresponding U'_i–plaque. Let $\gamma_i = \gamma|[t_{i-1},t_i]$, $1 \leq i \leq k$. Then $\ell(\gamma_i) \leq (1-\eta)\Delta < (1-\eta')\Delta$; so condition (2) on μ implies that $\gamma_i \subset P_{i-1} \cup P_i$. It follows that $(P_0, P_1, \ldots, P_k)$ is a plaque chain.

Since $d(x,y) < \epsilon < \mu/2$, the triangle inequality implies that $d(x_0,y) < \mu$ and this, by condition (1), implies that $y \in U_0$. Let Q_0 be the plaque of U_0 through y, $Q'_0 \supset Q_0$ the corresponding plaque of U'_0. By property (i), $p_y(x) \in Q'_0$. If $d(p_y(x),x) \geq \epsilon$, Lemma 13.3.9 implies that the Hausdorff distance between P'_0 and Q'_0 is at least ϵ, implying that the distance function

ρ on the set of plaques of U_0 gives $\rho(P_0, Q_0) \geq \epsilon$. Thus, x and y are $(0, \epsilon)$–separated by $\mathcal{U}_\Delta$ and we are done. Assume, then, that $d(p_y(x), x) < \epsilon$. In particular, we can assume that

$$d(p_y(\gamma(t)), \gamma(t)) < \epsilon, \quad 0 \leq t < r,$$
$$d(p_y(\gamma(r)), \gamma(r)) = \epsilon.$$

Since $\epsilon < \mu/2$, condition (3) implies that $p_y(\gamma(t_i))$ is a point of a plaque Q_i and that the segment $p_y(\gamma_i)$ has length at most $(1 - \eta')\Delta$. But $\gamma(t_i)$ is at distance less than $\mu/2$ from x_i and $d(p_y(\gamma(t_i)), \gamma(t_i)) \leq \epsilon < \mu/2$; so $d(p_y(\gamma(t_i)), x_i) < \mu$. Condition (2) on μ then implies that $p_y(\gamma_i) \subset Q_{i-1} \cup Q_i$. This guarantees that $(Q_0, Q_1, \ldots, Q_k)$ is a plaque chain. Another appeal to Lemma 13.3.9 shows that the plaques P'_k and Q'_k of U'_k containing $\gamma(r)$ and $p_y(\gamma(r))$, respectively, have Hausdorff distance at least ϵ, proving that x and y are $(\nu(R), \epsilon)$–separated by $\mathcal{U}_\Delta$. □

Exercise 13.3.17. For a unit speed flow φ and underlying foliation $\mathcal{F}$, modify the proof of Lemma 13.3.14 to prove that, for each $\eta \in (0, 1)$, there is a choice of atlas $\mathcal{U}_\Delta$ as in the lemma such that

$$(1 - \eta) h(\varphi) \leq \frac{1}{\Delta} h_{\mathcal{U}_\Delta}(\mathcal{F}).$$

Use this and Exercise 13.3.13 to prove

$$h(\varphi) = h_g(\mathcal{F}).$$

13.3.B. Remarks on foliated spaces. If $(M, \mathcal{F})$ is a compact, foliated metric space with a leafwise Riemannian metric $g = g_{\mathcal{F}}$, there should be a suitable notion of geometric entropy and a generalization of the theory just developed.

The definition we have given of geometric entropy might seem problematic in this context, but the reader should recall from Chapter 11 that any compact, foliated, metrizable space $(M, \mathcal{F})$ can be realized as a smooth–leaved lamination in separable Hilbert space $\mathbb{E}$. Thus, for each $x \in M$,

$$T_x(\mathcal{F}) \subset T_x(\mathbb{E}) = \mathbb{E}$$

is a finite–dimensional subspace; so the orthogonal complement $T_x^\perp(\mathcal{F})$ of $T_x(\mathcal{F})$ in $\mathbb{E}$ can be defined via the Hilbert inner product and is a closed, infinite–dimensional subspace. Identifying x with $0_x \in T_x^\perp$, we let $\mathcal{B}_{\mathcal{F}}^\perp(x, \epsilon)$ denote the ϵ–ball in $T_x^\perp(\mathcal{F})$ and set

$$B_{\mathcal{F}}^\perp(x, \epsilon) = \mathcal{B}_{\mathcal{F}}^\perp(x, \epsilon) \cap M.$$

Given a leafwise Riemannian metric g, one can now mimic the definition of $h_g(\mathcal{F})$ nearly word for word. The only difference is that the distance function d on M is not defined by g. For this distance, we are free to choose any metric for the compact space M, including the distance $d_{\mathbb{E}}$ which is the restriction to M of the Hilbert metric. The fact that the resulting geometric entropy does not really depend on d is a consequence, as usual, of the fact that all metrics on M are uniformly equivalent. The leafwise distance function $d_{\mathcal{F}}$ is induced by $g = g_{\mathcal{F}}$.

Remark. The definition in [**50**] does not use orthogonal projection, hence might seem preferable here. Nevertheless, orthogonal projection is used in [**50**, pp. 117–120] to prove the inequality in Corollary 13.3.15, and this would appear to be unavoidable.

The main technical difficulty in extending Theorem 13.3.2 to this context is posed by Lemma 13.3.11, which is a Lipschitz property. It seems that this could be sidestepped by restricting our attention to *Lipschitz* foliated spaces. Here is an outline.

Definition 13.3.18. A foliated chart (U, x, y), together with metrics d on U and ρ on the transverse space $Y = y(U)$, is Lipschitz if the homeomorphism

$$\varphi = (x, y) : U \to D \times Y$$

is bi–Lipschitz with respect to the metric d and the product metric $\lambda \times \rho$, where λ is the Euclidean metric on the p–disk D. The triple $(M, \mathcal{F}, d)$ is Lipschitz if every neighborhood of every point $w \in M$ contains a Lipschitz foliated chart that is a neighborhood of w. If a metric d exists so that the triple is Lipschitz, we call $(M, \mathcal{F})$ a Lipschitz foliated space.

It is straightforward to prove, for Lipschitz triples, that every foliated atlas admits a refinement by a regular foliated atlas of Lipschitz charts. One then mimics the proof of the inequality asserted in Proposition 13.3.12, using the above transverse metric ρ instead of the Hausdorff distance between plaques. Remark, however, that this inequality itself is indifferent to the choice of d and ρ. Thus, one replaces d with the metric $d_{\mathbb{E}}$ and ρ with the Hausdorff distance between plaques associated to $d_{\mathbb{E}}$, noting that Lemma 13.3.9 holds in this context. One now mimics the proof of Corollary 13.3.15 to obtain the second key inequality, obtaining the extension of Theorem 13.3.2. Since Lipschitz foliated spaces are a natural category generalizing smoothly foliated manifolds, this seems fairly satisfactory. Alternatively, motivated by these considerations, one can make the following definition.

Definition 13.3.19. The geometric entropy of arbitrary compact, foliated metric spaces $(M, \mathcal{F})$ with leafwise Riemann tensor g is

$$h_g(\mathcal{F}) = \sup_{\mathfrak{U}} \frac{1}{\operatorname{diam}_g \mathfrak{U}} h_{\mathfrak{U}}(\mathcal{F}).$$

Remark that finiteness of $h_{\mathcal{U}}(\mathcal{F})$and $h_g(\mathcal{F})$ are not guaranteed at this level of generality.

13.3.C. Examples. Here we sketch interesting examples, leaving nearly all of the details as exercises.

Exercise 13.3.20. Let $(M_i, \mathcal{F}_i)$, $i = 1, 2$, be compact, smoothly foliated manifolds of codimension one and assume that these have nonempty tangential boundary, that they are infinitesimally C^∞–trivial along $\partial_\tau M_i$ (Definition 3.4.1) and let $(M, \mathcal{F}) = (M_1, \mathcal{F}_1) \cup_\varphi (M_2, \mathcal{F}_2)$ be obtained by tangential gluing along certain boundary leaves by a diffeomorphism φ (Section 3.4). Let g be a Riemannian metric on M, $g_i = g|M_i$, $i = 1, 2$, and prove that $h_g(\mathcal{F}) = \max\{h_{g_i}(\mathcal{F}_i)\}_{i=1,2}$. Proceed as follows.

1. Given a regular foliated atlas $\mathcal{U}$ on M for $\mathcal{F}$, remark that this induces such an atlas $\mathcal{U}_i$ on M_i for $\mathcal{F}_i$, $i = 1, 2$, and prove that
$$h_{\mathcal{U}}(\mathcal{F}) = \max\{h_{U_i}(\mathcal{F}_i)\}_{i=1,2}.$$
(Hint. $S_{\mathcal{U}}(n, \epsilon) \leq S_{\mathcal{U}_1}(n, \epsilon) + S_{\mathcal{U}_2}(n, \epsilon)$ and $S_{\mathcal{U}_i}(n, \epsilon) \leq S_{\mathcal{U}}(n, \epsilon)$.)
2. Using the above step and Theorem 13.3.2, prove that
$$h_g(\mathcal{F}) \leq \max\{h_{g_i}(\mathcal{F}_i)\}_{i=1,2}.$$
3. The reverse inequality is easy, proving the result.

In particular, tangential gluing of foliated manifolds with zero entropy produces a foliated manifold with zero entropy.

It is an almost sacred tradition in foliation theory that, wherever possible, new concepts be illustrated by the Reeb foliation. Faithful to this, we pose the following problem.

Exercise 13.3.21. Show that the Reeb foliation of the three–sphere has zero entropy, proceeding as follows.

1. On a Reeb foliated solid torus $(M, \mathcal{F})$, choose a regular foliated atlas $\mathcal{U}$. Let $\Gamma_{\mathcal{U}}$ be the associated holonomy pseudogroup, S the (compact) transverse space on which it acts. Show that a metric ρ can be defined on S such that each component S_i is identified as a copy $[0, 1] \times \{i\}$ of the unit interval with the usual metric and so that every pseudogroup element is locally multiplication by a power of $1/2$.
2. Show that the pseudogroup on $[0, 1]$ generated by $t \mapsto t/2$ has zero entropy. Use this to show that $h_{\mathcal{U}}(\mathcal{F}) = 0$.

In light of Exercise 13.3.20, the Reeb foliation of S^3 has zero entropy.

The foliation in the following example has interesting dynamical properties. This is a mild modification of a construction due to M. W. Hirsch [**68**].

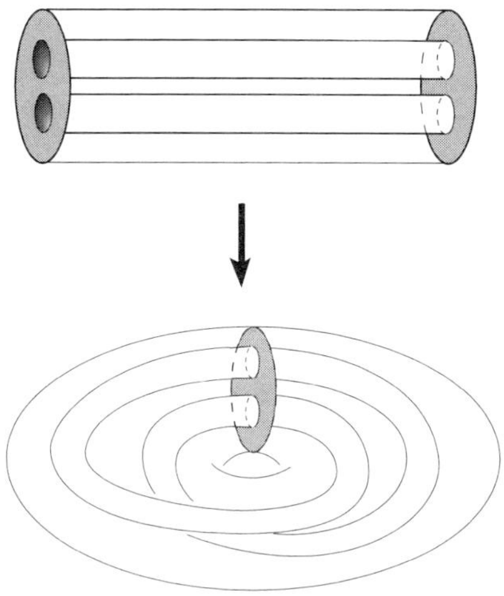

Figure 13.3.1. Forming the solid torus with wormhole drilled out

Example 13.3.22. Let P denote the compact surface (called a "pair of pants") obtained by deleting from S^2 the interiors of three disjoint disks, and let $N = P \times [0,1]$. Let $\varphi : P \to P$ be an orientation–preserving diffeomorphism leaving invariant one component of ∂P (called the "outer" boundary component) and interchanging the other two (the "inner" boundary components). The 3–manifold

$$T = N/\{(x,0) \equiv (\varphi(x),1)\}$$

is a solid torus with a wormhole drilled out that winds around twice longitudinally while winding once meridionally (Figure 13.3.1). Since

$$S^1 = [0,1]/\{0 \equiv 1\},$$

there is a fibration $\pi : T \to S^1$ with fiber P and we can parametrize these fibers as P_t, $0 \leq t \leq 1$. Here, of course, $P_0 = P_1$. Define imbeddings $h_i : [0,1] \to [0,1]$, $i = 0,1$, by

$$h_0(t) = t/2,$$
$$h_1(t) = (t+1)/2$$

and glue the inner toral boundary of T smoothly to the outer one so that each of the two inner boundary components of P_t is glued to the outer boundary component of $P_{h_0(t)}$ and $P_{h_1(t)}$, respectively (Figure 13.3.2). The resulting 3–manifold M is closed and, as in Example 3.3.1, the pairs of pants fit together to form the leaves of a smoothly foliated manifold $(M, \mathfrak{F})$.

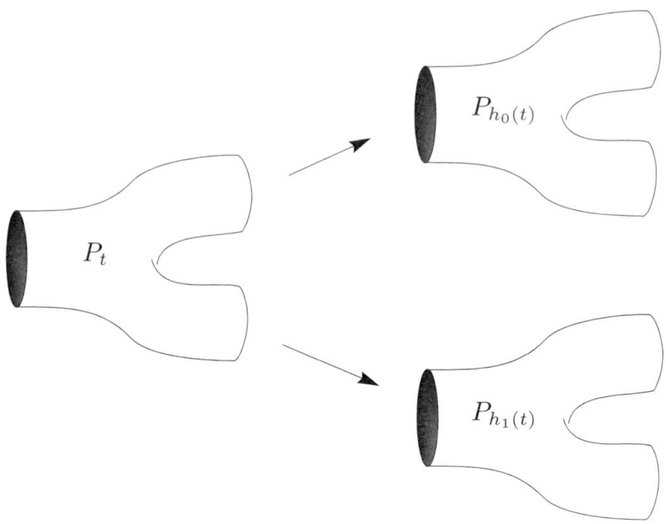

Figure 13.3.2. Gluing together the pairs of pants

It is interesting to visualize the leaves. Represent $t \in [0,1]/\{0 \equiv 1\}$ by its dyadic expansion $t = 0.t_1t_2 \cdots t_n \cdots$, where $t_i \in \{0,1\}$, $i \geq 1$. In this notation, the imbeddings h_i take the forms

$$h_0(0.t_1t_2\cdots) = 0.0t_1t_2\cdots,$$
$$h_1(0.t_1t_2\cdots) = 0.1t_1t_2\cdots.$$

Thus, if t is rational, write

$$t = 0.i_1 \cdots i_m t_1 \cdots t_r t_1 \cdots t_r \cdots,$$

where the expansion becomes periodic after the first m terms, $\tau = t_1 \cdots t_r$ being the minimal period. Thus, if $H = h_{i_1} \circ \cdots \circ h_{i_m}$, we see that $H(t') = t$, where t' has periodic expansion with minimal period τ. Geometrically, this means that the leaf containing P_t has a single cycle of pairs of pants, hence has genus one. Since there are no other cycles, the leaf has a Cantor set of ends. If t is irrational, the leaf containing P_t is a Cantor tree of genus zero, there being no cyclic connections (see Figure 4.2.2). The countably many leaves of genus one should be thought of as "negligible", hence "most of the leaves" are Cantor trees of genus zero. It is not hard to check that the genus zero leaves are exactly the ones with trivial holonomy. It is also easy to check that each leaf is everywhere dense in M.

We will see (Proposition 13.5.1) that the presence of a resilient leaf forces entropy to be positive. Coupled with the frequently cited theorem of Duminy, this proves that vanishing entropy implies the vanishing of $\mathrm{gv}(\mathcal{F})$. The following exercise illustrates Proposition 13.5.1 and gives a counterexample to the converse of Duminy's theorem.

Exercise 13.3.23. Let $(M, \mathcal{F})$ be the Hirsch foliation, as constructed in Example 13.3.1, and investigate its entropy and Godbillon–Vey class. Proceed as follows.

1. Show that $\mathcal{F}$ has a resilient leaf.
2. Show that there is a closed transversal $\Sigma \subset M$ and an identification of Σ with the unit circle $S^1 \subset \mathbb{C}$ such that the holonomy pseudogroup induced by $\mathcal{F}$ on Σ is generated by the local, one–to–one restrictions of $z \mapsto z^2$. By Exercises 13.1.4 and 13.2.6, this pseudogroup has nonzero entropy.
3. Deduce that $h_{\mathcal{U}}(\mathcal{F}) \neq 0$ for a suitable choice (hence every choice) of regular foliated atlas $\mathcal{U}$.
4. Show that $(M, \mathcal{F})$ is transversely affine, as defined in Exercise 1.3.20, hence that $\operatorname{gv}(\mathcal{F}) = 0$.

Let $(M, \mathcal{F})$ be a compact, smoothly foliated manifold with metric tensor g and associated norm $\|\cdot\|$. Suppose that $f : M \to M$ is a diffeomorphism that takes leaves to leaves. We say that $f : (M, \mathcal{F}) \to (M, \mathcal{F})$ is an expanding map if there is a constant $C > 1$ such that, for every vector $v \in T(\mathcal{F})$, $\|f_*(v)\| \geq C\|v\|$.

Exercise 13.3.24. Show that existence of an expanding diffeomorphism implies entropy zero. (Hint. Consult Exercise 13.3.7 and the proof of Corollary 13.3.4.) Generalize the notion of "expanding" to leafwise smooth homeomorphisms of foliated metric spaces, define geometric entropy as in Definition 13.3.19, and prove that entropy either vanishes or is infinite if there exists such an expanding map.

Exercise 13.3.25. Let Σ_n be the foliated space of Exercise 11.2.2. With respect to any leafwise metric tensor, show that the map f_n defines an expanding homeomorphism $\varphi : (\Sigma_n, \mathcal{F}) \to (\Sigma_n, \mathcal{F})$ by the formula

$$\varphi(z_1, z_2, \dots, z_k, \dots) = (f_n(z_1), f_n(z_2), \dots, f_n(z_k), \dots).$$

Give a direct proof that $\mathcal{F}$ has vanishing entropy.

13.4. Invariant Measures

We return to the general context of compact, foliated metric spaces. The term "entropy zero" will mean that all holonomy pseudogroup entropies vanish. Although this condition is formally similar to that of nonexponential growth of leaves, generally there is little relation between the two. For example, 1–dimensional leaves have at most linear growth, but there are many nonsingular flows with positive entropy. At the other extreme, there is the following.

Exercise 13.4.1. Construct a manifold carrying a foliation with all leaves of exponential growth and with holonomy pseudogroup consisting of local isometries. By Lemma 13.2.9, this foliation has entropy zero.

On the other hand, we will see that, for compact, smoothly (at least C^2) foliated manifolds of codimension one, if all leaves have quasi–polynomial growth the entropy will vanish (Exercise 13.5.8).

At any rate, Ghys, Langevin and Walczak [**50**] have established the following analogue of Plante's theorem (Theorem 12.3.1).

Theorem 13.4.2. *If $(M, \mathcal{F})$ is a compact, foliated space with zero entropy, then M has a nontrivial, holonomy–invariant measure μ.*

As in Chapter 12, we always assume that any representative measure $\mu_{\mathcal{U}}$ in the equivalence class μ is finite on the transverse space of plaques of $\mathcal{U}$.

In almost exactly the same way that Theorem 12.3.2 implies Theorem 12.3.1, Theorem 13.4.2 is a consequence of the following.

Theorem 13.4.3. *Let Γ be a finitely generated pseudogroup on a compact metric space X and let Γ_1 be a finite, symmetric generating set for Γ. If $h(\Gamma, \Gamma_1) = 0$, there is a Γ–invariant probability measure on X.*

The proof of Theorem 13.4.3 will proceed as a sequence of lemmas and definitions, a number of details being left as exercises. The hypothesis on entropy is needed for the first step.

Lemma 13.4.4. *If $h(\Gamma, \Gamma_1) = 0$, then*

$$\lim_{\epsilon \downarrow 0} \liminf_{n \to \infty} \frac{S(n+1, \epsilon)}{S(n, \epsilon)} = 1.$$

Proof. Assume not. Since

$$\frac{S(n+1, \epsilon)}{S(n, \epsilon)} \geq 1,$$

the only alternatives are that the limit does not exist or is > 1. In any event, we can select a sequence $\epsilon_k \downarrow 0$ such that there is a possibly infinite limit

$$\lim_{k \to \infty} \liminf_{n \to \infty} \frac{S(n+1, \epsilon_k)}{S(n, \epsilon_k)} > 1.$$

Thus, there is a number $b > 1$ such that, for each k there is N_k so large that

$$n \geq N_k \Rightarrow \frac{S(n+1, \epsilon_k)}{S(n, \epsilon_k)} > b.$$

But this implies that

$$S(n+1,\epsilon_k) > b^{n-N_k}S(N_k,\epsilon_k);$$

so $\operatorname{gr}(S(n,\epsilon_k)) = \operatorname{gr}(b^n)$, for all k. But then

$$\limsup_{n\to\infty}\frac{1}{n}\log S(n,\epsilon_k) \geq \liminf_{n\to\infty}\frac{1}{n}\log S(n,\epsilon_k) \geq \log b > 0.$$

Letting $k\to\infty$, hence $\epsilon_k \downarrow 0$, we conclude that $h(\Gamma,\Gamma_1) > 0$, contrary to hypothesis. □

Let $C_+(X)$ denote the space of nonnegative, continuous functions on X, topologized by the uniform norm $\|\varphi\| = \sup_{z\in X}\varphi(z)$. If $\varphi \in C_+(X)$ we define

$$\Lambda_{n,\epsilon}(\varphi) = \frac{1}{S(n,\epsilon)}\sup\Big\{\sum_{x\in A}\varphi(x) \mid A \subset X \text{ is } (n,\epsilon)\text{–separated}\Big\}.$$

The idea for the construction of a measure on X is to mimic the proof of the Riesz representation theorem, using these operators. They are almost, but not quite linear.

Lemma 13.4.5. *The operators $\Lambda_{n,\epsilon}$ satisfy the following properties.*

(1) *They are homogeneous. That is, $\Lambda_{n.\epsilon}(a\varphi) = a\Lambda_{n.\epsilon}(\varphi)$, for each nonnegative real number a.*
(2) *They are subadditive. That is, $\Lambda_{n.\epsilon}(\varphi_1+\varphi_2) \leq \Lambda_{n.\epsilon}(\varphi_1) + \Lambda_{n.\epsilon}(\varphi_2)$.*
(3) *If $\varphi_1 \leq \varphi_2$, then $\Lambda_{n.\epsilon}(\varphi_1) \leq \Lambda_{n.\epsilon}(\varphi_2)$.*
(4) *$\Lambda_{n,\epsilon}(1) = 1$ and $\Lambda_{n,\epsilon}(\varphi) \leq \|\varphi\|$.*
(5) *If the minimum distance between points of* $\operatorname{supp}\varphi_1$ *and points of* $\operatorname{supp}\varphi_2$ *is $> \epsilon$, then $\Lambda_{n.\epsilon}(\varphi_1+\varphi_2) = \Lambda_{n.\epsilon}(\varphi_1) + \Lambda_{n.\epsilon}(\varphi_2)$.*
(6) *If $\varphi,\psi \in C_+(X)$, then*

$$|\Lambda_{n,\epsilon}(\varphi) - \Lambda_{n,\epsilon}(\psi)| \leq 2\|\varphi-\psi\|.$$

In particular, the set $\{\Lambda_{n,\epsilon} \mid n \in \mathbb{Z}^+ \text{ and } \epsilon > 0\}$ is an equicontinuous family of functions on $C_+(X)$.

Properties (1) through (4) follow directly from the definition. For (5), simply note that, if A_1 and A_2 are (n,ϵ)–separated sets contained respectively in the supports of φ_1 and φ_2, then $A_1 \cup A_2$ is also (n,ϵ)–separated.

Exercise 13.4.6. Prove (6) in Lemma 13.4.5. (Hint. Define $\theta \in C_+(X)$ by

$$\theta(z) = \max\{\psi(z),\varphi(z)\}, \quad \forall\, z \in X,$$

and use properties (2), (3) and (4).)

Lemma 13.4.7. *There is a double sequence $\{n(k,p)\}_{k,p=1}^{\infty}$ of positive integers such that*

(i) *for each* $p \geq 1$, $n(k,p) \uparrow \infty$ *as* $k \uparrow \infty$;

(ii) *there are limits*

$$s_p = \lim_{k\to\infty} \frac{S(n(k,p)+1, 1/p)}{S(n(k,p), 1/p)},$$

$$1 = \lim_{p\to\infty} s_p.$$

Proof. By Lemma 13.4.4, the zero entropy condition implies that

$$\lim_{p\to\infty} \liminf_{n\to\infty} \frac{S(n+1, 1/p)}{S(n, 1/p)} = 1.$$

The existence of the double sequence $n(k,p)$ with the asserted properties follows immediately. □

Lemma 13.4.8. *Let* $n(k,p)$ *be as in* Lemma 13.4.7. *Then, for each* $p \geq 1$ *and each* $\varphi \in C_+(X)$ *the following limits exist*:

$$\Lambda_{p,0}(\varphi) = \lim_{k\to\infty} \Lambda_{n(k,p),1/p}(\varphi),$$

$$\Lambda_{p,1}(\varphi) = \lim_{k\to\infty} \Lambda_{n(k,p)+1,1/p}(\varphi).$$

Moreover, there is a strictly increasing sequence $\{p(r)\}_{r=1}^{\infty}$ *of positive integers such that the following limits exist, for every* $\varphi \in C_+(X)$:

$$\Lambda_0(\varphi) = \lim_{r\to\infty} \Lambda_{p(r),0}(\varphi),$$

$$\Lambda_1(\varphi) = \lim_{r\to\infty} \Lambda_{p(r),1}(\varphi).$$

Exercise 13.4.9. Prove Lemma 13.4.8. For this, observe that $C_+(X)$ has a countable dense subset and employ a diagonal process, using (4) and (6) in Lemma 13.4.5. Note that property (ii) of the sequence in Lemma 13.4.7 is not used. It will be needed subsequently.

Corollary 13.4.10. *For* $i = 0, 1$, *the operators* Λ_i *have properties* (1), (2), (3), (4) *of* Lemma 13.4.5. *In addition, if* $\varphi_1, \varphi_2 \in C_+(X)$ *have disjoint supports, then*

(5)′ $\Lambda_i(\varphi_1 + \varphi_2) = \Lambda_i(\varphi_1) + \Lambda_i(\varphi_2)$,

which is stronger than property (5) *of* Lemma 13.4.5.

Any operator $\Lambda : C_+(X) \to [0,1]$ with these properties will be called a *pseudo–integral* on X. For such operators, we have the following analogue of the Riesz representation theorem.

Lemma 13.4.11. *If* Λ *is a pseudo–integral on* X, *there is a regular Borel probability measure* μ *on* X, *uniquely determined by the condition that, on open subsets* $V \subseteq X$,

$$\mu(V) = \sup\{\Lambda(\varphi) \mid \varphi \in C_+(X), \varphi \leq 1, \operatorname{supp}\varphi \subset V\}.$$

For the proof, one attempts first to extend μ to an outer measure on the family of all subsets $A \subseteq X$ by

$$\mu(A) = \inf_{V \supseteq A} \mu(V). \tag{13.4.1}$$

Evidently, $\mu(A) \geq 0$ and it should be clear from (4) that $\mu(A) \leq 1$ and that $\mu(X) = 1$. Also, $\Lambda(0) = 0$ by (1); so $\mu(\emptyset) = 0$. It is also clear that $A \subseteq B \Rightarrow \mu(A) \leq \mu(B)$. Next, one defines $\mathfrak{M}_F$ to be the family of subsets $E \subseteq X$ such that

$$\mu(E) = \sup\{\mu(K) \mid K \subseteq E,\ K \text{ is compact}\}, \tag{13.4.2}$$

and then defines $\mathfrak{M}$ to be the class of all $E \subseteq X$ such that $E \cap K \in \mathfrak{M}_F$, for all compact subsets K. The following completes the proof of Lemma 13.4.11.

Lemma 13.4.12. *The family $\mathfrak{M}$ is a σ–algebra containing the Borel algebra $\mathfrak{B}$ and μ restricts to a regular measure on $\mathfrak{M}$, hence on $\mathfrak{B}$.*

Note that regularity is built in by equations (13.4.1) and (13.4.2) and, in turn, these equations are forced by regularity. The proof of Lemma 13.4.12 coincides with the first nine steps of W. Rudin's ten–step proof of the Riesz representation theorem [**112**, pp. 42–43]. Step I needs only the subadditivity (property (2)) of a pseudo–integral and step IV needs only the additivity in property (5)$'$. The monotonicity of Λ (property (3)) is used in step II. Otherwise, these steps rely only on definitions and preceding steps.

Step X in Rudin is to show that Λ extends to the integral associated to the measure μ. Since a pseudo–integral is only assumed to be subadditive, this step is not possible. Of course, Λ has some properties of the integral and we will need the following.

Lemma 13.4.13. *Let Λ be a pseudo–integral on X, μ its associated measure. If $V \subseteq X$ is open, $\varphi \in C_+(X)$, $\varphi \leq 1$ and $\varphi|V \equiv 1$, then $\Lambda(\varphi) \geq \mu(V)$.*

Proof. Indeed, for every $\psi \in C_+(X)$ with $\psi \leq 1$ and $\operatorname{supp}\psi \subset V$, it is clear that $\psi \leq \varphi$; so $\Lambda(\psi) \leq \Lambda(\varphi)$ by monotonicity of Λ. By Lemma 13.4.11, it follows that $\mu(V) \leq \Lambda(\varphi)$. □

We return to the pseudo–integrals Λ_0 and Λ_1 and let μ_0 and μ_1 be the respective Borel probability measures given by Lemma 13.4.11.

Lemma 13.4.14. *The measures μ_0 and μ_1 are identical.*

Proof. First we note that, since every (n, ϵ)–separated subset of X is obviously $(n+1, \epsilon)$–separated, we have

$$S(n, \epsilon)\Lambda_{n,\epsilon}(\varphi) \leq S(n+1, \epsilon)\Lambda_{n+1,\epsilon}(\varphi)$$

for every φ in $C_+(X)$. In particular, for the sequence $n(k, p(r))$,

$$\Lambda_{n(k,p(r)),1/p(r)}(\varphi) \leq \frac{S(n(k,p(r)+1),1/p(r))}{S(n(k,p(r)),1/p(r))} \Lambda_{n(k,p(r)+1),1/p(r)}(\varphi).$$

Taking two limits and appealing to (ii) in Lemma 13.4.7, we see that

$$\Lambda_0(\varphi) \leq \Lambda_1(\varphi).$$

By the definition of the measures on open sets we see that

$$\mu_0(V) \leq \mu_1(V),$$

for all open subsets $V \subseteq X$. By Equation (13.4.1),

$$\mu_0(B) \leq \mu_1(B),$$

for each Borel subset $B \subseteq X$. To obtain the reverse inequality, we note that

$$\mu_1(X \setminus B) = 1 - \mu_1(B) \leq 1 - \mu_0(B) = \mu_0(X \setminus B).$$

□

Henceforth we will denote by μ the measure induced on X by $\mu_0 = \mu_1$.

Lemma 13.4.15. *The measure μ on X is invariant by the pseudogroup Γ.*

Proof. Let $f \in \Gamma_1$ and let $\varphi \in C_+(X)$ have $\operatorname{supp}\varphi \subset \operatorname{im} f$. If $E \subset \operatorname{dom} f$ is an (n, ϵ)–separated set, then $f(E)$ is $(n+1, \epsilon)$–separated; so

$$\Lambda_{n,\epsilon}(\varphi \circ f) \leq \frac{S(n+1,\epsilon)}{S(n,\epsilon)} \Lambda_{n+1,\epsilon}(\varphi).$$

Applying this inequality to $(n, \epsilon) = (n(k, p(r), 1/p(r)))$ and taking two limits, we see that

$$\Lambda_0(\varphi \circ f) \leq \Lambda_1(\varphi).$$

Arguing as in the previous proof and appealing to the equality of μ_0 and μ_1, we conclude that

$$\mu(f^{-1}(B)) \leq \mu(B),$$

for each Borel subset $B \subseteq \operatorname{im} f$. Replacing $f \in \Gamma_1$ with $f^{-1} \in \Gamma_1$ and the Borel set B with $f^{-1}(B)$ gives the reverse inequality. □

The proof of Theorem 13.4.3, hence of Theorem 13.4.2, is complete.

13.5. Resilient Leaves

Recall the notion of "resilient leaf" for foliated spaces (Definition 12.2.39). It has the following elementary implication.

Proposition 13.5.1. *If the compact, foliated metric space $(M, \mathcal{F})$ has a resilient leaf F, then $\mathcal{F}$ has positive entropy.*

Proof. Let $x_0 \in F$ and $f_i : D \to B_i$, $i = 0, 1$, be as in the proof of Corollary 12.2.40. For $\epsilon > 0$ sufficiently small, $d(B_0, B_1) \geq \epsilon$. For each integer $n \geq 1$, consider the 2^n sequences $\iota = (i_1, i_2, \ldots, i_n)$ of 0's and 1's and the corresponding points $x_\iota = f_{i_1} \circ f_{i_2} \circ \cdots \circ f_{i_n}(x_0)$. If $\iota \neq \iota'$, let $r \geq 1$ be the smallest integer such that the corresponding terms i_r and i'_r of these two sequences are different, say $i_r = 0$ and $i'_r = 1$. Then $f^{-1}_{i_{r-1}} \circ \cdots \circ f^{-1}_{i_1}(x_\iota) \in B_0$ and $f^{-1}_{i_{r-1}} \circ \cdots \circ f^{-1}_{i_1}(x_{\iota'}) \in B_1$, hence are at distance at least ϵ apart. Thus, we have found a set of 2^n points that are (n, ϵ)–separated by $\{f_0^{\pm 1}, f_1^{\pm 1}, \mathrm{id}\}$. Since the transverse ball D, centered at x_0, can be chosen arbitrarily small, f_0 and f_1 can be assumed to be restrictions to D of elements of Γ_k, for some integer $k \geq 1$. It follows that, for all sufficiently small values of ϵ and all integers $n \geq 1$, $S_{\mathcal{U}}(nk, \epsilon) \geq 2^n$, hence $h_{\mathcal{U}}(\mathcal{F}) > 0$. □

By the result of Duminy on the vanishing of $\mathrm{gv}(\mathcal{F})$, we obtain the following.

Corollary 13.5.2. *If $(M, \mathcal{F})$ is a compact, C^2–foliated manifold of codimension one, then zero entropy implies that* $\mathrm{gv}(\mathcal{F}) = 0$.

The main result of this section will be the following special case of [**50**, Théorème 6.1].

Theorem 13.5.3. *Let $(M, \mathcal{F}, \pi)$ be a compact, connected, C^2–foliated bundle, where $\pi : M \to B$ has 1–dimensional fiber and $\partial B = \emptyset$. Then $\mathcal{F}$ has zero entropy if and only if there is no resilient leaf.*

The "only if" part of this theorem is given by Proposition 13.5.1. The reverse implication is more delicate.

Remark. Theorem 13.5.3 should be true without the requirement that $\mathcal{F}$ be transverse to a fibration. There seem to be technical difficulties, however, caused by the possibility that certain open, saturated sets may have nontrivial "octopus decompositions" (Definition 5.2.13). The reader who is interested in this problem should pay special attention to Exercise 13.5.17 and to the proof of Lemma 13.5.13 on pages 385–386.

For the first part of the proof, we will not use the hypothesis that $\mathcal{F}$ is transverse to a fibration. That hypothesis becomes useful in proving Lemma 13.5.9.

Accordingly, let $(M, \mathcal{F})$ be a compact, connected, C^2–foliated manifold of codimension one and without resilient leaves. Assume also that M is orientable, $\mathcal{F}$ is transversely orientable, and that $\partial_{\pitchfork} M = \emptyset$. The C^2–smoothness condition is required by the generalized Poincaré–Bendixson theory, especially the theory of levels (Section 8.3) and of open, saturated sets without holonomy (Chapter 9). The absence of transverse boundary is not strictly necessary, but avoids technical problems with some of the cited theory. This theory also requires orientability of M and transverse orientability of $\mathcal{F}$. Theorem 13.5.3 reduces to this case by passage to a finite cover.

We begin by recalling some facts from the theory of levels. A local minimal set of $\mathcal{F}$ is a minimal set of $\mathcal{F}|W$, where W is an open, connected, $\mathcal{F}$–saturated subset of M. In the absence of resilient leaves, a local minimal set in W is of one of the following two types:

1. a proper leaf;
2. W itself, each leaf of $\mathcal{F}|W$ being dense in W and without holonomy.

Indeed, exceptional local minimal sets have resilient leaves by Theorem 8.2.1 and it is elementary that, if W itself is a local minimal set with nontrivial holonomy, then W contains a resilient leaf (Exercise 8.2.6). As in Section 8.3, we denote the union of all local minimal sets by M_*. Recall that the leaves (if any) in $M \smallsetminus M_*$ are said to be at infinite level.

By Corollary 12.2.38, if $W \subseteq M$ is an open local minimal set of $\mathcal{F}$ without holonomy, the leaves of $\mathcal{F}|W$ all have the same growth type. By [**20**], in addition to polynomial and weakly polynomial types, this growth can be one of uncountably many quasi–polynomial types (Definition 12.2.15) dominating all polynomials, uncountably many nonexponential types (Definition 12.2.14) dominating all quasi–polynomials, or it can be exponential. Let $W_* \subset M_*$ denote the union of the open, local minimal sets in which the leaves have growth strictly dominating all polynomial types.

Choose a one–dimensional foliation $\mathcal{L}$ that is transverse to $\mathcal{F}$ (eventually, $\mathcal{L}$ will be the foliation by fibers of $\pi : M \to B$) and fix a biregular foliated atlas $\mathcal{U} = \{U_i\}_{i=1}^N$ for $\mathcal{F}$ and $\mathcal{L}$.

Denote the associated holonomy pseudogroup by Γ. The symmetric generating system, consisting of the holonomy cocycle and the identity, will be denoted by Γ_1. The transverse space X on which Γ acts is a finite collection of disjoint intervals $\{X_i\}_{i=1}^N$ with disjoint, compact closures $\overline{X}_i$ in the real line. Distances in X and $\overline{X}$ will be measured using the standard distance d in $\mathbb{R}$.

Let $\mathbf{P} \subseteq X$ denote the union of orbits of Γ that have (weakly) polynomial growth. By Proposition 12.2.35, the $\mathcal{F}$–saturation $\mathcal{F}(\mathbf{P})$ is exactly the union of leaves with (weakly) polynomial growth. If $\overline{\mathbf{P}}$ denotes the closure of $\mathbf{P}$

in $\overline{X}$, the saturation $\mathcal{F}(\overline{\mathbf{P}})$ is the closure in M of the union of leaves with (weakly) polynomial growth.

Lemma 13.5.4. *The set* $M_* \smallsetminus W_*$ *is exactly the saturation* $\mathcal{F}(\mathbf{P})$ *and the border* δW_* (Definition 5.2.4) *is a union of leaves with polynomial growth.*

Proof. A proper leaf either limits on an exceptional local minimal set or is totally proper, hence has finite depth (Proposition 8.3.16). Thus, in the absence of resilient leaves, proper leaves have polynomial growth (Exercise 12.2.42). Leaves at infinite level have growth dominating all polynomials (Corollary 12.2.43). By the definition of W_*, the first assertion follows. For the second, observe that δW_* is a union of semiproper leaves. By Lemma 8.3.23, these leaves cannot be at infinite level, hence they are proper leaves and have polynomial growth. □

Lemma 13.5.5. *The saturation* $\mathcal{F}(\overline{\mathbf{P}})$ *is exactly* $M \smallsetminus W_*$.

Proof. Clearly, $\mathcal{F}(\overline{\mathbf{P}}) \subseteq M \smallsetminus W_*$. For the reverse inclusion, let $x \in M \smallsetminus W_*$. Since M_* is dense in M (Theorem 8.3.7), there is a sequence $\{x_k\}_{k=1}^{\infty}$ of points of M_* converging to x in M. If infinitely many terms of this sequence are not in W_*, then we obtain a subsequence converging to x, each point of which is in $M_* \smallsetminus W_* = \mathcal{F}(\mathbf{P})$ (Lemma 13.5.4). If all but finitely many terms of the sequence lie in W_*, we can fix a transversal T through x and assume that each $x_k \in T \cap W_*$. Since $x \notin W_*$, either $x \in \delta W_* \subset \mathcal{F}(\mathbf{P})$ (Lemma 13.5.4) or there is a sequence $\{y_k\}_{k=1}^{\infty} \subset T \cap \delta W_*$ that converges to x and, again by Lemma 13.5.4, $y_k \in \mathcal{F}(\mathbf{P})$. We have proven that $M \smallsetminus W_*$ is contained in the closure of $\mathcal{F}(\mathbf{P})$. □

Corollary 13.5.6. *Let* I *denote a component of* $X \smallsetminus \overline{\mathbf{P}}$. *Then the saturation* $\mathcal{F}(I)$ *is an open saturated set* $W \subset M$, *each leaf of which is dense in* W *and without holonomy. Moreover, the* Γ*-orbits of the boundary points of* I *in* X *have polynomial growth.*

Fix ϵ sufficiently small. This will include the condition that

$$0 < \epsilon < d(\overline{X}_i, \overline{X}_j), \quad 1 \le i < j \le N.$$

For each integer $n \ge 0$, let $E_n \subset X$ be (n, ϵ)–separated by Γ_1 and of maximal cardinality $S_{\mathfrak{U}}(n, \epsilon)$. Let $S(n, \epsilon)$ denote the number of components of $X \smallsetminus E_n$ with both endpoints in E_n. Evidently,

$$S(n, \epsilon) \le S_{\mathfrak{U}}(n, \epsilon) \le S(n, \epsilon) + N,$$

where N is the number of components of X. Thus, entropy can be computed using $S(n, \epsilon)$.

We will decompose the number $S(n, \epsilon) = S_a(n, \epsilon) + S_b(n, \epsilon)$ and study the growth of each part separately. A component of $X \smallsetminus E_n$ with both endpoints in E_n is said to be of type (n, a) if it meets a leaf with (weakly) polynomial growth and, otherwise, it is of type (n, b). We denote by $S_a(n, \epsilon)$ the number of intervals of type (n, a) and, by $S_b(n, \epsilon)$, the number of type (n, b).

Lemma 13.5.7. *The growth type of $S_a(n, \epsilon)$ as a function of n is weakly polynomial.*

Proof. Remark that only finitely many components of $X \setminus \overline{\mathbf{P}}$ have length $\geq \epsilon/3$. By Corollary 13.5.6, endpoints in X of such components are in $\mathbf{P}$. Adding to these a finite subset of $\mathbf{P}$ that is $\epsilon/3$ dense in $\overline{\mathbf{P}}$, we obtain a finite set $Z \subset \mathbf{P}$. It is clear that the Γ–orbit of Z has weakly polynomial growth. That is, if we set

$$Z_n = \{x \in X \mid x = f(z), z \in Z, f \in \Gamma_n\},$$

then there is an integer $k \geq 0$, such that $\operatorname{card}(Z_n) \leq \operatorname{card}(Z) n^k$, for all sufficiently large integers n. If I is an interval of type (n, a), then there exists an element $f \in \Gamma_n$ such that $I \subset \operatorname{dom} f$ and the distance between the endpoints of $f(I)$ is at least ϵ. Since $I \cap \overline{\mathbf{P}} \neq \emptyset$, there is a point in $f(I) \cap \overline{\mathbf{P}}$. An elementary argument, left to the reader, then shows that $Z \cap f(I) \neq \emptyset$ and it follows that I contains a point of Z_n. This proves that $S_a(n, \epsilon) \leq \operatorname{card}(Z_n)$. □

Exercise 13.5.8. Adapt the argument, given so far, to show that, if every leaf has quasi–polynomial growth (Definition 12.2.15), then the entropy is zero. In fact, show that it is enough that the union of leaves with quasi–polynomial growth be dense in M.

For the remainder of the argument, we add the following condition.

Hypothesis. The transverse 1–dimensional foliation $\mathcal{L}$ has the fibers of a bundle $\pi : M \to B$ as leaves and $\partial B = \emptyset$.

The second condition guarantees that $\partial_{\pitchfork} M = \emptyset$. The following lemma will complete the proof of Theorem 13.5.3.

Lemma 13.5.9. *Under the above hypothesis, the growth type of $S_b(n, \epsilon)$ as a function of n is weakly polynomial.*

We break down the proof of Lemma 13.5.9 into a series of lemmas.

Lemma 13.5.10. *Without loss of generality, we can assume that the fibers of $\pi : M \to B$ are compact intervals.*

Proof. If the fibers are circles and no leaf is compact, the absence of resilient leaves implies that the foliation $\mathcal{F}$ has trivial holonomy. There is a continuous probability measure along the circle fibers that is holonomy invariant

(Lemma 9.2.16) and this provides a topological metric on X that is Γ–invariant. By Lemma 13.2.9, the entropy is zero. If there is a compact leaf, cut M apart along that leaf, obtaining a foliated interval bundle $(M', \mathcal{F}', \pi')$ where the base B' is a finite covering of B. Evidently, Lemma 13.5.9 holds for $(M, \mathcal{F})$ if and only if it holds for $(M', \mathcal{F}')$. □

Thus, we can and do choose the foliated atlas $\mathcal{U} = \{U_i\}_{i=1}^N$ such that the transversal X_i is a fiber, $1 \leq i \leq N$. There is no distinction now between X and $\overline{X}$. The generating set Γ_1 of Γ consists of orientation–preserving diffeomorphisms between components of X, as do the sets Γ_n.

If I is a component of $X \smallsetminus \overline{\mathbf{P}}$, let $\widetilde{I}$ denote the Γ–orbit of I. Say that the distance $\rho(I, J)$ between two components of $\widetilde{I}$ is n if that is the smallest nonnegative integer such that $f(I) = J$, for some $f \in \Gamma_n$. We define the growth function $g_I : \mathbb{Z}^+ \to \mathbb{Z}^+$ by

$$g_I(n) = \operatorname{card}\{J \in \widetilde{I} \mid \rho(I, J) \leq n\},$$

and denote its growth type by $\operatorname{gr}(\widetilde{I})$.

Lemma 13.5.11. *If I is a component of $X \smallsetminus \overline{\mathbf{P}}$, then $\operatorname{gr}(\widetilde{I})$ is polynomial.*

Indeed, $\operatorname{gr}(\widetilde{I})$ coincides with the growth of the Γ–orbit of an endpoint of $\overline{I}$. By Corollary 13.5.6, this is polynomial.

Corollary 13.5.12. *The number $I_b(n, \epsilon)$ of components of $X \smallsetminus \overline{\mathbf{P}}$ that contain intervals of type (n, b) has weakly polynomial growth in n.*

Proof. Let us say that a component of $X \smallsetminus \overline{\mathbf{P}}$ is large if its length is at least ϵ. There are only finitely many large components (not necessarily in the orbit of the same component I). By the previous lemma, the number $K(n)$ of components at ρ–distance at most n from some large one grows polynomially with n. If J is a component of $X \smallsetminus \overline{\mathbf{P}}$ that contains an interval (x, y) of type (n, b), then there is $f \in \Gamma_n$ defined on $[x, y]$ and such that $f(y) - f(x) \geq \epsilon$. Therefore, the interval $(f(x), f(y))$ is contained in a large component J' of $X \smallsetminus \overline{\mathbf{P}}$ and $\rho(J, J') \leq n$. That is, $I_b(n, \epsilon) \leq K(n)$ and the assertion follows. □

For a component I of $X \smallsetminus \overline{\mathbf{P}}$, denote by $S_b(n, \epsilon, I)$ the number of intervals of type (n, b) contained in I. Together with Corollary 13.5.12, the following lemma will establish Lemma 13.5.9 and complete the proof of Theorem 13.5.3.

Lemma 13.5.13. *There is a constant $c(\epsilon) > 0$, depending only on $\epsilon > 0$, such that $S_b(n, \epsilon, I) \leq c(\epsilon)n$, for every component I of $X \smallsetminus \overline{\mathbf{P}}$.*

Again, the proof of Lemma 13.5.13 will be divided into lemmas. First, we need some preparatory remarks.

Let I be a component of $X \setminus \overline{\mathbf{P}}$ and let W be the $\mathcal{F}$–saturation of I. There is a continuous measure μ along the leaves of $\mathcal{L}|W$ that is holonomy invariant, finite on compact sets and becomes unbounded at $\partial\widehat{W}$. This defines a continuous, $\widehat{\mathcal{F}}$–preserving flow

$$\Phi : \mathbb{R} \times \widehat{W} \to \widehat{W}$$

that leaves $\partial\widehat{W}$ pointwise fixed and has the leaves of $\mathcal{L}|W$ as nondegenerate flow lines (Theorem 9.2.1).

Definition 13.5.14. If I is a component of $X \setminus \overline{\mathbf{P}}$, then Γ_I denotes the collection of all $f \in \Gamma$ that are pure compositions of elements of Γ_1 such that $f(I) = I$. For each integer $n \geq 0$, $\Gamma_{I,n} = \Gamma_I \cap \Gamma_n$.

Lemma 13.5.15. *If $f \in \Gamma_I$, there is a unique real number $\tau(f)$ such that $\Phi_{\tau(f)} = f$.*

Proof. Choose arbitrary $x \in I$. If $x \leq f(x)$, set $\tau(x) = \mu[x, f(x)]$. If $x > f(x)$, set $\tau(x) = -\mu[f(x), x]$. Then $\Phi_{\tau(x)}(x) = f(x)$ and, since f preserves the continuous measure μ, this forces f to agree with $\Phi_{\tau(x)}$ in a neighborhood of x. Thus, $\tau(x)$ is a locally constant function of $x \in I$. By the connectivity of I, $\tau(x)$ is constant and we take $\tau(f) = \tau(x)$. □

Evidently, if $f, g \in \Gamma_I$, then $\tau(f \circ g) = \tau(f) + \tau(g)$.

Lemma 13.5.16. *For each component I of $X \setminus \overline{\mathbf{P}}$, there is a constant $K(I) > 0$ such that, if $f \in \Gamma_{I,n}$, then $|\tau(f)| \leq K(I)n$.*

Proof. Consider first arbitrary basic loops h at $I = (c, d)$ (Definition 8.1.7) of arbitrary length. These extend smoothly to basic loops $\overline{h}$ on $[c, d]$. By Lemma 8.1.16 (or, more properly, its proof), we can find $c < a < b < a' < d$ such that, independently of the choice of basic loop h, $a < h(b) < a'$. We then take $K(I) = \mu[a, a']$. Generally, $f \in \Gamma_{I,n}$ can be "factored" into a product of at most n basic loops by strategically inserting into the word defining f strings of the form $f_{i_m}^{-1} \circ \cdots \circ f_{i_1}^{-1} \circ f_{i_1} \circ \cdots \circ f_{i_m}$, all $f_{i_j} \in \Gamma_1$. The assertion follows. □

Exercise 13.5.17. Adapt the proof of Lemma 13.5.16 to foliations that are not necessarily transverse to fibrations. The problem is caused by the nucleus (Definition 5.2.13) of $\widehat{U}$, where U is the $\mathcal{F}$–saturation of I. You will need to use the fact that U supports a holonomy–invariant measure that can be viewed as a cohomology class on $\widehat{U}$.

Lemma 13.5.18. *For each $\epsilon > 0$, there is a constant $K(\epsilon) > 0$ such that, whenever I is a component of $X \setminus \overline{\mathbf{P}}$, whenever $y_1 < \cdots < y_s$ are points of*

I and whenever $f_1, \dots, f_{s-1}$ are elements of $\Gamma_{I,n}$ such that $f_i([y_i, y_{i+1}]) \subset I$ and $f_i(y_{i+1}) - f_i(y_i) \geq \epsilon$, then $s \leq K(\epsilon)n$.

Proof. If no component of $X \smallsetminus \overline{\mathbf{P}}$ has length greater than ϵ, the condition is vacuous and there is nothing to prove. Otherwise, there are only finitely many such components I, we find $K(I, \epsilon)$ for each and let $K(\epsilon)$ be the largest of these. Accordingly, fix one such component I of $X \smallsetminus \overline{\mathbf{P}}$, along with y_i and f_i as in the statement of the lemma.

Each f_i coincides with $\Phi_{\tau(f_i)}$ on (y_i, y_{i+1}) and $|\tau(f_i)| \leq K(I)n$, where $K(I) > 0$ is given by Lemma 13.5.16. Let $\eta > 0$ be such that, if J is a subinterval of I, if $\Phi_t(J)$ has length greater than or equal to ϵ, and if $|t| \leq 1$, then the length of J is greater than η. Clearly such a number η exists. Let Z be a finite, η–dense subset of I. If $k_i = [\tau(f_i)]$ denotes the integer between $\tau(f_i)$ and $-\tau(f_i)$ closest to $\tau(f_i)$, then Φ_{k_i} is defined on (y_i, y_{i+1}) and $\Phi_{k_i}(y_i, y_{i+1})$ has length greater than η. Hence (y_i, y_{i+1}) contains a point of $\Phi_{-k_i}(Z)$. Therefore, the number of intervals (y_i, y_{i+1}) is bounded by the cardinality of the set

$$\bigcup_{-[K(I)n] \leq k \leq [K(I)n]} \Phi_k(Z),$$

which in turn grows at most linearly in n. The existence of the constant $K(I, \epsilon)$, hence of $K(\epsilon)$ follows. □

Proof of Lemma 13.5.13. There are at most finitely many Γ–orbits $\widetilde{I}$ of components I of $X \smallsetminus \overline{\mathbf{P}}$ such that $\widetilde{I}$ contains a component of length $\geq \epsilon$. These are the only Γ–orbits that can contain intervals of type (n, b). For each of these, we find a constant $c(\widetilde{I}, \epsilon) > 0$ such that

$$S_b(n, \epsilon, I) \leq c(\widetilde{I}, \epsilon)n,$$

for every component I of $\widetilde{I}$ and all integers $n \geq 0$. We then take $c(\epsilon)$ to be the largest of these.

Let I be a fixed component of $X \smallsetminus \overline{\mathbf{P}}$ and let $x_1 < x_2 < \cdots < x_s$ be the points of $E_n \cap I$. Thus, the intervals (x_i, x_{i+1}) are exactly the intervals of type (n, b) in I. Let $I_0, I_1, \dots, I_\ell$ be the components of $\widetilde{I}$ of length at least ϵ. For a suitably large integer q, fix $g_j \in \Gamma_q$ such that $g_j(I_j) = I_0$, $1 \leq j \leq \ell$. If we suitably number the intervals I_j, we can also find $g_0 \in \Gamma_n$ such that $g_0(I) = I_0$. Let $\delta > 0$ maximize the Lipschitz constants for g_j^{-1}, $1 \leq j \leq \ell$. Finally, set $y_i = g_0(x_i)$, $1 \leq i \leq s$.

By assumption, for each $i = 1, 2, \dots, s$, there is $h_{i,j(i)} \in \Gamma_n$ such that $h_{i,j(i)}[x_{i-1}, x_i] \subset I_{j(i)}$ is an interval of length at least ϵ. The points

$$y_1 < y_2 < \cdots < y_s \text{ in } I_0$$

and the pseudogroup elements

$$f_i = g_{j(i)} \circ h_{i,j(i)} \circ g_0^{-1} \in \Gamma_{2n+q}$$

then satisfy the hypotheses of Lemma 13.5.18, with I replaced by I_0, ϵ by ϵ/δ and n by $2n+q$. This shows that $s \le K(\epsilon/\delta)(2n+q)$. Since q depends only on $\widetilde{I}$ and ϵ, there exists $c(\widetilde{I},\epsilon)$ such that $s \le c(\widetilde{I},\epsilon)n$. □

This concludes the proof of Theorem 13.5.3. To be sure that the smoothness hypothesis on $\mathfrak{F}$ is necessary for the conclusion, the authors of [**50**] offer an example which we give as an exercise.

Exercise 13.5.19. Let f be a homeomorphism of $[0,1]$ such that $f(x) < x$ in $(0,1)$. Let $[a,b] \subset (0,1)$ be an interval such that $f(b) = a$. Let φ_t be a flow on $[a,b]$, fixing the endpoints and such that $\varphi_t(x) > x$ for all $t > 0$ and all $x \in (a,b)$. Define a homeomorphism g by

$$g(x) = \begin{cases} 0 & \text{if } x = 0, \\ 1 & \text{if } x = 1, \\ f^n \circ \varphi_{2^n} \circ f^{-n} & \text{if } x \in f^n[a,b]. \end{cases}$$

Let Γ be the pseudogroup on $[0,1]$ generated by $\Gamma_1 = \{\mathrm{id}, f^{\pm 1}, g^{\pm 1}\}$.

(1) Show that Γ acts on $[0,1]$ without resilient orbit.

(2) Show that $h(\Gamma,\Gamma_1) > 0$.

(3) Can f and g be simultaneously Lipschitz?

At any rate, Lemma 8.1.1 shows that these homeomorphisms cannot be simultaneously C^2.

Bibliography

1. R. L. Adler, A. G. Konheim, and M. H. McAndrew, *Topological entropy*, Trans. Amer. Math. Soc. **114** (1965), 309–319.
2. L. V. Ahlfors, *Conformal Invariants: Topics in Geometric Function Theory*, McGraw–Hill, New York, NY, 1973.
3. L. V. Ahlfors and L. Sario, *Riemann Surfaces*, Princeton University Press, Princeton, 1960.
4. V. I. Arnol′d, *Small denominators I. Mappings of the circumference onto itself*, Translations Amer. Math. Soc. **46** (1965), 213–284.
5. O. Attie and S. Hurder, *Manifolds which cannot be leaves of foliations*, Topology **35** (1996), 335–353.
6. L. Auslander and R. E. MacKenzie, *Introduction to Differentiable Manifolds*, McGraw–Hill, New York, NY, 1963.
7. T. Banchoff, *Foliations of knot complements in the bicylinder boundary*, Bol. Soc. Brasil. Mat. **5** (1974), 31–43.
8. H. Becker and A. Kechris, *The Descriptive Set Theory of Polish Group Actions*, London Mathematical Society, Lecture Note Series, vol. 232, Cambridge Univ. Press, Cambridge, 1996.
9. H. Bohr, *Almost Periodic Functions*, Chelsea Publishing Company, New York, NY, 1947.
10. ———, *Collected Mathematical Works, Volume I*, Danish Mathematical Society, Copenhagen, 1952.
11. W. Boothby, *An Introduction to Differentiable Manifolds and Differential Geometry*, Academic Press, New York, NY, 1975.
12. R. Bott, *Lectures on characteristic classes and foliations*, Lecture Notes in Math., 279, Springer-Verlag, New York, NY, 1972, pp. 1–94.
13. R. Bott and L. Tu, *Differential Forms in Algebraic Topology*, Springer–Verlag, New York, NY, 1982.
14. N. Bourbaki, *Éléments de Mathématique, Livre VI*, Hermann, Paris, 1959.
15. C. Camacho and A. L. Neto, *Geometric Theory of Foliations*, Birkhäuser, Boston, Massachusetts, 1985.

16. A. Candel, *Uniformization of surface laminations*, Ann. Sci. École Norm. Sup. **26** (1993), 489–516.

17. J. Cantwell and L. Conlon, *Isotopies of foliated 3–manifolds without holonomy*, Adv. in Math. **144** (1999), 13–49.

18. ———, *Poincaré–Bendixson theory for leaves of codimension one*, Trans. Amer. Math. Soc. **265** (1981), 181–209.

19. ———, *Smoothing fractional growth*, Tôhoku Math J. **33** (1981), 249–262.

20. ———, *Nonexponential leaves at finite level*, Trans. Amer. Math. Soc. **269** (1982), 637–661.

21. ———, *Every surface is a leaf*, Topology **26** (1987), 265–285.

22. ———, *Endsets of exceptional leaves; a theorem of G. Duminy*, preprint.

23. ———, *Foliations and subshifts*, Tôhoku Math J. **40** (1988), 165–187.

24. ———, *The theory of levels*, Contemporary Mathematics: Index Theory of Elliptic Operators, Foliations, and Operator Algebras, Vol. 70, 1988, pp. 1–10.

25. ———, *Leaves of Markov local minimal sets in foliations of codimension one*, Publicacions Matemàtiques **33** (1989), 461–484.

26. ———, *Depth of knots*, Topology Appl. **42** (1991), 277–289.

27. ———, *Topological obstructions to smoothing proper foliations*, Contemporary Mathematics **161** (1994), 1–20.

28. R. Cogswell, *Leaves with growth dominating the quadratic function in foliations of codimension* 1, Thesis, Washington University, 1983.

29. L. Conlon, *Differentiable Manifolds; A First Course*, Birkhäuser, Boston, Massachusetts, 1993.

30. A. Connes, *Sur la théorie non commutative de l'integration*, Algèbres d'Opérateurs, Lecture Notes in Math. **725**, pp. 19-143, Springer-Verlag, New York, NY, 1979.

31. A. Connes, *A survey of foliations and operator algebras*, Proc. Symp. Pure Math., Amer. Math. Soc., 1982, pp. 521–628.

32. I. P. Cornfeld, F. V. Fomin, and Ya. G. Sinai, *Ergodic Theory*, Springer–Verlag, New York, NY, 1982.

33. W. de Melo and S. van Strien, *One-dimensional Dynamics*, Ergebnisse der Mathematik und ihrer Grenzgebiete, ser. 3, vol. 25, Springer–Verlag, Berlin, 1993.

34. G. de Rham, *Variétés Differentiables*, Hermann, Paris, 1960.

35. A. Denjoy, *Sur les courbes définies par les équations différentielles à la surface du tore*, J. Math. Pures Appl. **11** (1932), 333–375.

36. P. Dippolito, *Codimension one foliations of closed manifolds*, Ann. of Math. **107** (1978), 403–453.

37. J. Dugundji, *Topology*, Allyn and Bacon, Boston, Massachusetts, 1966.

38. G. Duminy, *L'invariant de Godbillon–Vey d'un feuilletage se localise dans les feuilles ressort*, preprint (1982).

39. W. G. Dwyer, D. B. Ellis, and R. H. Szczarba, *Foliations with nonorientable leaves*, Proc. Amer. Math. Soc. **89** (1983), 733–737.

40. D. B. A. Epstein, K. C. Millet, and D. Tischler, *Leaves without holonomy*, J. London Math. Soc. **16** (1977), 548–552.

41. D. Gabai, *Foliations and the topology of 3–manifolds*, J. Diff. Geo. **18** (1983), 445–503.

42. ______, *Foliations and genera of links*, Topology **23** (1984), 381–394.

43. ______, *Genera of the arborescent links*, Mem. Amer. Math. Soc. **59** (1986), 1–98.

44. ______, *Foliations and the topology of 3–manifolds* II, J. Diff. Geo. **26** (1987), 461–478.

45. ______, *Foliations and the topology of 3–manifolds* III, J. Diff. Geo. **26** (1987), 479–536.

46. L. Garnett, *Foliations, the ergodic theorem and Brownian motion*, J. Funct. Anal. **51** (1983), 285–311.

47. E. Ghys, *Une variété qui n'est pas une feuille*, Topology **24** (1985), 67–73.

48. E. Ghys, *Gauss-Bonnet theorem for 2-dimensional foliations*, J. Funct. Anal. **77** (1988), no. 1, 51–59.

49. E. Ghys, *Sur l'uniformisation des laminations paraboliques*, 73–91. In: *Integrable systems and foliations*, Papers from the colloquium held in honor of Pierre Molino's 60th birthday in Montpellier, May 22–26, 1995. Edited by Claude Albert, Robert Brouzet and Jean Paul Dufour. Progr. Math., 145. Birkhäuser Boston, Inc., Boston, MA, 1997.

50. E. Ghys, R. Langevin, and P. Walczak, *Entropie géométrique des feuilletages*, Acta Math. **160** (1988), 105–142.

51. H. Gluck, *Dynamical behavior of geodesic fields*, Lecture Notes in Math., 819, Springer–Verlag, 1980, pp. 93–105.

52. C. Godbillon, *Feuilletages, Études Géométriques*, Birkhäuser, Basel, 1991.

53. C. Godbillon and J. Vey, *Un invariant des feuilletages de codimension* 1, C. R. Acad Sci. Paris, Série A **273** (1971), 92–95.

54. S. E. Goodman, *Closed leaves in foliations of codimension one*, Comment. Math. Helv. **50** (1975), 383–388.

55. S. E. Goodman and J. F. Plante, *Holonomy and averaging in foliated sets*, J. Diff. Geo. **14** (1979), 401–407.

56. M. Greenberg and H. Harper, *Lectures on Algebraic Topology*, The Benjamin Cummings Publishing Co., Reading, Massachusetts, 1981.

57. B. Grunbaum and G. C. Shephard, *Tilings and Patterns*, W. H. Freeman, New York, NY, 1987.

58. A. Haefliger, *Variétés feuilletées*, Ann. Scuola Norm. Sup. Pisa **16** (1962), 367–397.

59. ______, *Some remarks on foliations with minimal leaves*, J. Diff. Geo. **15** (1980), 269–284.

60. P. R. Halmos, *Lectures on Ergodic Theory*, Chelsea Publishing Co., New York, NY, 1956.

61. G. Hector, *Sur un théorème de structure des feuilletages de codimension un*, Ph.D. thesis, Univ. de Strasbourg, 1972.

62. ______, *Feuilletages en cylindres*, Lecture Notes in Math., 597, Springer–Verlag, 1977, pp. 252–270.

63. ______, *Leaves whose growth is neither exponential nor polynomial*, Topology **16** (1977), 451–459.

64. ______, *Architecture des feuilletages de class C^2*, Astérisque **107–108** (1983), 243–258.

65. G. Hector and U. Hirsch, *Introduction to the Geometry of Foliations, Part B*, Vieweg and Sohn, Braunschweig, 1983.

66. S. Helgason, *Differential Geometry and Symmetric Spaces*, Academic Press, New York, NY, 1962.

67. M. Herman, *The Godbillon–Vey invariant of foliations by planes of T^3*, Lecture Notes in Math., 597, Springer–Verlag, 1977, pp. 294–307.

68. M. W. Hirsch, *A stable analytic foliation with only exceptional minimal sets*, Dynamical Systems, Warwick, 1974, Lecture Notes in Math., 468, Springer–Verlag, 1975, pp. 9–10.

69. ______, *Differential Topology*, Springer-Verlag, New York, NY, 1976.

70. M. W. Hirsch and S. Smale, *Differential Equations, Dynamical Systems, and Linear Algebra*, Academic Press, New York, NY, 1974.

71. K. H. Hofmann and P. S. Mostert, *Elements of Compact Semigroups*, Charles E. Merrill Books, Inc., Columbus, 1966.

72. S. Hurder, *The Godbillon measure for amenable foliations*, J. Diff. Geo. **23** (1986), 347–365.

73. ______, *Ergodic theory of foliations and a theorem of Sacksteder*, Dynamical Systems, College Park, MD, 1986–87, Lecture Notes in Math., 1342, Springer–Verlag, 1988, pp. 291–328.

74. T. Inaba, T. Nishimori, M. Takamura and N. Tsuchiya, *Open manifolds which are non–realizable as leaves*, Kodai Math. J. **8** (1985), 112–119.

75. H. Imanishi, *On the theorem of Denjoy–Sacksteder for codimension one foliations without holonomy*, J. Math. Kyoto Univ. **14** (1974), 607–634.

76. N. Jacobson, *Lie Algebras*, John Wiley and Sons, Inc., New York, NY, 1962.

77. T. Januszkiewicz, *Characteristic invariants of noncompact Riemannian manifolds*, Topology **23** (1984), 289–301.

78. J. P. Jouanolou, *Une preuve élémentaire d'un théorème de Thurston*, Topology **17** (1978), 109–110.

79. M. Kellum, *Uniformly quasi–isometric foliations*, Ergodic Theory and Dynamical Systems **13** (1993), 101–122.

80. S. Kobayashi and K. Nomizu, *Foundations of Differential Geometry, I*, Wiley (Interscience), New York, NY, 1963.

81. A. Kolmogorov and V. Tihomirov, *ε-entropy and ε-capacity of sets in functional space*, Amer. Math. Soc. Transl. (2) **17** (1961), 277–364.

82. N. Kopell, *Commuting diffeomorphisms*, Proc. Symp. Pure Math., XIV, Amer. Math. Soc., 1970, pp. 165–184.

83. G. Köthe, *Topological Vector Spaces I*, Springer–Verlag, New York, NY, 1969.

84. S. Lang, *Differential Manifolds*, Springer–Verlag, New York, NY, 1985.

85. R. Langevin and P. G. Walczak, *Entropy, transverse entropy and partitions of unity*, Ergodic Theory and Dynamical Systems **14** (1994), 551–563.

86. F. Laudenbach and S. Blank, *Isotopie de formes fermées en dimension trois*, Inv. Math. **54** (1979), 103–177.

87. O. Lehto and K. Virtanen, *Boundary behaviour and normal meromorphic functions*, Acta Math. **97** (1957), 47–65.

88. J. Milnor, *Morse Theory*, Princeton University Press, Princeton, 1963.

89. J. Milnor, *Singular Points of Complex Hypersurfaces*, Ann. of Math. Studies, vol. 61, Princeton University Press, Princeton, 1968.

90. J. Milnor and J. D. Stasheff, *Characteristic Classes*, Princeton University Press, Princeton, 1974.

91. P. Molino, *Riemannian Foliations*, Birkhaüser, Boston, Massachusetts, 1988.

92. C. C. Moore and C. Schochet, *Global Analysis on Foliated Spaces*, Mathematical Sciences Research Institute, vol. 9, Springer-Verlag, New York, NY, 1988.

93. S. Morita and T. Tsuboi, *The Godbillon–Vey class of codimension one foliations without holonomy*, Topology **19** (1980), 43–49.

94. J. Moser, *On the volume elements on a manifold*, Trans. Amer. Math. Soc. **120** (1965), 286–294.

95. R. Moussu and F. Pelletier, *Sur le théorème de Poincaré–Bendixson*, Ann. Inst. Fourier, Grenoble **24** (1974), 131–148.

96. M. Nadkarmi, *On the existence of a finite invariant measure*, Proc. Indian Acad. Sci. Math. Sci. **100** (1990), 203–220.

97. V. V. Nemytskii and V. V. Stepanov, *Qualitative Theory of Differential Equations*, Princeton University Press, Princeton, New Jersey, 1960.

98. T. Nishimori, *Behaviour of leaves of codimension one foliations*, Tôhoku Math J. **29** (1977), 255–273.

99. S. P. Novikov, *Topology of foliations*, Trans. Moscow Math. Soc. **14** (1965), 268–305.

100. R. Phelps, *Lectures on Choquet's Theorem*, Van Nostrand Mathematical Studies, vol. 7, American Book Company, New York, NY, 1966.

101. A. Phillips and D. Sullivan, *Geometry of leaves*, Topology **20** (1981), 209–218.

102. R. S. Pierce, *Existence and uniqueness theorems for extensions of zero-dimensional compact metric spaces*, Trans. Amer. Math. Soc. **148** (1970), 1–21.

103. D. Pixton, *Nonsmoothable, nonstable group actions*, Trans. Amer. Math. Soc. **229** (1977), 259–268.

104. J. F. Plante, *Asymptotic properties of foliations*, Comment. Math. Helv. **47** (1972), 449–456.

105. ———, *A generalization of the Poincaré–Bendixson theorem for foliations of codimension one*, Topology **12** (1973), 177–181.

106. ———, *On the existence of exceptional minimal sets in foliations of codimension one*, J. Diff. Eq. **15** (1974), 178–194.

107. ———, *Foliations with measure preserving holonomy*, Ann. of Math. **102** (1975), 327–361.

108. N. V. Quê and R. Roussarie, *Sur l'isotopie des formes fermées en dimension* 3, Inv. Math. **64** (1981), 69–87.

109. A. Ramsey, *Local product structure for group actions*, Ergodic Theory and Dynamical Systems **11** (1991), 209–217.

110. G. Reeb, *Sur Certaines Propriétés Topologiques des Variétés Feuilletées*, Hermann, Paris, 1952.

111. G. Reeb and P. Schweitzer, *Un théorème de Thurston établi au moyen de l'analyse nonstandard*, Differential Topology, Foliations and Gelfand–Fuks Cohomology, Rio de Janeiro, 1976, Lecture Notes in Math., 652, Springer–Verlag, 1978, p. 138.

112. W. Rudin, *Real and Complex Analysis*, 2nd ed., McGraw–Hill, New York, NY, 1974.

113. D. Ruelle and D. Sullivan, *Currents, flows, and diffeomorphisms*, Topology **14** (1975), 319–327.

114. H. Rummler, *Quelques notions simples en géométrie Riemanniennes et leurs applications aux feuilletage compacts*, Comment. Math. Helv. **54** (1979), 224–239.

115. R. Sacksteder, *Foliations and pseudogroups*, Amer. J. Math. **87** (1965), 79–102.

116. W. Schachermayer, *Une modification standard de la démonstration non standard de Reeb et Schweitzer (addendum à "un théorème de Thurston établi au moyen de l'analyse non standard")*, Differential Topology, Foliations and Gelfand–Fuks Cohomology, Rio de Janeiro, 1976, Lecture Notes in Math., 652, Springer–Verlag, 1978, pp. 139–140.

117. A. J. Schwartz, *A generalization of a Poincaré–Bendixson theorem to closed two-dimensional manifolds*, Amer. J. Math. **85** (1963), 453–458.

118. L. Schwartz, *Théorie des Distributions. Tome I*, Hermann, Paris, 1950.

119. S. Schwartzmann, *Asymptotic cycles*, Ann. of Math. **66** (1957), 270–284.

120. P. Schweitzer, *Surfaces not quasi-isometric to leaves of foliations of compact 3-manifolds*, Analysis and Geometry in Foliated Manifolds, Santiago de Compostela, 1994 (River Edge, NJ), World Sci. Publishing, River Edge, NJ, 1995, pp. 223–238.

121. M. Senechal, *Quasicrystals and Geometry*, Cambridge University Press, Cambridge, 1995.

122. M. Shub and D. Sullivan, *Expanding endomorphisms of the circle revisited*, Ergodic Theory Dynamical Systems **5** (1985), 285–289.

123. J. Sondow, *When is a manifold a leaf of some foliation?*, Bull. Amer. Math. Soc. **81** (1975), 622–624.

124. M. Spivak, *Differential Geometry, Volume I, Second Edition*, Publish or Perish, Inc., Houston, Texas, 1979.

125. J. Stallings, *On fibering certain 3–manifolds*, Topology of 3–manifolds; Proc. Top. Inst. of the University of Georgia, Prentice–Hall, Englewood Cliffs, NJ, 1961.

126. S. Sternberg, *Lectures on Differential Geometry*, Prentice–Hall, Englewood Cliffs, NJ, 1964.

127. D. Sullivan, *Cycles for the dynamical study of foliated manifolds and complex manifolds*, Inv. Math. **36** (1976), 225–255.

128. ———, *A homological characterization of foliations consisting of minimal surfaces*, Comment. Math. Helv. **54** (1979), 218–223.

129. D. Sullivan and R. Williams, *Homology of attractors*, Topology **15** (1976), 259–262.

130. I. Tamura, *Topology of Foliations: An Introduction*, Translations of Mathematical Monographs, vol. 97, American Mathematical Society, Providence, RI, 1992.

131. A. Tarski, *Cardinal Algebras*, Oxford Univ. Press, Oxford, 1949.

132. W. Thurston, *Non-cobordant foliations on S^3*, Bull. Amer. Math. Soc. **78** (1972), 511–514.

133. ———, *A generalization of the Reeb stability theorem*, Topology **13** (1974), 347–352.

134. ———, *The theory of foliations in codimension greater than one*, Comment. Math. Helv. **49** (1974), 214–231.

135. ———, *Existence of codimension–one foliations*, Ann. of Math. **104** (1976), 249–268.

136. D. Tischler, *On fibering certain foliated manifolds over S^1*, Topology **9** (1970), 153–154.

137. P. Tondeur, *Foliations on Riemannian Manifolds*, Springer–Verlag, New York, NY, 1988.

138. ———, *Geometry of Foliations*, Monographs in Mathematics, vol. 90, Birkhaüser Verlag, Basel, 1997.

139. N. Tsuchiya, *Growth and depth of leaves*, J. Fac. Sci. Univ. Tokyo **26** (1979), 473–500.

140. ———, *Leaves of finite depth*, Japan J. Math. **6** (1980), 343–364.

141. A. Verjovsky, *A uniformization theorem for holomorphic foliations*, The Lefschetz centennial conference, Part III (Mexico City, 1984), Contemp. Math. **58** (1987), 233–253.

142. S. Wagon, *The Banach-Tarski Paradox*, Cambridge Univ. Press, Cambridge, 1985.

143. P. Walczak, *Hausdorff dimension of Markov invariant sets*, J. Math. Soc. Japan **48** (1996), 125–133.

144. P. G. Walczak, *Dynamics of Foliations, Groups and Pseudogroups*, preprint.

145. G. Wallet, *Nullité de l'invariant de Godbillon–Vey d'un tore*, C. R. Acad Sci. Paris, Série A **283** (1976), 821–823.

146. P. Walters, *An Introduction to Ergodic Theory*, Springer–Verlag, New York, NY, 1982.

147. F. Warner, *Foundations of Differentiable Manifolds and Lie Groups*, Springer–Verlag, New York, NY, 1983.

148. J. H. C. Whitehead, *Convex regions in the geometry of paths*, Quart. J. Math. **3** (1932), 33–42.

149. R. Williams, *Expanding attractors*, Publ. Math. IHES **43** (1974), 169–203.

Index